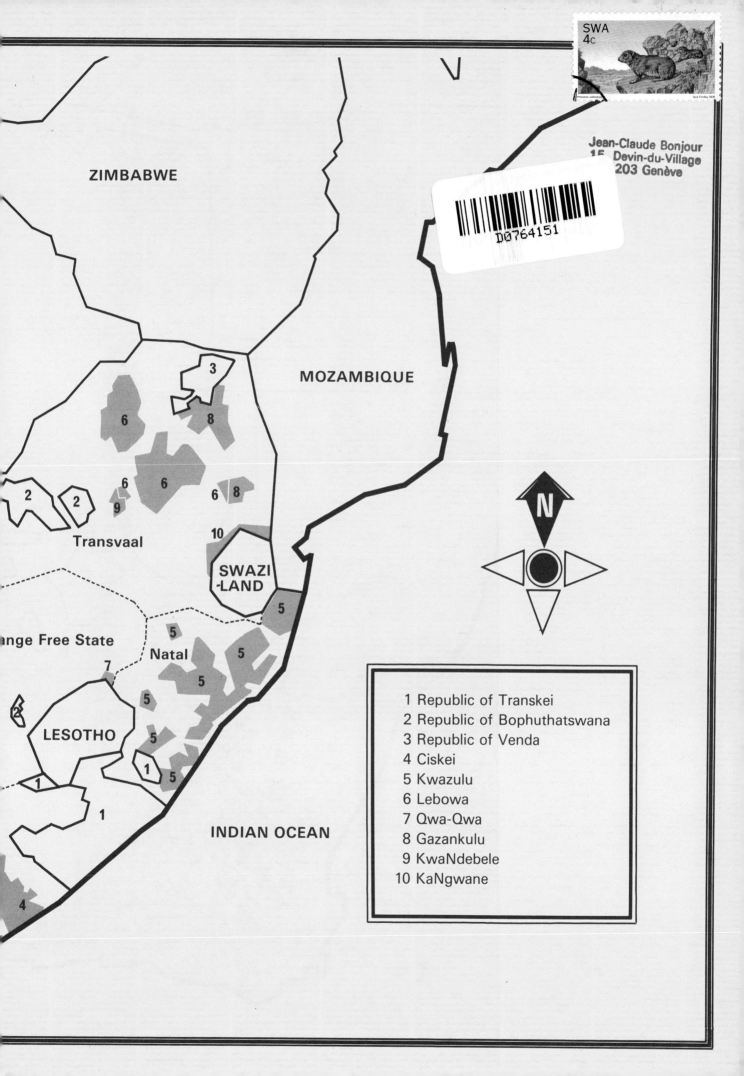

ZIMBABWE

MOZAMBIQUE

3

6

8

2 2

6 6

6 8

9

Transvaal

10

SWAZI
-LAND

5

5

ange Free State

Natal

5

5

7

5

2

LESOTHO

5

1

5

1

1

4

INDIAN OCEAN

1 Republic of Transkei
2 Republic of Bophuthatswana
3 Republic of Venda
4 Ciskei
5 Kwazulu
6 Lebowa
7 Qwa-Qwa
8 Gazankulu
9 KwaNdebele
10 KaNgwane

N

# THE RODENTS OF
# SOUTHERN AFRICA

# THE RODENTS OF SOUTHERN AFRICA

Notes on their identification,
distribution, ecology and taxonomy

G DE GRAAFF

MSc(Rand) DSc(Pret)

Assistant Director, National Parks Board of Trustees
formerly Associate Professor of Zoology, University of Pretoria

With colour plates by
R A R Black

BUTTERWORTHS

DURBAN · PRETORIA

BUTTERWORTH & CO. (SA) (PTY) LTD
© 1981

ISBN: 0 409 09829 9

**THE BUTTERWORTH GROUP**

**South Africa**
BUTTERWORTH & CO (SA) (PTY) LTD
152–154 Gale Street, Durban 4001

**England**
BUTTERWORTH & CO (PUBLISHERS) LTD
88 Kingsway, London WC2B 6AB

**Australia**
BUTTERWORTHS (PTY) LTD
586 Pacic Highway, Chatswood, Sydney, NWS 2067

**Canada**
BUTTERWORTH & CO (CANADA) LTD
2265 Midlands Avenue, Scarborough, Ontario MIP 451

**New Zealand**
BUTTERWORTHS OF NEW ZEALAND LTD
31–35 Cumberland Place, Wellington

**USA**
BUTTERWORTHS (PUBLISHERS) INC.
10 Tower Ofce Park, Woburn, Massachusetts 01801

Set in 10 on 12pt Bembo and designed by Dieter Zimmermann
(Pty) Ltd, Johannesburg
Cover design by Felix Calitz, Pretoria
Printed by Sigma Press (Pty) Ltd, Pretoria

This book is dedicated to my wife
*Alida*
and our children
*Ilse, Gerrit Hendrik* and *Linda*
who have made my life worthwhile.

May they never be denied the
privilege of experiencing God's
little creatures – lest they be
forgotten.

# Contents

# Foreword

The upsurge of interest in nature conservation and the severe plight of some of our vertebrate species has lead to an unprecedented interest in our larger mammals over the past few decades. Most research work on mammals was therefore devoted to the so-called endangered species. The necessity of studying these species before the total destruction of their habitat, as well as the glamour and public appeal attached to research of this nature, attracted a large number of research workers. Inevitably the smaller mammals were, and still are, neglected.

Because of their abundance and cosmopolitan distribution rodents are of extreme importance to man. They are the most numerous of all mammals and are significant members of nearly all terrestrial communities. Due to their wide range of adaptations they occur in every conceivable habitat and in many instances they are a major economic threat. Yet, in spite of its consequence to man and the salient role of rodents in nearly all ecosystems, the life histories of only a few rodent species have been adequately studied.

The last major work on rodents is Ellerman's *Families and genera of living rodents* which appeared in 1940. Since then different systems of classification have been proposed, mainly by Simpson (1945), Wood (1955) and Anderson (1967) and important contributions on other aspects of rodent biology were made by various authors from different parts of the world, including Africa.

In the four decades that passed since 1940 two major works on southern African mammals appeared, *viz.* Roberts (1951): *The mammals of South Africa;* and Ellermann, Morrison-Scott and Hayman (1953): *Southern African mammals.* Over the same period prominent contributions on the rodents of southern Africa, especially in relation to plague, were made by Davis in a number of articles. An identification manual of African rodents was compiled by Meester and Setzer with the collaboration of Misonne, Amtmann, Petter, Davis, Coetzee and De Graaff. Other important publications on rodent biology and taxonomy came from Meester, while Smithers made noteworthy contributions to our knowledge of the taxonomy, distribution and ecology of rodents in Botswana, Zimbabwe and Mozambique (the latter in collaboration with Lobão Tello).

Against this background De Graaff's work on *The rodents of southern Africa* is of major concern because no other work of this extent has been published on African rodents.

Gerrie de Graaff committed himself to rodent research about two decades ago. He has never looked back. He was a dynamic and inspiring teacher of Zoology at the University of Pretoria. He became an assistant-director of the National Parks Board and when he decided in 1972 to produce a comprehensive text exclusively devoted to southern African rodents, he displayed the same vigour and enthusiasm as in his teaching days.

The end result of his labours is a thorough manual on the rodents of southern Africa. It is not a systematic treatise, but rather a book that contains relevant information on rodents published by different people, including the author himself, over many generations. To put together the existing knowledge of South African rodents was an unenviable task because, with the possible exception of the marsupials, nowhere among the mammals is such diversity found within a single order.

However, the known facts of the 73 rodent species recognised in southern Africa has been elegantly knit together. Apart from the gross morphology and the structure of the skull and dentition, features such as distribution, habitat preference, diet, habits and breeding biology are emphasised throughout and it succeeds in representing the rodent as a living animal.

The value of the book is enhanced by identification keys and the splendid colour illustrations of virtually all the species. It is a compilation therefore that can be read profitably by the layman, the university student, the professional mammalogist and by those persons interested in the economic aspects of mammalian biology.

This book fulfills a great need and the author must be congratulated on having performed an enormous task which can most certainly be looked upon as a milestone in the history of African mammalogy.

F.C. Eloff
*Professor and Head: Department of Zoology*
*University of Pretoria*
*Pretoria*

# Preface

When I was a post-graduate student at the University of the Witwatersrand, Johannesburg, I was appointed as a CSIR (Council for Scientific and Industrial Research) research assistant to Prof. Raymond A. Dart, then Head of the Department of Anatomy of the Medical School, and presently the doyen of physical anthropologists in southern Africa. Much of my time from 1955 to 1958 was spent at the Bernard Price Institute for Palaeontological Research where Dart was studying the South African ape-men or australopithecines. Under his guidance and leadership I developed a keen interest in mammals and I embarked upon a detailed analysis of the small mammals which are found in the breccias containing fossils of *Australopithecus africanus* (from Makapansgat, Sterkfontein and Taung) and *Paranthropus robustus* (from Kromdraai and Swartkrans) respectively.

My appetite for research on small mammals and specifically rodents, was whetted. At the end of 1958 I joined the Department of Zoology at the University of Pretoria and was induced by the Head of the Department, Prof. F.C. Eloff, to extend my studies on South African rodents. In 1965 these activities culminated in a taxonomic revision of the molerats (Bathyergidae) of southern Africa. Preparing the work on the molerats made it clear to me that our knowledge of these and other rodents was scanty. My interest in rodents had therefore widened from palaeozoology to neozoology and more particularly to taxonomy.

I became a 'small-mammal' enthusiast. I had always had a rather diffuse concept of the order Rodentia in southern Africa, partly because of the absence of a comprehensive text which collated data scattered through a formidable array of textbooks, journals, magazines and other publications. Furthermore many books on the mammals of southern Africa tend to place the emphasis on the larger species. While our knowledge of larger mammals (especially bovids and carnivores) was expanding by leaps and bounds, there was still much to be learnt about the rodents, lagomorphs, bats and insectivores in southern Africa. I began to see that small mammals were interesting creatures and that they offered interesting, manageable, rewarding and important research possibilities. Several authorities have also urged that more research effort be focused on ecological studies of small mammals (see Meester 1954; Du Toit 1965; Davis 1966).

At the annual general meeting of the Mammal Research Institute at the University of Pretoria in November 1971, I discussed the situation with Dr R.H.N. Smithers, the then Director of National Museums and monuments in Zimbabwe. I mentioned the possibility of compiling a text on southern African rodents, which Smithers received enthusiastically. This book resulted from that conversation. I decided to produce a compendium of what is known about southern African rodents, thereby providing a review of available information for mentor, student and interested layman alike. The marshalling of data involved perusal of many sources scattered through the zoological literature over decades.

Simpson (1945) stated that rodents are inherently a difficult and relatively unsensational group of animals and, to many, of an unattractive nature which has discouraged and retarded research on both fossil and recent rodents. This situation has begun to change over the years and the whole approach has been transformed so that no part of mammalogy is likely to advance as rapidly as the study of rodents. So it is exciting to contribute to these new developments for the benefit of those interested in natural history.

This book is not intended to be a systematic treatise, but rather a book that can be used in teaching classes dealing with mammals and as a reference work for the layman or field worker, who wishes to learn something about interesting adaptations and ways of life of a very important order of mammals. It is assumed that the reader will be familiar with some basic principles of general zoology and its related general terminology. Hitherto, there has been no guide devoted exclusively to the rodents of southern Africa. This book may fill this void, especially if the upsurge in interest by the general public in matters pertaining to conservation and the environment is taken into consideration.

Where do we stand as far as our knowledge of southern African rodents is concerned? I have made a rough subjective assessment, placing the species in three categories which I call 'poorly known', 'moderately known' and 'fairly well known'. I would venture to say that some 5% would be fairly well known, 35% moderately known and 60% poorly known.

What little is known about the rodents is limited to their taxonomy, distribution and parasites, largely on account of their importance as vectors of disease and as competitors for man's agricultural produce. The gaps in our knowledge are greatest in the fields of physiology, reproduction, ethology and palaeontology. Another aspect concerns the relations with man as well as adequate descriptions of the preferred habitats of the different species. It is a wide-open field with plenty of opportunity for the interested person to make a worthwhile contribution to our knowledge of these animals which are, after all, as much part of the ecosystem as primary consumers as any of the larger herbivores. Rodents contribute significantly to the energy flow in nature and play a very important role in any biome in which they occur. In any event, the time has come for mankind to take cognizance of its co-passengers on this planet, and these include the much maligned group of mammals known as the Rodentia.

I have endeavoured to assemble a sample of information presently available (up till the middle of 1979) on the subject of southern African rodents. I fear that there will be many omissions, but I hope the publication of this book will stimulate those interested in natural history to collect additional data on the life histories of these animals. The present work contains information on all of the 73 rodent species known to occur in southern Africa. The limits of this area is south of a line drawn roughly across the continent from where the Cunene River in the west and the Zambezi River in the east enter the Atlantic and Indian oceans respectively. It therefore covers South Africa, Transkei, Bophuthatswana, Botswana, Lesotho, Swaziland, South West Africa/Namibia (SWA/Namibia), Zimbabwe and Mozambique (south of the Zambezi River). For an excellent treatise on the biogeography and ecology of southern Africa, see Werger (1978).

In preparing the accounts of the different species, I have of necessity relied heavily on the observations of others. This applies especially to general 'ecological' information where the published works of Smithers (1966, 1971, 1975) and Smithers & Lobão Tello (1976) have been of inestimable value. The same applies to data gleaned from the publications by Zumpt (1961a, 1966) and Theiler (1962) on the ectoparasitic arthropods which are known to be associated with these rodents. Formal acknowledgement of quoted information taken from these publications, as well as from a multitude of other publications, is made in the text through literature citations.

Identification keys have been included. While compiling these, the need of the non-specialist was kept in mind, but is often impossible to identify the species without reference to less obvious features, very often pertaining to the skull and dentition. These keys are intended to be applicable to species in southern Africa and may not hold good extralimitally.

The text is broken down into a discussion of all the recent rodents encountered in southern Africa. Each of the 73 species has been treated in a uniform way. Both the currently recognised scientific and common name (English and Afrikaans) are given, followed by a short paragraph explaining the derivation and meaning of the

generic and specific names. As could be expected, I have used the species as the basic category. The trinomial system is of value in the context of African mammalogy, but its overall position seems to be very fluid as far as rodents of southern African are concerned. The system of classification used in this book follows Simpson (1945) to a large extent, especially for suprageneric groupings. For alternative classifications, see Vaughan (1972) who has reviewed the work by Wood (1955) and Anderson & Jones (1967).

This is followed by an outline of the most important synonyms which have been given to an animal since its original description. I should stress that these lists of synonyms do not claim to be complete and serve only as an outline to help those who would like to read more about a particular species in the original and/or earlier literature. Only synonyms applicable to southern Africa have been included.

This introductory section is followed by a brief description of the gross morphology and phenotype of the species as well as parameters of size and mass for both males and females. I have often consulted the original descriptions of species and I have found them to be very detailed in the majority of cases. No attempt is made to improve on these and only the salient features are highlighted. The specialist falls back on the original descriptions in any case, if more detail is required. Lengths of the combined head and body (HB), tail (T), hindfoot (HF) and ear (E) were taken from specimen labels and catalogue cards. Dimensions are given for adult specimens only and the metric system is used throughout. It is well known that these measurements are taken by different people and that such data should be interpreted with care. However, trouble is taken to measure these animals when they are collected and these measurements are faithfully copied down, so they are there to be used in order to give one an indication of the order of magnitude which can be expected in natural populations. These measurements are expressed as a mean value (M), the size of the sample (N), the observed range and the coefficient of variability (% CV) of each set of measurements. Normal procedure takes a value of 3–10% indicative of an acceptable homogeneous statistical sample.

All the species are illustrated in colour by Mr Ray A.R. Black, because this book is also intended as an identification manual. Often even the most brilliant or enthusiastic student, or the interested layman has difficulty in identifying some closely related species. Actual comparison and study with named specimens will be required to lead to a satisfactory identification. I know that many of the larger museums in South Africa, will be glad to help in identifying difficult material.

After having briefly described the species, the salient features of structure and morphology of the skull and dentition are dealt with. Each genus is illustrated from its dorsal, lateral and ventral view, while the occlusal surfaces of the cheekteeth are also depicted. Every genus is portrayed in this way while only the upper molars of additional species have been illustrated.

The description is followed by an account of the species' geographical distribution throughout southern Africa and an accompanying map. The maps merely indicate the broad outlines of known distribution up to July 1979. The distributions have been plotted on the 1:4 000 000 map, a quarter degree General Index Chart (TSO Misc./1336 (1958)), printed by the Government Printer, Pretoria, for the Trigonometrical Survey. Every locality found in the literature on rodents or indicated on a specimen label in museum material or emanating from my own fieldwork has been plotted. In order to save space the maps have been reduced in size to show the basic outlines of distribution only. I intend to publish the detailed maps elsewhere. The same also applies to data on the actual localities where species have been recorded. To include pages and pages of localities with the respective quarter degree plotting units (referred to as the locus) is cumbersome in a book of this nature. To plot difficult or little known localities, much use was made of gazetteers published by Davis & Misonne (1964), Skead (1973) and Leistner & Morris (1976). I shall be glad to provide anyone interested with detailed information on localities indicated on the distribution maps. Information given in the publications of Pienaar (1964) for the Kruger National Park, Smithers (1971) for Botswana, Davis (1974) for southern Africa, Lynch (1975) for the Orange Free State (OFS), Smithers & Lobão Tello (1976) for Mozambique, Wilson (1975) for the Wankie National Park and Smithers (in litt.) for Zimbabwe, have also been incorporated in the distribution maps.

The discussions of the distribution of the 73 species are followed by notes on the preferred habitat of the animal, its diet, its habits, the predators with which it has to contend and its breeding. Reproduction is, of course, influenced directly or indirectly by the abovementioned factors. Another facet influencing the success of a species in nature, is the way in which it is adapted to ecto- and endoparasites. Notes on parasitology are presented; the association of the rodent with its parasites has potential consequences for man which are alluded to under a section called 'Relations with man'. Finally, I have given particulars about the prehistory of each species followed by notes on the taxonomy of the species.

The procedure outlined above has been followed for each of the six species of squirrels, the single species of porcupine, two species of canerats, one species of springhare, one species of dassie rat, five species of molerats, ten species of gerbils, eight species of dendromurines, eight species of otomyines, some five other species of cricetids, 23 species of murines and three species of dormice occurring in southern Africa.

Incomplete as these sections are bound to be, they may prove useful as a basis for further study. The emphasis today is on the study of a series of specimens rather than individual specimens and good collections representing both sexes, all ages, taken in all seasons and from as many localities over its range of distribution are required for each species. I have found that further organised collecting by trained persons everywhere is desirable to augment the already good collections available for study in South Africa. The preparation of study skins (properly provided with accurate information) is not difficult and interested amateurs could do much to ensure that valuable material is not wasted.

There are many problems yet to be solved about rodents in southern Africa. I hope that this account will stimulate a greater interest in studies of one of the most important groups of animals. Life may be seen as a parade too little understood, too little appreciated, in our day. If this book helps the reader to regain his contact with the amazing world of wild animals, it will have served its purpose.

*Acknowledgements*

One of the pleasures in doing research and preparing this work has been the friendly and kind co-operation of many individuals and institutions. This work was compiled, collated and written up in private time during evenings, over weekends and periods of leave. It also incorporates the results of field work undertaken by myself in association with others since 1954. It is well-nigh impossible to acknowledge all persons and sources which contributed directly or indirectly to a compilation of this kind, which has resulted from accumulated contacts. I make grateful acknowledgement to all these, so many that they cannot be thanked individually. I am obliged, however, to single out a number of persons whose support has played a vital role in my life. Credit must go in the first instance to my father and late mother who gave me their love and wise counsel and whose faith in me never wavered. In the second place I would like to single out my mentors at the University of the Witwatersrand and the University of Pretoria. Thirdly, I would like to thank colleagues at the Department of Zoology, University of Pretoria, especially Proff. F.C. Eloff and J.A.J. Nel who have unwittingly contributed much to shape my thinking over many years of contact at both academic and informal levels.

In Pretoria we are privileged to have a collection of rodents in the Transvaal Museum. Through the kind auspices of the Director and the professional officers in charge of the Mammal Department, this material was made available to me unreservedly. Actual handling of the specimens, gleaning information from the specimen labels and the efficient system of catalogue cards gave me a basis to work from. This was naturally augmented by an intensive study of the existing literature. I would therefore like to extend a very special word of thanks to the Director of the Transvaal Museum (Dr C.K. Brain) and to his Board of Trustees. I am extremely grateful for all the facilities extended to me, especially the magnificent library facilities. My appreciation is extended to Dr I.L. Rautenbach and Mr N.J. Dippenaar, curators of the Department of Mammals, for their unselfish co-operation and for allowing me free access to the collections in their care.

Towards the end of 1978 I completed a round trip to other national and provincial museums in the Republic of South Africa (RSA) in order to incorporate their data in this compilation. I would like to acknowledge the kind co-operation I received from those institutions who granted permission to use their facilities and collections. My thanks are due to the following organisations and persons who, in their capacities as directors and/or custodians have generously lent specimens to me or permitted me access to their collections. These include the Director of the McGregor Museum, Kimberley (Dr R. Liversidge), the Curator of the John Ellerman Museum of Zoology, University of Stellenbosch (Dr A. Channing), the Director of the South African Museum, Cape Town (Dr T.H. Barry), the Director of the Port Elizabeth Museum (Dr J.H. Wallace), the Director of the Albany Museum, Grahamstown (Mr B.C. Wilmot), the Director of the Kaffrarian Museum, King William's Town (Mr D. Comins), the Director of the East London Museum (Mr E.H. Bigalke), the Director of the Durban Museum (Mr P.A. Clancey) and the Director of the Natal Museum, Pietermaritzburg (Dr B.R. Stuckenberg). Finally, I am much obliged to the then Director of the National Museums in Zimbabwe (Dr R.H.N. Smithers) who provided me with a wealth of information on the rodents of Zimbabwe.

A word of recognition is also directed at the Department of Nature and Environmental Conservation of the Cape Provincial Administration. Through the kind auspices of the Assistant Director, Dr C.J. Loedolff, I was given access to the distribution data of rodents of the Cape Province compiled over a number of years by Mr C.T. Stuart.

My thanks are also due to Miss Jane Walker (Onderstepoort) who helped me with references pertaining to ticks and Miss M. Collins and Dr J.D.F. Boomker (both at that time attached to the Helminthology Department at Onderstepoort) for much appreciated help with information on helminths and nematodes associated with rodents in southern Africa. Mrs M. Avery of the South African Museum kindly supplied unpublished data on distribution during the Upper Pleistocene and Holocene of rodents in the southwestern Cape and the Cape Escarpment in the Kimberley area.

Sincere appreciation is due to Prof. E.D. Gerryts, Director of the Merensky Library at the University of Pretoria, who allowed me full use of the well-equipped library and its services.

I am also grateful to the Chief Director of the National Parks Board of Trustees for allowing me to do various spells of field work in the South African National Parks during the past two decades.

I should especially mention Prof. G.L. Maclean of the University of Natal, Pietermaritzburg, who generously gave his time to read the entire manuscript with patience, speed and efficiency and whose detailed suggestions and criticisms for improving the text were adopted with appreciation. Reading a manuscript is a thankless task and his willingness to have done so is gratefully acknowledged. To Prof. Maclean goes a large share of credit for accuracy and credibility of statements which I have made, although I assume unreserved responsibility for errors which may unwittingly have been included and perpetuated. An important lesson for any scientist is that the facts given to him could be wrong and that interpretations differ and change with time.

I would also like to mention the collaboration of Mr and Mrs N.J. Jensen of Pretoria who corrected grammar and syntax while the work was in early draft form. I have had unstinting help from Mr W. Massyn in preparation of the distribution maps – assistance which is appreciated very much indeed. The photographs of the skulls were taken by Mr W.A. de Beer. I would also like to thank Mrs G. Massyn who meticulously and efficiently typed the various drafts of the manuscript in such a competent way.

A special word of appreciation is also to be directed to Dr R.H.N. Smithers and Dr D.H.S. Davis. Dr Smithers was always full of encouragement and at times when I felt like abandoning the entire project, he offered words of encouragement. Dr Davis also read through a very large section of an early version of this book in manuscript form and graciously declined the co-authorship I offered him, which I feel would have given the book a more authoritative status.

I would also like to thank Dr C. de Jong who introduced me to my publisher. It is a pleasure to thank Mr J. Prinsloo and Miss S. Moolman who have been extremely helpful in guiding this book through the laborious channels leading to its final publication.

A final word of acknowledgement goes to my wife, Alida – without her encouragement, her willingness to take the good with the bad and her understanding during eight years of concerted literary effort, this book would not have been written.

# Colour plates

# Figures

# Maps

# An introduction to the rodents

Rodents are an extremely successful and well-defined assemblage of mammals, occurring throughout the world. They comprise 34 families, 354 genera and roughly 1 685 species (Vaughan 1972). They are the most numerous of all mammals, forming the largest mammalian order and they represent more than half the living species of mammals known to science. Admittedly, the validity of many subspecies, races and varieties is open to question and during the past two decades, taxonomy has focused its attention on the clarification of this uncertainty. The existing literature on rodents is vast and as a group they are familiar mammals, having played an important role in the history of man. Rodents derive their name from *rodere* which means 'to gnaw' in Latin; they all gather their food by gnawing. (For detailed and thorough treatments of the order, see Grassé (1955), Anderson & Jones (1967), Matthews (1971), Vaughan (1972) and Kingdon (1974) while Hebel & Stromberg (1976) discuss the detailed anatomy of the laboratory rat *Rattus norvegicus*.)

No other group shows such a wide range of adaptations for successfully colonising and inhabiting almost any type of habitat. Consequently, they are important components of nearly all terrestrial faunas. Rodents are mostly small mammals, predominantly herbivorous, with teeth modified for and adapted to gnawing and grinding. The dental formula is not known to exceed 1/1, 0/0, 2/1, 3/3 = 22. Their habits are extensive and varied – some are specialised for life underground (fossorial) while others are arboreal and some even lead a semi-aquatic life. Some species have the ability of gliding through the air, but the majority are predominantly terrestrial. Generally, the life-span and expectancy are short, but this is counterbalanced by the fact that rodents are extremely prolific. In many cases, local populations undergo marked, cyclical fluctuations in numbers.

## Global distribution

The geographical distribution of rodents is virtually cosmopolitan: they are absent from Antarctica and some small oceanic islands and have attained their maximum deployment and diversity in the Neotropical Region (South America, the West Indies and the Isthmus of Panama). In contrast, Madagascar and Australia have few indigenous rodents. Elsewhere, they are well represented in the remaining zoogeographic regions of the world known as the Afrotropical (= Ethiopian) (Africa south of the mid-Sahara and the southern portion of Arabia), the Palaearctic (Europe and Eurasia north of the Himalaya Range, as well as North Africa and the Middle East), the Oriental (India and the East Indies as well as the Philippines), the Australian (Australia, New Guinea and New Zealand) and the Nearctic (North America and the Greenland Ice Sheet) respectively.

## Features of the skull and lower jaw

Although there is much diversity in external form, rodents are remarkably uniform in structure. This applies especially to skull morphology and dentition (fig. 1).

The orbits communicate freely with the temporal fossae. The condyle of the mandible is elongated anterior-posteriorly and through the absence of a post-glenoid process admits propalinal motion of the jaw. A zygomatic arch is invariably present, the middle portion of which is formed by the jugal. The zygomatic arch varies in degree of development and is used as a distinguishing characteristic for taxonomic purposes. The lachrymal foramen is always within the orbital margin. The nasals are, with few exceptions, large and extend far forwards. The parietals are moderate and there is generally a distinct interparietal. Auditory bullae are always present and generally large.

The mandible always shows a narrow and rounded symphysial part supporting large incisors. The position of the angular process relative to the roots of the incisors is variable, originating lateral to the root in hystricognath rodents and ventral to the root in sciurognath (including myomorph) rodents.

### The zygomasseteric structure

According to Romer (1971), rodents are, without question, the most successful of all living mammals, and they are so numerous and varied that no two authors agree on their classification. The great diversity of rodents and the probable development of characteristics by parallel evolution is acknowledged by many, but consensus on how to resolve the problem of classifying rodents, especially above family level, has not yet been reached. It is generally accepted that the squirrels, the rats and mice, and the porcupines are representative of three distinct groups, considered as suborders – the Sciuromorpha (containing some 60 genera of living forms), Myomorpha (235 genera) and Hystricomorpha (57 genera) respectively. However, these subordinal groupings are uncertain and the numbers of genera and species assigned to them are ap-

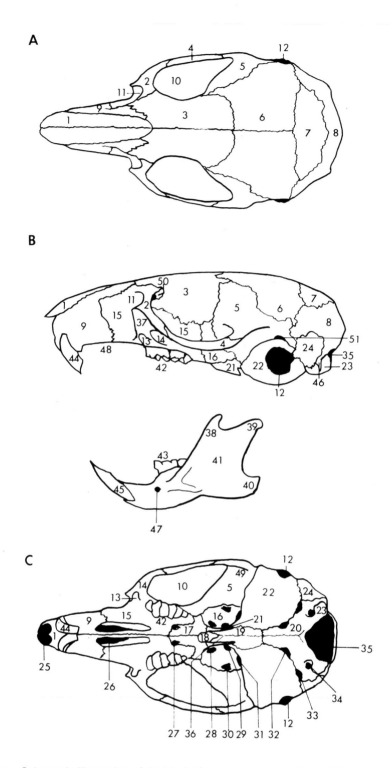

**Figure 1** Skull terminology. Schematic illustration of the skull of a rodent seen in dorsal (A), lateral (B) and ventral (C) views. Key to figure:

1 Nasal 2 Upper maxillary (zygomatic) process 3 Frontal 4 Jugal 5 Squamosal 6 Parietal 7 Interparietal 8 Supraoccipital 9 Premaxilla 10 Orbit 11 Infraorbital foramen 12 Auditory meatus (external) 13 Masseter knob 14 Lower maxillary (zygomatic) process 15 Maxilla 16 Alisphenoid 17 Palatine 18 Presphenoid 19 Basisphenoid 20 Basioccipital 21 Pterygoid 22 Auditory (tympanic) bulla 23 Exoccipital condyle 24 Mastoid 25 External nares 26 Anterior (incisive) palatal foramen 27 Posterior palatal foramen 28 Canalis nervi pterygoidei 29 Canalis basisphenoidicus 30 Foramen alare 31 Petrotympanic fissure 32 Canalis caroticus 33 Foramen jugulare 34 Canalis nervi hypoglossi 35 Foramen magnum 36 Mesopterygoid fossa 37 Zygomatic plate 38 Coronoid process 39 Condylar process 40 Angular process 41 Ascending ramus 42 Molars$^{1-3}$ 43 Molars $_{1-3}$ 44 Upper incisors 45 Lower incisors 46 Paroccipital process 47 Mental foramen 48 Diastema 49 Glenoid cavity 50 Lachrymal 51 Post-glenoid foramen.

proximations. According to Vaughan (1972) the terms *sciuromorph, myomorph* and *hystricomorph* have been used repeatedly in the past designating the major taxonomic divisions of rodents. These terms which are based on differences in skull morphology and the origin and insertion of different components of the masseter muscle are firmly entrenched in the mammalian literature. Although they have frequently been given subordinal rank, their taxonomic status is insecure. There are, however, many families and groups of families (both living and fossil), which do not readily fall into these categories and perhaps eight or more natural assemblages of rodent families occur. Wood (1955) recognises seven suborders. Rodents may, nevertheless, be squeezed somewhat arbitrarily into the classic framework by keeping the rat-like and porcupine-like suborders for important terminal groups, and using the sciuromorph category as a 'dumping-ground' for a number of primitive and variously specialised small groups.

As was indicated above, the structure of the cheekbone (zygomatic arch) and the origin as well as the insertion of the masseter muscle, has provided the classical basis of taxonomy whereby rodents are subdivided into hystricomorph (porcupine-like), sciuromorph (squirrel-like) and myomorph (rat- or mouse-like) units. This taxonomical approach has severe limitations and should at the most be seen as contrasting patterns of specialisations to solve the problems associated with mastication, but it has provided a platform to obtain some degree of order and sense in the often baffling diversity of species (fig. 2).

The zygomasseteric structure of the skull takes the configuration of the zygomatic arch, the maxillary process of the maxilla, the infraorbital canal, as well as the morphology of the masseter muscle into consideration. The zygomatic arch or zygoma consists of the squamosal process and maxillary process, connected by a third element, the jugal. It is the size and shape of the maxillary process in particular which are used in determining the major taxonomic relationships within the order.

In the rodents, the maxillary process surrounds a canal connecting the orbit with the anterior and lateral surfaces of the snout. This opening is called the infraorbital foramen (although it is situated in front of the orbit, and is therefore also known as the antorbital foramen or canal). In the majority of other mammals, this canal is small in diameter and accommodates a nerve and blood vessels. In some rodents, however, it is usually enlarged and also transmits portions of the masseter muscle complex. In rodents the masseter muscles consist of three parts. A superficial portion *(masseter superficialis)* is fairly uniform in all rodents arising far forward on the side of the face (on the maxilla) and often associated with a distinct tubercle (= masseter knob), running backwards and downwards inserting on the angular portion of the lower jaw. Two other sections of the masseter muscle, however, show variations. The middle layer *(masseter lateralis)*

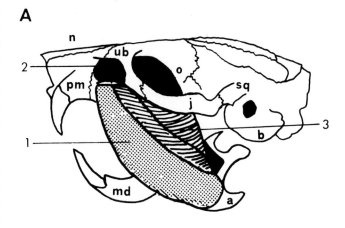

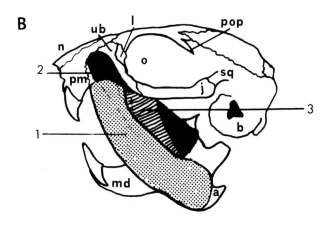

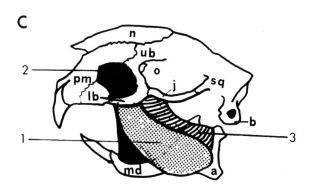

**Figure 2** Zygomasseteric patterns in rodents (highly schematic). A. The myomorph condition, B. The sciuromorph condition, C. The hystricomorph condition. 1 = *musculus masseter superficialis* 2 = *musculus masseter medialis* 3 = *musculus masseter lateralis.* Key: ub = upper branch of maxillary process; lb = lower branch of maxillary process; o = orbit; j = jugal; sq = squamosal; l = lachrymal; n = nasal; pm = premaxilla; md = mandible; b = bulla; a = angular process (not to scale).

originates from the under-edge of the zygomatic arch and inserts along the lower margin of the jaw. This muscle can have a clear anterior and posterior portion, especially in sciuromorphs. The deep layer *(masseter medialis)* also arises from the inner surface of the zygomatic arch, as well as on the rostrum (myomorphs, hystricomorphs) in front of the infraorbital canal, which it traverses and inserts higher up on the outer surface of the lower jaw.

In the Sciuromorpha, the infraorbital canal is small and primitive. There is no clear separation between upper and lower components of the maxillary process and this process forms a single surface to which the anterior head of the *masseter lateralis* is attached. The *masseter superficialis* attaches to the maxilla proper and has its origin on that bone. The *masseter medialis* is also attached to the inside of the maxillary process of the zygoma.

In the Myomorpha, the lower component of the maxillary process expands into a thin plate – the zygomatic plate – forming a vertical outer wall of the infraorbital canal, bridged dorsally by the upper component of the maxillary process. The foramen is enlarged and V-shaped: the broad, upper part transmitting strands of the *masseter medialis,* while the lower part accommodates the facial nerve. The anterior portion of the *masseter superficialis* is attached to the zygomatic plate, while the origin of the *masseter lateralis* has also shifted forward onto the lateral snout.

In the Hystricomorpha, the infraorbital canal is very big and the maxillary process has an upper and lower component as is the case in the myomorphs. It transmits a large portion of the *masseter medialis* which is greatly developed. This enlargement occurs at the expense of the smaller *masseter superficialis,* as well as the *masseter lateralis* muscles, which are smaller in extent.

*The dentition*

Rodents have a characteristic dentition. There is never more than one semi-circular scalpriform, persistently growing incisor in each side of the skull and in each lower jaw. Canine teeth and anteriorly situated premolars are always lacking and the toothless area between the incisors and the cheekteeth is known as the diastema. This gap is usually filled by a fold of skin from the upper lips, closing off the buccal cavity. This useful feature does not interfere with the free working of the incisors and prevents the filling of the buccal cavity with debris when the animal is gnawing at hard materials such as lead pipes or wood. The incisors grow throughout life from persistent pulps and are continually pushed out at the front of the premaxilla and lower jaw. In this way, portions of the incisors worn away by gnawing, are continually replaced. The gnawing action ensures that the upper incisors continually make contact with the lower incisors and these teeth are thus continually sharpened in a chisel-like fashion. Malocclusion of the incisors leads to uninterrupted growth and deformities, often with lethal consequences

for the animal. The outer surface of the incisors (often coloured yellow or orange) consists of hard enamel surrounding the inner dentine core. Many species show pronounced grooves on the outer surface. The incisors only have nerves near their roots where the teeth originate deeper in the skull or mandible. The lower incisors are often of greater diameter than the upper.

The cheektooth pattern found in rodents is more varied than in any other mammal group, while there are usually three or four (sometimes five) cheekteeth on each side of the skull. These cheekteeth, beyond the diastema, are placed in an unbroken series. They are primitively diphyondont, brachyodont, bunodont or lophodont and consist mostly of dentine, surrounded by an outer layer of harder enamel which usually forms folds or loops in the body of the tooth. The pattern on the occlusal surface is constant of diagnostic taxonomic value (fig. 3).

In many species, the molars have closed roots and the teeth eventually cease growing; in others, they have open roots and tend to grow throughout life, while a hard cement fills the space between the folds of enamel. High crowned, hypsodont (often open-rooted) teeth are encountered in species feeding on tough, hard vegetation, while low crowned molars are found in species with a mixed diet.

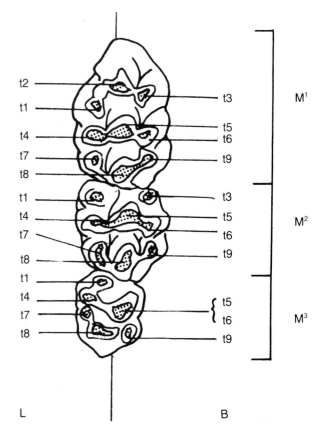

**Figure 3** Molar cusp pattern terminology. Modified and simplified after Misonne (1969). This is a left upper tooth row. L = lingual; B = buccal.

Chewing is important to rodents and the appropriate muscles are well developed, especially the internal and external pterygoid muscles, the temporalis muscle and the masseter muscle group. The skull and jaws have undergone marked modification for gnawing. The lower jaw is capable of an anterior-posterior movement, allowing considerable rotary motion on account of an elongate glenoid fossa. In conjunction with the temporal muscles, the masseter muscles pull the jaw forward and upwards and are therefore important masticatory muscles. They are usually better developed than the temporalis muscles and have their origin on the zygoma or the maxilla, anterior to the orbit, while their insertions are on the lower border of the jaw, below the molars.

## The postcranial skeleton

There are comparatively few specialisations in the postcranial skeleton. There are usually 13 thoracic and six lumbar vertebrae, their form varying in different groups. In saltatorial species, the lumbar transverse processes are long. The caudal vertebrae also show variety in structure and development.

The radius and ulna in the forearm are separate and the elbow joint permits free rotation of the forearm. The mobility of the front appendage is enhanced in many climbing rodents by the presence of a well-developed clavicle articulating with a narrow scapula with a long acromion process. The scaphoid and lunar elements are nearly always fused, while the centrale is free. An entepicondylar foramen on the humerus is usually absent. The hand usually has five digits, although the thumb (pollex) may often be vestigial or absent. The femur varies in form, but generally has a well-defined third trochanter. The tibia and fibulae are free (in sciurine and hystricine rodents) but fused distally in the murines. The foot is more variable in structure than the hand. The toes vary from three to five and as is the case in the fingers, are usually clawed. They usually walk on the entire surface of the foot or hand (plantigrade).

## The organ systems

Most species have three circumvallate papillae at the base of the tongue. Some species have cheekpouches. These are lined with hairy integument, open near the angles of the mouth and extend backwards behind the ears. The stomach may vary in form but is usually simple. There is a large caecum at the junction of the small and large intestines (except in the Gliridae), but it does not contain a spiral valve. The colon is elongated. The liver is lobed, but the lobes are variously subdivided in different species. The gallbladder may be absent.

The lungs may be without lobes or may show four lobes on the right and three lobes on the left lung. Vocal cords are present in the larynx, and rodents can make simple vocalisations.

The penis often possesses a penis bone (baculum). The testes are abdominally or inguinally situated and often descend into a distinct scrotum during the breeding season. The uterus is duplex (bicornuate) with the corna either opening separately into the vagina or united to form a corpus uteri. The placenta is discoidal and deciduate.

The brain is simple, usually with a smooth surface (furrowed in some species). Yet rodents have an excellent learning ability. Their visual acuity is moderate (poor in molerats) and their field of vision is fairly large because the eyes tend to be placed on the sides of the head. Their senses of hearing and smell are acute. Some species can hear in the ultrasonic range (up to 100 kHz). The sense of touch is intensified by the presence of guard hairs in the fur, augmenting the well-developed vibrissae. The integument is generally thin and the panniculus (a sheet of muscle underlying the skin) is rarely much developed.

There is great variety in the structure of the pelage of rodents. The pelage may consist of a single type of hair, but more often different specialised kinds of hair occur on different areas of the body, such as whiskers (vibrissae), quills, spines, bristles, guard hairs, wool and underfur. However, there is not a very clear definition of this terminology and the relationships of the one to the other are intergrading and vague. Rosevear (1969) discusses this problem at greater length and is paraphrased in the following paragraphs. Taking the pelage of the Brown House Rat *Rattus norvegicus* as a reference type, four categories of hair on the dorsal surface of the animal are to be found:

*Category 1:* 'Underfur'. Extremely fine and almost invariably widely wavy; of uniform diameter, except for a tapering tip, sometimes pigmented distally.

*Category 2:* 'Sub-bristle'. A straight hair with a long, often flat shaft (stipes), not quite as fine as the underfur; expanding distally to an elongated blade (lamella), shallowly concavo-convex from front to back, often pigmented, tapering to a point of subcircular secondary contour hair.

*Category 3:* 'Gutter hair'. A relatively broad bristle openly channelled concavo-convexly virtually throughout its entire length. There is no distinction into stipe and lamella, while distally they sometimes sharpen to a point. The distal end is often pigmented, the tip often darkly so. When present, these hairs are always the broadest and most conspicuous component of the fur, often distinguishable with the naked eye. The number, breadths and lengths of these bristles impart the harshness or spring to the pelage and when present in quantity, form the primary contour element. In *Acomys*, they attain a breadth of 0,5 mm and except for a little, scattered, short underfur, form the whole coat.

*Category 4:* 'Guard hair'. Long bristles, almost circular in section, tapering gradually to a base almost imperceptibly stouter than the underfur, hence with little demarcation into stipe and lamella. The tip is heavily pigmented distally and projects beyond the coat contour and sometimes almost to the base. Distally, they often also have fine, colourless tips. They are never a major element in the pelage and serve a tactile function. According to Rosevear (1969), the categories referred to above are distinct. The body is given its contour by any one, or combination of the first three categories, the distal portions and tips of which form the visible coat. The belly fur bears the same characteristics as that of the back (described above), except that it is of different colour, shorter and lacks guard hairs. Juveniles may lack certain elements found in adults and all coats are subject to moults.

It has generally been acknowledged in the past that rodents have no sweat glands. A survey of the literature by Green and Rodgers *(pers. comm.)* showed that sweat glands of *Rattus* had already been described in 1896 and many publications on sweat gland histology, the mechanism of sweating and even the anyalysis of sweat of laboratory rats follows. No reference was found to sweat glands in any of the southern African rodents. Green and Rodgers are subsequently attempting to establish the presence of sweat glands in southern African species. Sweat glands were found in *Thryonomys, Pedetes, Xerus, Mystromys, Tatera, Desmodillus, Saccostomus, Otomys, Aethomys, Rhabdomys, Lemniscomys, Praomys, Mus minutoides* and *Acomys,* restricted to small areas of the integument. Well-developed eccrine sweat glands were found in the naked pads of the digits as well as the palmar and plantar soles of all the Muridae, Sciuridae and the majority of Cricetidae examined. In *Pedetes,* apocrine sweat glands occur on the hands, while in *Thryonomys,* sweat glands are present in the relatively naked areas behind the ears.

## The fossil record

The origin of rodents is obscure and *Paramys,* a species appearing in the late Palaeocene, is already a typical, primitive true rodent. Romer (1971) presumed that rodents arose from basal, insectivorous, placental stock, but transitional forms are unknown. It is clear, however, that once they appeared, they almost immediately occupied an important niche in terrestrial environments and have continually and increasingly flourished from the Eocene to the present (Colbert 1969). For reviews dealing with the fossil record of the African hysticomorphs and murids respectively, see Wood (1974) and Misonne (1969).

## Classification

It is commonly agreed that rodents are characterised by teeth uniquely adapted to gnawing with the aid of continuously growing upper and lower incisors. There, unfortunately, the agreement ends as Nel (1969) has laconically observed. The scope of the order does not present difficulties, but the relative position and relationships of infraordinal taxa are still hotly debated. Parallelism and convergence occur to an amazing degree in rodents, geographically, and presumably phylogenetically, far removed. To this source of confusion one must also add the extreme paucity of fossil forms with which the living members can be compared.

The literature on the classification of rodents is voluminous. Brandt (1855), Tullberg (1899), Miller & Gidley (1918), Allen (1939), Ellerman (1940, 1941), Winge (1941), Simpson (1945), Stehlin & Schaub (1951), Ellerman, Morrison-Scott & Hayman (1953), Schaub (1953b), Lavocat (1956b), Wood (1955, 1958, 1959), Wood & Patterson (1959), Vandebroek (1966), Anderson (1967) and Lavocat (1974) have all attempted to define a satisfactory rodent classification. The order Rodentia has been sliced taxonomically in many ways and if one considers the different bases for the systems proposed by Stehlin & Schaub and that of Wood '. . . the amount of agreement between the systems is more remarkable that the disagreement' (Anderson 1967). A comparison of Anderson's classification (1967) based on Simpson's (1945) with that of Wood's (1955) classification, is included to emphasise the lack of uniformity but also to point out the considerable agreement at superfamily level.

## The role of rodents

Rodents are of great importance to mankind and knowledge of these creatures is of great value. Rodents often compete directly with man. They thrive on plants and crops developed and domesticated by man, thereby competing directly with man for available food (Myllymaki 1975). In many cases they are pests, feeding on and destroying agricultural crops and the damage they do to food stores is well known, where food is consumed or spoiled in silos, warehouses and individual homes. They often carry parasites that transmit diseases to which man is susceptible (Arata 1975; Gear 1954). It is well known that rats and mice, together with fleas, have played an important role in the history of mankind, not only as living reservoirs of the plague bacillus (Davis 1948, 1953), but also for many other contagious diseases. Rats and other rodents spread leptospirosis (through their urine), toxoplasmosis, rabies, trichinosis, murine typhus, relapsing fever, plague, helminths and salmonellosis to mention but a few (Hobson 1969). Vast sums of money are lost through damage inflicted by rodents on wood products, insulation and a wide variety of objects which are not safe from their incessant gnawing.

On the other hand, some rodent species are used for their fur and as a source of food and protein, especially in more primitive societies. For additional information, see Hanney (1975). Others are used extensively in research. As experimental methodology in medicine and biology advanced, so did the increasing need for laboratory animals. Freye *(in Grzimek 1969)* states that an animal was involved in every great medical or biological discovery and in 90% of these cases, the animal was a rodent. In the National Institute of Health in Maryland, USA, an important medical research institute, some 800 000 mice, 300 000 rats and 200 000 guinea pigs and hamsters were used in 1965 (Freye *in* Grzimek 1969). In May 1973 it was reported to the British House of Commons that in excess of 5,5 million experiments on living animals took place in the country annually and that 95% of the species used were rodents. It is clear that the phenomenal progress of medical research would not have been possible without the use of rodents as experimental animals (Hanney 1975). Rodents have consequently become biological reagents for man.

Finally, it is feasible that natural populations of rodents play an important part in the destruction of harmful insects and their eggs, as well as in the eradication of weeds and other unwanted plants by consuming the seeds or the plants themselves. Rodents also form the staple diet for many other mammals, birds and reptiles, thereby occupying a key position in the intricate balance of nature.

## A blueprint for survival

Hanney (1975) has succinctly described rodents as a group using a 'blueprint' for success. Hanney's remarks are briefly paraphrased in the following sentences. The blueprint refers in the first instance to the everlasting and self-sharpening incisors which are situated in a sturdy skull where an efficient jaw mechanism has evolved. Consequently the individual rodent can utilise diverse kinds of food, satisfying the energy requirements needed for an active lifestyle. Thirdly, a rodent has excellent sensory organs as well as an adequately developed brain, which are housed in a robustly constructed body with a full complement of effective accessories (hands, digits) showing infinite adjustability. All these characteristics finally lead to a rapid production rate and the consequent highly adapted and successful biological populations encountered in nature.

# Two systems of classification of living rodents

**WOOD (1955)**

Suborder SCIUROMORPHA
  Superfamily Aplodontoidea
    Family Aplondontidae
  Superfamily Sciuroidea
    Family Sciuridae
? SCIUROMORPHA
  Superfamily Ctenodactyloidea
    Family Ctenodactylidae

Suborder THERIDOMYOMORPHA
  Superfamily Anomaluroidea
    Family Anomaluridae
? SCIUROMORPHA or THERIDOMYOMORPHA
    Family Pedetidae

Suborder CASTORIMORPHA
  Superfamily Castoroidea
    Family Castoridae

Suborder MYOMORPHA
  Superfamily Muroidea
    Family Cricetidae
    Family Muridae
  ? Muroidea
    Family Spalacidae
    Family Rhizomyidae
  Superfamily Geomyoidea
    Family Heteromyidae
    Family Geomyidae
  Superfamily Dipodoidea
    Family Zapodidae
    Family Dipodidae
? MYOMORPHA
  Superfamily Gliroidea
    Family Gliridae
    Family Seleviniidae

Suborder CAVIOMORPHA
  Superfamily Octodontoidea
    Family Octodontidae
    Family Echimyidae
    Family Ctenomyidae
    Family Abrocomidae
  Superfamily Chinchilloidea
    Family Chinchillidae
    Family Capromyidae
  Superfamily Cavioidea
    Family Caviidae
    Family Hydrochoeridae
    Family Dinomyidae
    Family Heptaxodontidae
    Family Dasyproctidae
    Family Cuniculidae
  Superfamily Erethizontoidea
    Family Erethizontidae

Suborder HYSTRICOMORPHA
  Superfamily Hystricoidea
    Family Hystricidae
  Superfamily Thryonomyoidea
    Family Thryonomyidae
    Family Petromyidae

Suborder BATHYERGOMORPHA
  Superfamily Bathyergoidea
    Family Bathyergidae

**ANDERSON (1967), modified version
of Simpson (1945)**

'SCIUROMORPHA'
  Superfamily Aplodontoidea
    Family Aplodontidae
  Superfamily Sciuroidea
    Family Sciuridae
  Superfamily Geomyoidea
    Family Geomyidae
    Family Heteromyidae
  Superfamily Castoroidea
    Family Castoridae
  Superfamily Anomaluroidea
    Family Anomaluridae
    Family Pedetidae

'MYOMORPHA'
  Superfamily Muroidea
    Family Cricetidae
    Family Spalacidae
    Family Rhizomyidae
    Family Muridae
  Superfamily Gliroidea
    Family Gliridae
    Family Platacanthomyidae
    Family Seleviniidae
  Superfamily Dipodoidea
    Family Zapodidae
    Family Dipodidae

'HYSTRICOMORPHA'
  Superfamily Hystricoidea
    Family Hystricidae
  Superfamily Erethizontoidea
    Family Erethizontidae
  Superfamily Cavioidea
    Family Caviidae
    Family Hydrochoeridae
    Family Dinomyidae
    Family Heptaxodontidae
    Family Dasyproctidae
  Superfamily Chinchilloidea
    Family Chinchillidae
  Superfamily Octodontoidea
    Family Capromyidae
    Family Myocastoridae
    Family Octodontidae*
    Family Ctenomyidae
    Family Abrocomidae
    Family Echimyidae
    Family Thryonomyidae
    Family Petromyidae
  Superfamily Bathyergoidea
    Family Bathyergidae
  Superfamily Ctenodactyloidea
    Family Ctenodactylidae

*Including the family Ctenomyidae.
Anderson considered the Ctenomyidae as a separate family.

# Rodentia: general suprageneric classification scheme

The general suprageneric classification scheme of living rodents followed in this book shows southern African taxa in CAPITALS (mostly after Simpson (1945)). However, the Cricetomyinae, Dendromurinae, Petromysci-nae and Otomyinae are here placed under the Cricetidae and not with the Muridae, and the Bathyergidae have been split into two subfamilies.

**SCIUROMORPHA**

Aplodontoidea
Aplodontidae
SCIUROIDEA
SCIURIDAE
SCIURINAE
Petauristinae

Geomyoidea
Geomyidae
Geomyinae
Heteromyidae
Perognathinae
Dipodomyinae
Heteromyinae

Castoroidea
Castoridae
Castorinae

**SCIUROMORPHA**
*incertae sedis*

**HYSTRICOMORPHA**

HYSTRICOIDEA
HYSTRICIDAE
HYSTRICINAE
Atherurinae

Erethizontoidea
Erethizontidae
Erethizontinae
Chaetomyinae

Cavioidea
Caviidae
Caviinae
Dolichotinae
Hydrochoeridae
Hydrochoerinae
Dinomyidae
Dasyproctidae
Cuniculinae
Dasyproctinae

Chinchilloidea
Chinchillidae

OCTODONTOIDEA
Capromyidae
Octodontidae
Ctenomyidae
Abrocomidae
Echimyidae
Echimyinae
Dactylomyinae
THRYONOMYIDAE
PETROMURIDAE

**HYSTRICOMORPHA**
*incertae sedis*
BATHYERGOIDEA
BATHYERGIDAE
BATHYERGINAE
GEORYCHINAE

**MYOMORPHA**

MUROIDEA
CRICETIDAE
CRICETINAE
Nesomyinae
Lophiomyinae
Microtinae
GERBILLINAE
CRICETOMYINAE
DENDROMURINAE
PETROMYSCINAE
OTOMYINAE

Spalacidae
Rhizomyidae
MURIDAE
MURINAE
Phloeomyinae
Rhynchomyinae
Hydromyinae

GLIROIDEA
GLIRIDAE★
Muscardininae
GRAPHIURINAE
Platacanthomyidae
Seleviniidae

Dipodoidea
Zapodidae
Sicistinae
Zapodinae
Dipodidae
Dipodinae
Cardiocraniinae
Euchoreutinae

**?Hystricomorpha or**
**?Myomorpha**
*incertae sedis*

Ctenodactyloidea
Ctenodactylidae

★ = Muscardinidae

9

# *Checklist of southern African rodents*

ORDER      # Rodentia Bowdich, 1821

| | |
|---|---|
| Suborder | SCIUROMORPHA Brandt, 1855 |
| Superfamily | SCIUROIDEA Gill, 1872 |
| Family | SCIURIDAE Gray, 1821 |
| Subfamily | SCIURINAE Baird, 1857 |
| Genus | *Xerus* Ehrenberg, 1833 |
| | *X. inauris* (Zimmermann, 1780). Cape Ground Squirrel |
| | *X. princeps* (Thomas, 1929). Kaokoveld Ground Squirrel |
| Genus | *Heliosciurus* Trouessart, 1880 |
| | *H. rufobrachium* (Waterhouse, 1842). Red-legged Sun Squirrel |
| Genus | *Funisciurus* Trouessart, 1880 |
| | *F. congicus* (Kuhl, 1820). Striped Tree Squirrel |
| Genus | *Paraxerus* Major, 1893 |
| | *P. cepapi* (A. Smith, 1836). Bush Squirrel |
| | *P. palliatus* (Peters, 1852). Red Bush Squirrel |
| | |
| Suborder | SCIUROMORPHA *incertae sedis* |
| Superfamily | ANOMALUROIDEA Gill, 1872 |
| Family | PEDETIDAE Owen, 1847 |
| Genus | *Pedetes* Illiger, 1811 |
| | *P. capensis* (Forster, 1778). Springhare |
| | |
| Suborder | HYSTRICOMORPHA Brandt, 1855 |
| Superfamily | HYSTRICOIDEA Gill, 1872 |
| Family | HYSTRICIDAE Burnett, 1830 |
| Subfamily | HYSTRICINAE Murray, 1866 |
| Genus | *Hystrix* Linnaeus, 1758 |
| | *H. africaeaustralis* Peters, 1852. Cape Porcupine |
| Superfamily | OCTODONTOIDEA Simpson, 1945 |
| Family | THRYONOMYIDAE Pocock, 1922 |
| Genus | *Thryonomys* Fitzinger, 1867 |
| | *T. swinderianus* (Temminck, 1827). Greater Canerat |
| | *T. gregorianus* (Thomas, 1894). Lesser Canerat |
| Family | PETROMURIDAE Tullberg, 1899 |
| Genus | *Petromus* A. Smith, 1831 |
| | *P. typicus* A. Smith, 1831. Dassie Rat |
| | |
| Suborder | HYSTRICOMORPHA *incertae sedis* |
| Superfamily | BATHYERGOIDEA Osborn, 1910 |
| Family | BATHYERGIDAE Waterhouse, 1841 |
| Subfamily | BATHYERGINAE Roberts, 1951 |
| Genus | *Bathyergus* Illiger, 1811 |
| | *B. suillus* (Schreber, 1782). Cape Dune Molerat |
| | *B. janetta* Thomas & Schwann, 1904. Namaqua Dune Molerat |
| Subfamily | GEORYCHINAE Roberts, 1951 |

| | |
|---|---|
| Genus | *Heliophobius* Peters, 1846 |
| | *H. argenteocinereus* Peters, 1846. Silvery Molerat |
| Genus | *Cryptomys* Gray, 1864 |
| | *C. hottentotus* (Lesson, 1826). Common Molerat |
| Genus | *Georychus* Illiger, 1811 |
| | *G. capensis* (Pallas, 1778). Cape Molerat |
| | |
| Suborder | MYOMORPHA Brandt, 1855 |
| Superfamily | MUROIDEA Miller & Gidley, 1918 |
| Family | CRICETIDAE Rochebrune, 1883 |
| Subfamily | CRICETINAE Murray, 1866 |
| Genus | *Mystromys* Wagner, 1841 |
| | *M. albicaudatus* (A. Smith, 1834). White- tailed Rat |
| Subfamily | GERBILLINAE Alston, 1876 |
| Genus | *Gerbillurus* Shortridge, 1942 |
| | *G. paeba* (A. Smith, 1836). Hairy-footed Gerbil |
| | *G.*sp.aff. *paeba* vide Davis (1975) |
| | *G. tytonis* (Bauer & Niethammer, 1959). Dune Hairy-footed Gerbil |
| | *G. vallinus* (Thomas, 1918). Brush-tailed Hairy-footed Gerbil |
| | *G. setzeri* (Schlitter, 1973). Setzer's Hairy-footed Gerbil |
| Genus | *Tatera* Lataste, 1882 |
| | *T. leucogaster* (Peters, 1852). Bushveld Gerbil |
| | *T. afra* (Gray, 1830). Cape Gerbil |
| | *T. brantsii* (A. Smith, 1836). Highveld Gerbil |
| | *T. inclusa* Thomas & Wroughton, 1908. Gorongoza Gerbil |
| Genus | *Desmodillus* Thomas & Schwann, 1904 |
| | *D. auricularis* (A. Smith, 1834). Short-tailed Gerbil |
| Subfamily | CRICETOMYINAE Roberts, 1951 |
| Genus | *Cricetomys* Waterhouse, 1840 |
| | *C. gambianus* Waterhouse, 1840. Giant Rat |
| Genus | *Saccostomus* Peters, 1846 |
| | *S. campestris* Peters, 1846. Pouched Mouse |
| Subfamily | DENDROMURINAE Allen, 1939 |
| Genus | *Dendromus* A. Smith, 1829 |
| | *D. mesomelas* (Brants, 1827). Brants' Climbing Mouse |
| | *D. mystacalis* Heuglin, 1863. Chestnut Climbing Mouse |
| | *D. melanotis* A. Smith, 1834. Grey Pygmy Climbing Mouse |
| | *D. nyikae* Wroughton, 1909. Nyika Climbing Mouse |
| Genus | *Steatomys* Peters, 1846 |
| | *S. pratensis* Peters, 1846. Fat Mouse |
| | *S. parvus* Rhoads, 1896. Tiny Fat Mouse |
| | *S. krebsii* Peters, 1852. Krebs' Fat Mouse |
| Genus | *Malacothrix* Wagner, 1843 |
| | *M. typica* (A. Smith, 1834). Large-eared Mouse |
| Subfamily | PETROMYSCINAE Roberts, 1951 |
| Genus | *Petromyscus* Thomas, 1926 |
| | *P. collinus* (Thomas & Hinton, 1925). Pygmy Rock Mouse |
| | *P. monticularis* (Thomas & Hinton, 1925). Berseba Rock Mouse |
| Subfamily | OTOMYINAE Thomas, 1897 |
| Genus | *Otomys* Cuvier, 1823 |
| | *O. irroratus* (Brants, 1827). Vlei Rat |
| | *O. angoniensis* Wroughton, 1906. Angoni Vlei Rat |
| | *O. saundersae* Roberts, 1929. Saunders' Vlei Rat |
| | *O. laminatus* Thomas & Schwann, 1905. Laminate Vlei Rat |
| | *O. unisulcatus* F. Cuvier, 1829. Bush Karoo Rat |
| | *O. sloggetti* Thomas, 1902. Rock Karoo Rat |

| Genus | *Parotomys* Thomas, 1918 |
| | *P. brantsii* (A. Smith, 1834). Brants' Whistling Rat |
| | *P. littledalei* Thomas, 1918. Littledale's Whistling Rat |
| Family | MURIDAE Gray, 1821 |
| Subfamily | MURINAE Murray, 1866 |
| Genus | *Zelotomys* Osgood, 1910 |
| | *Z. woosnami* (Schwann, 1906). Woosnam's Desert Rat |
| Genus | *Thamnomys* Thomas, 1907 |
| Subgenus | *Grammomys* Thomas, 1915 |
| | *T. (G.) dolichurus* (Smuts, 1832). Woodland Mouse |
| | *T. (G.) cometes* Thomas & Wroughton, 1908. Mozambique Woodland Mouse |
| Genus | *Dasymys* Peters, 1875 |
| | *D. incomtus* (Sundevall, 1847). Water Rat |
| Genus | *Pelomys* Peters, 1852 |
| Subgenus | *Pelomys* Peters, 1852 |
| | *P. fallax* Peters, 1852. Creek Rat |
| Genus | *Aethomys* Thomas, 1915 |
| Subgenus | *Aethomys* Thomas, 1915 |
| | *A. (A.) chrysophilus* (De Winton, 1897). Red Veld Rat |
| | *A. (A.) nyikae* (Thomas, 1897). Nyika Veld Rat |
| | *A. (A.) silindensis* Roberts, 1938. Selinda Veld Rat |
| Subgenus | *Micaelamys* Ellerman, 1941 |
| | *A. (M.) namaquensis* (A. Smith, 1834). Namaqua Rock Mouse |
| | *A. (M.) granti* (Wroughton, 1908). Grant's Rock Mouse |
| Genus | *Thallomys* Thomas, 1920 |
| | *T. paedulcus* (Sundevall, 1846). Tree Rat |
| Genus | *Lemniscomys* Trouessart, 1881 |
| | *L. griselda* (Thomas, 1904). Single-striped Mouse |
| Genus | *Rhabdomys* Thomas, 1916 |
| | *R. pumilio* (Sparrman, 1784). Striped Mouse |
| Genus | *Praomys* Thomas, 1915 |
| Subgenus | *Mastomys* Thomas, 1915 |
| | *P. (M.) natalensis* (A. Smith, 1834). Multimammate Mouse |
| | *P. (M.) shortridgei* (St Leger, 1933). Shortridge's Mouse |
| Subgenus | *Myomyscus* Shortridge, 1942 |
| | *P. (Myomyscus) verreauxii* (A. Smith, 1834). Verreaux's Mouse |

| Genus | *Rattus* Fischer, 1803 |
|---|---|
| | *R. rattus* (Linnaeus, 1758). House Rat |
| | *R. norvegicus* (Berkenhout, 1769). Brown House Rat |
| Genus | *Mus* Linnaeus, 1758 |
| Subgenus | *Mus* Linnaeus, 1758 |
| | *M. (M.) musculus* Linnaeus, 1758. House Mouse |
| Subgenus | *Leggada* Gray, 1837 |
| | *M. (L.) minutoides* A. Smith, 1834. Pygmy Mouse |
| Genus | *Acomys* I. Geoffroy, 1838 |
| | *A. subspinosus* (Waterhouse, 1838). Cape Spiny Mouse |
| | *A. spinosissimus* Peters, 1852. Spiny Mouse |
| Genus | *Uranomys* Dollman, 1909 |
| | *U. ruddi* Dollman, 1909. Rudd's Mouse |
| Superfamily | GLIROIDEA Simpson, 1945 |
| Family | GLIRIDAE Thomas, 1897 |
| Genus | *Graphiurus* Smuts, 1832 |
| Subgenus | *Graphiurus* Smuts, 1832 |
| | *G. (G.) ocularis* (A. Smith, 1829). Smith's Dormouse |
| Subgenus | *Claviglis* Jentink, 1888 |
| | *G. (C.) platyops* Thomas, 1897. Rock Dormouse |
| | *G. (C.) murinus* (Desmarest, 1822). Woodland Dormouse |

**ANALYSIS**

| Suborder | Super-families | Families | Sub-families | Genera | Species |
|---|---|---|---|---|---|
| Sciuromorpha | 1 | 1 | 1 | 4 | 6 |
| Sciuromorpha *incertae sedis* | 1 | 1 | — | 1 | 1 |
| Hystricomorpha | 2 | 3 | 1 | 3 | 4 |
| Hystricomorpha *incertae sedis* | 1 | 1 | 2 | 4 | 5 |
| Myomorpha | 2 | 3 | 7 | 27 | 57 |
| | 7 | 9 | 11 | 39 | 73 |

# Key to the suborders and their families in southern Africa

1 Lower jaw specialised by outward distortion of angular portion by a specialised limb of *masseter lateralis superficialis* ........................................ 2

   Lower jaw normal, not as described above ............ 3

2 Infraorbital foramen much enlarged for muscle transmission, wider below than above; zygomatic plate below the infraorbital foramen, not tilted upwards; fibula well developed, not fused with tibia; cheekteeth 4/4, occlusal surfaces with infolds and islands of enamel; eyes well developed, ear pinnae present; pelage modified to quills, bristles or springy hairs; animal poorly adapted to fossorial mode of life .................................... HYSTRICOMORPHA
(HYSTRICIDAE, Porcupine;
THRYONOMYIDAE, Canerats;
PETROMURIDAE, Dassie Rat)
Page 45

   Infraorbital foramen small, not transmitting muscle; zygomatic plate below infraorbital foramen; fibula reduced and fused with tibia; cheekteeth 4/4, occlusal surfaces simple with enamel round edges; eyes suppressed, ear pinnae absent; pelage soft; animal adapted to fossorial mode of life ..........................
.....................HYSTRICOMORPHA *incertae sedis*
(BATHYERGIDAE, Molerats)
Page 63

3 Infraorbital foramen not transmitting muscle with zygomatic plate tilted upwards; jugal bone long, usually extending to the lachrymal; skull with postorbital processes; cheekteeth cuspidate, upper molars typically with series of transverse ridges with cusps at corners; cheekteeth 4/4 or 5/4; tail busy; fibula not fused to tibia ........... SCIUROMORPHA
(SCIURIDAE, Squirrels)
Page 15

   Infraorbital foramen always enlarged for muscle transmission; frontals without postorbital processes
.............................................................. 4

4 Zygomatic plate narrow and below infraorbital foramen; premolars not suppressed, normally as large as the molars; molars simplified in occlusal pattern; cheekteeth 4/4, rootless; external form modified for bipedal and saltatorial way of life with long hindlimbs and tail; skull specialised with mastoids inflated, frontals wide and zygoma thickened ............
SCIUROMORPHA *incertae sedis*
(PEDETIDAE, Springhare)
Page 37

   Zygomatic plate tilted upwards and broadened alongside the infraorbital foramen, the latter well open for muscle transmission; cheekteeth rooted, either 3/3 or 4/4; animals not modified for saltatorial or bipedal way of life; skull not particularly specialised; fibula fused with tibia................................ 5

5 Infraorbital foramen flattened by zygomatic plate, the latter more or less tilted upwards; premolars suppressed, cheekteeth not exceeding 3/3; jugal usually short; tail usually not bushy ......................
MYOMORPHA
(CRICETIDAE, White-tailed Rat,
Gerbils, Giant Rat, Pouched
Mouse, Climbing Mice, Fat Mice,
Large-eared Mouse, Rock Mice,
Vlei Rats, Karoo Rats,
Whistling Rats)
Page 83
(MURIDAE, typical Rats and Mice)
Page 163

   Infraorbital foramen not flattened by zygomatic plate, the latter not tilted upwards; cheekteeth 4/4; jugal usually long; tail bushy ........ MYOMORPHA
(GLIRIDAE, Dormice)
Page 245

# Sciuromorpha
# SCIURIDAE: Squirrels

# SUBORDER Sciuromorpha Brandt, 1855

The squirrels and their kin are an interesting assemblage of rodents. The family comprises some 50 genera and slightly fewer than 400 species, found in a diversity of habitats throughout the world, excepting Australia, New Zealand and New Guinea and adjoining territories, the island of Malagasy, southern South America, the polar regions and the true desert areas like the Sahara and Arabia. They are active and conspicuous animals and generally well known to man. Like many other mammals, they have undergone marked adaptive radiation during their evolution; arboreal, semi-terrestrial, terrestrial, semi-fossorial and 'flying' species have developed. Most, however, are predominantly arboreal. The suborder also includes the superfamilies Aplodontoidea (mountain beavers, North America), Geomyoidea (pocket gophers, North America) and Castoroidea (beavers, North America and Europe).

Approximately four extinct genera are known (Anderson & Jones 1967). Palaeontological data show that the squirrels were established in Eurasia and North America by the Miocene, '... after which time they probably also invaded Africa, South America was invaded much later, in the Pleistocene' (Kingdon 1974). The Eurasian origin of squirrels is also shown by the presence in that part of the world of some 33 relatively diverse genera, mostly occurring in southeast Asia (Kingdon 1974).

The head of a squirrel is relatively small and there is no clearly defined neck region separating the body from the head. The eyes are generally well developed, implying keen vision with short ears and face region. The pelage of the body may be soft but often consists of harsh, bristly hair, while the well-developed tail is usually covered with long, bristlelike hairs without a terminal tuft. The tail is never prehensile as is the case in some South American monkeys. It plays an important role in balancing the animal and is also used in communication between individuals. The 'language' of tail waving has broad similarities in many species (Kingdon 1974).

The hindlimbs are generally better developed than the forelimbs. The latter has four digits (the thumb or pollex being rudimentary and usually provided with a flattish nail) while the feet usually have five digits. All the digits are clawed, adapted for climbing or digging, depending on whether the species is predominantely arboreal or terrestrial.

Structural features in the skull indicate that they are close to the generalised ancestral stock from which all rodents evolved. The skull is strongly built with a small but pronounced postorbital process on the frontal bone, a feature unique to the squirrels. The jugal is long, usually in contact with the lachrymal and is not supported ventrally by a backward continuation of the maxillary process. The infraorbital foramen is small and rounded and does not transmit muscle. The palate between the molar rows is broad.

The incisors have an enamel covering confined to their front surfaces. The molars are rooted, brachydont or hypsodont, and may consist of one or two premolars above, followed by the usual three molars, while only one premolar and three molars are present in the lower jaw. The presence of these premolars is a primitive condition. The total number of grinding teeth may therefore be 20 or 22. The occlusal surfaces of the molars are characterised by wavy crossbands of enamel, forming prominent cusps and ridges.

As is to be expected in agile animals with a good climbing ability, the clavicles are well developed. The tibia and fibula are never fully fused.

Squirrels are mostly diurnal and the arboreal species make their abodes in hollow trees. They feed mostly on seeds, nuts and fruits. They often forage on the ground, to collect their vegetable food, usually eating it on the spot. Some terrestrial species, e.g. *Xerus*, often take insects as do the Funisciurina, a subtribe of the Funambulini. When eating, the food is grasped between the hands while the animal sits on its haunches. The tail is usually tilted up over the back and the vernacular name, squirrel, is probably derived from the Greek *skioura (skia* = shade, *oura* = tail) indicating that the animal often sits in the shade of its own tail.

Some eight genera occur in Africa, of which four genera and six species occur in southern Africa, predominantly in the tropical and temperate parts. Following Moore (1959), two genera *(Funisciurus* and *Paraxerus)* are representatives of the Tribe Funambulini, the African and Asian tree squirrels. The third genus, *Xerus*, is a member of the Tribe Xerini, the African ground squirrels, while the fourth genus, *Heliosciurus,* the sun squirrels, is classified under the Tribe Protoxerini. Indigenous squirrels are absent from the southwestern winter rainfall region. Moore (1959) used the number of transbullar septa in the auditory bulla as important evidence for the classification of squirrels. *Funisciurus* has one septum, while *Xerus* has three septa and *Heliosciurus* has from zero to two and a half septa (usually only one).

The North American species *Sciurus carolinensis* occurring in Cape Town and environs, was introduced by Cecil John Rhodes, probably soon after the start of the 20th century (Bigalke 1939). As is so often the case with alien introductions, it has multiplied to such an extent that it has virtually overrun the Cape Peninsula, attaining pest proportions in domestic gardens and orchards as well as in indigenous patches of forest on Table Mountain. The species figured on the Cape Vermin List during the years 1918–1922 when rewards were paid out for no fewer than 11 188 specimens (Bigalke 1939). See Davis (1950) for additional data on the status of this exotic species in the southwestern Cape.

SUPERFAMILY **Sciuroidea** Gill, 1872

FAMILY **Sciuridae** Gray, 1821

SUBFAMILY **Sciurinae** Baird, 1857

**Key to the southern African sciurid genera**
(Modified after Roberts (1951), Meester, Davis & Coetzee (1964) and Amtmann (1975))

1 Fur bristly; ears small; third finger of hand normally longest; claws on hand and foot adapted to digging; a pale white stripe down each side of the body; lachrymal enlarged; palate well over half of occipitonasal length; four upper molars; two pairs of mammae ......
*Xerus*
(Ground Squirrels)
Page 18
Fur soft; ears conspicuous; fourth finger of hand normally longest; claws on hand and foot adapted for climbing; lateral body stripe present or absent; lachrymal not enlarged; palate normally clearly less than half of occipitonasal length; four or five upper molars; two or three pairs of mammae ................ 2
2 Four upper molars; occlusal pattern unspecialised; three pairs of mammae; no lateral body stripe ...........
*Heliosciurus*
(Sun Squirrel)
Page 24
Five upper molars; occlusal pattern specialised, molars tending to become flatcrowned with isolated deep folds in adult; females with either two or three pairs of mammae ............................................. 3
3 Cheekteeth of upper and lower jaw tend to become flat-crowned in adult; with both white and dark lateral stripes along body; two pairs of mammae .......
*Funisciurus*
(Striped Tree Squirrel)
Page 27
Cheekteeth of lower jaw cuspidate in adult (although upper ones flatcrowned); no lateral body stripe; three pairs of mammae ................. *Paraxerus*
(Tree Squirrels)
Page 30

GENUS *Xerus* Ehrenberg, 1833

The Cape Ground Squirrel *Xerus inauris* was first mentioned by Pennant in 1780, who referred to it as an 'earless dormouse' in his *History of Quadrupeds*. In the same year Zimmermann described it as *Sciurus inauris* from 'Kaffirland' some 100 miles north of the Cape of Good Hope, based on Pennant's information.

In 1792 Kerr referred to specimens from an area in the Karoo to the north of the Sneeuberge (probably near Graaff-Reinet) as *Sciurus capensis* after Gmelin had referred to this species as *Sciurus dschinschicus* in 1788. Since then the species has been described under numerous names, resulting in a long list of synonyms. (For additional detailed information, see Sclater (1901), Ellerman (1940), Roberts (1951) and Ellerman *et al.* (1953).) In 1834 Smith proposed the generic name *Geosciurus*, which was revived and upheld during the first four decades of the 20th century, but it was demoted to subgeneric rank by Ellerman in 1940. In *Xerus* I include *Geosciurus* (Moore

1959), an interpretation adopted by Amtmann (1975). Two monotypic species *(X. inauris* and *X. princeps)* occur in southern Africa.

Because of similar habits, general overall size and appearance, these rodents are often confused with small suricates (i.e. carnivores) known as 'meerkats' *(Suricata* sp.) in South Africa. This is reflected in the Afrikaans vernacular name of 'waaierstertmeerkat'. The squirrel can rapidly be identified by its incisor teeth, the absence of canines and the resultant diastema.

---

**Key to the southern African species of *Xerus***
(Modified after Meester *et al.* (1964) and Amtmann (1975))

1 Longitudinal flank stripe; white or buffy; incisors white; hairs of tail black at base with another black band adjoining white tip; orbit diameter normally less than 1/3 of occipitonasal length.........*X. inauris*
(Cape Ground Squirrel)
Page 19

Longitudinal flank stripe; white or buffy; incisors orange; three blackish bands on hairs of tail; orbit diameter normally more than 1/3 of occipitonasal length.............................................. *X. princeps*
(Kaokoveld Ground Squirrel)
Page 22

---

# *Xerus inauris* (Zimmermann, 1780)

**Cape Ground Squirrel**

**Grondeekhoring**

**Erdmannetjie**

**Waaierstertmeerkat**

The generic name is derived from the Greek *xeres* = dry, arid, indicating the preferred habitat of the animal. The specific name refers to a combination of the Latin *in* = not, without and *auris* = ear, in reference to the very small ear pinnae of the species.

## Outline of synonymy

1780 *Sciurus inauris* Zimmermann, *Geogr. Geschichte* 2: 344. 'Kaffirland, 100 miles north of the Cape of Good Hope.'

1788 *S. dschinschicus* Gmelin, *Linn. Syst. Nat.* 1:151. Probably from South Africa.

1792 *S. capensis* Kerr, *Linn. Anim. Kingd.:* 266. Near Graaff-Reinet, Cape Province (CP).

1793 *S. namaquensis* Lichtenstein, *Cat. Rer. Nat.:* 2 Orange River, Namaqualand, CP.

1801 *S. ginginianus* Shaw, *Gen. Zool.* 2:147. Gingi, East Indies = South Africa.

1801 *Myoxus africanus* Shaw, *Gen. Zool.* 2:172. Sneeuberge, CP.

1817 *Sciurus albovittatus* Desmarest, *Nouv. Dict. N.H.* 10:110. Cape of Good Hope.

1820 *S. levaillantii* Kuhl, *Beitr. Zool.* 2:67. 'In Africa meridionalis.'

1832 *S. setosus* Smuts, *Enum. Mamm. Cap.:* 33. Southern CP.

1834 *Geosciurus capensis* A. Smith, *S. Afr. Quart. Journ.:* 2, pts 1–3.

1862 *Xerus setosus* Layard, *Cat. Mamm. S. Afr. Mus.:* 47.

1882 *X. capensis* Jentink, *Notes Leyd. Mus.* 4:48.

1909 *Geosciurus capensis* Thomas, *Ann. Mag. nat. Hist.* (8) 3:473.

1923 *G. c. namaquensis* Thomas & Hinton, *Proc. zool. Soc. Lond.:* 483.

1939 *G. inauris inauris* (Zimmermann). *In* Allen, *Bull. Mus. comp. Zool. Harv.* 83:293.
*G. i. namaquensis* (Lichtenstein). *In* Allen, *Bull. Mus. comp. Zool. Harv.* 83:293.

1940 *Xerus (Geosciurus) inauris inauris* (Zimmermann). *In* Ellerman, *The families and genera of living rodents.*

1951 *Geosciurus inauris* (Zimmermann). *In* Roberts, *The mammals of South Africa.*

1964 *Xerus inauris* (Zimmermann). *In* Meester *et al., An interim classification of southern African mammals.*

The measurements in table 1 compare with values obtained by Smithers (1971) from a much larger sample of animals in Botswana. The mass for males, however, seems to be on the low side when compared to Smithers' figures (M = 659 g, N = 40, range 511–1 022 g).

TABLE 1
Measurements of male and female *Xerus inauris*

| | Parameter | Value (mm) | N | Range (mm) | CV (%) |
|---|---|---|---|---|---|
| Males | HB | 243 | 29 | 200–300 | 8,5 |
| | T | 208 | 29 | 182–228 | 6,0 |
| | HF | 63 | 30 | 57–70 | 5,5 |
| | E | 10 | 10 | 10–12 | 6,7 |
| | Mass: | 569 g | 3 | 546–600 g | — |
| Females | HB | 240 | 56 | 185–292 | 9,5 |
| | T | 217 | 48 | 190–260 | 8,5 |
| | HF | 65 | 55 | 56–76 | 6,9 |
| | E | 11 | 14 | 10–13 | 9,8 |
| | Mass: | 580 g | 8 | 540–650 g | 7,0 |

## Identification

This animal is readily distinguishable in the field by its plain cinnamon dorsal coloration, a white lateral stripe from shoulder to hip and small ears. The eyes are prominent, head broad, with a dull whitish line above and below each eye. The front of the face, the hands, feet and belly are white. The hair is coarse with little underfur, the skin dark. The claws are black, well developed, slightly curved and adapted to digging. The pollex is rudimentary with a flat nail. The long hairs of the tail are white, with a black band near the base and a broader subterminal black band near the tip (plate 1).

A whining shriek is uttered when the animal is alarmed. Females are slightly heavier than males, especially when lactating. There are two pairs of abdominally and inguinally situated mammae.

Albinism also occurs in *X. inauris* and white squirrels have been observed on a farm in the western Orange Free State (OFS) in 1973 by Straschil (1974). Subsequently their numbers have increased to 12–13 individuals and they have split into different colonies inhabiting the same burrows as do normally coloured ones. According to Straschil, the white fur is softer and less bristly than usually found in the case of *X. inauris,* while the skin and claws are pink and the eyes red.

## Skull and dentition

The skull is stout with the bony palate extending well behind the molars. The bullae are inflated, rounded, with three transbullar septa. The lachrymals are enlarged and the postorbital processes small but clear. The squamsosal extends to the base of the postorbital process of the frontal. Below the infraorbital foramen is a well-developed masseter knob (fig. 4). The diameter of the orbit is usually less than one third of the occipitonasal length of the skull. A single premolar (Pm⁴) is present in each jaw

followed by three molars which tend towards hypsodonty. The $M^3$ is smaller and simpler than the $M^1$ and $M^2$ (fig. 5). The incisors are white, opisthodont and not grooved.

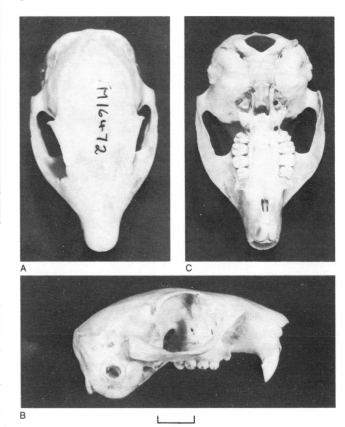

**Figure 4** Dorsal (A), lateral (B) and ventral (C) views of the skull of Cape Ground Squirrel *Xerus inauris.* Note: the palate extending well beyond molars; the rounded bulla with three transbullar septa; the well-developed masseter knob and large postorbital processes.

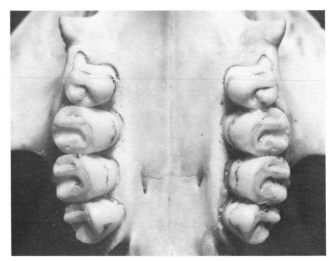

**Figure 5** Cheekteeth of the Cape Ground Squirrel *Xerus inauris.* A single $Pm^4$ is followed by $M^{1-3}$ tending towards hypsodonty.

## Distribution

The Cape Ground Squirrel extends from the OFS to the northeastern and northwestern Cape Province into Great Namaqualand and SWA/Namibia. From the western Transvaal it ranges northwards into Botswana and the Kalahari. It therefore occurs in the Southern Savanna Grasslands west of the Great Escarpment and in the South West Arid (map 1).

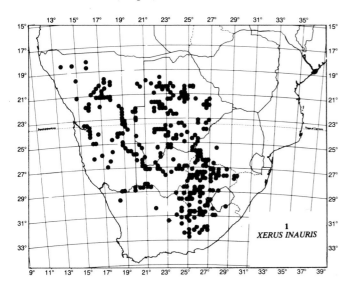

1
XERUS INAURIS

## Habitat

Smithers (1971) has analysed 236 records in Botswana, which indicate a preference for open country with hard and often stony ground. *X. inauris* also occurs in *Acacia* scrub on fringes of pans or floodplains where the soil may be hard, consolidated or sandy. Their occurrence in fairly loose sand, as is indicated by the presence of *Terminalia* trees ('Sandgeelhout'), signals the use of sandy soils, provided it is not too loose to hinder the construction of burrows. Amtmann (1975) states that it is common in calcareous tufa soils on plains. In the Kalahari Gemsbok National Park, Cape ground squirrels occur in the dry river beds, on pans (a feature characteristic of the Kalahari) and on flat calcrete areas adjoining the riverbeds. In the central Kalahari, Botswana, they feed and excavate their burrow systems on pans (Parris 1976). Under these circumstances they may live in colonies of up to 30 individuals (Smithers 1971).

## Diet

Cape ground squirrels feed on bulbs (which they dig out), while they also subsist on fleshy roots and seeds. Occasionally they may take insects and birds eggs. In the Nossob River, Kalahari Gemsbok National Park, I have observed them foraging up to 200 m or more from their burrows. The water requirements are apparently derived from metabolic processes and in captivity they readily take to maize, sunflower seed, groundnuts and acorns. Experiments by Bolwig (1958) indicate that Cape ground squirrels are more efficient than the white laboratory rat *Rattus norvegicus* at water retention in the kidneys and colon.

## Habits

The Cape Ground Squirrel is purely terrestrial and is never found in trees. Although colonies number up to 30 individuals, this figure is generally lower, six to ten individuals being more common. Colony size is, however, difficult to assess with accuracy.

The burrows are 10–15 cm in diameter, of varying length and with several openings. Neighbouring burrows frequently interconnect. In any warren, disused passages are often found and the excavated soil forms low, flattened mounds near the entrances. When disturbed on the surface, the squirrels run to their burrows. They are inquisitive, however, and slowly emerge again to determine the source of disturbance. They often sit up on their haunches, viewing their surroundings. When feeding, the conspicuous tail is folded over the back. Sandbathing occurs frequently (Straschil 1975).

They are strictly diurnal and they emerge only well after sunrise. They do not hibernate or store up food supplies. They do not emerge on cold or overcast days (Smithers 1971). The burrows are often shared by the Suricate *Suricata suricatta* and occasionally the Yellow Mongoose *Cynictis penicillata* (Smithers 1971).

Cape ground squirrels tame readily and make gentle and amusing pets.

## Predators

It could be expected that the predators of squirrels are many, yet little exact data seems to be available. According to Dean (1977) they are taken by the Spotted Eagle Owl *Bubo africanus* at Barberspan in the Transvaal, while Pitman & Adamson (1978) report them as being preyed upon by the Giant Eagle Owl *Bubo lacteus*. It is likely that they are preyed upon by diurnal raptors, snakes and mongooses.

## Reproduction

The female has one inguinal and one abdominal pair of mammae. Two to six young form a litter, one to three being the average. In the western Transvaal, gravid females have been taken in August and October at Ventersdorp and Lichtenburg respectively with two embryos each. They breed throughout the year (Smithers 1971) and become sexually mature at approximately six months of age (Walker 1975).

## Parasites

According to Neitz (1965) three cases of rabies have been reported from Cape ground squirrels in the OFS up to 1965. The plague bacillus *Yersinia pestis* has also been associated with these rodents and its sporadic incidence has been indicated by laboratory tests.

Helminth parasites of *Xerus inauris* are poorly known. *Catenotaenia geosciuri* is the only plathyhelminth hitherto on record (Ortlepp 1938a). Nemathelminths include *Enterobius polyoon* (Le Roux 1930a), *Streptopharagus geosciuri* (Le Roux 1930b), *Rictularia aethechini* (Le Roux 1930c), *Oesophagostomum xeri* (Ortlepp 1922a) and *Physaloptera capensis* (Ortlepp 1922b).

The mesostigmatic mite *Haemolaelaps casalis* is also known to occur in *X. inauris* (Zumpt 1961a). Adult individuals of the argasid tick *Ornithodoros zumpti*, have been recorded by Theiler (1962) as are the following adult (A) or immature (I) ixodid ticks: *Haemaphysalis houyi* (A, I), *H. leachii leachii* (A, I), *H. l. muhsami* (A, I), *Rhipicephalus appendiculatus* (I), *R. pravus* (I), *R. simus* (I) and *R. theileri* (A, I). The presence of *Hyalomma truncatus* (I) has also been observed (Walker *pers. comm.*).

Fleas are represented by three families. Pulicidae include *Echidnophaga bradyta*, *E. gallinacea*, *Pulex irritans*, *Ctenocephalides connatus*, *C. felis*, *Synosternus caffer*, *Xenopsylla hipponax*, *X. philoxera*, *X. versuta*, *X. cryptonella* and *X. erilli*. Hystrichopsyllidae are represented by *Dinopsyllus ellobius*, while a third family, the Chimaeropsyllidae, is represented by *Demeillonia granti* and *Chiastopsylla numae* form *rossi* (Zumpt 1966).

### Relations with man

Cape ground squirrels are said to be destructive in maize fields, where the cobs are torn from the stalks in order to get to the grain. The impact of this destruction does not appear to be extensive and generally they are tolerated by farmers.

During plague epidemics, they may contract the disease themselves, and may be involved as a vector in the spread of the disease on account of their association with fleas, especially with *Dinopsyllus ellobius* and *Xenopsylla philoxera* as well as *Pulex irritans*, the only flea with man as primary host. Furthermore, they are reservoirs for enterobiasis, strongyliosis and other helminth infections. The ticks they harbour transmit *(inter alia)* the neurotropic strain of the spirochaete *Borellia tillae* (via *Ornithodoros zumpti)*, rickettsias (via *Haemaphysalis leachii)*, biliary fever, East Coast fever and tick-bite fever (via *Rhipicephalus appendiculatus)* (Theiler 1962).

### Prehistory

Zeally (1916) reported *X. inauris* (= *X. capensis)* as a fossil from the Upper Pleistocene in breccias at the Bulawayo waterworks reserve in Zimbabwe, yet this species does not occur in this region today. Apart from this, the genus has not yet been found in fossiliferous breccias in southern Africa. A fossil species, *X. janenschi*, was described by Dietrich (1941) from Lower Pleistocene deposits in Tanzania.

### Taxonomy

The taxonomy of this species presents no difficulties in southern Africa. The single species *inauris* without any subspecies, is found. Apart from *X. princeps* (see below), two other species are found extralimitally (as here defined) in Africa. *Xerus rutilus* (the Unstriped Ground Squirrel) is found in northeastern Sudan, Ethiopia, Somalia and Kenya as well as in the northeastern areas of Uganda and Tanzania, while *X. erythropus* (Geoffroy's Ground Squirrel) ranges from Senegal and southern Mauritania to the Blue Nile area in Sudan, and southwards to southwestern Kenya and the Poko-Uele region of Zaïre.

# *Xerus princeps* (Thomas, 1929)

**Kaokoveld Ground Squirrel**
**Kaokoveld Waaierstertmeerkat**
**Kaokoveld Grondeekhoring**

The derivation of the generic name has been explained under *Xerus inauris*. The species name is derived from the Latin *princeps* = first or primary and the application may refer to the larger than average size, brighter coloration and more profusely ringed tail of this species in contrast to the somewhat smaller, drabber *inauris*.

### Outline of synonymy

1929 *Geosciurus princeps* Thomas, *Proc. zool, Soc. Lond.*: 106. Otjitundua, Kaokoland, SWA/Namibia.
1939 *G. inauris princeps* Thomas. *In* Allen, *Bull. Mus. comp. Zool. Harv.* 83:294.
1940 *Xerus (Geosciurus) princeps* Thomas. *In* Ellerman, *The families and genera of living rodents.*
1964 *X. princeps* Thomas. *In* Meester *et al., An interim classification of southern African mammals.*

#### TABLE 2
#### Measurements of male and female *Xerus princeps*

|  | Parameter | Value (mm) | N | Range (mm) | CV (%) |
|---|---|---|---|---|---|
| Males | HB | 250 | 4 | 225–280 | — |
|  | T | 233 | 4 | 223–250 | — |
|  | HF | 67 | 4 | 63–71 | — |
|  | E | 14 | 3 | 13–15 | — |
|  | Mass: See text | | | | |
| Females | HB | 247 | 6 | 225–260 | 5,0 |
|  | T | 238 | 7 | 210–250 | 7,2 |
|  | HF | 65 | 8 | 62–71 | 5,8 |
|  | E | 14 | 4 | 13–15 | — |
|  | Mass: See text | | | | |

I can find no data on the mass of animals made into study skins in southern African museum collections and neither have masses been recorded in the published literature.

## Identification

This 'splendid' ground squirrel (as it was described by Thomas), was originally collected at Otjitundua in the central Kaokoveld and '... is a very fine discovery' due to Captain Shortridge's fifth Percy Sladen and Kaffrarian Museum expedition to the Kaokoveld in SWA/Namibia. In colour and appearance, this species is much like *X. inauris,* although Thomas, in his original description (1929) referred to it as 'Finer, more brightly coloured, with longer and more profusely ringed tail...' The general colour dorsally was described as more profusely mixed with grey than in *inauris,* and the head and outer ridge of the hips strongly grizzled with white-tipped hairs (plate 1).

Each individual hair of the tail has three black bands and four white ones in contrast to two black bands in *X. inauris.* The overall size of *X. princeps* also tends to be somewhat larger, especially the length of hindfoot and tail.

## Skull and dentition

The skull of *X. princeps* is similar in structure to that of *X. inauris.* The diameter of the orbit is normally more than one third of the occipitonasal length. The skull is slightly larger than that of *X. inauris.* In *X. inauris,* the nasals are broad anteriorly and rapidly narrow backwards, while in *princeps* they are parallel-sided, the front being little broader than the back (fig. 6). The incisors show a yellow to orange hue. The occlusal patterns of the upper molars do not differ from that of *X. inauris* (fig. 7).

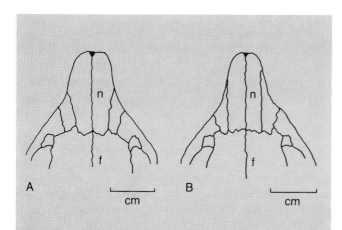

**Figure 6** Nasal bones (n) seen from above in skulls of (A) the Cape Ground Squirrel *Xerus inauris* and (B) the Kaokoveld Ground Squirrel *X. princeps.* In *X. inauris* they taper on approaching the frontals (f) but are parallel in *X. princeps.*

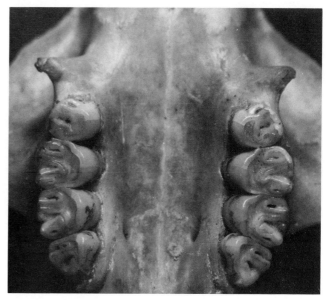

**Figure 7** Cheekteeth of the Kaokoveld Ground Squirrel *Xerus princeps.*

## Distribution

In southern Africa the Kaokoveld Ground Squirrel occurs exclusively in SWA/Namibia, ranging from Great Namaqualand and Damaraland northwards to the Kaokoveld and across the Cunene River into Angola (map 2). It is a typical representative of the South West Arid biotic zone (Rautenbach 1978a).

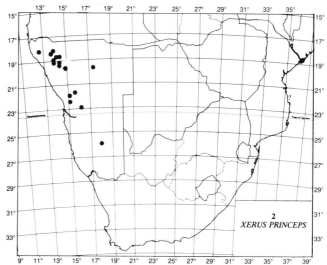

2
*XERUS PRINCEPS*

## Habitat

Shortridge (1934) states that it is only found in rocky and hilly ground, while Roberts (1951) reports them to occur in the plains as well. They utilise crevices amongst rocks to a greater extent than *X. inauris.*

## Diet

No information available.

23

## Habits

Like *X. inauris,* this species is diurnal and lives in colonies. The social structure within a colony is not known.

## Predators and reproduction

Nothing is known about these aspects of the animal's ecology.

## Parasites

The pulicid flea *Ctenocephalides connatus* has been recorded from *Xerus princeps* (Zumpt 1966). Like *X. inauris,* it is also a principal host of this insect. Species of *Ctenocephalides* are poor plague vectors and are consequently regarded as of minor importance (De Meillon, Davis & Hardy 1961).

## Relations with man and prehistory

Unknown.

## Taxonomy

The single, monotypic species *princeps* occurs in southern Africa, implying that no subspecies have been described. I have seen specimens in the Transvaal Museum, Pretoria, and the Kaffrarian Museum, King William's Town, and am not entirely convinced that *princeps* is, in fact, different from *inauris*. Although I accept the difference described by Thomas (e.g. brighter coloration, slightly larger size, the difference in nasal structure), I wonder whether this would stand up to multivariate analysis. In addition, there is a very close overlap in the distribution of the two species (see maps). What factors keep them separate as two species? This is an interesting taxonomic problem waiting to be unravelled.

GENUS *Helidsciurus* Trouessart, 1880

This species resembles the Bush Squirrel *Paraxerus cepapi,* in the absence of the white lateral stripe, presence of claws adapted for climbing and three pairs of mammae, but differs in having one premolar only. It is purely tropical in range, extending south of the Zambezi Valley as far south as the Tropic of Capricorn. As in many other species, little information about its ecology is available.

The genus *Helioschurus* occurs exclusively in Africa and, according to Kingdon (1974) must have evolved on this continent. It has been grouped with the giant squirrels, *Protoxerus* and *Epixerus,* in the Tribe Protoxerini (Moore 1959). Three species occur in Africa. *Helioschurus ruwenzorii* is a montane squirrel with a limited distribution in eastern Zaïre and western Rwanda and Burundi into western Uganda. A second species, *H. gambianus,* ranges from Sierra Leone across Africa to the Ethiopian Highlands and southwards into Angola, Zaïre and western Tanzania. The third species, which concerns us here, *H. rufobrachium,* also ranges across tropical Africa from west to east and southwards to Mozambique and Zimbabwe in lowland and montane forests.

The taxonomy of *Helioschurus* has been difficult to assess and many of the described forms were either assigned to *gambianus* (originally described as *Sciurus gambianus* by Ogilby in 1835) if they were small and lacked red on the legs, or to *rufobrachium* if they had red legs. In addition, they were assigned to other species (e.g. *punctatus* and *maculatus,* now both relegated to subspecific rank under *gambianus* and *rufobrachium* respectively) if they did not conveniently fall into either of these categories (Rosevear 1969). In 1927 Ingoldby '... expressed the view that *Helioschurus* as a genus was highly susceptible to local ecological conditions and that there was in fact only one species *gambianus,* throughout Africa and that all the various forms which had been described were solely climatic races of this, the outcome of given combinations of temperature, humidity and so forth' (Rosevear 1969). Ellerman (1940) accepted this interpretation and listed 44 races all as subspecies of *gambianus*. A year before this Allen (1939), however, placed 47 races in eight species including *bongensis, gambianus, multicolor, mutabilis, punctatus, rhodesiae, rufobrachium* and *undulatus*. In 1975, Amtmann placed 26 races under *rufobrachium* and 22 under *gambianus* relegating all the other species as listed by Allen to subspecific rank.

# Heliosciurus rufobrachium (Waterhouse, 1842)

**Red-legged Sun Squirrel**

**Rooipootsoneekhoring**

The generic name is derived from a combination of two Greek words, *helios* = sun and *skioura* = shade. It likes to bask in the sun and it is implied that its own tail is available if shade is required. The species name is derived from the Latin *rufo* = red and *brachium* = forearm, referring to the reddish coloration of the front limb.

## Outline of synonymy

1842 *Sciurus rufobrachium* Waterhouse, *Ann. Mag. nat. Hist.* (1) 10:202. Fernando Po.

1852 *S. mutabilis* Peters, *Monatsber. K. Preuss. Akad. Wiss. Berlin:* 273. Boror, Mozambique.

1853 *S. maculatus* Temminck, *Esquisses Zoologiques sur la Côte de Guine:* 130. Probably Gold Coast.

1867 *S. aubryi* Milne-Edwards, *Rev. Zool.* (2) 19:228. Gabon.

1867 *Macroxus rufobrachiatus* var. *waterhousii* Gray, *Ann. Mag. nat. Hist.* (3) 20:328. Ashanti, Gold Coast.

1867 *M. shirensis* Gray *Ann. Mag. nat. Hist.* (3) 20:327. Shire River, Zimbabwe.

1867 *M. isabellinus* Gray, *Ann. Mag. nat. Hist.* (3) 20:329. Lower Niger.

1900 *Sciurus (Heliosciurus) libericus* Miller, *Proc. Wash. Acad. Sci.* 2:633. Mount Coffee, Liberia.

1902 *S. aschantiensis* Neumann, *Sitzb. Ges. Naturf. Freunde, Berlin:* 175. Gold Coast.

1939 *Heliosciurus rufobrachium* (Waterhouse). *In* Allen, *Bull. Mus. comp. Zool. Harv.* 83:296.

1941 *Heliosciurus (Heliosciurus) rufobrachium* (Waterhouse). *In* Ellerman, *The families and genera of living rodents.*

1951 *Heliosciurus mutabilis* (Ingoldby). *In* Roberts, *The mammals of South Africa.*

1953 *H. (H.) gambianus* Ogilby. *In* Ellerman *et al., Southern African mammals.*

1964 *H. rufobrachium* (Waterhouse). *In* Meester *et al., An interim classification of southern African mammals.*

1975 *H. rufobrachium* (Waterhouse). Amtmann, *in* Meester & Setzer (eds), *The mammals of Africa: an identification manual.*

## TABLE 3
## Measurements of male and female *Heliosciurus rufobrachium*

|  | Parameter | Value (mm) | N | Range (mm) | CV (%) |
|---|---|---|---|---|---|
| Males | HB | 237 | 6 | 229–246 | 3,3 |
|  | T | 273 | 7 | 240–287 | 6,2 |
|  | HF | 55 | 7 | 44–60 | 9,8 |
|  | E | 17 | 4 | 16–17,5 | 3,7 |
|  | Mass: No data available | | | | |
| Females | HB | 230 | 6 | 220–243 | 4,0 |
|  | T | 287 | 3 | 270–292 | 1,3 |
|  | HF | 54 | 6 | 45–59 | 9,1 |
|  | E | 17 | 5 | 16–20 | 9,5 |
|  | Mass: No data available | | | | |

## Identification

This species is notoriously variable in dorsal colouring. Individuals may vary from speckled grey to light or dark red. The base of each individual hair is always dark brown, followed by a lighter coloured band, another dark band, and a white to yellow band with reddish to black tips. The ventral surface is more uniform, but may also vary in coloration. The tail is banded transversely with a series of narrow, indistinct white bands (plate 1).

Smithers (1966) reports that during the months of September to February, some individuals are an overall pale reddish-brown colour, with buffy grizzling, while others show all grades of change from this condition to the grizzled greyish buff colour more normally encountered in the dry months of March to August.

## Skull and dentition

The skull is typically sciurid with one premolar ($Pm^4$), making a total of four cheekteeth. The infraorbital foramen is a narrow canal no more than half the length of the maxillary tooth row. A supraorbital notch is closed on the margin of the orbit and persists as a foramen piercing the frontal. The frontoparietal suture is absent in adults. A temporal foramen is present in the squamoso-parietal suture dorsal to the posterior glenoid foramen (Moore 1959). A well-developed masseter knob occurs in typical *Heliosciurus* (Ellerman 1940). The postorbital processes are well developed (fig. 8). The number of trans-bullar septa varies considerably, from 0 to 2,5 septa per bulla, with a mode of one septum in some 47 skulls of a sample of 84 skulls looked at in detail by Moore (1959). The upper cheekteeth have their main ridges strongly convergent internally. The anterior cusp of $Pm^4$ is very well developed, projecting forwards (fig. 9).

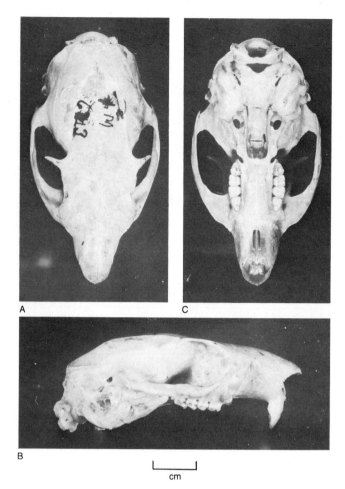

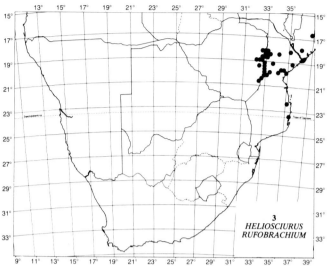

## Distribution

In southern Africa, this species occurs in Mozambique and southeastern Zimbabwe. Elsewhere it occurs in Zambia, Malaŵi and further north to southeastern Sudan and Kenya and westwards to Senegal (map 3).

**3**
**HELIOSCIURUS**
**RUFOBRACHIUM**

**Figure 8** Dorsal (A), lateral (B) and ventral (C) views of skull of the Red-legged Sun Squirrel *Heliosciurus rufobrachium*. The bony palate terminates at level of M³. A well-developed masseter knob occurs. Postorbital processes conspicuous.

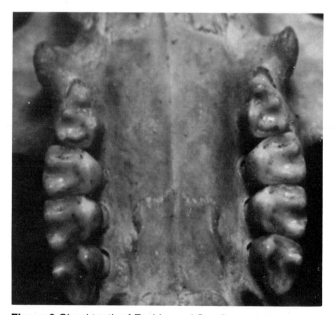

**Figure 9** Cheekteeth of Red-legged Sun Squirrel *Heliosciurus rufobrachium*. The single Pm⁴ has its anterior cusp well developed, projecting forwards.

## Habitat

It can be found in savanna with large trees, showing its preference for heavily wooded areas. They are common on the escarpment of the eastern border of Zimbabwe in the mist belt of the montane evergreen forest with a high annual rainfall. In Mozambique, they occur in tropical and riverine forest thickets and in riparian conditions in Miombo savanna (Smithers & Lobão Tello 1976). They tend to move about at medium height in trees, but also move from lower levels to canopy with ease (Kingdon 1974). T.S. Jones *(in* Rosevear 1969) states that it is a common inhabitant of mangrove swamps, both *Rhizophora* (the Red Mangrove) and *Avicennia* (the White Mangrove).

## Diet

They consume fruits, nuts and fresh green shoots of trees and *Phoenix* palms and are reported as taking birds' eggs (Smithers 1966). Insects are also taken and Kingdon (1974) records one animal from southern Tanzania whose stomach was filled exclusively with grasshoppers during the early rains. Small birds and even geckos are occasionally included in their diet. Feeding is mostly accomplished in trees, but they will descend to ground-level if attracted.

## Habits

This species is diurnal and is reputed to like basking in the hot, tropical sun. According to Roberts (1951), this habit probably accounts for changes of colour of the hairs as mentioned above. It frequents large trees, in the holes of which it rears its young in nests lined with leaves. It is most active in mid-morning and late afternoon.

## Predators

Kingdon (1974) describes them as quick-fleeing when frightened. They advance through branches in a fast but continuous rush from tree to tree, often gaining height as they do so. They sometimes sunbathe on exposed trunks. Consequently, they are possibly more vulnerable to hawks than some of the more timid and cryptic squirrels. Kingdon remarks that on casual assessment, these squirrels appear to be more common and that they expose themselves more than other sciurid genera – '... perhaps their more numerous young are a response to heavier predation'.

## Reproduction

One to three young are born per litter, usually from June to October (Smithers 1966), although five young have been recorded. Nests are made in holes in tree trunks or hollow logs. Kingdon (1974) states that they may even nest in beehives, irrespective of whether the hives are occupied or not. Ellerman *et al.* (1953) mentions the complete or nearly complete suppression of the baculum in the male. Kingdon (1974) writes that the timing of reproduction may be determined by the males who show signs of a sexual cycle. As their testes reach their largest size, the squirrels develop two large anal glands. In Uganda males showed this condition in August and February respectively.

## Parasites

To the best of my knowledge there is no information available on the ticks and worms associated with this species. Three laelaptid mites have, however, been collected, *viz. Ugandolaelaps protoxera, Hirstionyssus heliosciurus* and *H. liberiensis.* A fourth species, the trombiculid *Schoengastia lucassei sciuri* is also known (Zumpt 1961a).

The flea *Libyastus* is a parasite of squirrels and *L. infestus* is known from *H. rufobrachium* (Zumpt 1966).

## Relations with man and prehistory

Unknown.

## Taxonomy

Roberts (1951) and earlier authors regard *mutabilis* as the prior name for southern African representatives of *Heliosciurus,* but Ellerman *et al.* (1953) included it in *gambianus.* Meester *et al.* (1964) point out that Rosevear (1963) suggests that *Heliosciurus* includes two species, *gambianus* and *rufobrachium* and that the southern African forms belong to the latter.

*H. r. mutabilis* (Mozambique), *H. r. chirindensis, H. r. smithersi* and *H. r. vumbae* (Zimbabwe) have been described as subspecies in our area. *H. r. mutabilis* includes *beirae* and *shirensis* and it is possible that *vumba* and *smithersi* are synonyms of *chirindensis* (Amtmann 1966). Twenty-two other subspecies in the rest of Africa are at present tentatively accepted. These include *arrheni, semlikii, rubricatus, pasha* and *medjianus* (Congo), *aubryi* (Gabon and parts of Central African Republic), *benga* (Rio Muni), *caurinis* and *occidentalis* (Guinea-Bissau), *shindi, keniae, leakeyi* and *daucinus* (Kenya), *dolusus* (Mafia and Zanzibar Islands), *hardyi* (Ivory Coast), *isabellinus* (Togo), *leonensis* (Sierra Leone), *maculatus* (Liberia), *nyansae* and *undulatus* (Tanzania), *obfuscatus* (Nigeria) and *rufobrachium* (Fernando Po).

Apart from *rufobrachium* (with 26 subspecies referred to above), the genus contains two additional species: the Mountain Sun Squirrel *H. ruwenzorii* (with four subspecies) and the Gambian Sun Squirrel *H. gambinus* (with 22 subspecies) which occur in the eastern Congo and in West and West Central Africa respectively.

GENUS *Funisciurus* Trouessart, 1880

The genus *Funisciurus* was named by Trouessart in 1880 based on the genotype *Sciurus isabella* Gray, 1862.

In 1959 *Funisciurus* and *Paraxerus* were united in a sub-tribe Funisciurina by Moore. Some authorities (e.g. Kingdon 1974) treat *Paraxerus* as a subgenus of *Funisciurus.* However, as is stated by Kingdon, only very much more detailed study will give a satisfactory picture of the remarkable evolutionary radiations of the funisciurine squirrels. Kingdon quotes Thomas (1909) who stated that *Funisciurus* '... would seem to be the representative of *Paraxerus* in the West African forest region'. Although this seems to be a somewhat oversimplified statement of the situation, Kingdon accepts this statement to be true in broad outlines.

# *Funisciurus congicus* (Kuhl, 1820)

**Striped Tree Squirrel**

**Kuhl's Squirrel**

**Western Striped Squirrel**

**Ovamboland Gestreepte Eekhoring**

The generic name is derived from the Latin *funis* = rope, pointing to the agility of these animals on slender branches reminiscent of tightrope walkers. The species name *congicus* = pertaining to the Congo. They can thus be called the rope squirrels of the Congo.

## Outline of synonymy

1820 *Sciurus congicus* Kuhl, *Beitr. Zool.* 2:55. Northern Angola.

1843 *S. praetextus* Wagner, Schreber's *Säugeth.* Suppl. 3:216. No locality.

1901 *Funisciurus congicus* (Kuhl). *In* Sclater, *Fauna of South Africa.*

1926 *F. c. oenone* Thomas, *Proc. zool. Soc. Lond.:* 297. Cunene Falls.

1938 *F. c. damarensis* Roberts, *Ann. Transv. Mus.* 19:236. Tsumeb district. (= *Heliosciurus (sic) c. damarensis.*)

### TABLE 4
Measurements of male and female *Funisciurus congicus*

| | Parameter | Value (mm) | N | Range (mm) | CV (%) |
|---|---|---|---|---|---|
| | HB | 153 | 13 | 141–165 | 5,0 |
| | T | 162 | 13 | 150–175 | 5,5 |
| Males | HF | 39 | 13 | 38–42 | 3,1 |
| | E | 17 | 19 | 15–19 | 5,4 |
| | Mass: | 105 g, 115 g | 2 | — | — |
| | HB | 149 | 12 | 137–161 | 6,8 |
| | T | 154 | 11 | 142–175 | 5,6 |
| Females | HF | 38 | 14 | 36–42 | 5,9 |
| | E | 16 | 11 | 14–18 | 10,25 |
| | Mass:. | 110 g | 1 | — | — |

## Identification

The white stripe along the flank from the shoulder to the haunch is obvious, with a pronounced darker stripe below it. The tail is long and thin. The fur is soft. The colour can be described as yellow-buff to yellow-brown dorsally, darker than the flanks which are greyer. The ventral surface is white from chin to anus. The hands and feet are pale-yellowish, the digits white, covered with hair virtually concealing the sharp curved claws characteristic of arboreal squirrels. There is a light ring around the eye, which does not extend to the muzzle (plate 1). The pelage is short and soft. Pectoral mammae are absent, but an abdominal and an inguinal pair are usually present. The baculum is very small in this genus (Ellerman 1941).

## Skull and dentition

The weakly ridged skull is strongly built, but in contrast to the Ground Squirrel *Xerus inauris,* the ovate bullae are not as rounded and the palate hardly extends behind the back of the molars. The rostrum is somewhat elongated, while the masseter knob is not very prominent. The postorbital processes are moderately developed (fig. 10).

The incisors are coloured orange, not grooved. Premolars $^3$ and $^4$ occur in the maxilla, while the $Pm_3$ is absent in the mandibular tooth row. The cheekteeth are more specialised than in almost any other squirrel (Ellermann 1940) and they tend to become flat-crowned (or nearly so) in adults (fig. 11).

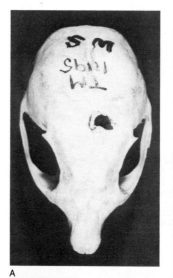
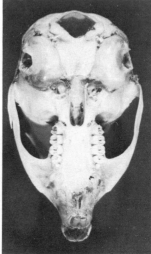
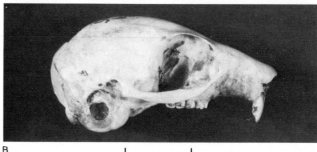

**Figure 10** Dorsal (A), lateral (B) and ventral (C) views of the skull of the Striped Tree Squirrel *Funisciurus congicus.* Note: palate not extending beyond M³, bulla ovate, masseter knob poorly developed. Both Pm³ and Pm⁴ are present. Skull sturdily built, weakly ridged and with poorly developed postorbital processes.

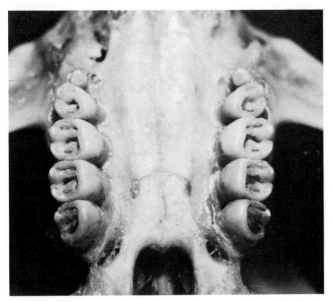

**Figure 11** Cheekteeth of the Striped Tree Squirrel *Funisciurus congicus*. The Pm³ is peglike with the Pm⁴ slightly smaller than the M¹. The tooth rows are parallel with occlusal surfaces of teeth pronouncedly ridged.

## Distribution

In southern Africa they occur only in western Ovambo in the north of SWA/Namibia (map 4) and range extralimitally northwards to Zaïre. In the southern African context they are confined to the South West Arid (Rautenbach 1978a).

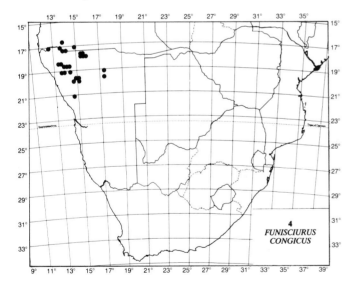

**4**
**FUNISCIURUS CONGICUS**

## Habitat

They live in forest belts such as are found between Ondongua and Ukuambi in western Ovambo.

## Diet

Golding (1938 *in* Rosevear 1969) found that their tastes are catholic and in West Africa this genus may include

insects as an appropriate part of their natural diet (Rosevear 1969). Viljoen *(pers. comm.)* has also found this squirrel feeding on Mopane worms *Gonimbrasia belina* in SWA/Namibia.

## Habits

The Striped Tree Squirrel or Kuhl's Squirrel is predominantly arboreal, occasionally venturing on to the ground. They are possibly more given to excited chattering than the Sun Squirrel *(Heliosciurus)*, being quarrelsome and highly excitable. According to Rosevear (1969) their notes have a staccato quality which can be described as a click or a cluck. Rosevear (1969) also refers to work by Cansdale who noticed differences in the ability of *Heliosciurus* and *Funisciurus* to smell insects hidden in debris. *Funisciurus* traced the insects without hesitation, while *Heliosciurus* was apparently unaware of their presence.

## Predators and reproduction

Unknown.

## Parasites

The flea *Libyastus vates* occurs on this squirrel. In fact the type series, the only known species of this flea, was collected from this squirrel in Angola (Zumpt 1966). Members of this flea species are rarely collected, so that nothing definite can be said about their host relationship (Zumpt 1966). Adults of the tick *Rhipicephalus sanguineus* have also been recorded from this mammal; this tick transmits various babesias, rickettsias and bubonic plague (Theiler 1962).

## Relations with man and prehistory

Unknown.

## Taxonomy

Roberts (1951) listed two subspecies: *Funisciurus congicus oenone* (Thomas, 1926) from the Kaokoveld and *F. c. damarensis* (Roberts, 1938) from the Tsumeb district in SWA/Namibia. These two subspecies were also upheld by Ellerman *et al.* (1953) and were listed by Meester *et al.* (1964). Amtmann (1975), following Amtmann (1966) states that the extralimital subspecies *congicus* (which includes *poolii* from northern Angola, *flavinus* from southwestern Angola, *interior* from southern Congo and *olivellus* from northern Angola) of earlier authors are invalid. This remark also applies to *oenone* and *damarensis*. *Funisciurus congicus* is therefore interpreted as a monotypic species. Extralimitally, eight other species of *funisciurus* are known. These include the Mountain Tree Squirrel *F. carruthersi*, Gray's Four-striped Tree Squirrel *F. isabella*, Leconte's Four-striped Tree Squirrel *F. lemniscatus*, Bocage's Tree Squirrel *F. bayonii*, De Winton's Tree Squirrel *F. substriatus*, the Orange-headed Tree Squirrel *F. leucogenys*, Cuvier's Tree Squirrel *F. pyrrhopus* and Thomas' Tree Squirrel *F. anerythrus*.

# GENUS *Paraxerus* Major, 1893

The genus *Paraxerus* was erected by Forsyth Major in 1893, 13 years after Trouessart had described another group of African squirrels as *Funisciurus* in 1880. The close relationship between these two genera was understood by Thomas (1909) who stated that *Funisciurus* '... would seem to be the representative of *Paraxerus* in the West African forest region'. Consequently, these two genera were united in a subtribe Funisciurina by Moore in 1959. Some authorities (e.g. Kingdon 1974) have accepted *Paraxerus* as a subgenus of *Funisciurus*, and although Amtmann (1975) treated them as separate genera, an interpretation followed by me.

*Paraxerus* species all have sharp claws, enabling them to climb with agility in their arboreal habitat. They are consequently known as African tree squirrels and the southern African genera *Paraxerus* and *Funisciurus* are but two of the six genera and subgenera which make up the Tribe Funambulini as interpreted by Moore (1959). Both *Paraxerus* and *Funisciurus* are also representatives of the subtribe Funisciurina as described by Moore. The tribe is characterised by the presence of one or two transverse bony septa in the auditory bulla. The bushy tail (often highly coloured and ornamental) is usually shorter than the head-body length. A pair of pectoral mammae are present, as well as inguinal and abdominal pairs. Mostly tropical in distribution. The comparative eco-ethology of southern African tree squirrels has been discussed by Viljoen (1978b) and her work makes it abundantly clear that the different squirrels are specifically adapted to their differing environments '... from the small and light-coloured *F. congicus* and *P. cepapi* to the yellowish *P. p. tongensis* and the saturated red and large *P. p ornatus* of the dense, moist "evergreen" forest'.

---

**Key to the southern African species of *Paraxerus***
(Modified after Meester *et al.* (1964))

1 Ventral surface whitish; head, back and tail dull grey, (not rufous); skull length ± 39–45 mm in adults ............................................... *P. cepapi*
(Bush Squirrel)
Page 31

Ventral surface red or orange; tail deep red to orange; head rufous; skull length ± 45–52 mm in adults ............................................... *P. palliatus*
(Red Bush Squirrel)
Page 34

---

# *Paraxerus cepapi* (A. Smith, 1836)

**Bush Squirrel**

**Yellow-footed Squirrel**

**Smith's Bush Squirrel**

**Geelpooteekhoring**

The name is derived from the Greek *para-* = a prefix meaning near and *xerus,* the sciurid genus preferring dry areas (*xeres* = dry, arid). A closeness to the Cape Ground Squirrel is, therefore, implied. The Greek word *kepos* indicates a garden and may refer to the fact that the Bush Squirrel or Yellow-footed Squirrel (I prefer the latter name) is often found in gardens.

This species was first encountered by Sir Andrew Smith near the Marico River in the western Transvaal. The type specimen is housed in the British Museum (Natural History), London (Sclater 1901). It is the commonest and most widely distributed of the southern African arboreal squirrels and is often seen scampering across a road or sitting in a tree.

This species was studied extensively by Viljoen (1975) who had a thorough look at some aspects of the ecology, reproductive physiology and ethology of the Yellow-footed Squirrel in the Naboomspruit area of the Transvaal. It is a penetrating study of the biology of this rodent and a valuable addition to the literature of African sciurids.

## Outline of synonymy

1836 *Sciurus cepapi* A. Smith, *App. Rpt. Exped. Explor. S. Afr.:* 43. Marico River, western Transvaal.
1843 *S. cepate* Gray, *List Spec. Mamm. B.M.:* 140. *Lapsus* for *cepapi*.
1893 *Paraxerus cepapi* Major, *Proc. zool. Soc. Lond.:* 189.
1897 *Funisciurus cepapi* Thomas, *Proc. zool. Soc. Lond.:* 933.

Plate 1

Striped Tree Squirrel
*(Funisciurus congicus)*

Bush Squirrel
*(Paraxerus cepapi)*

Cape Ground Squirrel
*(Xerus inauris)*

Kaokoveld Ground Squirrel
*(Xerus princeps)*

Red Bush Squirrel
*(Paraxerus palliatus)*

Red-legged Sun Squirrel
*(Heliosciurus rufobrachium)*

Plate 2

Lesser Canerat
*(Thryonomys gregorianus)*

Greater Canerat
*(Thryonomys swinderianus)*

Cape Porcupine
*(Hystrix africaeaustralis)*

## TABLE 5
### Measurements of male and female *Paraxerus cepapi*

|        | Parameter | Value (mm) | N  | Range (mm) | CV (%) |
|--------|-----------|-----------|----|-----------|--------|
| Males  | HB        | 176       | 8  | 165–182   | 3,3    |
|        | T         | 170       | 8  | 157–185   | 4,5    |
|        | HF        | 43        | 8  | 41–46     | 3,1    |
|        | E         | 17        | 8  | 15–19     | 6,5    |
|        | Mass:     | 190 g     | 28 | 76–242 g* | —      |
| Females| HB        | 177       | 8  | 156–180   | 4,7    |
|        | T         | 175       | 8  | 160–195   | 5,2    |
|        | HF        | 41        | 8  | 38–45     | 4,7    |
|        | E         | 17        | 8  | 15–20     | 8,0    |
|        | Mass:     | 195 g     | 24 | 130–265 g*| —      |

*Rautenbach (1978b)

Although I could find no data on the mass of animals made into study skins in museum collections, Viljoen (1975) lists some figures. The average adult mass of both sexes from Naboomspruit was 214,1 g (52 specimens). In much smaller samples (five and six) the animals had average masses of 242,9 g (five specimens) and 231,4 g (six specimens) from Potgietersrust and Rustenburg respectively.

## Identification

Black tips of the individual hairs give the animals a grizzled, grey-yellow appearance. The fur is short, soft and close. The tail is darkened by the presence of three black bands on the individual hairs, while the limbs are a more richly coloured buffy-yellow. The claws are curved and sharp. The back of the ears is dark brown, while the eyebrows, sides of the face, chin, throat, breast and inner surfaces of legs are white. The rest of the ventral surface is a dull white. The bushy tail is distinctly barred with narrow, transverse buffy bands, and its length is shorter than the combined length of the head and body (plate 1).

Coloration in this species is geographically variable. It may be white below *(P. c. sindi* – Tete), more rusty-coloured dorsally *(P. c. cepapoides* – Beira) or pale grey dorsally *(P. c. phalaena* – Ovambo). Smithers (1971) states that material from Botswana falls into three distinct colour groups: those from the northwestern part are darker in contrast to those from the southern part of the eastern sector which are a buffy yellow, while specimens from eastern Okavango are a buffy grey. Cases of albinism have been reported from the Kruger National Park (Pienaar, Rautenbach & De Graaff 1980).

According to Viljoen (1975) this species moults twice annually, the spring moult commencing in September and the autumn moult in January. Both moults move from front to back over the body. Juveniles commence their first moult when they are about 41 days old and the process is completed between the 110th and 120th day.

The voice consists of a volley of high-pitched chattering whinnies, varied with croaking notes (Astley Maberly 1963).

## Skull and dentition

The various races differ in skull-size. The bony palate extends only to the back surface of the third molar. The anterior palatine foramina are short and set far forward. The nasals are narrowest at the middle, at least in the typical form, but this varies geographically. The rostrum is not particularly elongated. One or two transverse bony septa occur in the auditory bulla.

The anterolateral edge of the zygomatic arch continues forward as a sharp ridge on to the rostrum but stops short of the premaxilla and the arch rises no higher than the lower edge of the lachrymal (Moore 1959). Furthermore, the anterolateral part of the nasal overshoots the nasal process of the premaxilla, so that the premaxilla makes an indentation in the lateral margin of the nasal (fig. 12).

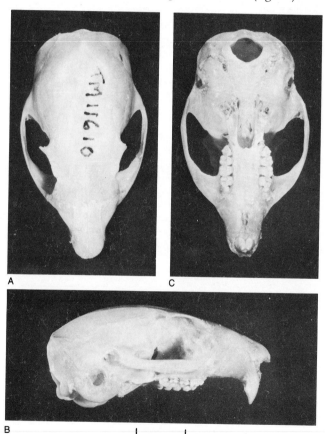

**Figure 12** Dorsal (A), lateral (B) and ventral (C) views of the skull of the Bush Squirrel *Paraxerus cepapi*. Note: palate not extending beyond M³, while anterior palatine foramina are set far forward. Masseter knob small. Bulla with one or two transverse septa. Both Pm³ and Pm⁴ are present and postorbital processes are large.

31

Two premolars occur in the maxillary tooth row. Pm$^3$ is small and peglike while Pm$^4$ shows no prominent anterior cusp. M$^1$ and M$^2$ each show three clear depressions, while M$^3$ shows only two depressions, the second being rather broad (fig. 13).

The lower molars are ridged more or less transversely; three depressions are separated by four ridges. With age, the teeth are simplified to two-lobed structures.

The incisors tend towards pro-odonty.

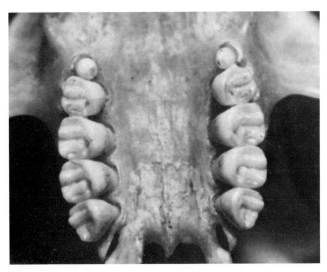

**Figure 13** Cheekteeth of the Bush Squirrel *Paraxerus cepapi*. Note: presence of peglike Pm$^3$ while Pm$^4$ is devoid of a forwardly directed anterior cusp. Tooth rows arch buccally slightly resulting in wide palate.

## Distribution

The Yellow-footed Squirrel ranges from the eastern Transvaal lowveld and southern Mozambique to the western Transvaal, eastern and northern Botswana and northern SWA/Namibia, to the eastern lowveld of Mozambique and throughout Zimbabwe (map 5). It is therefore a Southern Savanna Woodland species, stopping abruptly on the boundary of the South West Arid (Davis 1974).

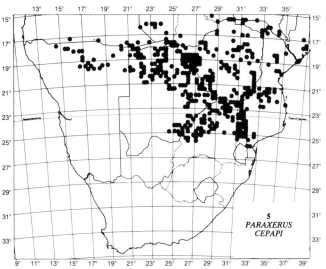

**5**
*PARAXERUS CEPAPI*

## Habitat

The species tends to inhabit savanna rather than dense forests and is frequently encountered in wooded valleys along watercourses. Smithers (1971) says that in Botswana their principal habitat requirements are mostly mopane woodland *(Colophospermum mopane)* and riverine woodland (including *Acacia nigrescens, A. erioloba, Ficus* spp., *Diospyros mespiliformis, Kigelia pinnata* and *Combretum imberbe)*. *Acacia* woodland (with a variety of species) also harbours this species in large areas of the northeastern and eastern parts of Botswana. This also applies to mixed woodland with stands of *Acacia* in a predominantly *Terminalia-Combretum* association.

In the central Transvaal, the Yellow-footed Squirrel occurs where the dominant tree species is the Tambotie *Spirostachys africana* (Viljoen 1975).

## Diet

The Yellow-footed Squirrel subsists on berries, seeds and pods and often feeds on the ground. The food is masticated thoroughly so that identification of food in stomach contents is well-nigh impossible. Smithers (1971) records that it eats dry or green fruits of wild figs *Ficus* spp., Blinkblaar *Ziziphus mucronata, Acacia* spp. and the Mopane *Colophospermum mopane*. It also takes fresh green shoots of *Acacia* spp., forbs and grass *Cynodon dactylon*. Astley Maberly (1963) records that it feeds on the acid fruits of the Marula *Sclerocarya caffra*, and scuffles about on the ground for bulbous roots. It is also stated that it will take birds' eggs and nestlings, but such items seem to be rare in the normal diet. Feeding habits of *Paraxerus cepapi* have been noted by Shortridge (1934), Roberts (1951) and Dobroruka (1970).

Viljoen (1975, 1977a) observed yellow-footed squirrels feeding on a wide variety of plant foods. The seeds of *Acacia tortilis* are taken as are the seeds of *A. giraffae*. They also consume the gum of another *Acacia* species, *Acacia robusta*. Flowers and seeds of some *Aloe* species are eaten as are the leaves of the *maculatae* group of aloes. The flowers of the Tambotie *Spirostachys africana*, two species of *Justicia*, two species of *Commelina*, *Senecio radicans* and *Loranthus zeyheri* are also items of diet, as are berries of *Ehretia rigida, Boscia rehmanniana, Euclea undulata, Solanum capsicastrum, Diospyros* sp. and *Ziziphus mucronata*. The seeds of the grasses *Urochloa panicoides, Panicum maximum, Dactyloctenium aegyptium* and *Monechma fimbriatus* are also consumed when available. Altogether some 38 different plant species were recorded in the diet of these squirrels. Viljoen (1975) also states that the fruit of *Mimusops zeyheri* and *Burkea africana* is also eaten. Lichen and bark were chewed throughout the year.

Viljoen (1977a) also noted that insects play an important part in the diet of this squirrel, especially at certain times of the year in certain localities. These insects include aphids, ants, termites, flies, beetles and roaches.

## Habits

These squirrels are not shy when undisturbed. For protection they often lie flat on horizontal branches and are not easily detected. They make nests of grass and dry leaves in hollow tree trunks. They may store up quantities of food underground and in crevices. Although mainly arboreal, they spend a fair amount of time on the ground. They are active little animals and climb vertical tree trunks with great speed. Interspecific competition between these squirrels and the Crested Barbet *(Trachyphonus vaillantii)* has been reported by Viljoen (1973).

They live singly, in pairs or in small family parties, but there is much variation in group size. Viljoen (1975) reports two per group before the breeding season, to as many as 12 a little while later (two adult males, three adult females and seven young, persumably of three litters). Viljoen suggests that the group is maintained chiefly through a characteristic smell which they have as a result of allo-grooming and occupancy of the same nest. Occasionally, they also mark each other by anal-dragging. The group feeds communally by day and members alert each other in alarm situations. When alarmed by a ground predator (identified by Viljoen as Black-footed Jackal, Slender Mongoose, Banded Mongoose and Black-footed Cat), the whole group often displays mobbing behaviour, all vocalising simultaneously, clicking harshly while tailflicking. They emit a high-pitched whistle at raptors which usually has the effect of sending juveniles headlong into their nest holes. They are diurnal and inactive during rainy weather. When threatened, they keep a tree-trunk or branch between them and the intruder. They are vocal animals and keep up a continual noisy chatter during their active and playful escapades.

## Predators

According to Astley Maberly (1963), yellow-footed squirrels frequently mob tree snakes *(Dispholidus typus)* and other snakes (including large mambas) by perching close to them and scolding incessantly, while whisking their tails '. . . though there is evidence that the mamba will catch and devour squirrels when the opportunity occurs'. Apart from that, their principal enemies seem to be hawks, owls, genets, wild cats and pythons. Astley Maberly also relates that many are killed by Black children using dogs and sticks. Kemp & Kemp (1977) have observed a foraging Southern Ground Hornbill *Bucorvus cafer* chasing yellow-footed squirrels in the Kruger National Park. Worden & Hall (1978) have observed a White-faced Owl *Otus leucotis* feeding its chicks with portions of a dismembered yellow-footed squirrel.

## Reproduction

The female usually has three pairs of mammae, although the pectoral pair may be underdeveloped. They breed throughout the year with an increase in litters from October to April (i.e. the summer months) with a peak in November and December in Botswana (Smithers 1971). Stevenson-Hamilton (1950) states that two young are born in a litter and that they breed in spring or early summer in the lowveld of the Transvaal. As many as four young have been recorded.

The Naboomspruit population studied by Viljoen (1975) were in oestrus in the first week of September and the young were born in about the first week of November. She determined that the gestation period in captive litters was approximately 55 days. A postpartum oestrus is manifest after four days when conditions are favourable. A captive female had six litters during the year; with a gestation period of 55 days, this would necessitate conception soon after parturition.

Viljoen also noted regression of the testes in February and regeneration in June; peak testis development occurred from August to January. Sperm counts in the epididymis varied from $5 \times 10^6$ per single epididymis in March to $474 \times 10^6$ in January.

Viljoen (1975) has observed that the young develop quickly and leave the nest for the first time at 21 days. They are weaned at 4–6 weeks and attain sexual maturity at ten months.

## Parasites

The endoparasites of yellow-footed squirrels are poorly known. *Streptobacillus moniliformis,* a bacterium causing rat-bite fever, developed in a person who was bitten while handling this squirrel immediately after its arrival from South Africa at Hamburg, West Germany (Neitz 1965). According to Neitz, nothing is known about the prevalence of this form of rat-bite fever in South Africa. Apart from this record, no additional information is available on other pathogens which may be found in this species.

This paucity of information also applies to the parasitic worms. In the Naboomspruit population, studied by Viljoen (1975), only 4% of the stomachs and 7% of the small intestines contained parasites. She also recorded that all the caeca and 48% of the large intestines contained numerous very small nematodes. These internal parasites still await identification. The only other record available in the literature refers to the nematode *Physaloptera africana* which has been recorded from this squirrel by Ortlepp (1937).

The ectoparasitic trombiculid chiggers *Schoengastia katangae* and *Schoutedenichia (S.) paraxeri* are known from extralimital records (Lubumbashi) (Zumpt 1961a). The laelaptid mite *Hirstionyssus transvaalensis* as well as the acarinid *Sarcoptiform hypopi* have also been recorded by Viljoen (1975, 1977b) from yellow-footed squirrels collected near Naboomspruit and the Limpopo River.

Ticks include adult *Haemaphysalis leachii muhsami* and three species of *Rhipicephalus* (immature *appendiculatus,* adult *muehlensi* and adult *pravus*) (Theiler 1962). *Rhipicephalus theileri, R. simus* and *Haemophysalis* cf. *zumptii* were recorded by Viljoen (1977b). Fleas include the families Pulicidae and Hystrichopsyllidae. The former is represented by *Xenopsylla brasiliensis, X. zumpti* (Zumpt 1966) and *Ctenocephalides felis* as recorded by Viljoen (1975) and the latter family by *Listropsylla prominens* (Zumpt 1966). The sucking family louse *Neohaematopinus heliosciuri* was also collected by Viljoen (1975).

## Relations with man

Yellow-footed squirrels are regarded as a great delicacy by the Mashona of eastern Zimbabwe (Sclater 1901). In the lowveld of the Transvaal, they are formidable pests in gardens and orchards. Stevenson-Hamilton (1950) states that they delight in mangoes, which are frequently half-gnawed before they are ripe enough to be picked and are, therefore, spoilt. They also take oranges and in fact any cultivated fruit.

They harbour *Streptobacillus moniliformis* organisms which cause a form of rat-bite fever. Nothing is known about the prevalence of this fever in South Africa (Neitz 1965). The ticks they harbour transmit biliary fever *(Babesia canis),* tick-bite fever *(Rickettsia conorii)* and Q-fever *(Coxiella burnetii) (Haemaphysalis leachii).* *Rhipicephalus appendiculatus* and *R. muhlensi* both transmit East Coast fever *(Theileria parva)* (Theiler 1962).

Stevenson-Hamilton (1950) remarks that, in spite of their mischievous propensities, they are attractive and amusing little animals, which tame fairly easily. He found that they are useful in detecting snakes among garden trees and creepers, commencing a loud and continued alarm chatter, should one of these reptiles be encountered.

## Prehistory

These squirrels have not been found as fossils in southern African deposits.

## Taxonomy

Apart from *P. cepapi* and *P. palliatus* (see below) some nine other species occur in Africa, but extralimital to this work. These include *alexandri* (Alexander's Bush Squirrel), *boehmi* (Boehm's Bush Squirrel), *flavivittis* (Striped Bush Squirrel), *lucifer* (Black and Red Bush Squirrel), *vexillarius* (Swynnerton's Bush Squirrel), *cooperi* (Cooper's Green Squirrel), *poensis* (Small Green Squirrel), *vincenti* (Vincent's Bush Squirrel) and *ochraceus* (Huet's Bush Squirrel). With the exception of *poensis* and *cooperi,* which occur in West Africa, the others all occur in Central and East Africa. Too many subspecies within *P. cepapi* have probably been recognised in the past.

*Paraxerus cepapi chobiensis* includes *maunensis, kalararicus* and *tsumebensis* of earlier authors (Meester *et al.* 1964), while the taxonomic status of *carpi, cepapoides, phalaena* and *sindi* need revision. Smithers (1971) proposes the revival of the name *P. c. maunensis,* as well as *P. c. kalaharicus.*

# *Paraxerus palliatus* (Peters, 1852)

**Red Bush Squirrel**

**South African Red Squirrel**

**Rooieekhoring**

The derivation of *Paraxerus* has been explained under *Paraxerus cepapi.* The species name is derived from the Latin *palliatus* = cloaked, which refers to the striking coloration of the animal, reminding one of a cloak worn by fine gentlemen in former years.

*Paraxerus palliatus* is a coastal lowland species and their red bellies are most characteristic of the *palliatus* squirrel group (Kingdon 1974). Kingdon has also suggested the possibility of hybridisation between *palliatus* and an evolving *cepapi* and he attempted to define many intermediate forms with varying characteristics of the parent stocks.

Within southern African as defined in this book, two subspecies will be referred to. The first is *Paraxerus palliatus ornatus,* the Ngoye Red Squirrel, the largest of the subspecies of *palliatus,* which occurs in the 2 900 ha Ngoye Forest in KwaZulu. The second subspecies is *P. p. tongensis,* the Tonga Orange Squirrel, which occurs among the dry forests of northern KwaZulu.

Meester (1976) has listed this species as rare in the *South African Red Data Book.*

## Outline of synonymy

*(tongensis)*

1852 *Sciurus palliatus* Peters, *Monatsber. K. Preuss. Akad. Wiss. Berlin:* 273. Mozambique.

1873 *Macroxus annulatus* Gray, *Ann. Mag. nat. Hist.* (4) 12:265. Zanzibar Island, Tanzania.

1906 *Funisciurus palliatus* Thomas, *Ann. Mag. nat. Hist.* (7) 18:297. Zanzibar Island, Tanzania.

1939 *Paraxerus palliatus palliatus* (Peters). *In* Allen, *Bull. Mus. comp. Zool. Harv.* 83:302.

1914 *Paraxerus bridgemani* Dollman, *Ann. Mag. nat. Hist.* 14:152. Mozambique.

1926 *P. b. auriventris* Roberts, *Ann. Transv. Mus.* 11:250. Magudi, Mozambique.

1931 *P. sponsus tongensis* Roberts, *Ann. Transv. Mus.* 14:220. Magusi Forest, Natal.

1939 *P. s. bridgemani* Dollman. *In* Allen, *Bull. Mus. comp. Zool. Harv.* 83:303.

1951 *P. bridgemani bridgemani* Dollman. *In* Roberts, *The mammals of South Africa.*

1975 *P. palliatus tongensis* Roberts. Amtmann, *in* Meester & Setzer (eds), *The mammals of Africa: an identification manual.*

*(ornatus)*

1864 *Sciurus ornatus* Gray, *Proc. zool. Soc. Lond.:* 13. Ngoye Forest, Natal.

1899 *Funisciurus palliatus* W.L. Sclater, *Ann. S. Afr. Mus.* 1:185.

1905 *Sciurus palliatus ornatus* Gray. *In* Thomas & Schwann, *Proc. zool. Soc. Lond.*: 266.

1939 *Paraxerus p. ornatus* (Gray). *In* Allen, *Bull. Mus. comp. Zool. Harv.* 83:302.

### TABLE 6
Measurements of male and female *Paraxerus palliatus*

|        | Parameter | Value (mm) | N | Range (mm) | CV (%) |
|--------|-----------|-----------|----|-----------|--------|
| Males  | HB        | 195       | 10 | 180–219   | 6,5    |
|        | T         | 193       | 9  | 170–209   | 6,2    |
|        | HF        | 47        | 9  | 43–53     | 5,9    |
|        | E         | 19        | 8  | 17–22     | 8,9    |
|        | Mass:     | 182 g     | 1  | —         | —      |
| Females| HB        | 192       | 8  | 175–205   | 5,6    |
|        | T         | 186       | 6  | 170–205   | 8,0    |
|        | HF        | 47        | 8  | 44–53     | 7,7    |
|        | E         | 17        | 7  | 14–20     | 15,2   |
|        | Mass: No data available |  |  |  |  |

## Identification

According to Dollman (1914) the most conspicuous feature of *P. p. tongensis* is its orange-bordered tail. The dorsal surface of the body and tail is pale yellow speckled with grey. The individual hairs of the body have a slate-black base, pale lemon-yellow middle, white subterminal ring and dark tip. The nose, sides of face, back of hands and feet and ventral surface of the body are yellowish-orange. The dorsal and ventral surfaces of the tail are coloured like that of the back, with the tips of the individual hairs a dark auburn colour (plate 1).

In *P. p. ornatus* the dorsal colour is grey-brown, grizzled with pale buff or yellow. The belly is orange-red and the tail is a pronounced wine-red. The individual hairs of the tail show three black bands with a ferruginous tip.

## Skull and dentition

In comparison with *P. cepapi,* the skull of *P. palliatus* is slightly shorter and the cranium is smaller. The nasals are parallel-sided (fig. 14). Two premolars precede the three molars in the maxilla and the length of the entire upper tooth row is geographically variable. A peculiar tooth pattern has been a characteristic held to separate the genera (fig. 15). In *Paraxerus* (either *cepapi* or *palliatus*) the molars in the lower jaw are cuspidate in adults while in *Funisciurus* both the upper and lower molars are more or less flat-crowned in adults (Amtmann 1975).

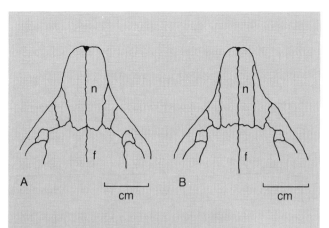

**Figure 14** Nasal bones (n) seen from above in skulls of (A) the Bush Squirrel *Paraxerus cepapi* and (B) the Red Bush Squirrel *P. palliatus*. In *P. cepapi* they are narrowest in the middle, but of even width in *P. palliatus*.

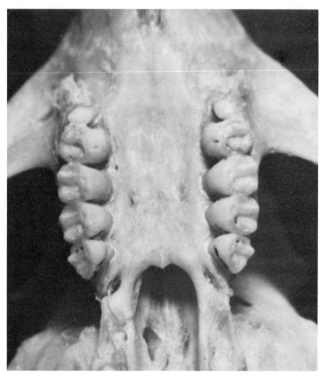

**Figure 15** Cheekteeth of the Red Bush Squirrel *Paraxerus palliatus.*

## Distribution

*Paraxerus palliatus* with its various subspecies is known from the Ngoye and Manguzi forests in KwaZulu and also occurs in southeastern Mozambique to Beira and inland to the Chirinda Forest near Melsetter in Zimbabwe. It therefore inhabits the Southern Savanna Woodland zone (map 6). Extralimitally, the species occurs on the eastern seaboard northwards to Somalia.

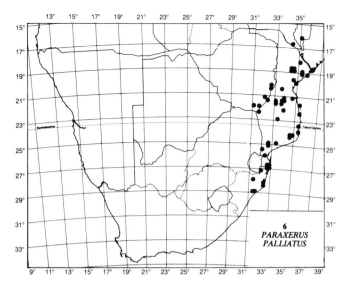

6 PARAXERUS PALLIATUS

## Habitat

Its preferred habitat is littoral forest, as well as evergreen, montane vegetation.

*P. p. tongensis* lives in coastal and dune forests on the Mozambique plain and extends southwards to the well-known Maphelane dune, just south of the St Lucia Estuary, Natal. *P. p. ornatus* prefers moist, montane, evergreen habitat in Ngoye Forest with an annual precipitation of 1 300 mm (Viljoen 1978a). Indeed, Viljoen is of the opinion that habitat is the single most important factor modifying behaviour, colour and population structure of the tree-squirrels *Paraxerus cepapi, P. palliatus ornatus* and *P. p. tongensis* in the RSA.

## Diet

There are few data in the available literature, but it is likely that they eat bark, seeds and nuts of various trees. Viljoen (1978a) says of *P. p. ornatus* in the Ngoye Forest, that they are fond of seeds and nuts and forage for fruits at or just before sunrise. They often bury these items either in holes in trees or on the forest floor. When food is scarce during or just after winter, they may feed on the scatter-hoarded reserves. Under experimental conditions, Viljoen *(pers. comm.)* has found *P. p. ornatus* to concentrate on larger sized kernels, while *Paraxerus cepapi* prefers fleshy fruits.

## Habits

Red bush squirrels are diurnal, living in pairs or small groups in the thick underbush of forests where there is a dense mat of forest debris (Smithers 1966). *P. p. ornatus* maintains contact by emitting a soft murmer. Viljoen (1978a) states that it is an extremely vocal animal, using bird-like trills and clicks both to advertise territories and in their mobbing alarm displays when human intruders or snakes are seen. When a forest bird of prey appears, the squirrels emit a low-pitched warning bark as they flee for cover among branches; they never bolt into a hole. They can disappear completely among the branches and twigs in a matter of seconds, so there is no need for them to seek refuge in holes.

Viljoen records that they sleep in holes at night, lining the holes with leaves and wood-fibres gnawed from the interior of the hole.

The alarm call of *P. p. tongensis* is an irregular, repeated 'chuck-chuck' (Smithers 1966).

## Predators, reproduction and prehistory
Unknown.

## Parasites

It was implied above that we know little about the overall biology of *P. palliatus* and this certainly applies to our knowledge of the associated ecto- and endoparasites. No viruses, protists, worms or mites have been recorded.

The only flea hitherto recorded is the ceratophyllid *Libyastus infestus selindae* (on specimens from Mount Selinda in Zimbabwe). All species of this flea genus are parasites of squirrels belonging to the genera *Paraxerus, Funisciurus* and *Helioschiurus* (Zumpt 1966).

## Predators, reproduction and prehistory
Unknown.

## Relations with man

Not much is known, for these animals are not often seen, let alone collected. Scatter-hoarded seeds are often found on forest floors and indications are that these seeds are not always recovered by the squirrels. These seeds have effectively been planted to the benefit of the forest and Viljoen (1978a) states that many people regard squirrels as indispensable for the generation of tree life in certain habitats. From a conservation point of view, this sounds plausible and if it is accepted that conservation of natural forests has some value for mankind, it would imply that squirrels fulfil an important role.

## Taxonomy

This species has suffered under different schemes of classification at some time or other. The arrangement as proposed by Amtmann (1975) is followed in this book. Eleven subspecies are listed, of which six occur in southern Africa *(auriventris, bridgemani, ornatus, sponsus, swynnertoni* and *tongensis)*. Other subspecies occur extralimitally in Malaŵi *(palliatus)*, southeastern and eastern Tanzania *(freri, suahelicus)*, eastern Kenya *(tanae)* and southern Somalia *(barawensis)*. Amtmann (1975) notes that too many subspecies are probably recognised and that *bridgemani* (= *tongensis* as understood in the present work) could well be a separate species. There is room here for a proper reassessment of the taxonomy of this species.

# Sciuromorpha
*incertae sedis*
# PEDETIDAE: Springhare

# SUBORDER Sciuromorpha *incertae sedis*
# SUPERFAMILY Anomaluroidea Gill, 1872
# FAMILY Pedetidae Owen, 1847

The Springhare is placed with the sciuromorphs, following the early arrangements by Tullberg (1899), Weber (1928) and Winge (1941), for want of a better place to put them systematically. This uncertainty is due to the absence of sufficient morphological and palaeontological evidence of their tribe relationship within the rodents. There are sciuromorph resemblances in the dentition, though the skull is markedly different. An affinity with the sciurid anomalurids is debatable and Simpson (1945) has placed the Pedetidae within the superfamily Anomaluroidea. Yet there is a wide gap between the two families and Ellerman (1940) suggested that there is virtually no alternative to the classification of *Pedetes* as a superfamily distinct from all others in living rodents. Alston (1876) placed this anomalous rodent in a special subfamily of jerboas (Dipodidae) among the myomorphs, but the supposed resemblances between jerboas and springhares are artificial and adaptive rather than cladistic. Cockrum (1962) also placed it provisionally under the Myomorpha. Thomas (1896) put the family in the hystricomorphs as did Roberts (1951). Miller & Gidley (1918) pursued the idea of Alston and placed it as a family under the superfamily Dipodoidea. Romer (1971), in his classic work on *Vertebrate Paleontology* interpreted it as a family, but did not assign it to any particular superfamily or suborder. This is also the interpretation adhered to by Misonne (1974).

# GENUS *Pedetes* Illiger, 1811

These rodents, with their kangaroo-like appearance and saltatorial gait, are commonly seen at night over vast areas of the South African countryside. This holds true when travelling by car and the eyes of the animal show a characteristic reddish glow when reflecting the headlights of the vehicle.

# *Pedetes capensis* (Forster, 1778)

## Springhare

## Springhaas

The generic name is derived from the Greek *pedetes* = a leaper or a dancer, referring to the springhare's saltatorial gait. The specific name *capensis* is the Latin for 'of the Cape'.

The animal was first described as *Yerbua capensis* by Forster in 1778 antedating the name *Mus cafer* (Pallas 1779). Confusion often arises when the Springhare is referred to as *M. cafer*, Pallas, 1778, but pp. 1–70 of Pallas' *Novae Species Quadropedum e Glirium Ordine* were published in 1778, the rest in 1779.

## Outline of synonymy
1778 *Yerbua capensis* Forster, Oefvers. K. Svenska Vet. Akad. Förh. Stockholm 39:108. Cape of Good Hope.
1779 *Mus cafer* Pallas, Nov. Spec. Quad. Glir. Ord. 87. Cape of Good Hope.
1826 *Pedetes cafer* A. Smith, Descr. Cat. S. Afr. Mus. 26.
1829 *Helamys capensis* F. Cuvier, Hist. Nat. Mamm. pt. 59. *Helamis* (variant).
1834 *P. typicus* A. Smith, S. Afr. Quart. Journ. 2:169. Eastern CP. (Renaming of *capensis*.)

## TABLE 7
### Measurements of male and female *Pedetes capensis*

|         | Parameter | Value (mm) | N  | Range (mm) | CV (%) |
|---------|-----------|------------|----|------------|--------|
| Males   | HB        | 393        | 20 | 370–396    | 3,8    |
|         | T         | 414        | 20 | 335–468    | 6,6    |
|         | HF        | 152        | 20 | 115–170    | 7,6    |
|         | E         | 74         | 20 | 67–84      | 5,3    |
|         | Mass: Up to 4 kg | | | | |
| Females | HB        | 372        | 18 | 272–400    | 5,5    |
|         | T         | 411        | 18 | 300–450    | 7,4    |
|         | HF        | 145        | 18 | 127–160    | 4,5    |
|         | E         | 74         | 15 | 67–85      | 5,6    |
|         | Mass: Up to 2,8 kg* | | | | |

*Butynski (1973)

### Identification

Springhares are kangaroo-like in appearance with short front limbs and powerful hind legs, and are adapted to a saltatorial way of life. The head is blunt, with ears long and narrow, thinly haired apically, naked inside with a well-developed tragus which probably keeps sand out of the meatus while the animal is burrowing. The ears are brownish. The eyes are well developed. The vibrissae are long, while longish hairs also occur on the eyelids.

The hands are clawed, adapted for burrowing while the hindfeet have four relatively long toes with straight claws. The third toe is the longest, while the hallux is absent. The sole of the foot, heel and base of each toe are naked. A large, rounded naked palmar pad is found on the hand, near the root of the thumb with another smaller oval one at the root of the fifth digit. The pollex is not reduced – a rare feature in the order Rodentia.

The tail is about as long as the head-body length, bushy, heavily haired, tawny in colour with the distal third dark brown to black, and a thick, black brush terminally.

The pelage of the body is soft and furry, tending to be long and straight, cinnamon-buff dorsally with black-tipped hairs decreasing in number towards the tail. The bases of the hairs on the back and tail are dark grey. The ventral colour is buffy-white with no definite underfur. The abdominal hairs are uniform in colour without any black coloration. Whitish hairs extend upwards in front of the thighs onto the flanks (plate 8).

The overall coloration of these animals is variable geographically. Specimens from the OFS are paler than those from the eastern Cape, while individuals from Botswana have a greater amount of white on the upper parts of the forelimbs, with the sides of the face paler (Smithers 1971), and the white of the underparts of the tail extending over the top behind the black tip to varying degrees.

The fibula is reduced and fused with the tibia.

The spoor is characteristic because of the elongated hindfoot and enlarged third toe with its pointed triangular claw. The adjoining toes are much shorter. The tarsus is as long as the distance from the knee to the ankle. The droppings are bean-sized, flattish, irregular square and about 15–18 mm in length, usually slightly longer than broad.

Its voice is a series of low grunts.

### Skull and dentition

The skull is stoutly built, the infraorbital foramen very large. The rostrum is deep and broad. The large orbits are also conspicuous, the well-defined frontals longer than the parietals. The nasals are broad. No sagittal or occipital crests occur. The zygomatic region is conspicuously thickened anteriorly. The tympanic bullae are well developed, with inflated mastoids. A squamosal process projects backwards over the mastoid posterior to the bullae.

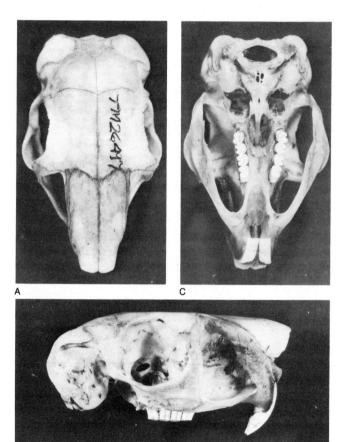

**Figure 16** Dorsal (A), lateral (B) and ventral (C) views of the skull of the Springhare *Pedetes capensis*. Note the deep rostrum, the conspicuous orbit, infraorbital foramen and nasals; mastoid inflated with squamosal process projected over it; short palate (reaching only to level of M²); marked palatine depression and wide pterygoid fossae are characteristic; frontals longer than the parietals.

40

The palate is short and extends backwards only to the middle of the M¹ or M² of the upper tooth row. A marked palatine depression where the anterior palatine foramina normally occur, is situated in front of the tooth rows and extends forwards to the incisors. The pterygoid fossae are wide (fig. 16).

In the mandible the condyle projects beyond the level of the short angular process and the coronoid consists of a thin ridge. The bulging roots of the lower incisors give the lower jaw a knobbly appearance.

Both upper and lower incisors are smooth, thick and strong. The single premolar (not reduced in size) and the three molars are double ovals with an indentation on the outer side (upper) and inner side (lower), while the teeth are generally simplified in crown pattern (fig. 17). The molars are open-rooted, so the teeth grow continuously.

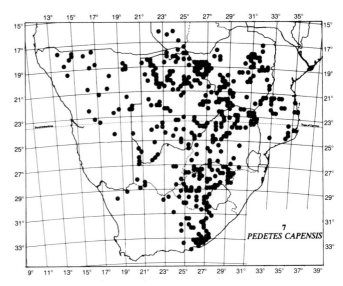

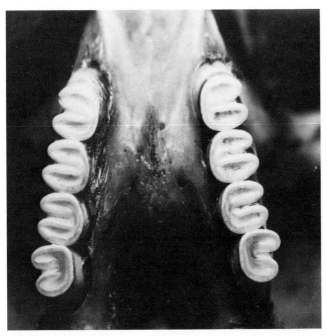

**Figure 17** Cheekteeth of the Springhare *Pedetes capensis*. Pm⁴ not reduced in size. All teeth show a simplified crown pattern. Tooth rows diverge posteriorly.

## Distribution

Springhares occur over most of southern Africa. They range from the Cape Province into SWA/Namibia (except the coastal edge of the Namib and to a lesser extent the Great Namaqualand) to Angola in the west and to Kenya in the east. They are unknown in the coastal parts of Transkei, but do occur in the highlands of Natal (map 7). They are widespread and common in Botswana, although absent from extensive areas where the ground is hard or rocky (Smithers 1971). They therefore occur in the Namib, South West Arid, Southern Savanna Woodland and Southern Savanna Grassland biotic zones (Rautenbach 1978a).

## Habitat

They seem to prefer open sandy ground, often covered with light woodland, in arid or semi-arid country where the vegetation is scanty or covered with open scrub. Generally they are absent from mopane scrub, heavy woodland and tall grassveld. According to Smithers (1971), they tend to occur in areas under heavy grazing pressures, either by domestic stock or wildlife.

In Zimbabwe the Springhare is widely distributed because of the predominantly sandy soils. The effect of springhare populations was studied by Choate and Hill *(pers. comm.)* in the Victoria Falls National Park and the Wankie National Park. The perennial grass cover was often overgrazed by larger ungulates resulting in a lower herbaceous layer. The short grasses where then grazed down to the crown by hares and springhares. This defoliation led to loss in plant vigour and finally the destruction of tufts by springhares digging for underground parts of *Bothriochloa insculpta*, *B. glabra*, *Heteropogon contortus* and *Cynodon dactylon*. As a result of springhare activity, as well as utilisation by larger ungulates, there were signs of breakdown in perennial grass cover, and pioneer grasses *Aristida scabrivalvis*, *Eragrostis viscosa* and *Chloris virgata* then invaded. There is, therefore, a case of springhares causing grassland succession. Choate drew attention to the fact that springhares would also move into medium to tall unutilised grassland where feeding was selectively on *Imperata cylindrica* rhizomes only.

## Diet

Springhares are predominantly vegetarian. They take stems and rhizomes of grasses *(Chloris gayana, Cynodon dactylon, Eragrostis trichophora* and *Oryza longistaminata)* (Wilson 1975) as well as growing and green shoots of other plants, including various species of Cucurbitaceae (Kingdon 1974). Locusts and beetles are also taken on occasion. They are common in the Nylsvley Nature Reserve in the central Transvaal where they consume the

41

roots of *Cynodon dactylon*, *Lannea edulis*, *L. disolor*, *Elephantorhiza obliqua*, *Rhynchosia monophylla*, the leaf bases of *Brachiana nigropedata*, *Eragrostis chloromelas* and the culms of *Eragrostis pallens* (Jacobsen 1977).

## Habits

Springhares are terrestrial and entirely nocturnal. At night the eyes reflect red in artificial light. They are usually solitary and do not form communal warrens. They are excellent burrowers and tree roots may be excavated to 60 cm depth in the soil (Jacobsen 1977). Tunnels slope downwards with loose earth thrown out at the entrance. A little way down, the tunnels are often closed by loose soil, beyond which they may twist and turn, coming to the surface again at a small outlet without surrounding soil (Roberts 1951). The burrows are 20–25 cm in diameter and may reach one metre below the surface. Aggregations of burrows may occur over an area of one hectare. A pair may have several burrows, occupied on successive days.

Springhares emerge at dusk, supposedly by a great leap into the air (although Smithers (1971) states that this is not substantiated). They can travel far at night to feed, returning night after night to a favourite feeding area (Kingdon 1974), but do not habitually use beaten paths. They often move on all fours while feeding. When running at full speed, the tail is raised and slightly curved upwards. In flight they tend to hop with great agility by means of long jumps, covering as much as two metres at a leap. However, they cannot attain the speed of true hares.

Kingdon (1974) described the Springhare as a single balanced cantilever with the body balanced on a fulcrum between the hip joints, with the toes and the tail acting as a counter-poise. The animal's design is subject to environmental limitations such as rough ground or anything more than the lightest vegetation over which it has difficulty in progressing.

They are alert animals with acute vision, hearing and smell. Kicking seems to be their only defence. When captured young, they tame easily and become interesting pets and in captivity they show partiality to oats (Haagner 1920) and may live up to six years or more. The tail is used as a support when sitting up to feed. They do not defaecate in their burrows. They sleep on their haunches with head and forelimbs between the thighs and tail coiled around head and body.

Disused burrows are often taken over by other mammals such as mongooses. They seldom emerge from their burrows during rainy weather and there is circumstantial evidence of seasonal and other fluctuations in numbers of different populations (Ansell 1960; Pienaar 1964).

## Predators

The major predators seem to be the smaller carnivores like genets *(Genetta* sp.) and mongooses like the Grey Mongoose *Myonax pulverulentus*. Kingdon (1974) points out that the Serval *Felis serval*, the Caracal *F. caracal*, the Wild Cat *F. lybica*, the Ratel *Mellivora capensis* and the Black-backed Jackal *Canis mesomelas* all share the Springhare's habitat. The Mongoose *Herpestes* has also been reported killing springhares. Springhare remains were recovered from stomach contents of five black-backed jackals in the Wankie National Park (Wilson 1975). In the Kalahari Gemsbok National Park Labuschagne *(pers. comm.)* observed *P. capensis* being taken as prey by the Cheetah *Acinonyx jubatus,* while Mills & Mills (1978) recorded it in the diet of the Brown Hyaena *Hyaena brunnea*.

## Reproduction

The female has two pairs of pectoral mammae, closely spaced. The young are born in chambers in the burrows, without nesting material, after a gestation period estimated to be two months (Roberts 1951). One young is born at a time, probably annually. Pregnant females may evidently be taken at any month of the year and at birth the well-haired young are large in relation to the size of the adult. According to Coe (1969), specimens from Kenya indicate that young are produced throughout the year. Shortridge (1934) records pregnant females in SWA/Namibia in November and April, while Smithers (1966) stated that in Zimbabwe and Zambia breeding occurs from November to February. Coe states that if the assumption of continuous breeding in Kenya is correct, there would seem to be a distinct specific difference between *P. capensis* of the south and *P. capensis surdaster* of the north. Birthweight may be around 250 g and the eyes open on the second day. Juveniles tend to remain in the burrows for a long time before emerging to the surface.

Other aspects of springhare biology that have been studied include breeding (Van der Horst 1935), newborn young (Jones 1941; Hediger 1950; Coe 1967 (on *P. c. surdaster)*), and placental type (Mossman 1957). Coe (1969) has written about the anatomy of the reproductive tract, and about breeding in *P. c. surdaster*.

## Parasites

A fair amount of information is available. The organism causing bubonic plague *(Yersinia pestis)* occurs sporadically in natural populations of springhares in the OFS, Transvaal, Western Province and SWA/Namibia. Their susceptibility to this micro-organism has also been tested in the laboratory (Neitz 1965).

Nematode parasites include *Dermatoxys veligara* (Mönnig 1927), *Hyracofilaria hyracis* (Onderstepoort records), *Oesophagostomum susannae* (Le Roux 1929a), *Physaloptera capensis* (Ortlepp 1922b) and *Trichuris pedetei* (Verster 1960).

Three species of mites, all of the family Laelaptidae, have been recorded. These include *Radfordilaelaps capensis*, *Androlaelaps marshalli* and *Histionyssus santos-diasi*

(Zumpt 1961a). Other arthropod parasites include fleas and ticks. The following pulicid fleas are known to be associated with *Pedetes: Echidnophaga bradyta, E. gallinacea, Delopsylla crassipes* (on extralimital subspecies *surdaster), Ctenocephalides connatus, C. felis, Synosternus caffer, Pariodontis riggenbachi riggenbachi* and *Xenopsylla cryptonella* (Zumpt 1966). The following ticks have been recorded: *Amblyomma hebraeum* (immatures), *Haemaphysalis leachii leachii* (immatures and adults), *Rhipicephalus appendiculatus* (adults), *R. capensis* (adults), *R. sanguineus* (adults), *R. simus* (immatures) and *R. tricuspis* (adults) (Theiler 1962), immature specimens of *R. capensis* (Walker *pers. comm.)* and immature *R. evertsi* on the extralimital subspecies *surdaster.*

## Relations with man

Springhares may seriously damage groundnut and other crops like maize (digging planted seeds from the ground) and wheat. Kingdon (1974) states that where agriculture has invaded their habitat, they can take to crop-raiding which includes sweet potatoes and pumpkins.

The flesh is pale and firm, has little flavour, but seems to make good eating. Springhares are prized as food by Bushmen. Silberbauer (1965) records that a single band of Bushmen may kill up to 227 specimens in one year. Butynski (1973) estimates that the total number killed by Bushmen in Botswana is in excess of 346 000 annually and that the number killed by Botswana hunters amounts to 2,2 million per year. This represents some 2,2 million kg of springhare meat annually, the equivalent of 20 000 head of cattle. Other indigenous people of southern Africa, like the Ndebele, often catch them with a long, supple stick with a hook or 'burred' seedpod attached to the end. This is pushed down the passage and twisted to embed the 'burr' in its fur, whereupon the animal can be dragged out (Smithers 1971). An animal provides about 60% usable carcass. Quinn (1959) also lists this species as a food item of the Pedi in Sekukuniland (Lebowa).

Because of the various parasites they harbour, springhares play a role in the transmission to man and domestic stock of bubonic plague, rickettsiasis, babesiasis, theileriosis and toxicosis paralysis.

## Prehistory

The Pedetidae is known from Lower Miocene deposits in Africa as a rather specialised fossil, *Parapedetes namaquensis,* described from Elizabethfeld south of Lüderitz in SWA/Namibia (Stromer 1922; Hamilton & Van Couvering 1977). A larger version of the living form, *Megapedetes,* is a Miocene fossil known from East Africa (Kingdon 1974). *Megapedetes* is less specialised than *Parapedetes* (MacInnes 1957). A fossil species *P. gracilis* was identified by Broom (1930, 1934) from the australopithecine locality at Taung in the northwestern Cape. *P. hagenstadi* was reported by Dreyer & Lyle (1931) from Florisbad in the OFS. Fossil specimens of *P. capensis* are also known from Pleistocene deposits in Bulawayo, Zimbabwe (Zeally 1916).

## Taxonomy

Too many subspecies have probably been recognised and described. Misonne (1974) includes some 12 different subspecies within *P. capensis.* The status of the southern African forms *capensis* and *cafer* (both from the 'Cape of Good Hope'), *albaniensis* (Albany District), *orangiae* (OFS), *salinae* (northern Transvaal), *damarensis* (Damaraland) and *fouriei* (Ovambo) is open to question and is in need of revision. This also applies to the status of the extralimital *angolae* (central Angola), *larvalis* and *currax* (Kenya), and the Tanzanian forms *taborae* and *dentatus.* Misonne (1974) also lists the extralimital *P. surdaster* as a subspecies of *P. capensis.*

# Hystricomorpha
## HYSTRICIDAE: Porcupine
## THRYONOMYIDAE: Canerats
## PETROMURIDAE: Dassie Rat

Plate 3

Common Molerat
(*Cryptomys hottentotus*)

Cape Molerat
(*Georychus capensis*)

Silvery Molerat
(*Heliophobius argenteocinereus*)

Namaqua Dune Molerat
(*Bathyergus janetta*)

Cape Dune Molerat
(*Bathyergus suillus*)

Plate 4

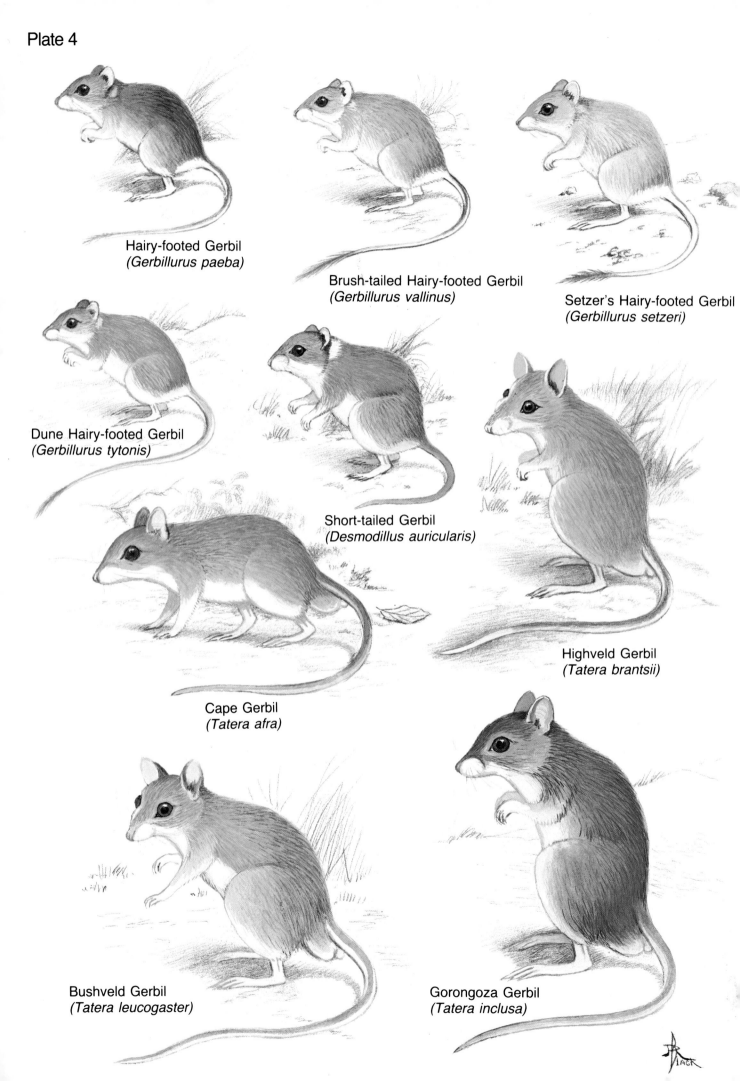

Hairy-footed Gerbil
*(Gerbillurus paeba)*

Brush-tailed Hairy-footed Gerbil
*(Gerbillurus vallinus)*

Setzer's Hairy-footed Gerbil
*(Gerbillurus setzeri)*

Dune Hairy-footed Gerbil
*(Gerbillurus tytonis)*

Short-tailed Gerbil
*(Desmodillus auricularis)*

Highveld Gerbil
*(Tatera brantsii)*

Cape Gerbil
*(Tatera afra)*

Bushveld Gerbil
*(Tatera leucogaster)*

Gorongoza Gerbil
*(Tatera inclusa)*

# SUBORDER Hystricomorpha Brandt, 1855

The Hystricomorpha have been recognised as one of the major groups of rodents by all authorities (Ellerman 1940). The 'Hystricomorpha', originally proposed by Brandt in 1855 (together with the 'Sciuromorpha' and 'Myomorpha' constituting the three classical 'waste-paper baskets' of rodent classification), include animals showing great structural diversity. The suborder contains the largest of the contemporary rodents (e.g. the South American Capybara, *Hydrochoerus hydrochaeris,* which may weigh 50 kg) and includes 17 currently accepted families. Common features occur in the skull. In all cases the infraorbital foramen through which the deeper portion of the masseter muscle (= *masseter medialis)* passes *en route* to the outside of the lower jaw below the cheek-teeth, is conspicuously enlarged. This arrangement of the jaw muscles typifies the suborder. The jugal bone of the enlarged zygomatic arch is not supported by a zygomatic process of the maxillary. The angular portion of the dentary arises from its outer wall rather than from its lower border. This formation is for the insertion of a specialised limb of the *masseter lateralis superficialis* muscle connecting the skull with the dentary. All hystricomorphs have a dental formula of $1^1/1$ $C^0/0$ $P^1/1$ $M^3/3 = 20$.

The great diversity of rodents and the probable occurrence of parallel evolution in different groups are acknowledged by most workers (Anderson & Jones 1967). This has resulted in different classifications of the hystricomorphs by Simpson (1945), Stehlin & Schaub (1951) and Wood (1955), to mention but three schemes. There is no consensus as yet on classifying rodents above the level of families. Anderson & Jones (1967) state that, considering the different bases for the two systems of Stehlin & Schaub on the one hand and that of Wood on the other, the amount of agreement is more remarkable that the disagreement.

The hystricomorphs present interesting and often baffling zoogeographical problems because of striking resemblances between African and South American representatives. The Afrotropical and Neotropical zoogeographical regions are therefore the centres of abundance for the hystricomorphs (Vaughan 1972). This is especially true for the South American continent where three superfamilies occur. The Cavioidea is divided into the following families: the Caviidae (the familiar Guinea Pig and related forms, some five genera and 12 species), the Hydrochoeridae (the Capybara, the largest living rodent, one genus and two species), the Dinomyidae (the Paca-rana, one genus and one species), the Heptaxodontidae (two genera, two species, rodents which became extinct in Recent or Sub-recent times in the West Indies) and the Dasyproctidae (the agouti's and paca's, four genera and some 11 species). The second South American superfamily, the Chinchilloidea contains but a single family, the Chinchillidae (the well-known chinchillas, three genera and eight species). The third superfamily is known as the Octodontoidea which contains seven families of which two occur in Africa (the Thryonomyidae and Petromuridae). The others are the Capromyidae (the hutias, one genus, some seven living species from the West Indies), the Myocastoridae (the Nutria, one genus and one species), the Octodontidae (tuco tuco's and the like, six genera and some 33 species), and the Echimyidae (rat-sized spiny rats, 14 genera and 43 species). It can thus be seen that a total of some 39 genera and 122 species occur in the neotropics.

In the Neartic Region another superfamily, the Erethizontoidea, occurs. These are the well-known New World porcupines of which there are four genera and eight species, ranging across the North-American Continent from the Arctic to Mexico.

Four superfamilies are represented in the Afrotropical and Palaearctic regions. Hystricoidea include the single family Hystricidae, the Old World porcupine, of which there are four genera and some 15 species. These range across Africa, from the Cape to southern Europe and eastwards to southern Asia and the Philippines. The second superfamily represented in Africa is the Octodontoidea, which, as was indicated above, includes the family Thryonomyidae or canerats, with one genus and six species, the members of which are known as canerats. The second octodontoidid family is the Petromuridae, which contains the Dassie Rat, with one genus and one species, having a limited distribution in western southern Africa. A third African hystricomorph superfamily, the Bathyergoidea, contains the African molerats, a group of unusual, highly fossorial rodents. The Bathyergidae with five genera and 20 species occupy much of West, Central and East Africa, southwards to the Cape of Good Hope. The fourth superfamily, Ctenodactyloidea (the gundis) also occurs in Africa, especially in North Africa, with four genera and eight species.

The Afrotropical Region (even with some help from the Palaearctic and Oriental regions) is not as rich in hystricomorphs (15 genera, 50 species) when compared with the Neotropical Realm.

Of this formidable array of taxa above the generic level, only the Hystricoidea, the Octodontoidea (part) and the Bathyergoidea are represented within the geographical limits defined for this book.

---

**Key to the southern African hystricomorph families**
(Modified after Roberts (1951))

1 Size large (600 mm); body covered with long spines; skull with dorsal profile oval, orbits placed far back, nasals very broad; molars four, with wavy enamel patterning, occlusal surfaces flat; upper incisors smooth .............................. Family Hystricidae
(Porcupine)
Page 48

Size smaller (<600 mm HB length); no spines on body ...................................................... 2

2 Length of HB 300–600 mm; hair bristly, also covering tail which is less than half the HB length; body compact, legs short; molars four with infoldings of enamel; upper incisors with two grooves on anterior surface .......................... Family Thryonomyidae
(Canerats)
Page 54

HB length not exceeding 200 mm; tail about the same length, covered with scattered long hairs, not close-set as in squirrels; skull flattened; molars four, in double transverse sections; upper incisors not grooved .......................... Family Petromuridae
(Dassie Rat)
Page 60

---

SUPERFAMILY **Hystricoidea** Gill, 1872

FAMILY **Hystricidae** Burnett, 1830

SUBFAMILY **Hystricinae** Murray, 1866

The Old World porcupines are terrestrial animals with thickset bodies. Conspicuous are the well-developed spines and quills. The skull has a marked pneumatic inflation of the nasal bones, giving the skull a curved outline. An interorbital constriction is lacking. A very large infraorbital foramen transmits the medial masseter. The zygomatic arch is simple, the auditory bullae not enlarged and the paroccipital processes not particularly long. The occipital region of the skull is strongly ridged, providing attachments for powerful neck muscles.

The four hypsodont semi-rooted molars have re-entrant folds of enamel that, with wear, become islands on the occlusal surfaces. The upper molars throughout the family have one inner and three outer folds (Ellerman 1940), which become isolated as islands almost immediately on the occlusal surface. The lower teeth show the reverse pattern. The molars become rooted only later in life (Rosevear 1969). The incisors are plain and the deciduous dentition is shed relatively late in life. Hystricids eat a diversity of plant matter, often damaging crops. Fairly extensive burrows are dug which may be used as dens. In the Hystricinae the quills are conspicuously marked with black and white bands which may act as a visual signal to deter predators in combination with rattling of the quills (Vaughan 1972). Apart from flattened spines and cylindrical quills some genera, e.g. *Hystrix,* have developed long, wiry hairs forming a conspicuous nuchal crest. In all cases the thumb is reduced, leaving the hand with four functional clawed digits. The hindfoot has five digits, the two middle ones being slightly longer than the outer ones.

The Hystricidae, consisting of two subfamilies, the Atherurinae (brush-tailed porcupines) and the Hystricinae (crested porcupines) range from across tropical Asia to the southern tip of Africa. The Atherurinae, erected by Lyon in 1907, are predominantly forest animals and are extralimital to this work. The Hystricinae, erected by Murray in 1866, are better adapted to savanna conditions (Kingdon 1974). They may be differentiated by the dorsal pelage which consists of long, stout, cylindrical, bicoloured quills in *Hystrix* while *Atherurus* has a dorsal pelage of short, flat spines though a few unicoloured quills may occasionally be present. The nasals are very large in *Hystrix*. Although these two subfamilies are widely accepted today, Ellerman (1940) questioned the proposal by Lyon to divide the family, stating that there appear to be too many essential characters common to both for this division to be maintained.

# GENUS *Hystrix* Linnaeus, 1758

Two genera of porcupines occur in Africa. The Brush-tailed Porcupine *Atherurus africanus* occurs in the forests from Gambia to the southern parts of Zaïre and eastwards to western Uganda and environs. It is extra-limital to this work and will not be discussed in detail. The genus *Hystrix* consists of two species. The Crested Porcupine *Hystrix cristata* inhabits the northern half of Africa to the Mediterranean coast including northern Zaïre,

Uganda, Kenya, northern and central Tanzania as well as Somalia and Ethiopia. It is therefore extralimital. The South African Porcupine *Hystrix africaeaustralis* occurs in the southern half of Africa showing some sympatric overlap with *H. cristata* in Central and eastern Africa. In the broad area of sympatry no intermediate forms have been discovered (Kingdon 1974).

# *Hystrix africaeaustralis* Peters, 1852

**Cape Porcupine**

**Kaapse Ystervark**

*Hystrix* was the Greek and Latin name for a porcupine. The species' name refers to the Latin *afer* = pertaining to Africa and *auster* = south wind. It is therefore the animal hailing from the south of Africa.

## Outline of synonymy

1852 *Hystrix africaeaustralis* Peters, *Reise nach Mossam-bique. Säugeth.:* 170. Querimba Coast, Mozambique.

1858 *H. capensis* Grill, *Oefvers. K. Svenska Vet. Akad. Förh. Stockholm* (2) 2:19. Salt River, near Knysna, CP.

1910 *H. stegmani* Müller, *Arch. f. Naturgesch.* 76 (1):186. Kissenji, Tanzania.

1910 *H. africaeaustralis prittwitzi* Müller, *Sitzb. Ges. Naturf. Freunde, Berlin:* 311. Tabora, Tanzania.

1936 *H. a. zuluensis* Roberts, *Ann. Transv. Mus.* 18:240. White Umfolozi River, KwaZulu.

1951 *H. a. capensis* Grill. *In* Roberts, *The mammals of South Africa.*

## TABLE 8
Measurements of male and female *Hystrix africaeaustralis*

|  | Parameter | Value (mm) | N | Range (mm) | CV (%) |
|---|---|---|---|---|---|
| Males | HB | 655 | 3 | 649–674 | — |
|  | T | 117 | 2 | 105–130 | — |
|  | HF | 99 | 4 | 89–102 | — |
|  | E | 40 | 4 | 39–41 | — |
|  | Mass: May reach 18 kg | | | | |
| Females | HB | 701 | 6 | 630–805 | 10,6 |
|  | T | 85 | 2 | 85 | — |
|  | HF | 103 | 5 | 93–114 | 6,9 |
|  | E | 46 | 5 | 45–48 | 2,9 |
|  | Mass: May reach 22,6 kg | | | | |

Measurements of museum specimens seem to have been irregularly taken for this species and some of the values given in the table are evidently based on too heterogeneous or too small a sample. The figures do indicate that porcupines are formidable rodents weighing in excess of 20 kg.

## Identification

They are stout, heavily built animals with blunt, rounded heads and small eyes situated far back, and coats of thick cylindrical spines covering the body behind the shoulders, sparsely intermingled with ordinary hairs. It is the largest rodent in southern Africa (plate 2).

The general colour is a dark brown to black, with a white patch on the side of the neck. The vibrissae on the greyish-brown bristled face are long and stout. The forepart of the body, covered with bristles elongating along the nape of the neck and the occipital region of the skull to form long wiry hairs forming a conspicuous nuchal crest. These bristles and the crest are erectile. The chin and forepart of the throat are nearly naked. Towards the hinder part of the back and haunches the bristles become longer, stronger and sharper, forming the true spines and quills implanted in the skin in transverse grooves, about 3 cm apart. Each groove contains five to eight quills or spines. The spines of the back are banded brown and white, the brown bands being broader; the tips are white. The sharper and stiffer defensive quills mixed with the spines at the sides of the back are black-tipped and white at the base. The quills near the midline of the body above and near the tail are hollow and open at their ends – known as *Rasselbechern* in German. These 'rattle quills' produce a sound which may differ in *cristata* and *africaeaustralis* serving as an ethological barrier between these two species where they occur sympatrically (Corbet & Jones 1965). The ears are inconspicuous. The tail is short and well hidden. The legs are short, covered with bristles, the digits clawed, the soles naked. The pollex on the forefoot is rudimentary, while a hallux is present on the hindfoot. The female has three laterally placed pairs of mammae. Albinistic individuals occur sporadically. Clavicles are present, though small. The centrale in the hand is not free, a characteristic known also in the Cuniculidae, the paca's of South America (Dasyproctidae *sensu* Anderson 1967) (Ellerman 1940).

## Skull and dentition

The skull is conspicuously arched and domed, on account of the pneumatisation which has occurred in the well-developed nasals. The zygomatic arch is strong but simple and the jugal element does not make contact with the lachrymal. The lachrymal has an upwardly directed process. Sagittal and occipital crests are prominent. There is no interorbital constriction. The palate is straight and wide and extends behind the molars. The infraorbital foramen is well developed, though small. The palatine foramina are small, lying far forwards. The bullae and paroccipital processes are relatively small. The zygomatic plate is poorly differentiated and lies below the infraorbital foramen (fig. 18).

Some twisting of the dentary can be seen. When viewed from below, there is a distinct longitudinal groove between the dental and angular portions which has given rise to the concept of the hystricognath *vs* sciurognath jaw types in rodents. The coronoid is low, and the condyle rounded and oblong.

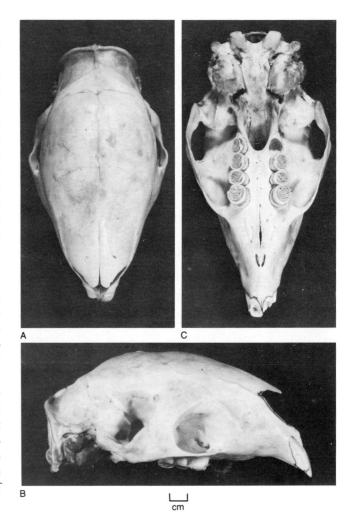

**Figure 18** Dorsal (A), lateral (B) and ventral (C) views of the skull of the Cape Porcupine *Hystrix africaeaustralis*. Note the conspicuously domed skull, the slender zygomatic arch and the prominent sagittal and occipital crests. An interorbital constriction is absent. The infraorbital foramen is well developed, though small. The zygomatic plate is poorly differentiated.

In both the upper and lower jaws, four cheekteeth are present. The large molariform $Pm^4$ is shed and replaced late in life. It erupts close against the first molar which either decays or becomes smaller than the second molar (Roberts 1951). All the teeth have three outer folds on the buccal side (outer side) and one inner fold lingually (inner side). In the lower molars this arrangement is reversed. The teeth are hypsodont and the patterns of enamel which remain as islands on the occlusal surfaces become obliterated only after considerable wear. These molars do not grow continuously (fig. 19). The incisors are well developed and smooth.

## Distribution

The Cape Porcupine is widely distributed in southern Africa, though not always abundant. Its presence is

50

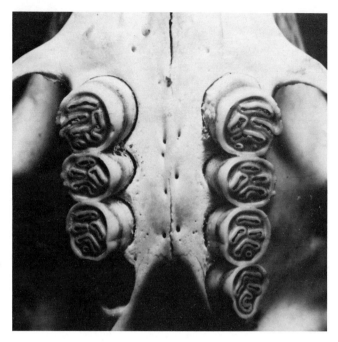

**Figure 19** Cheekteeth of the Cape Porcupine *Hystrix afri-caeaustralis*. The large molariform Pm$^4$ is shed and replaced late in life, erupting close against the M$^1$ which becomes smaller than the M$^2$. The teeth tend towards hypsodonty.

largely dependent on food supply and available shelter. It ranges from the Zambezi Delta southwards into the Transvaal, KwaZulu and the OFS and along the coast to the Cape Peninsula. It occurs also in SWA/Namibia, Zimbabwe, Botswana and Mozambique (map 8).

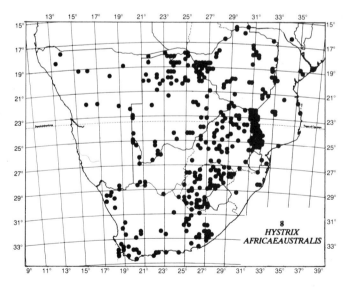

8
HYSTRIX
AFRICAEAUSTRALIS

### Habitat

During daylight porcupines hide in caves or natural crevices with narrow entrances which are often overlain by rock or boulders. They may also conceal themselves in scrub or tall grass near cultivated areas while feeding

there. Their warren, used over many years, usually has a number of exits, which they either dig themselves or take over from other species (often the Aardvark *Orycteropus afer*). They inhabit all types of country apart from swampy areas, very moist forests and barren desert areas, as long as there is suitable rock or scrub cover. They therefore tend to be common in hilly, rocky country.

### Diet

Porcupines are mainly vegetarian and consume roots, bulbs, berries, fruits and the bark of trees. Local farmers in Natal state that porcupines are omnivorous and traps can be baited with meat and bones (Maclean *pers. comm.*). In the Kruger National Park, they ring-bark trees such as the Wild Mango *Cordyla africana* (Pienaar *et al.* 1980), *Erythrina* spp., *Ficus* spp., *Spirostachys africanus* (Pienaar 1964) as well as *Trichilia emetica* along the Limpopo River. This activity is also witnessed in the Umfolozi Game Reserve in Natal, where many Tambotie trees are disfigured 12–15 cm above the soil with resultant secondary growth. At Nylsvley Nature Reserve in the Transvaal they also eat the bark of young *Strychnos pungens* (Jacobsen 1977). They are wasteful feeders, usually foraging alone or in small parties. Feeding is done at night and therefore these animals are not often seen. Their normal diet (in undeveloped areas) consists of indigenous bulbs and roots. Bulbs of *Crinum macowani*, tubers of *Jatropha zeyheri* and the roots of *Elephantorrhiza obliqua*, *Albizia tanganyicensis*, *Lannea edulis*, *L. discolor* and *Bauhinia macrantha* are additional feeding records from Nylsvley (Jacobsen 1977). Kingdon (1974) states that it eats the roots, tubers, rhizomes and bulbs of the following plants: *Sanservieria*, *Combretum*, *Commiphora*, *Colophospermum*, *Aloe*, *Ficus petersi*, *Gladiolus* and species of *Cyperus*. The fruits of *Kigelia*, *Sclerocarya* and *Strychnos* are also eaten. They seem to forage singly or in pairs during the breeding season.

### Habits

Porcupines are active at night and may nightly move as far as 16 km from den to feeding area using well-defined paths. They live singly, in pairs or in small groups. They are wary animals with a keen sense of smell and have no climbing or jumping ability.

The quills can be raised when annoyed or attacked and if cornered the porcupine charges backwards with the erected quills directed to the rear; the stout quills of the body and tail detach easily and when lodged in the flesh, can cause irritating and festering wounds in the jaws and feet of lions and other carnivores which frequently attack them. The quills tend to work their way deeper into the flesh when not removed, causing severe discomfort and pain and may even penetrate the lungs or liver of the attacker. A predator injured in this way is a pitiful sight. Some years ago we came across a lioness near Dankbaar in the Kalahari Gemsbok National Park which had been thus tormented and for humane reasons she was put to death.

The sounds they emit have been described as piping calls and pig-like grunts and while they wander along, their presence is often betrayed by the rattling of the quills. The ordinary pace is a shuffling plantigrade gait, which on pursuit breaks into a clumsy gallop.

The presence of porcupines is indicated by dropped quills, lying about the entrance to their abodes.

They are efficient and adept at digging. They can dig their own burrows and Anderson & Jones (1967) records one burrow extending 18 m from the main entrance to a chamber measuring 1,2 × 1,2 × 0,46 m in size, some 1,5 m below the surface of the soil. This system had two or three less-used escape exits.

The droppings are shaped like large date stones, woody and fibrous in texture. The burrows are often littered by gnawed bones carried there by the porcupines on which they hone their incisors. This habit was commented upon by Alexander (1956) as well as by Dart (1957), suggesting the possible role of porcupines in the accumulation of bones associated with remains of early man.

## Predators

Their chief enemies seem to be the larger cats (lions and leopards) and hyaenids. In the Kalahari Gemsbok National Park they are eaten by the Brown Hyaena *Hyaena brunnea* (Mills & Mills 1978). FitzSimons (1962) states that on occasion they are swallowed by pythons *Python sebae*.

## Reproduction

Two young form a normal litter, but it may vary between one and four. The gestation period varies between six and eight weeks and the female has three laterally situated pairs of mammae. The young are born in grass-lined chambers in burrows with their eyes open and come into this world during the summer months (Shortridge 1934). Near full-time fetuses in Rhodesia (November) weighed 300 g (Smithers 1971). A gravid female with three young was taken in February in the Maria van Riebeeck Nature Reserve, south of Pretoria. Two embryos were in the left uterine horn and one in the right horn. In the Kruger National Park mating has been observed in May and one to three young are born in autumn and early winter (Pienaar *et al.* 1980).

There may be a considerable difference in the development of sibling fetuses, as is shown by two taken from the same female in November at weights 294 g and 107 g respectively. The bristles of the young are soft and pliable and they accompany the female on foraging explorations only after the quills have hardened at about 14 days of age. It is said that the males make good parents.

## Parasites

The sarcosporid *Sarcocystis* sp. is apparently enzootic in the porcupine (Viljoen 1921). Mites have been taken from *Hystrix africaeaustralis* and the mesostigmatid *Rhinophaga hystrici* (inhabiting the nasal and frontal sinuses of hystricids), the trombidiformid *Psorergates hystrici* and the sarcoptiformid *Rhyncoptes recurvidens* have been recorded (Zumpt 1961a; Till 1957). The ticks include both argasids as well as ixodidids. Immatures of *Argas brumpti* occur (Theiler 1962) as well as adult specimens of the following Ixodidae: *Rhipicentor nuttalli, Hyalomma truncatum, Rhipicephalus appendiculatus, R. capensis, R. pravus, R. reichenowi, R. sanguineus,* and *R. simus* as well as both adults and immatures of *Haemaphysalis leachii leachii* (Theiler 1962).

Fleas are represented by Pulicidae and Hystrichopsyllidae. The pulicidids include *Echidnophaga gallinacea, E. larina, Ctenocephalides canis, C. felis, Pariodontis riggenbachi* (highly host specific) and *Xenopsylla cheopis.* The hystrichopsyllid flea is *Dinopsyllus lypusus* (Zumpt 1966).

## Relations with man

Porcupines do extensive damage to cultivated crops like carrots, groundnuts, pumpkins, maize, cotton and potatoes. They are wasteful foragers consuming only part of the bounty and leaving the rest to decay. They are therefore actively hunted by a variety of methods, chiefly by tracking down with dogs and clubbing to death. The dogs have to be trained to do this kind of driving as the porcupines are apt to halt suddenly in their tracks '...with the object of impaling the pursuer on the long, sharp quills, and thus inexperienced dogs may lose their eyes in this way' (Roberts 1951). A rap over the head is an effective way of killing porcupines. Shooting, snaring and trapping are also employed. When caught in traps they may twist off the appendage so that trapping is not always effective.

On the other hand, they are useful animals in devouring *Homeria* spp., bulbs which are poisonous to cattle, and in retarding bush encroachment by eating the bark of *Acacia* spp.

It is said that the meat makes excellent eating and some African tribes treat the meat as a delicacy. Roberts (1951) states that in some places certain tribes enter porcupine burrows armed with spears – a method requiring courage.

Porcupine quills are often used by man for ornamentation and as a talisman. Kingdon (1974) records having been approached by Ugandan schoolboys in search of quills as a talisman to ensure that they would pass their examinations. In other cases quills were thought to provide protection against diseases, including smallpox and febrile convulsions.

The protozoan *Sarcocystis* sp. (of uncertain systematic position) is found regularly in skeletal muscles and myocardium of domestic ruminants, pigs and horses and appears to be enzootic in *H. africaeaustralis* (Neitz 1965). The laelaptid *Rhinophaga hystrici* inhabits the nasal passages of porcupines and monkeys but nothing is known about the pathogenicity of these mites (Zumpt 1961a). The ticks associated with porcupines carry a variety of micro-organisms pathogenic to man and domestic stock, which includes babesiasis, rickettsiasis and theilerioses. The flea *Echidnophaga gallinacea* is a severe pest of domestic poultry and the porcupine also carries the common cat and dog flea *Ctenocephalides canis,* as well as the important flea vector in the spread of bubonic plague *Xenopsylla cheopis.*

### Prehistory

In Africa, the genus *Hystrix* ranges from Pleistocene to Recent times. In South Africa, *Hystrix* cf. *africaeaustralis* is known from the Wonderwerk Cave in the Kuruman district (Cooke 1941), as well as from Makapansgat, Sterkfontein, Swartkrans and Kromdraai (Greenwood 1955; Maguire 1978). From the *Australopithecus* layers the Makapansgat breccias have also yielded a fossil species *H. major,* as well as an extinct Giant Porcupine *Xenophystrix crassidens. Hystrix major* has subsequently been renamed *H. makapanensis* as the name *major* was a junior homonym of *H. major* Gervais 1859 (Maguire 1978). Apart from

Makapansgat, single doubtful isolated teeth from Sterkfontein, Swartkrans and Kromdraai A might belong to this taxon. The extralimital *Hystrix cristata* has also been reported by Greenwood (1955) from Sterkfontein in the Transvaal.

### Taxonomy

Porcupines of the genus *Hystrix* are found from southern Europe *(H. cristata)* to the Cape *(H. africaeaustralis),* as well as in Asia *(H. indica (=leucura)).* Corbet & Jones (1965) revised the genus in Africa, based largely on characteristics of the skull. *H. africaeaustralis* (including *stegmanni*) occurs over virtually the whole of the southern half of the continent, its northern limit being approximately the Congo River; *cristata* ranges northwards to the Mediterranean and the Red Sea (Rosevear 1969).

The nominate race in southern Africa, *H. a. africaeaustralis* is taken to include the subspecies *capensis* and *zuluensis* of Roberts (1951) and earlier authors. However, it is possible that at least *zuluensis* may be valid as a subspecies (Meester *et al.* 1964) if the narrower and more pointed muzzle as well as the sutures of the nasals with the adjoining bones are taken into consideration.

SUPERFAMILY # Octodontoidea Simpson, 1945

Apart from the Hystricoidea discussed above, the Octodontoidea occur as a second superfamily of the Hystricomorpha in southern Africa. Two families are usually incorporated under the Octodontoidea: the Thryonomyidae and the Petromuridae (canerats and dassie rats respectively). Most related forms occur in South America, but both the thryonomyids and the petromurids are purely African in their distribution. Therefore, in spite of striking morphological resemblances, some doubts have been expressed about whether the Neotropical and Afrotropical groups are in fact related, as the generally accepted classification would imply (Rosevear 1969). The resemblances may be due to evolutionary convergence, but Kingdon (1974) says that an ancient common ancestor is also possible.

### Key to the southern African octodontoid families
(After Meester *et al.* (1964))

1 Cheekteeth complex, outer side of upper molars with more than one fold; bullae small, paroccipital processes enlarged; four hindtoes; tail not bushy; incisors thick, upper ones heavily three-grooved; skull massive and powerfully ridged.....................
Family Thryonomyidae
(Canerats)
Page 54

Cheekteeth simple, outer side of upper molars with one re-entrant fold; bullae enlarged, but paroccipital process not enlarged; five hindtoes; tail more or less bushy; incisors not heavily grooved; skull not heavily ridged........................... Family Petromuridae
(Dassie Rat)
Page 60

# Thryonomyidae Pocock, 1922

# *Thryonomys* Fitzinger, 1867

This family includes two species in Africa, occupying different ecological niches: the larger *Thryonomys swinderianus* ranges from West Africa to East Africa and southwards to the Transvaal, Natal and eastern Cape; *T. grego-* *rianus* (a smaller species) is known from Mount Selinda, eastern escarpment of Zimbabwe, in our area, but also occurs in Malaŵi, Zambia, East Africa generally, and Cameroon.

---

**Key to the southern African species of**
**_Thryonomys_**

Skull (frontal region) considerably arched anteriorly; outermost of the three grooves of the upper incisors on midline of tooth; tail more than twice length of hind-foot ............................................. *T. swinderianus*
(Greater Canerat)
Page 54

Skull (frontal region) not arched anteriorly; outer groove of incisor nearer outside edge of tooth; tail shorter, or scarcely longer, than hindfoot ..................
*T. gregorianus*
(Lesser Canerat)
Page 58

---

# *Thryonomys swinderianus* (Temminck, 1827)

**Greater Canerat**

**Groter Rietrot**

The name is coined from the Greek *thyron* = a rush and *mys* = mouse with reference to its common association with waterside vegetation. The species is named after the late Prof. Van Swinderen of Groningen in the Netherlands.

This animal was first described by Temminck in 1827 as *Aulacodus swinderianus* on a specimen from Sierra Leone in West Africa. In 1852 Peters described specimens from Tete, Macanga and Boror in Mozambique as *A. variegatus*. Ellerman *et al.* (1953) point out that this name was stillborn since Peters realised before publication that it was a synonym of *swinderianus,* in the synonym of which he publishes it. Another interesting anecdote on the species name has been pointed out by Thomas in 1894. It should correctly be spelt *swinderenianus* but no-one seems to have followed this apart from W.L. Sclater in 1899, and neither did Thomas himself do so in his later writings (Rosevear 1969). The use of the name *Aulacodus* also became inadmissible as it was previously used by Estricht in 1822 for a coleopterous genus. The type specimen is in the Rijksmuseum voor Natuurlijke Historie in Leyden.

**Outline of synonymy**

1827 *Aulacodus swinderianus* Temminck, *Monogr. Mamm.* I. 248. Sierra Leone.

1852 *Aulacodus variegatus* Peters, *Reise nach Mossambique. Säugeth.*: 138. Tete, Macanga, Sena and Boror, Mozambique.

1864 *A. semipalmatus* Heuglin, *Nov. Act. Acad. Leop. Dresden* 31:6. Central Africa.

1894 *A. swinderenianus* Thomas, *Ann. Mag. nat. Hist.* (6) 13:202.

1896 *Triaulacodus swinderianus* Lydekker, *Geog. Hist. of Mamm.:* 91. Substitute for *Aulacodus,* preoccupied.

1897 *Thryonomys calamophagus* De Beerst, *Bull. Mus. d'Hist. Nat. Paris* 160. Nyasa, Central Africa.

1899 *Thryonomys swinderenianus* W.L. Sclater, *Ann. S. Afr. Mus.:* 1:234.

1910 *T. swinderianus* Jentink, *Zool. Jahrb. Jena:* 28.

1922 *T. s. raptorum* Thomas, *Ann. Mag. nat. Hist.* (9) 9:392. Lagos, Nigeria.

## TABLE 9
### Measurements of male and female *Thryonomys swinderianus*

| | Parameter | Value (mm) | N | Range (mm) | CV (%) |
|---|---|---|---|---|---|
| Males | HB | 465 | 2 | 429–483 | — |
| | T | 195 | 4 | 182–209 | — |
| | HF | 82 | 8 | 72–94 | 8,0 |
| | E | 35 | 5 | 33–40 | 8,0 |
| | Mass: | 5–6,5 kg | | | |
| Females | HB | 448 | 3 | 427–466 | — |
| | T | 172 | 5 | 162–187 | 6,8 |
| | HF | 77 | 7 | 72–81 | 4,3 |
| | E | 29 | 6 | 25–32 | 10,1 |
| | Mass: | 2,4 kg | | | |

Information on the weights of these animals is meagre. Individuals weighing 9 kg (20 lbs) have been recorded (Rosevear 1969). Males are considerably heavier than females.

### Identification

The pelage is deep brown, tinged with yellow. The hairs of the coat are harsh and spiny yet supple (plate 2). On the dorsal surface gutter hairs occur, 30–35 mm in length, thick and pithy so that the gutter channel is obscured distally. These and other bristle-like hairs emerge from the skin in four to five widely spaced parallel lines. These hairs grow in tufts, from three to five out of common root pores. The skin is easily torn and loosely attached to the body. Chin and throat are white, underparts of body whitish. The head is small, with squarely cut muzzle. The small ears are obscured by fur. The tapering, scaly tail is covered with shorter bristles. Legs strong, soles of feet naked with well-developed claws. Pollex on hand poorly developed, fifth digit rudimentary though its claw as well developed as those of digits 2, 3 and 4. Hallux on hindfoot lacking, fifth digit small.

Cases of albinism have been reported from the Malelane area in the Kruger National Park (Pienaar *et al.* 1980).

### Skull and dentition

The distinctive skull is stoutly built, the muzzle broad and arched. The frontal region is slightly inflated. Supraoccipital and sagittal crests are well developed and prominent. The antorbital foramen is very large and elliptical with a special basal groove for nerve transmission. The tympanic bulla is well developed with strong exoccipital processes. The anterior nasal opening is large. The rostrum is short and deep. The zygomatic arches are strong with the jugal elements large. The palate is nar-

row, the front portion in the diastema at a markedly different level from that between the molar rows. The palate terminates about the level of M³. The anterior palatinal foramina are level with M¹. Paroccipital processes are strong and well developed (fig. 20).

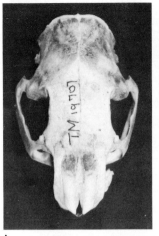
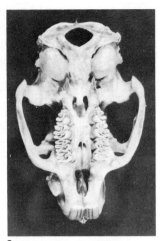
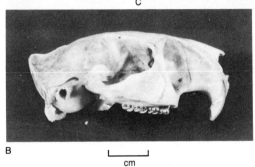

**Figure 20** Dorsal (A), lateral (B) and ventral (C) views of the skull of the Greater Canerat *Thryonomys swinderianus*. Note: large infraorbital foramen, short and deep rostrum, well-developed incisors and strong zygomatic arch, conspicuous paroccipital process and sagittal and supraoccipital crests. Intertemporal gutter clearly visible. Diastemic part of palate lies at a different level than bony palate between molars. Anterior palatine foramina large.

A characteristic indentation ('intertemporal gutters') can be seen near the posterior part of the orbit where the squamosal joins the frontal and is continued down the inner wall of the orbit. Rosevear (1969) states that this is an obvious feature in skulls from southern and eastern Africa.

The foramen leading to the cribriform plate as seen through the foramen magnum, is broad below and narrow above (Rosevear 1969).

The incisors are broad, yellow or orange anteriorly; upper incisors have three grooves on the inner side, the outermost deepest. Rosevear (1969) states that it probably has the most powerfully built incisors of any African rodent – yet these are very brittle (at least in prepared skulls) because of the large volume of dentine in relation to the thin surround of enamel (Rosevear 1969). The lower incisors are smooth. The incisors are opisthodont. The rooted premolar and molars show infoldings of enamel, single on the inside, double on the outside in the maxillary set and reversed in the mandibular set. The third molar erupts late in life. The $M^2$ is usually the largest tooth and the Pm the smallest (fig. 21).

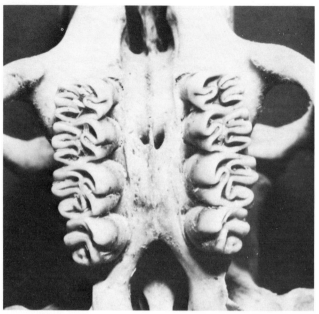

**Figure 21** Cheekteeth of the Greater Canerat *Thryonomys swinderianus*. Enamel infoldings single on the inside, double on the outside (reversed in lower molars). $Pm^4$ is smallest tooth, $M^2$ the largest while $M^3$ erupts late in life.

## Distribution

The Greater Canerat ranges throughout West and Central Africa south of the Sahara to southern Africa where it occurs mainly in warmer parts. In South Africa it occurs in the Eastern Province, Natal and almost the entire Transvaal, while it is also known from northern Botswana and eastern SWA/Namibia, as well as along the Cunene River (map 9).

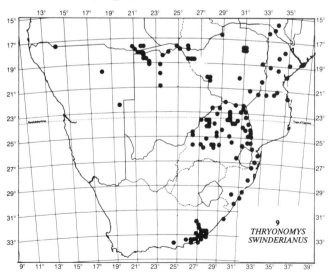

9
THRYONOMYS
SWINDERIANUS

## Habitat

It is found in forest belts and open woodlands wherever there is tall and matted grass or reeds growing in damp or wet places. It likes a semi-aquatic environment in marshes and reedbeds. It therefore occurs frequently on fringes of rivers or swamps, usually on dry ground which is subjected to seasonal flooding. It is common wherever the right surroundings prevail, which includes a good grass cover which they use for both shelter and food (Kingdon 1974).

## Diet

According to Kingdon (1974), canerats eat roots and shoots of grasses *(Setaria, Echinochloa, Hyparrhenia)*, reeds, sedges and other semi-aquatic vegetation. They will also eat nuts and fruits that have fallen to the ground, and the bark of several trees. The bark of shrubs or small bushes like Bugweed *Solanum mauritianum,* is also stripped off (Marinier 1978). Although they are primarily vegetarian, they may also consume an occasional smaller species of rodent (Rosevear 1969).

## Habits

Despite the fact that canerats have well-developed claws, they do not appear to burrow to any extent. They are crespuscular to nocturnal animals with increased activity towards sunset and sunrise. In some areas they may be partly diurnal.

Distinct runs are found, leading to available water. These runs are often cluttered with small piles of cut stems. They are excellent swimmers and fond of playing in the water to which they often take when pursued. The senses of hearing and smell are well developed, but sight is not.

They are rather timid animals, lying up during the day and when flushed, take cover in dense vegetation where dogs cannot penetrate. They are exceedingly agile animals despite their appearance, and difficult to trap. If insufficient cover is available, they excavate shallow burrows (FitzSimons 1919).

The call has been described as a slow staccato whistle (Rosevear 1969) while others (Astley Maberly 1963) describe it as a booming grunt. Kingdon (1974) states that canerats whistle and utter low hooting grunts, thumping their hindlegs when alarmed. More recently, Cox (1978) investigated the syntactics of auditory signals of the canerat under experimental conditions and she found that the animal uses two types of recognisable signals which she calls discrete and graded (continuous). It was also determined that the Greater Canerat uses both mechanical auditory signals (hindfoot stamping) and true vocalisations which consisted of three syllable types. The discrete signals were named phonetically: a *boom*, with or without hindfoot stamping and a *growl* (Type IV syllables). A *boom* was given as a warning signal, while a *growl* was uttered as a threat signal during agnostic behaviour. The graded or continuous signals were also phonetically named the *squeak* (Type I syllable), the *wheet* (Type II syllable) and the *quirr* (Type IV syllable). It is concluded from the type and range of these signals that the Greater Canerat probably has a well-defined social organisation. Greater Canerat droppings are similar to those of porcupines, but smaller. They appear to exist singly, but at times become gregarious. In the dry season in East Africa, females have been reported to segregate from the males who become solitary (Kingdon 1974).

## Predators

In sugar estates, the Python *Python sebae* is frequently protected as a means of checking their increase. Leopards are very partial to canerats (Kirby *in* Sclater 1901; Kent *in* Shortridge 1934) while they are also taken by smaller carnivores, e.g. servals *Felis serval* (Sclater 1901) and the Ratel *Mellivora capensis* (Pienaar *et al.* 1980), hawks and eagles. According to Roberts (1951) wild dogs *(Lycaon pictus)* sometimes hunt canerats in the absence of larger game. They are also taken by the Giant Eagle Owl *Bubo lacteus* (Pitman & Adamson 1978).

## Reproduction

Little is known about their reproduction and available information varies. Rosevear (1969) states that two to four young are born from mid-October to the beginning of January. Smithers (1971) records newborn litters in June and August (Okavango) and that young individuals were taken in August and November (Zimbabwe) while a gravid female with three fetuses was collected in November. Stevenson-Hamilton (1950) refers to newborn litters during spring and early summer in the eastern Transvaal, confirming Rosevear's remarks. Kingdon (1974) reports that they breed at the age of one year and between two and six young are born at a time. Two litters a year are possible (Ajayi 1971). The young are born in an advanced state in shallow depressions, lined with grass or shredded reeds and they are hidden under a thin top layer of reed vegetation (Shortridge 1934). Canerats may occasionally breed in aardvark *Orycteropus afer* or porcupine *Hystrix africaeaustralis* burrows. The eyes are open at birth. The young are covered with dense, soft bristles and are soon able to run, after a gestation period of three months (Kingdon 1974). The tail is longer relative to the head-body length in young than in adults.

The female has three laterally situated mammae, enabling the young to suckle while she is lying on her belly (Roberts 1951).

## Parasites

Cestodes associated with *T. swinderianus* include *Monezia (Fuhrmanella) transvaalensis, Raillietina (R.) mahonae, R. (R.) trapezoides* and *R. (R.) thryonomysi* (Baylis 1935; Collins 1972). The nematodes include *Acheilostoma moucheti* (unpublished Onderstepoort records), *Physaloptera afri-*

cana, *Trichuris vondwei, Longistriata (L.) spira* and *Trachypharynx natalensis* (Ortlepp 1937, 1938b, 1939, 1962b). Mites are paucily known and Zumpt (1961a) lists the trombidiform *Trombicula sicei* as an example. Ticks, as recorded by Theiler (1962) include immatures (I) and adults (A) of *Ixodes aulacodi (A), I. ugandanus (A), Amblyomma hebreum (I), A. nutalli (I), A. splendidum (I), A. variegatum (I, A), Rhipicephalus appendiculatus (I), R. senegalensis (A), R. simpsoni (I, A)* and *R. simus (I, A)*. The pulicid flea *Ctenocephalides felis* has been recorded by Zumpt (1966), as has the calliphorid fly *Chrysoma inclinata* (Zumpt 1961b).

## Relations with man

The Greater Canerat is a pest of any garden near its habitat and it often raids farmlands, damaging crops of groundnuts, pumpkins and sweet potatoes. Canerats eat maize and sugarcane, felling the stems of individual plants by cutting their bases. Roberts (1951) points out that if these animals are not destroyed, great damage is done to sugarcane and on many sugar estates the python is protected as a means of checking their numbers. Like porcupines they occasionally strip bark off young trees in orchards. They are difficult to trap, for they readily tear away the ensnared limb. Rautenbach (1978b) has also observed damage in wheat fields which appear to be temporarily inhabited.

They are often driven out by fire and are hunted in Africa by canerat drives for their tasty white, veal-like flesh. Shortridge (1934) states that canerats form an important source of meat for the Okavango tribes who hunt them with spears and dogs. Astley Maberly (1963) also states that Blacks prize canerats highly as excellent food and he finds it surprising that these animals are not bred for food in South Africa. Ajayi (1971) has suggested field management of canerats on a sustained yield basis in view of their highly acceptable protein content.

Canerats carry tapeworms of the genus *Raillietina* which are important to poultry and related birds. Severe infestations can lead to high mortality. Ticks can transmit rickettsiases and babesiases, while the chiggers (Trombiculidae) can transmit scrub-itch or trombidiasis to man. The cat flea *Ctenocephalides felis,* occurs frequently on dogs and man and may act as an important vector of the numerous pathogens in the domestic dog. It is, however, not an effective vector of plague. The calliphorid fly *Chrysoma* can cause myiasis in man and traumatic myiasis has in fact been recorded in a canerat from Komatipoort (Zumpt 1961b).

## Prehistory

From fossil evidence it is clear that the genus *Thryonomys* existed in the central Sahara during the Pleistocene. The fossils were described as *T. logani* from a locality approximately 800 km from the Niger River (Romer & Nesbit 1930). Apparently they existed in Africa from the Upper

57

Miocene (Walker 1975). Another fossil form, *T. arkelli,* has been found in Pleistocene deposits in Sudan (Bate 1947). *T. swinderianus* has been recorded as a fossil from Kenya and Tanzania by Hopwood (1931).

Other thryonomyid-like genera are also known from the Lower Miocene deposits in SWA/Namibia. These include *Neosciuromys africanus* and *Phiomyoides humilis* (Stromer 1922 and 1926). Two incomplete mandibles of *Neosciuromys africanus (= Phythinylla fracta)* were reported on by Hendey (1978) from Arrisdrift in SWA/Namibia, where they were part of an assemblage of fossil vertebrates of Miocene Age. This taxon has since been recognised as a junior synonym of *Paraphiomys pigotti* (Andrews, 1914) by Lavocat (1973). Another species close to *Phiomys andrewsi* has been found under similar circumstances and was mentioned by Stromer (1926). *Paraphiomys stromeri (= Apodecter stromeri)* has also been found in these deposits (Hendey 1978). In addition, two incomplete lower incisors have been found and they have tentatively been attributed to *Bathyergoides* sp., the genus recorded by Stromer and also in East Africa by Lavocat (1973).

In addition to *Bathyergoides, Paraphiomys pigotti, Phiomyoides humilis,* and *Paraphiomys stromeri* referred to above, Hamilton & Van Couvering (1977) also found *Diamantomys luederitzi* and *Pomonomys dubius* in the fossiliferious locality referred to as the Diamond Fields south of Lüderitz in SWA/Namibia, originally described by Stromer.

## Taxonomy

Only one genus, *Thryonomys,* is now recognised in this family. Thomas (1922) split off from this genus a group of mainly East African representatives *(Choeromys)* on characteristics not now regarded as valid. *Thryonomys* is today divided into two species 'groups': the '*Thryonomys* group' and the '*Choeromys* group', corresponding to the former two genera. The former includes *swinderianus* and the latter *gregorianus.*

Roberts (1951) and earlier authors place the South African material in the subspecies *variegatus,* regarded as a synonym of the nominate race by Ellerman *et al.* (1953).

According to Misonne (1974), *T. swinderianus* is understood to include *angolae* Thomas, 1922 (from Angola), *logani* Romer & Nesbit, 1930 (the fossil form from the central Sahara), *raptorum* Thomas, 1933 (from Nigeria) and *variegatus* Peters, 1852 (from Mozambique).

# Thryonomys gregorianus (Thomas, 1894)

**Lesser Canerat**

**Kleiner Rietrot**

The derivation of *Thryonomys* has been given under *T. swinderianus.* This species was named after Dr J. W. Gregory who obtained the first specimens on an expedition to East Africa.

The species was first described by Thomas in 1894, based on material (a skull only) from Kenya. In 1922 he proposed the genus *Choeromys* for representatives of canerats from East Africa based on certain obvious differences between *Thryonomys* and *Choeromys.* Subsequently, it has been accepted that these differences are scarcely of generic or even subgeneric importance (Rosevear 1969) and *Thryonomys* is today merely divided into two species groups. *T. gregorianus* is a representative of the '*Choeromys*' group.

## Outline of synonymy

1894 *Aulacodus gregorianus* Thomas, *Ann. Mag. nat. Hist.* (6) 13:202. Kenya.

1897 *Thryonomys sclateri* Thomas, *Proc. zool. Soc. Lond.:* 432. Nyika Plateau, Malaŵi.

1907 *T. harrisoni* Thomas & Wroughton, *Ann. Mag. nat. Hist.* (7) 19:384. Sudan.

1912 *T. gregorianus pusillus* Heller, *Smithson. misc. Collns.* 59 (16):17. Kenya.

1918 *T. rutshuricus* Lönnberg, *Stockholm Vet. Ak. Handl.* (2) 58:78. Central Africa.

1922 *T. harrisoni congicus* Thomas, *Ann. Mag. nat. Hist.* (9)9:390. Zaïre.

1932 *Choeromus sclateri* St Leger, *Proc. zool. Soc. Lond.:* 982.

1941 *T. gregorianus gregorianus* Thomas. *In* Ellerman, *The families and genera of living rodents.*

1951 *T. (Choeromys) sclateri* Thomas. *In* Roberts, *The mammals of South Africa.*

1953 *Choeromys harrisoni sclateri* Thomas. *In* Ellerman *et al., Southern African mammals.*

1964 *T. gregorianus* Thomas. *In* Meester *et al., An interim classification of southern African mammals.*

## Identification

The individuals occurring in southern Africa are very much the same in colour and general appearance as *Thryonomys swinderianus,* but their overall size is smaller (plate 2). The tail is also stated to be more reduced in '*Choeromys*' (i.e. shorter or more or less as long as the hindfoot), but in this characteristic, *T. (C.) sclateri* as listed by Roberts (1951), has a tail nearly as long as in the typical group.

## Skull and dentition

In contrast to *T. swinderianus*, the frontal region is more or less flat. The opening from the cerebral to the olfactory fossa is narrow below and broad above, i.e. shield-shaped. The outer groove of the incisor is situated closer to the outer edge of the tooth, rather than on the midline as is the case in *T. swinderianus* (fig. 22).

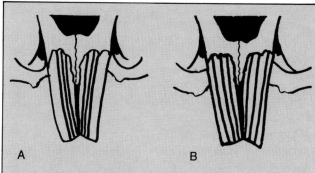

**Figure 22** Upper incisors of the Greater Canerat *Thryonomys swinderianus* (A) and the Lesser Canerat *T. gregorianus* (B). In *T. swinderianus* the outermost of the three grooves is situated in the midline of the tooth. In *T. gregorianus* the outermost of the three grooves is situated nearer the outside edge of the tooth than the midline.

## Distribution

In southern Africa it is known to occur only at Mount Selinda and vicinity in Zimbabwe, as well as in Mozambique (map 10). Northwards it occurs in Malaŵi, along the East Coast of Africa to southern Sudan and westwards to Cameroon.

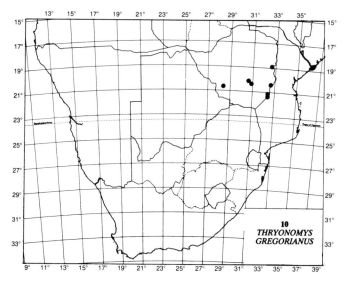

**10**
**THRYONOMYS GREGORIANUS**

## Habitat

The Lesser Canerat seems to be confined to the higher altitudes on the eastern tropical escarpment. Virtually nothing is known about the ecology of this species and much of the information presented here has been gleaned from Kingdon (1974). In contrast to *T. swinderianus*, it prefers dry land in moist savannas. In long, rank grass it has little need for added shelter. During the dry season, however, it uses natural shelters including crevices, termitaries and disused aardvark holes.

## Diet

They feed on the stems of grass species which include *Pennisetum*, *Setaria* and *Hyparrhenia*. In montane areas the bracken fern *Pteridium* is also taken. They also debark certain trees and shrubs.

## Habits

Lesser canerats are primarily nocturnal, but occasionally active by day, particularly during the rainy season when cover, shade and moisture are abundant (Kingdon 1974).

## Predators

Their predators include man, leopards, servals and other cats, pythons, viverrids and some eagles, e.g. *Aquila rapax nipalensis* in Zaïre (Chapin 1932). In Zimbabwe *T. gregorianus* is preyed upon by Mackinder's Eagle Owl *Bubo capensis mackinderi* (Gargett & Grobler 1976).

## Reproduction

Kingdon (1974) stresses that breeding data are remarkably scarce. In southern Kenya pregnant females have been taken in November and December. They are reported to be seasonal breeders in northern Uganda, bearing litters at the end of the dry season in February and March.

## Parasites

Zumpt (1961a) records the occurrence of the laelaptid mite *Laelaps muricola* and states the infestation as '. . . certainly only accidental'. Ticks are known from one record only, adults of *Rhipicephalus simpsoni*, reported by Theiler (1962).

## Relations with man

They raid cultivated lands and can cause considerable damage to groundnuts, sweet potatoes, maize and pumpkins.

## Prehistory

There is no information available.

## Taxonomy

The name *Choeromys* was used as a genus by Allen (1939) and as a subgenus by Roberts (1951) for *gregorianus* and related species. Ellerman *et al.* (1953) regard *Choeromys* as a synonym of *Thryonomys*. The form *sclateri* is regarded as a good species by Roberts and as a subspecies of *C. harrisoni* by Allen, but Ellerman *et al.* (1953) regard all the lesser canerats as probably conspecific. This interpretation is also upheld by Misonne (1974), certainly as far as *congicus* Thomas, 1922 (Zaïre), *harrisoni* Thomas &

Wroughton, 1907 (Sudan), *pusillus* Heller, 1912 (Kenya), *rutshuricus* Lönnberg, 1917 (Central Africa) and *sclateri* Thomas, 1897 (Malawî) are concerned. The affinities of the other extralimital species, *logonensis* Jeannin, 1936 and *camerunensis* Monard, 1949, both from Cameroon, should be reassessed.

FAMILY # Petromuridae Tullberg, 1899

GENUS ## *Petromus* A. Smith, 1831

This family includes but a single species, *Petromus typicus*. A second species *(P. cunealis)* and some ten subspecies have been described based on minor colour differences, but these are no longer accepted as valid. Structurally, the dassie rats as they are known, show certain specialisations enabling them to seek shelter in narrow rock crevices. Vaughan (1972) lists some of these specialisations which include a dorso-ventrally flattened skull and a high flexibility of the ribs allowing the body to be flattened without injury. The type specimen of *P. typicus* is in the British Museum (Natural History) London.

The Dassie Rat has been listed as rare, with a limited distribution, in the *South African Red Data Book* by Meester (1976).

# *Petromus typicus* A. Smith, 1831

**Dassie Rat**

**Dassierot**

The name is carried from the Greek *petro* = rock and *mys* = mouse, emphasising the preferred habitat of the animal. The species name is derived from the Greek *typikos* = typical. In other words, this is a typical rat of the rocks in areas where they occur.

**Outline of synonymy**

1831 *Petromys typicus* A. Smith, *S. Afr. Quart. Journ* 1 (5):11 (misprint for p. 2). Mountains of Little Namaqualand near mouth of Orange River, CP.

1925 *P. t. tropicalis* Thomas & Hinton, *Proc. zool. Soc. Lond.*: 241. Karibib, SWA/Namibia.

1926 *P. cunealis* Thomas, *Proc. zool. Soc. Lond.*: 307. Cunene Falls.

1936 *P. typicus marjoriae* Bradfield, *Auk* 53:131. Khan River, SWA/Namibia.

1938 *P. t. quinasensis* Roberts, *Ann. Transv. Mus.* 19:240. Guinas Lake, SWA/Namibia.

1938 *P. t. windhoekensis* Roberts, *Ann. Transv. Mus.* 19:240. Neudamm, SWA/Namibia.

1938 *P. t. kobosensis* Roberts, *Ann. Transv. Mus.* 19:240. Kobos, SWA/Namibia.

1938 *P. t. barbiensis* Roberts, *Ann. Transv. Mus.* 19:241. Helmeringshausen, SWA/Namibia.

1938 *P. t. ausensis* Roberts, *Ann. Transv. Mus.* 19:241. Aus, SWA/Namibia.

1938 *P. t. namaquensis* Roberts, *Ann. Transv. Mus.* 19:241. Warmbad, SWA/Namibia.

1946 *P. t. karasensis* Roberts, *Ann. Transv. Mus.* 20:314. Great Karas Mountains, SWA/Namibia.

1946 *P. t. cinnamomeus* Roberts, *Ann. Transv. Mus.* 20:314. Ariamsvlei, SWA/ Namibia.

### TABLE 10
### Measurements of male and female *Petromus typicus*

| | Parameter | Value (mm) | N | Range (mm) | CV (%) |
|---|---|---|---|---|---|
| Males | HB | 171 | 14 | 154–210 | 7,7 |
| | T | 140 | 9 | 125–150 | 4,6 |
| | HF | 30 | 13 | 28–35 | 5,8 |
| | E | 13 | 13 | 11–15 | 7,1 |
| | Mass: 170,212 g | | 2 | — | — |
| Females | HB | 150 | 9 | 137–190 | 10,3 |
| | T | 140 | 7 | 116–168 | 10,0 |
| | HF | 33 | 9 | 31–36 | 4,0 |
| | E | 14 | 9 | 11–16 | 9,3 |
| | Mass: 251,262 g | | 2 | — | — |

Data on the weight of these animals are meagre. The figures given above (170 g, 212 g) are obviously not taken from adult males. The females are much heavier (251 g, 262 g). When looking at long series of study skins in museums, sexual dimorphism between smaller males and larger females is not evident, as these figures would suggest.

## Identification

These animals are squirrel-like in appearance (plate 7). However, the tail is not as bushy and is covered with scattered, long hairs. The tail is shorter than the head-body length and its scaly nature is disguised by its hirsute covering. In life, the hair near the tail base is iridescent but soon loses its lustre when the animal is dead. The fur is soft and springy due to the absence of underfur. The nose is yellowish (tawny-yellow) as is the area around the eyes, while the rest of the dorsal surface is grey, becoming a dull chestnut towards the hinder portion of the body and hindlegs.

The dorsal hairs are ashy grey at their bases. The tail thickens at the base, being reddish in colour merging into dark brown to black distally, with individual hairs longer than those of the body. The body hairs grow in clusters varying between three and five in number as is found in other hystricomorphs (cf. *Thryonomys*). The ventral surface ranges from yellowish to dirty white. The colour is geographically variable. Specimens from Neudamm (Windhoek) are darker, while those from the Namib area are lighter. Individuals from O'Kiep are a rich light brown and in these cases the tails are pronouncedly darker (blackish) in contrast to the rest of the body.

The ears are short, rather wider than high, blackish, covered with a few hairs. The vibrissae are long and black.

The soles of the feet are naked. There are four clawed toes (pollex short and rudimentary) with swollen pads on the forefeet and five toes on the hindfeet (hallux short).

The dorsal surface of the appendages shows a yellowish coloration. The feet are narrow.

Generally, the hand is flattened, the clavicles are imperfectly developed and the overall size of the animal corresponds to newborn dassies *Procavia capensis*.

## Skull and dentition

The infraorbital canal is wide as in other hystricomorphs. The skull is broad posteriorly, not heavily ridged, bullae well developed (inflated) with the paroccipital process not enlarged. There is no interorbital constriction. The anterior palatal foramina are deep and long and penetrate between the molars (fig. 23).

The incisors are yellow, smooth and narrow, and opisthodont.

The premolar and molars are rooted, high crowned (hypsodont), with deep infoldings of enamel lingually (on the inside) in the maxilla and buccally (on the outside) in the lower jaw. This creates the impression that each tooth consists of a double section, oblique to the length of the tooth row. Smaller and opposite folds which are obliterated with wear, occur especially in young animals (fig. 24).

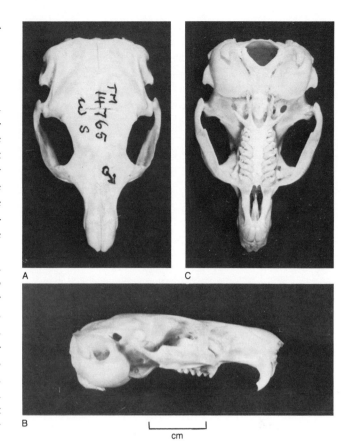

**Figure 23** Dorsal (A), lateral (B) and ventral (C) views of the skull of the Dassie Rat *Petromus typicus*. Note flattened character of skull and inflated bulla. Interorbital constriction slight. Posterior portion of skull well developed, as are the zygomatic arches.

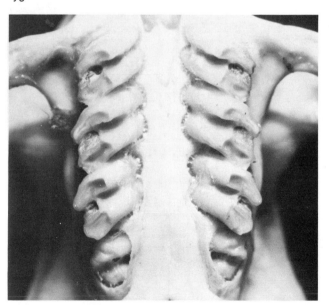

**Figure 24** Cheekteeth of the Dassie Rat *Petromus typicus*. The anterior palatine foramina penetrate between premolars. Hypsodont teeth show deep infoldings of enamel lingually, creating the impression that each tooth consists of a double section.

61

## Distribution

Dassie rats occur in the northwestern Cape Province (Little Namaqualand), northwards into SWA/Namibia and Angola. Eastwards from Namaqualand they are encountered all along the Orange River up to Kakamas (map 11). It is therefore an inhabitant of the South West Arid and also occurs marginally in the Southern Savanna biotic zone.

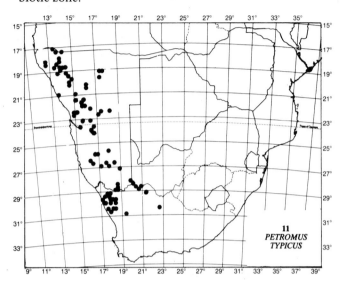

**11 PETROMUS TYPICUS**

## Habitat

Dassie rats are confined to rocky hills and mountainous areas where they live in narrow rock crevices and in among large boulders. Their coloration blends superbly with their surroundings.

## Diet

They subsist entirely on soft vegetable matter which includes leaves of shrubs and to a large extent on flowers of the many species of Compositae (daisy family) which grow in their immediate vicinity. Stomach contents have also revealed berries and seeds. They may occasionally climb shrubs to feed.

## Habits

Dassie rats favour rock shelters from which they emerge at daylight to forage. Feeding activity is greater towards sunrise and sunset. They often tend to sun themselves near their shelters and such areas are often stained yellowish from their urine (Roberts 1951). It builds its nest from twigs and characteristically collects dried leaves of the Kokerboom *Aloe dichotoma* (Dieckmann *pers. comm.*).

Each individual or pair seems to occupy its own crag or hole. They are playful when undisturbed and seem to be fairly docile animals. They are extremely agile and are more rock runners than rock jumpers. when jumping they tent to flatten their bodies to a most remarkable degree because of the flexible thorax.

Roberts (1951) has described their voice as a '. . . complaining whining whistling note'.

Their characteristic elongated droppings are larger than those of associated murids (approximately the size of a blow-fly pupa). These pellets are not deposited in large accumulations but occur scattered near their dwellings.

These animals are apt to be overlooked, even where they occur in large numbers. They can be trapped by baiting with mealies. Dieckmann *(pers. comm.)* has caught them at the same trap sites as *Aethomys namaquensis* and *Mus musculus*.

The skin is very thin and the tails tend to break off easily when the animals are captured.

## Predators

Unknown, but certainly include different species of birds of prey.

## Reproduction

Information is meagre. Gravid females have been collected in September (near Brukaros) and in November (at Karibib) and Shortridge (1934) states that breeding seems to take place during hot weather. The young are born in an advanced state of development, well covered with hair. One or two seems to be the normal litter size.

The testes of the male are inconspicuous. The mammae in the female consist of two laterally situated pairs below the shoulders, while a third pair (often non-functional) occurs farther back. The mammae are small and difficult to detect.

## Parasites

Information is meagre. Mönnig (1931) has recorded the presence of the nematode *Acanthoxyuris shortridgei* while Zumpt (1966) has listed the occurrence of the pulicid flea *Echidnophaga gallinacea*.

## Relations with man

Dassie rats act as vectors for fleas referred to above, and they in turn can spread plague *(Yersinia pestis)*.

The flesh of this animal is described as having a pleasant, aromatic smell and taste.

## Prehistory

A Dassie Rat has been found as a fossil at Taung, whence Broom (1939) described a fossil species referred to as *Petromus minor*.

## Taxonomy

Roberts (1951) recognises two species, *typicus* and *cunealis*, but Ellerman *et al.* (1953) synonymises them. Some 14 subspecies have been described and on subspecies level, this family is clearly in need of taxonomic revision. It is unlikely that 14 valid forms could occur in the fairly restricted range of the species (Meester *et al.* 1964). Ellerman *et al.* (1953) list some 12 subspecies *(typicus, tropicalis, cunealis, marjoriae, guinasensis, windhoekensis, kobosensis, barbiensis, ausensis, namaquensis, karasensis* and *cinnamomeus)* while Lundholm (1955b) describes two more, *greeni* and *pallidior*.

# Hystricomorpha
*incertae sedis*
# BATHYERGIDAE: Molerats

# Hystricomorpha *incertae sedis*
# Bathyergoidea Osborn, 1910

The group of rodents popularly known as molerats to be dealt with in this section, is of doubtful relationship and I have followed Simpson (1945) in treating them as a separate superfamily. This idea was first proposed partially by Tullberg (1899) when he divided the rodents into two major tribes, the Hystricognathi and the Sciurognathi, on the basis of the inflection of the angle of the lower jaw. Under the former he proposed two subtribes, the Bathyergomorphi and Hystricomorphi respectively and under the sciurognaths he proposed the Myomorphi and Sciuromorphi. This outline was basically followed by Ellerman (1940, 1941) in his monumental work on *The families and genera of living rodents*. In 1951, Roberts raised the molerats to subordinal rank, the Bathyergomorpha, equivalent to the Hystricomorpha, Sciuromorpha, Dipodomorpha and Myomorpha. There have been two general tendencies in connection with the molerats – to isolate them or to place them with the hystricognaths (Wood 1955). Tullberg (1899) isolated them as did Miller & Gidley (1918). Schaub (1953a) called them Palaeotrogomorpha *incertae sedis* and stated that they are so highly specialised '. . . that closer affinities are entirely unknown'. Wood (1955) suggests that these forms separated from the Paramyidae in the early or middle Miocene and have had an independent existence in southern Africa ever since. He consequently also treats them as an isolated group.

The Molerat has a small infraorbital foramen, not transmitting any part of the masseter muscle. The anterior part of the zygomatic arch is not modified into a zygomatic plate. The masseter is, however, complex and highly differentiated. Wood (1955) summarises the position as follows: the *masseter mediatis anticus* has extended upward inside the zygoma and arises from the upper part of the median wall of the orbit, while the *masseter lateralis superficialis* has moved partly onto the anterior face of the zygoma.

The Bathyergoidea consist of a single family, the Bathyergidae. They occur in Africa only and are highly fossorial rodents, occupying much of Africa. They occur in all the countries of southern Africa, and range northwards to Tanzania, Kenya and Ethiopia in the east, across the continent to Ghana in West Africa.

# FAMILY Bathyergidae Waterhouse, 1841

This family of rodents is endemic to Africa, occurring from the Cape to some 10° north of the equator. Although termed molerats, they are neither moles (a separate order of insectivorous mammals), nor rats (a completely different family of rodents).

Molerats are all fossorial with reduced eyes and ear pinnae. The legs and tail are short. Generally the pelage is well developed, although there is a tendency for reduction of the pelage within the family culminating in the nearly naked Sand Puppy *Heterocephalus* found in Ethiopia, Somalia and Kenya.

The skull is stoutly built with large incisors, which in life project outside the mouth. The infraorbital canal is secondarily reduced in size, not transmitting the deep (anterior) masseter medialis muscle. The small, upright zygomatic plate is situated below the infraorbital foramen. The cheekteeth are usually four in number, rooted, tending towards hypsodonty and the occlusal surfaces are relatively simple. According to Thomas (1909) and Roberts (1951), these cheekteeth consist of two premolars and two molars in the southern African genera, but homology is uncertain, chiefly because of the absence of pertinent embryological data. According to Tullberg (1899) the malleus and incus are fused as in the Hystricoidea and Ctenodactyloidea, but unlike the remainder of the order.

The lower jaw is hystricomorph in character, the superficial *masseter lateralis* muscle is inserted on the angular process of the mandible, thereby modifying the shape of the jaw outwards.

The fore- and hindfeet have five digits each and naked soles. The hind claws are always short and concave below. The carpal bones, the radiale (scaphoid) and intermedium (lunar) are separate, a feature unique to Bathyergidae and Ctenodactylidae (Tullberg 1899).

The peculiar nature of the jaw-muscles, the variability in the number of cheekteeth in the different genera and the variable size of the infraorbital foramen indicate that the bathyergids are an isolated group within the rodents (Ellerman 1940). They have been placed in the Sciuromorpha (Romer 1958), the Mymorpha (Thomas 1896), the Hystricomorpha (Winge 1941; Ellerman 1940; Simpson 1945 – *incertae sedis;* Landry 1957) and in a separate group (Weber 1928; Roberts 1951; Wood 1955). At present they are best considered as hystricomorphs (De Graaff 1979) (especially when internal anatomical features are considered) and final agreement on this question must await accumulation of further evidence (De Graaff 1975).

Roberts (1951) has divided the family into two subfamilies, the Bathyerginae and the Georychinae. In southern Africa, *Bathyergus* is placed within the former, while the other southern African genera *(Heliophobius, Georychus* and *Cryptomys)* are referred to in the latter. According to Ellerman *et al.* (1953) there is much to be said for such a classification, as the genus *Bathyergus* is a relatively generalised form, whereas the excessively long upper incisors rooted in the pterygoidal region of the skull in the other genera seem to be unique within the rodents.

**Key to the southern African bathyergid subfamilies**

1 Size large, HB length 150–250 mm or more; claws of forefeet long and adapted to burrowing; upper incisors grooved on outer surface, their roots based above the molars............................Bathyerginae
(Dune Molerats)
Page 66

Size smaller as a rule, HB length not exceeding 215 mm as a rule; claws of forefeet small, less obviously adapted to burrowing; upper incisors not grooved on outer surfaces, their roots based well behind the molars in the pterygoid region............Georychinae
(Cape Molerat and Common Molerats)
Page 71

SUBFAMILY **Bathyerginae** Roberts, 1951

GENUS *Bathyergus* Illiger, 1811

Two species, *B. suillus* and *B. janetta* occur in southern Africa. On the basis of structural features, some authorities think that this genus should be classified in a family by itself as is pointed out by Walker (1975).

**Key to the southern African species of** *Bathyergus*
(Modified after Meester *et al.* (1964))

1 Colour cinnamon to light brown to tawny; darker brown mid-dorsal band from nape of neck to hindquarters varied and not clearly defined .................
*Bathyergus suillus*
(Cape Dune Molerat)
Page 67

Colour dark slate-grey; distinct dark dorsal band from nape of neck to hindquarters; overall size smaller than *suillus*...............................*B. janetta*
(Namaqua Dune Molerat)
Page 70

# *Bathyergus suillus* (Schreber, 1782)

## Cape Dune Molerat

## Kaapse Duinmol

The Greek word *bathys* = deep, while *ergo* = work. *Bathyergus* therefore implies the one who works below the surface of the soil. The species name is derived from the Latin *suillus* = pertaining to a pig. It must have reminded Schreber of some sort of ground hog.

The earliest mention of the Cape Dune Molerat was made by Abbé N. L. de la Caille who visited the Cape in 1750 while making certain astronomical determinations. Masson, a gardener from the Royal Botanical Gardens at Kew, also visited South Africa between 1772 and 1776 and he also referred to the existence of this species. The first satisfactory description was transmitted to Prof. Allamand of Leyden in the Netherlands by Col. Gordon who commanded the Dutch forces at the Cape prior to the first British occupation of the Cape in 1795 (Sclater 1901).

## Outline of synonymy

1782 *Muis suillus* Schreber, Schreber's *Säugeth.:* 715. Cape of Good Hope.

1788 *M. maritimus* Gmelin, *Linn. Syst. Nat.* 13:140. Cape of Good Hope.

1788 *Marmota africana* Thunberg, *Resa uti Europa, Africa, Asia,* etc. I:293. Cape of Good Hope.

1826 *Georychus maritimus* A. Smith, *Descr. Cat. S. Afr. Mus.:* 28.

1832 *Bathyergus maritimus* Smuts, *Enum. Mamm. Cap.:* 48.

1842 *Orycterus maritimus* Waterhouse, *Ann. Mag. nat. Hist.* (1) 8:81.

1926 *B. s. intermedius* Roberts, *Ann. Transv. Mus.* 11:261. Klaver, CP.

### TABLE 11
Measurements of male and female *Bathyergus suillus*

| | Parameter | Value (mm) | N | Range (mm) | CV (%) |
|---|---|---|---|---|---|
| Males | HB | 281 | 39 | 240–330 | 5,2 |
| | T | 50 | 39 | 30–70 | 11,7 |
| | HF | 51 | 37 | 45–65 | 6,6 |
| Mass: May attain 750 g | | | | | |
| Females | HB | 256 | 45 | 204–300 | 7,3 |
| | T | 47 | 45 | 27–65 | 10,3 |
| | HF | 46 mm | 43 | 42–55 | 5,7 |
| Mass: May attain 500 g | | | | | |

## Identification

The Cape Dune Molerat is the largest known bathyergid. The claws on digits 2, 3 and 4 of the forefeet are big, well developed, curved and adapted to digging. The claws on the hindfoot are not as developed.

The coloration is cinnamon dorsally, sometimes with a clearer mid–dorsal dark brown band. The flanks are paler. Individual hairs are black at their bases, slaty-grey along their shafts and their distal tips are light yellow or brown. The hair on the ventral surface is usually short, resulting in a grey ventral coloration. There is a buffy to white area surrounding the eyes and the area around the muzzle and throat is usually white. Some specimens (e.g. from Klaver) show a conspicuous white spot on the forehead. This white patch occurs in both sexes in approximately 40% of individuals (De Graaff 1965) (plate 3).

The species varies somewhat in colour geographically. Along the west coast they are paler in colour and of lighter hue. In contrast, specimens along the coast from Cape Town to Knysna are darker. These colour differences become apparent only when a number of specimens from different localities are compared simultaneously. This gradual colour change may be correlated with humidity; Lamberts Bay on the west coast has an annual rainfall of only 200–300 mm, while the Knysna area experiences 800–1 000 mm annually.

The fur is soft, thick and woolly, without any trace of sheen. Piebald varieties occur sporadically.

Males are not significantly larger than females. Individuals of both sexes from the area north of the Berg River and south of the Olifants River tend to be slightly smaller than those around Cape Town. The female has two pairs of pectoral and one pair of inguinal mammae.

## Skull and dentition

The skull is rather elongate, sturdy, more or less flattened, tapering with a small angle downwards to the muzzle. The nasals are narrow. The interorbital constriction is not pronounced. The zygomatic arches tend to be wide; the jugal has a fairly broad posterior root. A zygomatic plate is scarcely developed. There is a well-developed sagittal crest between the frontals and occipital region which is strongly ridged and well developed (fig. 25). Small anterior palatine foramina occur placed far back; the narrow palate extends well behind the last molar. The lower jaw is stockily built and the symphysis between the two rami firmly ankylosed. The angular portion is curved outwards.

The pro-odont upper incisors are heavily grooved (smooth in the lower incisors), white and rooted at a point above the anterior cheekteeth. The molars which do not grow continuously are simple and hypsodont. Re-entrant folds occur in juveniles only. They are somewhat oval in section, the central portion of dentine surrounded by a ring of enamel (fig. 26). The last molar is the smallest element of the tooth row, while the maxillary set is

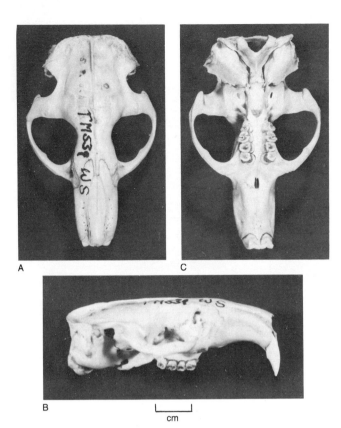

A                               C

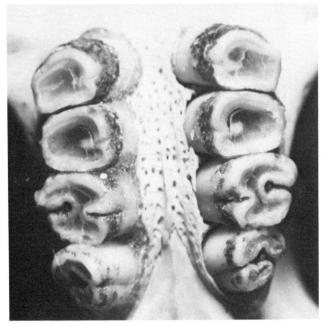

B

cm

**Figure 25** Dorsal (A), lateral (B) and ventral (C) views of the skull of the Cape Dune Molerat *Bathyergus suillus*. The skull is elongate with well-developed sagittal crest; infraorbital foramen small; muzzle narrow and anterior palatine foramina short. The palate is narrow extending behind M³; incisor grooved, pro-odont.

**Figure 26** Cheekteeth of the Cape Dune Molerat *Bathyergus suillus*. The teeth are hypsodont, somewhat oval in shape with re-entrant folds in young animals only. Pm⁴ is about as large as M¹ while M³ is the smallest tooth.

usually shorter in its total length than the mandibular tooth row. It has been suggested by Thomas (1909) that the cheekteeth consist of Pm³, Pm⁴, M¹ and M², but it is more likely to consist of Pm⁴, M¹, M² and M³ (De Graaff 1979).

## Distribution

The Cape Dune Molerat occurs predominantly in the southwestern Cape, ranging northwards to Lamberts Bay along the coast and inland to Travellers' Rest (about 80 km from the coast) and the area south of the Olifants River and north of the Berg River. It occurs abundantly on the Cape Flats and thence eastwards along the coast to the vicinity of Knysna. Sclater (1901) states that it may perhaps occur as far east as the Bathurst area (map 12). It therefore occurs in the South West Cape and Forest biotic zones.

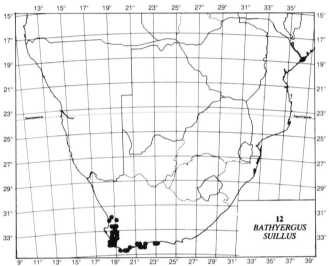

12
*BATHYERGUS SUILLUS*

## Habitat

The Cape Dune Molerat occurs wherever loose, coastal sand dunes are present. They also occur along banks of the larger rivers and their tributaries in the Western Province. All known localities are below 300 m above sea level. They are not found on rocky mountain slopes.

## Diet

Virtually nothing is known about food or feeding habits of this species in the wild. It is stated that they feed on roots and bulbs (Dreyer 1910; Sclater 1901). The sense of touch is apparently delicately adapted to its mode of feeding. Food is located within the soil by means of tunneling. It is uncertain whether *B. suillus* stores food in special food chambers (De Graaff 1965).

## Habits

Cape dune molerats live in small colonies of two to five individuals in tunnel systems. When captured they show aggression, but soon calm down under captive conditions. When on the surface of the soil, there is some ran-

68

dom movement, without real fleeing. This is undoubtedly correlated with poor eyesight.

The tunnels are dug by means of the large claws and are occupied permanently. The presence of these animals in an area is given away by the large mounds, representing the excess soil removed from the tunnels. These mounds are pushed up through side tunnels which branch off the main tunnel at an angle of approximately 45° and below the mounds the tunnel is airtight, soil being rammed into a compact plug. The tunnels of *B. suillus* are 12,5–20 cm in diameter, undulating in depth below the surface from 15–20 cm to a depth of one metre. In the winter rainfall area where they occur, these systems are often flooded and new tunnels are usually dug above the water level. In contrast to other bathyergid genera like *Cryptomys* and *Georychus,* the tunnel system of *Bathyergus suillus* has remarkably few side tunnels. The mounds are large, 60–75 cm in diameter and 35–50 cm in height (De Graaff 1965). The burrow systems made by *B. suillus* were also reported upon by Schulz (1978).

It seems that the animals are more active during the rainy season and extention to the tunnel system may be as much as 2,5–3 m a day. Any damage to the system (by means of flooding, sagging of the roofs, mechanical opening, etc.) is quickly repaired, for they are extremely sensitive to air currents.

The nest is built in a special rounded nesting chamber, 40 cm in diameter, constructed of roots, leaves, bracts and twigs of the prevailing vegetation.

The animals orientate themselves by means of echolocation within the tunnels and also utter a snorting type of grunt.

## Predators
Few predators are known, but the Molesnake *Pseudaspis cana* does enter their tunnels to take its toll (De Graaff 1965).

## Reproduction
Two pairs of pectoral and one pair of inguinal mammae are present. Mating occurs towards the middle of winter and pregnant females have been taken in August and September. Three to four young seem to form the normal litter, but the length of the gestation period is unknown. Smears indicate the presence of ripe spermatozoa during July and August. The testes descend during the breeding season, as in rats.

## Parasites
Information on the parasites associated with the bathyergids is meagre. For a more detailed discussion of parasites of the Bathyergidae, see De Graaff (1964).

The nemathelminth worms, *Longistriata (L.) bathyergi, Libyostrongylus bathyergi* and *Heterakis macrospiculum,* have all been taken from *Bathyergus suillus* collected at Strand-

fontein (Ortlepp 1939). To this can be added an unidentified species of *Trichuris* (Ortlepp *pers. comm.*) also from animals taken near Strandfontein on the Cape Flats.

The laelaptid mites, *Haemolaelaps bathyergus* and *Macronyssus bacoti,* occur as ectoparasites on *Bathyergus suillus.* The listrophorid mite *Listrophoroides bathyergians* is also recorded by Zumpt (1961a). As far as I could assess from the available literature, the hystrichopsyllid flea *Dinopsyllus ingens* is the only one hitherto known to be associated with *Bathyergus suillus* (Zumpt 1966), and for which it is taken as the type host (De Meillon, Davis & Hardy 1961). A suckling louse (Anoplura, Haematopinidae) was taken off females of *B. suillus* from an unknown locality representing *Proenderleinellus lawrensis* (Bedford 1932). Finally, both adults and immatures of *Ixodes alluaudi* and *Haemaphysalis leachii muhsami* occur on the Cape Dune Molerat (Theiler 1962).

## Relations with man
Roberts (1951) states that the skin has not much value, for it is thin and the fur short. On the other hand, the meat of the Cape Dune Molerat is regarded as a delicacy when baked in an anthill with salt and pepper. In the Citrusdal district, these molerats are eaten in stews and some families trap up to four or five animals per household per week. This often forms their sole source of protein, apart from fish (De Graaff 1965).

In the areas where they occur, railway gangers are instructed to watch out that these animals do not tunnel underneath the tracks, as a number of these tunnels in close proximity causes the lines to sag, increasing the possibility of derailments (Roberts 1951).

An important agricultural activity is the production of wheat. When fields in the Boland of South Africa are prepared for wheat planting, the tunnel systems are not destroyed so that the mounds which the animals throw up, are eventually covered by the germinating wheat. During the reaping season, the blades of combine harvesters cut through these mounds, thereby considerably shortening the life expectancy of the blades. Wheat farmers consequently regard these animals as pests and employ diverse methods to eradicate them from their properties.

Furthermore they damage gardens, golf links, bowling greens and tennis courts.

Cape dune molerats may be eradicated by chemical methods (poisoning, etc.) or mechanical methods. Various types of traps are available commercially, all effective to a certain degree, depending on the ability of the trapper. In addition, they may be dug out with considerable physical exertion, especially since tunnel systems may exceed 100 m in length. Finally, they may be shot. A tunnel system is opened (preferably near a new mound) and usually the animal appears within a few minutes to investigate. I have had limited success with this method.

## Prehistory

In former years it was accepted that the earliest African bathyergid known came from the Oligocene-Miocene boundary in SWA/Namibia (Stromer 1924) and was named *Bathyergoides neotertianius*. Subsequent research has shown this fossil rodent to be a phyomyidid canerat (Romer 1971).

*Bathyergus suillus* remains are known as fossils which are associated with the neanderthaloid Saldanha Man. These have been found at Elandsfontein, a few miles to the west of Hopefield in the southwestern Cape. These *Bathyergus* skulls and fragments are not more (if anything) than subspecifically distinct from the living *suillus*. This site has yielded the only evidence hitherto where the extant genus is known to be represented in the fossil or subfossil state.

## Taxonomy

In contrast to the views of Roberts (1951) and Ellerman *et al.* (1953) I regard *B. suillus* as monotypic with *intermedius* of Roberts as a synonym.

# *Bathyergus janetta* Thomas & Schwann, 1904

## Namaqua Dune Molerat
## Namakwa Duinmol

The derivation of *Bathyergus* has been given under *B. suillus*. The origin of the specific name *janetta* could not be traced.

During 1903, Mr C. H. B. Grant collected natural history specimens in Namaqualand for the British Museum in London. Namaqualand has been almost entirely neglected by collectors with the exception of a few specimens collected there by Dr Andrew Smith about 1830 '...and the little set collected by Dr R. Broom at Port Nolloth in 1897... being the only mammals that the museum has ever received from that country' (Thomas & Schwann 1904). Among the specimens collected by Grant was a Namaqua Dune Molerat which Thomas and Schwann thought worthy of specific distinction and called it *Bathyergus janetta*. The type specimen, a female, was collected at Port Nolloth on 3 August 1903 and is housed in the British Museum (Natural History), London.

This species has been included in the *South African Red Data Book* where its status is described as rare, undoubtedly coupled to its limited distribution (Meester 1976).

## Outline of synonymy

1904 *Bathyergus janetta* Thomas & Schwann, *Proc. Zool. Soc. Lond.*: 180. Port Nolloth, CP.

1939 *B.j. inselbergensis* Shortridge & Carter, *Ann. Transv. Mus.* 32:290. Ezelfontein, Kamiesberg, CP.

1939 *B. suillus janetta* Thomas & Schwann. *In* Allen, *Bull. Mus. comp. Zool. Harv.* 83:426.

1946 *B. janetta plowesi* Roberts, *Ann. Transv. Mus.* 20:315. Oranjemund, SWA/Namibia.

### TABLE 12
Measurements of males and female *Bathyergus janetta*

|  | Parameter | Value (mm) | N | Range (mm) | CV (%) |
|---|---|---|---|---|---|
| Males | HB | 205 | 10 | 170–235 | 7,7 |
|  | T | 47 | 10 | 41–52 | 6,4 |
|  | HF | 41 | 10 | 38–43 | 3,0 |
|  | Mass: No data available | | | | |
| Females | HB | 190 | 13 | 170–230 | 8,1 |
|  | T | 44 | 13 | 38–52 | 6,7 |
|  | HF | 37 | 13 | 34–45 | 7,8 |
|  | Mass: No data available | | | | |

## Identification

The dorsal area is coloured seal-brown for a breadth of approximately 50 mm from the nape of the neck to the hind-quarters. This band contrasts markedly with the drab grey of the shoulders and sides. The ventral surface is darkish, with the head black. The chin often shows an obvious white patch. A thin white stripe frequently occurs along the top of the nose, between the eyes. The pelage around the external orifice of the ear is white and the lighter hue of the flanks commences just posterior to the ears. The limbs are dark slaty-grey proximally, while the hands and feet are usually off-white. The fringing hairs of the tail are dull white with the central hairs of the tail pale brown. In the female the mammae consist of two inguinal and one pectoral pair. The males are significantly larger than the females (plate 3).

In size, *B. janetta* is much smaller than *B. suillus*. The mean head-body length of *janetta* is 205 mm (males) compared to 281 mm in *suillus* (De Graaff 1965).

## Skull and dentition

The skull of *B. janetta* is less heavily ridged than that of *B. suillus* (see above), while the muzzle width (compared to the width of the brain case) is proportionally greater in the former. The animals are much smaller than *B. suillus*. The palate does not extend beyond the molars to the same extent as in *B. suillus*. Also, in *B. suillus* the bullae are normal, whereas in *B. janetta* they appear rather swollen. The upper molars are like those of *B. suillus,* but smaller (fig. 27).

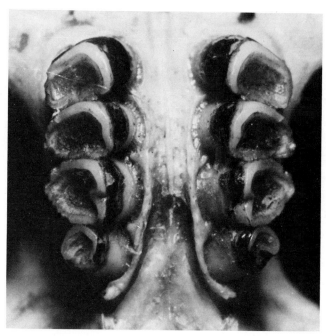

**Figure 27** Cheekteeth of the Namaqua Dune Molerat *Bathyergus janetta*.

## Distribution

Namaqua dune molerats are common in the white dune soils near Port Nolloth and extend northwards to the

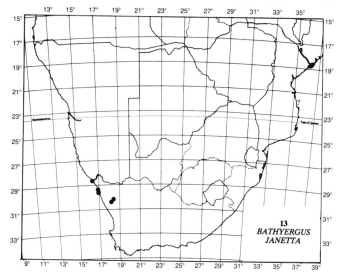

north bank of the Orange River near Oranjemund. Southeastwards from Port Nolloth, they are known from Ezelfontein (Kamiesberg, approximate altitude 1 350 m above sea level) and they are also encountered in the hilly area between Springbok and Kamaggas (map 13).

### Habitat

Sand dunes along the west coast of the Cape Province appear to be their favoured habitat. As indicated above, they also occur inland where sand dunes extend up valleys as a result of prevailing strong westerly winds. *B. janetta* occupies more diverse habitats than *B. suillus* which appear to be limited to soft soils of rivers and tributaries and dunes immediately alongside the coast.

### Diet

No specific information is available, but like other members of the family, they subsist mainly on bulbs and roots.

### Habits

Namaqua dune molerats live in small colonies in tunnel systems. The geographical distribution is determined by the physical condition of the soil and availability of food supply. Mounds of soil are thrown up at irregular distances to clear the passages of soil and these animals seldom emerge above ground. It is uncertain whether they hoard food supplies in special underground chambers.

### Predators, reproduction, parasites, relations with man and prehistory

No information available.

### Taxonomy

As here understood, *B. janetta* is a monotypic species which is in contrast to Roberts (1951), who interprets this species as polytypic (i.e. recognising *janetta*, *inselbergensis* and *plowesi* as subspecies). On the other hand, Ellerman *et al.* (1953) regard *janetta* as only subspecifically distinct from *suillus* (following Shortridge (1934) and Allen (1939) in this respect).

# SUBFAMILY Georychinae Roberts, 1951

This subfamily of molerats was erected by Roberts in 1951. He referred all the remaining southern African genera (*Georychus, Cryptomys* and to which *Heliophobius* would also belong) to it, restricting the typical subfamily Bathyerginae to *Bathyergus*. In the Georychinae the upper incisors are never grooved and they extend backwards into the pterygoid region, a unique characteristic of these bathyergids. The genera in this subfamily have foreclaws which are not enlarged (cf. *Bathyergus*) because the major share of digging is done by means of the incisors rather than by the front appendages. Finally, compared to *Bathyergus,* the angular process of the mandible is not much drawn backwards in these forms. Ellerman *et al.* (1953) point out that if subfamilies are admitted in this family, the aberrant Sand Puppy *Heterocephalus* of East Africa should probably also be regarded as the type of a third subfamily.

---

**Key to the southern African genera of Georychinae**

(Modified after Ellerman (1940) and Ellerman *et al.* (1953))

1 Cheekteeth at full dentition 6/6 (the teeth are usually not simultaneously in position, for the anterior premolars are shed before the posterior molars are cut); hairs of pelage about 20–25 mm in length; palate not extending behind tooth row ............... *Heliophobius* (Silvery Molerat) Page 72

Cheekteeth at full dentition 4/4; pelage usually 20 mm in length; palate extending behind toothrow .... 2

2 Cheekteeth simplified to ring-pattern in adult. The posterior tooth is cut early in life; face not contrastingly coloured ................................. *Cryptomys* (Common Molerat) Page 74

Cheekteeth retaining one inner and one outer fold to old age. Posterior tooth cut late in life; black cap on head, white ring round ear, cheeks black, nose white – face prettily coloured ....................... *Georychus* (Cape Molerat) Page 79

---

GENUS *Heliophobius* Peters, 1846

This molerat was first described by Peters in 1846 in the *Berichter der Preussische Akademie für Wissenschaft* after acquiring a specimen from Tete on the Zambezi River in Mozambique.

# *Heliophobius argenteocinereus* Peters, 1846

**Silvery Molerat**

**Silwerkleurige Knaagdiermol**

The generic name is compounded from the Greek *helios* = the sun and *phobos* = fright. The specific name is derived from the Latin *argenteus* = silvery and *cinereus* = grey-coloured. The name therefore refers to the silver to ash-coloured molerat which is afraid of the sun.

I have not had experience with this species in the field and what follows is a compilation of data from the literature, especially Ellerman (1940). However, I have had a look at the limited number of study skins available in southern Africa and these vindicate the observations (where applicable) which follow.

**Outline of synonymy**

1846 *Heliophobius argenteocinereus* Peters, *Monatsber. K. Preuss. Akad. Wiss. Berlin:* 259. Tete on the Zambezi, Mozambique.
1906 *H. robustus* Thomas, *Ann. Mag. nat. Hist.* (7) 17:179. Mpika, Zambia.
1917 *H. angonicus* Thomas, *Ann. Mag. nat. Hist.* (8) 20:314. Malaŵi.

These animals are poorly represented in southern African museums and I could not compile any statistical parameters.

**Identification**

The species has a fairly long pelt, silky to the touch. Individual hairs are 20–25 mm in length compared to *Cryptomys* with a much shorter pelage (6–9 mm).

Like other bathyergids, the pinnae of the ears are absent, while the tail is vestigial (plate 3).

The claws on the forefeet are not excessively lengthened. Ellerman (1940) remarks that the hindfoot differs from that of *Bathyergus* in that the second digit is the main component rather than the third (as in the forefoot), though the hallux is slightly longer than the fifth digit.

## Skull and dentition

The infraorbital foramen is small. The palate is very narrow and does not extend behind the tooth row.

A remarkable feature of *Heliophobius* is its dentition which consists of 6/6 in its full complement. However, these teeth are very infrequently simultaneously in position, for the anterior premolars are shed before the posterior molars erupt late in life. The animal appears to be erupting teeth more or less throughout life. In 50 skulls available for study, Ellerman (1940) came across only one where all six teeth were in place together (one side of the jaw only). The usual number in place at once appears to be either 5/4 or 4/4 and occasionally 4/5. The teeth show one external and one internal fold, but the occlusal surface soon wears down to a simple ring-pattern.

Thomas (1909) has suggested that the cheekteeth could be annotated as follows:

$$6/6 = Pm\frac{2.3.4.}{2.3.4.} \quad M\frac{1.2.3.}{1.2.3.}$$

This arrangement seems unlikely, however, and for an alternative interpretation, see De Graaff (1979).

As is the case in *Georychus* and *Cryptomys*, the roots of the upper incisors are situated in the pterygoid region of the skull. The incisors, which stand out in front of the closed lips, are white.

The angular portion of the mandible is not produced far backwards.

## Distribution

In southern Africa, it occurs in Mozambique, in the Tete and Zambezi districts, ranging northwards to Tanzania, southern Zaïre and southern Kenya (map 14), and is therefore an inhabitant of the Southern Savanna Woodland biotic zone.

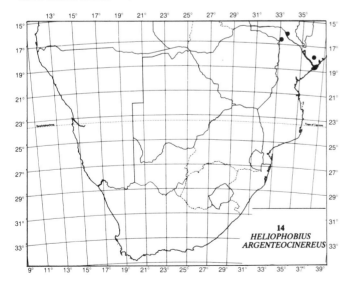

**14**
*HELIOPHOBIUS*
*ARGENTEOCINEREUS*

## Habitat

According to Kingdon (1974) *Heliophobius* occupies the drier and warmer habitats at lower altitudes (750–1 500 m with an annual precipitation of 250–600 mm). It can withstand long dry seasons and '... favours well-drained, sandy soils on rocky hillsides on open plains or in *Combretum* or *Brachystegia* woodland'. In Mozambique, it occurs in alluvium and sandy ground.

## Diet

The food consists mainly of tubers and bulbs which include *Dolichos* and *Vigna* as principal elements (Jarvis & Sale 1971).

## Habits

The presence of heaps of soil, at intervals, disclose the presence of these animals. Their burrows are about 200 mm below the surface of the soil where most of the edible tubers are found. Kingdon (1974) states that it does not feed on the surface, although it occasionally emerges to collect grass to line its nest.

The average length of four burrows measured by Jarvis & Sale (1971) was 47 m. They discovered bolt holes that went down 460–540 mm which the animals use if danger threatens. These are difficult to find, because they are blocked with earth as soon as the animals have entered them.

Kingdon (1974) states that in captivity these animals are aggressive when kept together and fight at the least provocation. This behaviour is also seen in *Cryptomys* which I have kept in captivity. In the wild, burrow systems of *Heliophobius* are occupied for much of the year by single animals.

## Predators

Skulls of this species are not infrequently found in owl pellets.

## Reproduction

Their behaviour during the breeding season is unknown and bearing in mind their asociality and aggression, should be an interesting study. Copley (1950) says that they produce one litter annually at the beginning of the rainy season. There are one to four young per litter.

## Parasites

As is to be expected from a cryptic animal such as *Heliophobius*, the list of parasites associated with it is brief. I could not trace any mention of arthropods hitherto recorded as ectoparasites. However, the cestode *Inermicapsifer madagascariensis* has been reported from a Silvery Molerat in Kenya (Baylis 1949) and Tanzania (Ortlepp 1961), while *Catenotaenia lobata* from *Heliophobius argenteocinereus marungensis* was collected at Dahome (Joyteux & Baer 1927).

## Relations with man

Not much is known. Baylis (1949) reported on a new human cestode infection which was found in Kenya. This tapeworm, *Inermicapsifer madagascariensis (= I. arvicanthides),* is a tapeworm of rats and it seems as if the Silvery Molerat has also become a reservoir. That this anoplocephalid tapeworm has been transmitted to man, is also verified by Ortlepp (1961).

**Prehistory**
No information available.

**Taxonomy**
Ellerman (1940) lists two species of *Heliophobius:* the nominate species *H. argenteocinereus* and *H. spalax,* a population which occurs east of Kilimandjaro. The subspecies of *H. argenteocinereus* listed, include *albifrons* Gray, 1864 (Tanzania), *marungensis* Noack, 1807 (southeastern Zaïre), *emini* Noack, 1894 (Tanzania), *robustus* Thomas, 1906 (northeastern Zambia), *kapiti* Heller, 1909 (Kenya), *mottoulei* Schouteden, 1913 (Zaïre) and *angonicus* Thomas, 1917 (northeastern Zambia). This is also the position accepted by De Graaff (1975) in his bathyergid contribution to *The mammals of Africa,* edited by Meester & Setzer. Kingdon (1974) states that *H. argenteocinereus* is a somewhat variable species and that the races described may not have much validity. He is of the opinion that *H. spalax* is almost certainly '… no more than a local race of *H. argenteocinereus* as there is no evidence of sympatry'.

GENUS *Cryptomys* Gray, 1864

As presently understood, three species of *Cryptomys* occur in Africa: *C. ochraceocinereus* (ranging from northern Uganda into Sudan and westwards to northern Zaïre, Nigeria and eastern Ghana), *C. mechowi* (southern Zaïre, Nyika plateau (Malaŵi), northwestern Zambia and northern and central Angola), and *C. hottentotus* occurring throughout southern Africa, excluding the western Karoo and northwestern Cape (De Graaff 1975). It also occurs in Zambia, Malaŵi, parts of Zaïre and as far north as Tanzania.

# *Cryptomys hottentotus* (Lesson, 1826)

**Common Molerat**

**Hottentot Molerat**

**Hottentot Knaagdiermol**

The generic name is derived from the Greek *kryptos* = hidden and *mys* = mouse, with reference to the subterranean habits of the animal. Why Lesson associated the early Hottentot inhabitants of the Cape with this species, is not clear. It is probably attributed to the fact that both were encountered at the Cape of Good Hope.

This molerat was first described by Lesson in 1826 as *Bathyergus hottentotus* from the vicinity of Paarl, Cape. Lesson acted as apothecary on board the French ship *Coquille* during its voyage around the world and he thus had the opportunity to encounter this molerat during the vessel's sojourn at the Cape.

As here understood, *Cryptomys hottentotus* is a geographically variable polytypic species with an extensive distribution pattern in southern Africa. At the most five subspecies can be admitted in southern Africa as here defined, including *bocagei, damarensis, darlingi, hottentotus* and *natalensis* (De Graaff 1975).

**Outline of synonymy**
1826 *Bathyergus hottentotus* Lesson, *Voyage autour du Monde sur la Coquille.* Zool. 1:166. Paarl, CP.
1827 *B. caecutiens* Brants, *Het Geslacht der Muizen:* 37. Cape of Good Hope. (Knysna *vide* Roberts 1951.)
1829 *B. ludwigii* A. Smith, *Zool. J.* 4:439. (Cape Town *vide* Roberts (1951).)
1838 *Bathyergus damarensis* Ogilby, *Proc. zool. Soc. Lond.:* 5. Damaraland, CP.
  *Synonyms*
  1898 *Georychus lugardi* De Winton, *Ann. Mag. nat. Hist.* (7)1:253. Northeastern Botswana.
  1909 *G. micklemi* Chubb, *Ann. Mag. nat. Hist.* (8)3:35. Western Zambia.
  1946 *Cryptomys ovamboensis* Roberts, *Ann. Transv. Mus.* 20:315. Ondonga, SWA/Namibia.
1843 *Georychus holosericeus* Wagner. *In* Schreber's *Säugeth.* Suppl. 3:373. Graaff-Reinet, CP.
1895 *Georychus darlingi* Thomas, *Ann. Mag. nat. Hist.* (6) 16:239. Salisbury, Zimbabwe.
  *Synonyms*
  1896 *Georychus nimrodi* De Winton, *Proc. zool. Soc. Lond.:* 808. Matabeleland, Zimbabwe.

1907 *Georychus beirae* Thomas & Wroughton, *Proc. zool. Soc. Lond.:* 780. Beira, Mozambique.

1946 *Cryptomys zimbitiensis* Roberts, *Ann. Transv. Mus.* 20:315. Zimbiti, Mozambique.

1897 *Georychus bocagei* De Winton, *Ann. Mag. nat. Hist.* (6) 20:323. Angola.

*Synonym*

1933 *C. kubangensis* Monard, *Bull. Soc. Neuchatel Sci. Nat.* 57:58. Angola.

1899 *Georychus exenticus* Trouessart, *Cat. Mamm. Viv. Foss.:* 1338. Error for *caecutiens.*

1906 *Cryptomys hottentottus* (sic) *talpoides* Thomas & Schwann, *Proc. zool. Soc. Lond.:* 166. Knysna, CP.

1909 *Georychus jorisseni* Jameson, *Ann. Mag. nat. Hist.* (8) 4:466. Waterberg, Transvaal.

1913 *G. albus* Roberts, *Ann. Transv. Mus.* 4:100. Type locality unknown. Possibly Wynberg, CP.

1913 *Cryptomys natalensis* Roberts, *Ann. Transv. Mus.* 4:94. Wakkerstroom, Transvaal.

*Synonyms*

1913 *Georychus jamesoni* Roberts, *Ann. Transv. Mus.* 4:95. Johannesburg, Transvaal.

1913 *Georychus arenarius* Roberts, *Ann. Transv. Mus.* 4:96. Pretoria, Transvaal.

*Synonyms*

1913 *Georychus pretoriae* Roberts, *Ann. Transv. Mus.* 4:99. Pretoria, Transvaal.

1917 *G. palki* Roberts, *Ann. Transv. Mus.* 6:5. Vaal River, Transvaal.

1913 *Georychus mahali* Roberts, *Ann. Transv. Mus.* 4:108. Pretoria, Transvaal.

1917 *Georychus komatiensis* Roberts, *Ann. Transv. Mus.* 5:272. Carolina, Transvaal.

*Synonyms*

1917 *Georychus stellatus* Roberts, *Ann. Transv. Mus.* 5:272. Komatipoort, Transvaal.

1917 *G. rufulus* Roberts, *Ann. Transv. Mus.* 5:272. Tzaneen, Transvaal.

1951 *Cryptomys komatiensis zuluensis* Roberts. *In The mammals of South Africa:* 396. St Lucia, Natal.

1926 *Cryptomys montanus* Roberts, *Ann. Transv. Mus.* 11:260. Pretoria, Transvaal.

1926 *Cryptomys junodi* Roberts, *Ann. Transv. Mus.* 11:260. Masiyeni, Mozambique.

1926 *Cryptomys melanoticus* Roberts, *Ann. Transv. Mus.* 11:260. Leydsdorp district, Transvaal.

1929 *Cryptomys langi* Roberts, *Ann. Transv. Mus.* 13:119. Howick district, Natal.

1946 *Cryptomys natalensis streeteri* Roberts, *Ann. Transv. Mus.* 20:316. Hectorspruit, Transvaal.

1917 *G. vandami* Roberts, *Ann. Transv. Mus.* 5:273. Leydsdorp, Transvaal.

*Synonyms*

1917 *Georychus natalensis pallidus* Roberts, *Ann. Transv. Mus.* 5:278. Soutpansberg district, Transvaal.

1939 *Cryptomys natalensis nemo* Allen, *Bull. Mus. comp. Zool. Harv.* 83:429.

1917 *C. vryburgensis* Roberts, *Ann. Transv. Mus.* 5:274. Vryburg, CP.

1924 *Cryptomys cradockensis* Roberts, *Ann. Transv. Mus.* 10:73. Cradock, CP.

1924 *C. transvaalensis* Roberts, *Ann. Transv. Mus.* 10:73. Pretoria district, Transvaal.

1924 *C. bigalkei* Roberts, *Ann. Transv. Mus.* 10:73. Glen, OFS.

1926 *Cryptomys orangiae* Roberts, *Ann. Transv. Mus.* 11:259. Glen, OFS.

1926 *C. vetensis* Roberts, *Ann. Transv. Mus.* 11:259. Vet River, OFS.

1946 *Cryptomys holosericeus valschensis* Roberts, *Ann. Transv. Mus.* 20:316. Bothaville, OFS.

## TABLE 13
Measurements of male and female *Cryptomys hottentotus*

| | Parameter | Value (mm) | N | Range (mm) | CV (%) |
|---|---|---|---|---|---|
| Males | HB | 133 | 271 | 105–185 | 8,1 |
| | T | 19 | 222 | 10–30 | 16,0 |
| | HF | 23 | 238 | 12–33 | 10,9 |
| | Mass: | 134 g | 21 | 112–145 g | 7,6 |
| Females | HB | 129 | 248 | 100–164 | 5,1 |
| | T | 18 | 202 | 10–32 | 16,0 |
| | HF | 22 | 246 | 12–38 | 12,6 |
| | Mass: | 119 g | 26 | 98–153 g | 11,6 |

## Identification

This is a geographically variable species, in both size and colour. It is small to medium sized, cinnamon-buff to clay-coloured dorsally, slightly paler below. Individual hairs usually have dark slaty bases with fawn-coloured tips. Individuals tend to be smaller in the southwestern Cape and larger further north. A white occipital patch is variable in occurrence: the patch is usually absent in the southern populations and reaches its maximum development in specimens from the northwestern Cape, SWA/Namibia and Botswana. In one subspecies, *Cryptomys hottentotus damarensis,* a pronounced colour polymorphism may occur. Individuals in different colour phases may occur within the same tunnel system (De Graaff 1972). Occasional albino specimens are known from study collections while in some populations buffy-yellowish individuals are encountered (plate 3).

The claws on the hands and feet are not enlarged and the second and third digits tend to be subequal in length.

The pelage is short and well developed, silky to the touch and juveniles are usually darker than adults.

There is no significant difference in size between males and females. The mammae consist of two pectoral and one inguinally situated pairs.

## Skull and dentition

In contrast to the Cape Dune Molerat *Bathyergus suillus*, the skull is much smaller. Seen from the lateral aspect, the skull has a convex curvature dorsally and is not as robust and sturdily built. Sagittal and lambdoid crests are but faintly developed. It presents a far more 'rounded' appearance. The bullae are well developed (fig. 28).

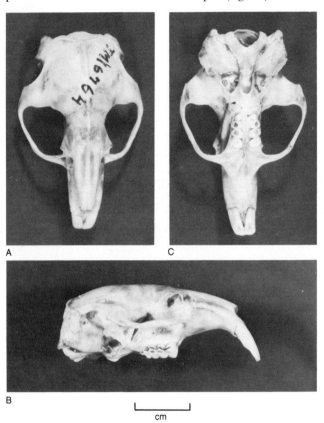

**Figure 28** Dorsal (A) lateral (B) and ventral (C) views of the skull of the Common Molerat *Cryptomys hottentotus*. Crests are but faintly developed; infraorbital foramen small; pulp cavities of incisors lie behind molar tooth row. Upper incisors plain. Muzzle strongly built with zygomatic arches sturdy; interorbital constriction clearly visible.

The braincase is reasonably broad with a narrow interorbital constriction. The nasals and muzzle are slender. The infraorbital foramen is more or less elliptical in shape, small and does not normally transmit any of the masseteric musculature as in hystricomorphs.

In the lower jaw, the symphysis between the two rami is not rigidly enclosed, allowing a certain freedom of movement between the two lower incisors. The upper incisors are not grooved and their pulp cavities are sit-

uated in the pterygoid region of the skull. These teeth are not as broad as those of *Bathyergus* and may attain a length of 38 mm. The lower incisors are also smooth on their anterior surfaces. The upper molars are simple structures, decreasing in size from front to back, consisting of a central portion of dentine surrounded by a ring of well-developed enamel (fig. 29).

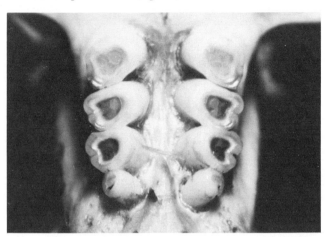

**Figure 29** Cheekteeth of the Common Molerat *Cryptomys hottentotus*. These teeth (Pm$^4$, M$^{1-3}$) are simple structures decreasing in size from front to back, consisting of a dentine core surrounded by a ring of enamel.

## Distribution

Of the rodents, this species exhibits one of the most extensive distribution patterns in southern Africa. It ranges from Namaqualand and the southwestern Cape Province to the Eastern Province, northwards over large stretches of the eastern Karoo and northwestern Cape into the OFS, western, northern and eastern Transvaal. It occurs in SWA/Namibia and Ovambo, the Kalahari areas of Botswana and Ngamiland into Zimbabwe as far as Mount Selinda in the east. It extends from the eastern Transvaal into the southern and northern parts of Natal (as well as Swaziland) to the mouth of the Zambezi River in the north (map 15).

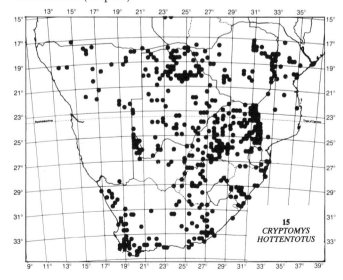

15
*CRYPTOMYS HOTTENTOTUS*

76

## Habitat

Bearing in mind the wide geographical distribution, it is expected that the Common Molerat occurs in diverse habitats from the drier and arid northwest to areas of high rainfall and humidity. It occurs in a diversity of soil types: loose sandy soils (weathered granites, e.g. at Pretoriuskop in the Kruger National Park) to stony soils and hills (e.g. the hills surrounding Pretoria) to montane and escarpment conditions (e.g. at Wakkerstroom). There is a tendency, however, for greater dispersal in loose, sandy soils (especially alluvial soils associated with major rivers and streams).

Botanically speaking, the Common Molerat frequents karroid veldtypes, coastal rhenosterbushveld, coastal forests, thornveld, mopaneveld, savanna and pure grassveld, as well as temperate and transitional forests, scrub and bushveld.

## Diet

The Common Molerat subsists on bulbs, tubers, fleshy rootstocks and bulbous grass roots. It is especially partial to the blue tulip *Moraea* (Iridaceae) which is poisonous to cattle. In the wild, it eats *Citrullus* (Cucurbitaceae), *Pseudogaltonia* (Liliaceae), *Homeria* (Iridaceae), *Cyperus* (Cyperaceae) and occasionally the leaves of certain species of *Aloe*. It is very fond of garden plants, including dahlias and agapanthus, while they readily take to cultivated crops including potatoes and carrots. In the Nylsvley Nature Reserve in the central Transvaal they are common in the ecotonal zone between savanna and grassland where they feed on bulbs, corms and roots of *Cyperus* spp., *Gladiolus* spp. and even the roots of *Aloe* spp. in times of stress (Jacobsen 1977).

## Habits

Like the other genera of this family, the Common Molerat throws up mounds and develops extensive tunnel systems (Hickman 1979) with many side-tunnels leading off from the main tunnel, which is approximately 150 mm below the surface of the soil. The diameter of the tunnel varies between 50–60 mm. The incisors are used for digging and the species makes extensive tunnels when the soil is damp and easily worked. Nests are usually situated at a deeper level than the rest of the tunnel system and are constructed from vegetation collected on the surface of the ground (De Graaff 1962).

Normally, these animals do not appear above ground, although they may do so at night, as is evidenced by the fact that they are often found in pellets of the Barn Owl *Tyto alba*. The gait on the surface is wobbly and it seems that they forage after sunset. It is said that they construct special chambers in which to hoard food. Transporting round objects like bulbs is facilitated by the fact that the lower jaws can move independently, thereby ensuring a better grip on the food item.

Common molerats live in colonies, varying in size: the males are polygamous and in one instance six females and one male were collected from a single tunnel system (Bothma *pers. comm.*). Individuals coming from different tunnel systems fight readily. On the surface, *Cryptomys hottentotus* adopts an aggressive stance when threatened, facing its persecutor with jaws agape. This apparent fearlessness probably stands this molerat in good stead when underground, since any available burrow is attractive to other rodents and snakes (Hanney 1975).

The sense of hearing is well developed, but sight is poor (Eloff 1958). Specimens can be trapped in the field with eyelids virtually fused: others may walk about on the soil surface with eyes closed, opening them only when excited or aroused. Similarly, the sense of smell seems to be poorly developed. In contrast, however, the tactile sense seems to be well developed – large vibrissae are present on the face while stiffened, bristle-like hairs occur on the outer surfaces of the fore- and hindfeet, serving a tactile function.

Other animals are often encountered sharing the nests or tunnel systems. The Running Frog *Kassina senegalensis* is frequently found in tunnels bordering pans and vleis (Eloff 1952). Lizards *(Mabuia* sp.) and scorpions *(Opisthopthalmus pictus, O. wahlbergi, O. carinatus* and *Parabuthus granulatus)* occur in occupied tunnels of *Cryptomys* in the Kalahari Gemsbok National Park (De Graaff 1972). Larvae of certain beetles, *Gonopus* sp. (Coleptera) are often found in the entwined vegetation of the nest structure.

When agitated, these animals emit grunts, while normal communication is by means of high-pitched squeaks.

## Predators

Skulls of this species are often encountered in the pellets of the Barn Owl *Tyto alba* (De Graaff 1960b; Dean 1977). Other bird predators include the Marsh Owl *Asio capensis* (Dean 1977), the Giant Eagle Owl *Bubo lacteus* (Pitman & Adamson 1978) and the Grey Heron *Ardea cinerea*. Snakes are probably their chief enemies and the Molesnake *Pseudaspis cana* and the Shield-nosed Snake *Aspidelaps scutatus* are known to take them (De Graaff 1965). FitzSimons (1962) says that molesnakes in search of molerats '. . . push the forepart of the body down through a molehill into the runway below and patiently wait in this position for the return of the unwary prey'. Domestic dogs and cats often molest them, while they are also eaten by the Black-backed Jackal *Canis mesomelas* and the Silver Jackal *Vulpes chama*. Mills & Mills (1978) recorded them as prey items for the Brown Hyaena *Hyaena brunnea* in the Kalahari Gembsbok National Park. Ratels *Mellivora capensis,* are inclined to dig them out while the Small-spotted Genet *Genetta genetta* hunts them (Pienaar *et al.* 1980).

## Reproduction

Two pairs of mammae are situated pectorally, one pair inguinally. The number of young per litter varies. Five and three fetuses were found in gravid females during April and July (SWA/Namibia and Zimbabwe) while pregnant females have been taken during February to March in Pretoria, and in the western Transvaal (March). In the Transvaal, monthly pregnancy rates recorded by Rautenbach (1978b) are February (11%), March (55%) and August (11%). Lactating females were recorded during January and April. Smithers (1971) recorded pregnant females during February and July. Rautenbach states that the samples are too small to be conclusive, but the inference is that parturition occurs throughout the year, with a mean number of fetuses per female being 1,75 (N = 8).

## Parasites

The tape worm *Inermicapsifer madagascariensis*, commonly found in southern African rodents, was also encountered in *C. hottentotus*, which I trapped near Shingwidzi in the Kruger National Park (De Graaff 1964).

Somewhat more information is available on the ectoparasitic arthropods. Zumpt (1961a) has listed four families of mites which have been taken from *Cryptomys*. This include the myobiid *Radfordia rotundata* and the trombiculids *Schoutedenichia crocidurae*, *Gahrliepia nana* and *Acomatacarus polydiscum*. The laelaptids include four species of *Haemolaelaps (capensis, natalensis, bathyergus* and *eloffi)*, while *Androlaelaps marshalli* and *Macronyssus bacoti* have also been identified. The family Ascaidae is represented by *Myonyssoides capensis*, also reported by Bedford (1932).

Eight species of fleas are associated with *Cryptomys*. These are the pulicid species *Procaviopsylla creusae*, *Xenopsylla philoxera*, *X. piriei* and *X. georychi*, the hystrichopsyllids *Ctenophthalmus edwardsi*, *C. ansorgei* and *?Dinopsyllus zuluensis*, and the chimaeropsyllid *Cryptopsylla ingrami*, a flea specific to this molerat (Zumpt 1966).

Very few ticks are known to be associated with *C. hottentotus* which is not surprising when the fossorial way of life of the host is heeded. Theiler (1962) lists only one species, both adult and immature individuals of *Ixodes alluaudi*, found on this molerat. The anoplurid *Proenderleinellus hilli* and also been taken from this rodent (Bedford 1932).

## Relations with man

There is no doubt that the tunnel systems increase the porosity of the soil (Eloff 1954). Economically undesirable plant genera like *Homeria* and *Cyperus* are kept in check by the Common Molerat in natural areas. Agriculturally favourable grasses like *Themeda* (rooigras) often spread and germinate in the cool, moist tunnels. On the other hand, this molerat is a serious pest in vegetable gardens and small-holdings with cultivated lands of potatoes. They spoil more than they can devour, gnawing off small portions and discarding the rest which rapidly rots.

## Prehistory

The genus *Cryptomys* also occurs in Pleistocene breccias (De Graaff 1960a). *Cryptomys robertsi* (an extinct species) occurs in all the fossiliferous localities hitherto analysed and must have been a versatile and successful species. Lavocat (1973) established a new genus and species *Paracryptomys mackennae* on the basis of material collected at Lüderitz (SWA/Namibia) representing Lower Miocene mammals housed in the collection of the American Museum of Natural History (Hamilton & Van Couvering 1977).

## Taxonomy

*Cryptomys hottentotus* is a very variable species in southern Africa. Presently five subspecies are recognised: *C. h. hottentotus* (Lesson, 1826), *C.h. damarensis* (Ogilby, 1838), *C. h. bocagei* (De Winton, 1897), *C. h. darlingi* (Thomas, 1895) and *C. h. natalensis* (Roberts, 1913) (De Graaff 1975). *C. hottentotus*, occurring in the Cape Province, the OFS, the western, central and northern Transvaal and southern Zimbabwe, includes *holosericeus*, *caecutiens*, *ludwigii*, *exenticus*, *talpoides*, *cradockensis*, *jorriseni*, *albus*, *vandami (= pallidus, nemo)*, *bigalkei*, *vryburgensis*, *transvaalensis*, *orangiae*, *vetensis* and *valschensis* of previous authors. *C. h. damarensis*, resident in the northwestern Cape, Botswana, central SWA/Namibia and eastern Zimbabwe, is taken to include *lugardi*, *ovamboensis* and *micklemi* of earlier authors. *C. h. bocagei*, found in western Angola and ranging southwards across the Cunene River into northern SWA/Namibia, includes *kubangensis* as a synonym.

*C. h. darlingi* occurs in Zimbabwe (Mashonaland) and ranges southeastwards to Mount Selinda and Mozambique and southwestwards to Matabeleland in Zimbabwe. The Mozambique forms *beirae* and *zimbitiensis* as well as *nimrodi* (described from the vicinity of Bulawayo) appear to be synonyms of *darlingi*.

Finally, *C. h. natalensis*, residing in the Witwatersrand/Pretoria area and ranging eastwards to the southeastern Transvaal and Natal and from thence northwards into southern Mozambique, is taken to include *jamesoni*, *arenarius*, *anomalus (= pretoriae, palki)*, *komatiensis*, *mahali*, *montanus*, *junodi*, *langi*, *zuluensis*, *rufulus*, *stellatus*, *melanoticus* and *streeteri* as synonyms.

Apart from the subspecies mentioned above, *Cryptomys hottentotus* also includes yet another extralimital subspecies in Zambia and the Katampa Province of Zaïre. It has been listed as *C. h. amatus* and includes *molyneuxi* as a synonym.

Further north in Africa, and extralimital to this work, two other species of *Cryptomys* occur. The Ochre Molerat *Cryptomys ochraceocinereus* (Heuglin, 1864) (with *ochraceocinereus* and *oweni* as subspecies) ranges from northern Uganda and Sudan westwards to Nigeria and eastern Ghana. *Cryptomys mechowi* (Peters, 1881), known as the Giant Molerat, occurs in southern Zaïre, on the Nyika Plateau in Malaŵi and in northwestern Zambia and central Angola. *C. m. mellandi* is recognised as a separate subspecies, but *mechowi* is taken to include *ansorgei* and *blainei* as synonyms.

# GENUS *Georychus* Illiger, 1811

This monotypic genus, endemic to southern Africa, occurs in the coastal regions of the southern and southwestern Cape. Relict populations of this molerat occur in Natal (Nottingham Road vicinity), as well as in the eastern Transvaal (near Belfast and Ermelo).

# *Georychus capensis* (Pallas, 1778)

## Cape Molerat
## Blesmol

The generic name is derived from the Greek *georychos* = throwing up earth, while the specific name *capensis* is the Latin reference to the Cape. As is the case in the genera *Bathyergus* and *Cryptomys*, *Georychus* also pushes up mounds of earth which results from its tunnelling activities, thereby announcing its presence in the area.

This animal was first referred to as *Mus capensis* by Pallas in 1778, the type locality being given as the Cape of Good Hope. According to Sclater (1901), Masson made the first allusion to the species as the 'blesmol', while Buffon (1776, 1782) supplied early descriptions of it. In 1785 Sparrman supplied notes on the habits and occurrence of this species, followed by Thunberg in 1788 describing and naming it *Marmota capensis*. In 1811 Illiger proposed the erection of a new genus for this species, *viz. Georychus*. Eventually, A. Smith (1826) seems to be the first authority to refer to this molerat as *G. capensis,* as it has been known ever since.

Roberts (1913) divided the genus into two species based mainly on colour variations. Specimens from Belfast, Transvaal, he referred to as *G. yatesi*. Prior to this, Thomas & Schwann described a new subspecies of *G. capensis* in 1906 and referred to it as *G. c. canescens*. In 1940 Ellerman relegated *G. yatesi* to subspecific rank and this interpretation was also followed by Roberts in 1951. This polytypy of *Georychus* at species level was not accepted by De Graaff (1965, 1975) who studied the taxonomy of southern African molerats in greater depth. According to Shortridge (1934) the type specimen of *G. capensis* obtained from the Cape Flats near Cape Town, no longer exists.

## Outline of synonymy

1778 *Mus capensis* Pallas, *Nov. Spec. Quad. Glir. Ord.:* 76. Cape of Good Hope.

1788 *Marmota capensis* Thunberg, *Resa uti Europa, Africa, Asia, etc.* I:293. Cape of Good Hope.

1811 *Georychus* Illiger, *Prodr. Syst. Mamm. (= Mus capensis* Pallas).

1826 *Georychus capensis* A. Smith, *Descr. Cat. S.A. Mus.* I.

1832 *Bathyergus capensis* Smuts, *Enum. Mamm. Cap.:* 108.

1834 *Mus buffoni* Cuvier, *Ann. Sci. Nat.* (2) 1:196. Cape of Good Hope.

1844 *Fossor leucops* Lichtenstein, Forster's *Desc. Anim. Iter. ad. Maris Austr. Terras Suscepto:* 364, Cape of Good Hope.

1858 *Georychus capensis* Grill, *Oefvers. K. Vet. Akad. Förh. Stockholm:* 19; Layard 1862, *Cat. Spec. Coll. S.A. Mus.* 1.; Moseley 1879, *Notes by a Naturalist on the Challenger:* 620.

1906 *G. c. canescens* Thomas & Schwann, *Proc. zool. Soc. Lond.:* 1:165. Knysna, CP.

1913 *Georychus yatesi* Roberts, *Ann. Transv. Mus.* 4:92. Belfast, Transvaal.

1939 *Georychus capensis capensis* (Pallas). *In* Allen, *Bull. Mus. comp. Zool. Harv.* 83:431.

## TABLE 14
### Measurements of male and female *Georychus capensis*

| | Parameter | Value (mm) | N | Range (mm) | CV (%) |
|---|---|---|---|---|---|
| Males | HB | 184 | 11 | 153–200 | 6,1 |
| | T | 28 | 11 | 20–40 | 7,0 |
| | HF | 32 | 11 | 26–35 | 7,7 |
| Mass: Only three weights are known: 245 g, 310 g, 360 g | | | | | |
| Females | HB | 177 | 25 | 143–204 | 6,8 |
| | T | 27 | 17 | 23–38 | 12,9 |
| | HF | 29 | 23 | 25–35 | 8,8 |
| Mass: Four weights are known: 124 g, 173 g, 195 g, 326 g | | | | | |

## Identification

The most characteristic feature of this species is the fact that the face is prettily marked. The muzzle, the area around the eyes and ears, and a conspicuous frontal to occipital patch (invariably present) are white on a background of black. The rest of the body is buffy to buff-orange, with variable tipping of brown to the hairs. The hands, feet and tail are also white (plate 3).

The pelage is thick and woolly; complete or partial albinism is encountered, ranging from pure white individuals to light grey or creamy orange.

Geographical variation in coloration is evident: specimens from the vicinity of Knysna are more drab, while those east of Port Elizabeth tend to be most brightly coloured.

Juveniles are a decidedly darker slate-grey dorsally.

The size of the Cape Molerat is geographically variable. Specimens from the vicinity of Worcester tend to be larger than those of the Cape Peninsula.

The Cape Molerat shows no unique specialisations externally. It also digs with the aid of its incisors and consequently the claws are not enlarged. The arrangement of the digits conforms closely to that of *Bathyergus*.

## Skull and dentition

The skull is larger than that of *Cryptomys*, and its profile is arched. The jugal bone dovetails into the zygoma. The palate is narrow, not wider than the occlusal width of the

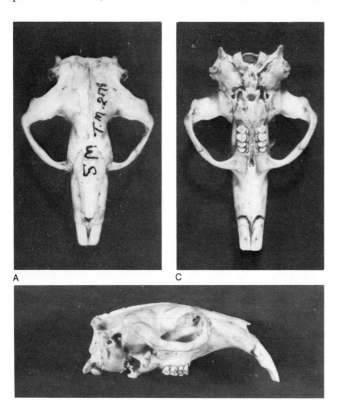

**Figure 30** Dorsal (A), lateral (B) and ventral (C) views of the skull of the Cape Molerat *Georychus capensis*. The arched skull is robust with a sagittal crest in old individuals; muzzle slender; jugal dovetails into the zygomatic branch of the maxilla; infraorbital foramen is small; palate narrow with zygomatic arches wide. Incisors smooth and their pulp cavities lie in pterygoid region.

molar teeth. The skull is robust and sagittal crests are often present in old individuals. In the living animal the muzzle does not protrude to such a marked extent as in *Bathyergus*. The infraorbital foramen is small. The sphenolateral foramen is obliterated by the roots of the upper incisors. The bullae are well developed (fig. 30).

The upper incisors are smooth, rooted in the pterygoid region behind the molars, thereby possibly checking the development of the posterior molars. The lower, ungrooved incisors have a limited degree of movement because of the unfused symphysis of the two jaw rami.

Each of the upper cheekteeth has one narrow inner and outer fold of enamel, which persist until the molars are well worn. *Georychus* is thus the only member of the family without simplified rounded (ovate) cheekteeth in the adult (fig. 31). The lower cheekteeth have one outer fold persistent, and one inner fold tending to disappear with age. Males tend to have wider zygomatic arches than females.

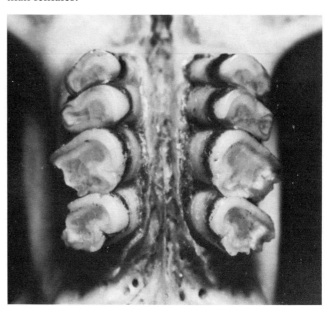

**Figure 31** Cheekteeth of the Cape Molerat *Georychus capensis*. Single inner and outer folds of enamel jutting into core of dentine persist well into old age.

## Distribution

The Cape Molerat is confined to the RSA, from the Cape Peninsula to the mountains around Tulbagh and Worcester. Ellerman *et al.* (1953) state that specimens are known from Citrusdal and Nieuwoudtville, the latter probably its most northerly distribution in the western Cape. From Cape Town it extends eastwards along the coast to Port Elizabeth and Pondoland in Transkei, and inland to Nottingham Road in Natal. A few specimens have been collected at Belfast and Ermelo on the Transvaal highveld (map 16).

The Cape Molerat thus occurs sympatrically with the larger *Bathyergus suillus* and the smaller *Cryptomys hotten-*

*totus* in the Western Province and southern coastal belt to the Eastern Province.

It therefore occurs in four biotic zones, *viz.* the South West Arid, the South West Cape, the Forest and in the Southern Savanna Grassland.

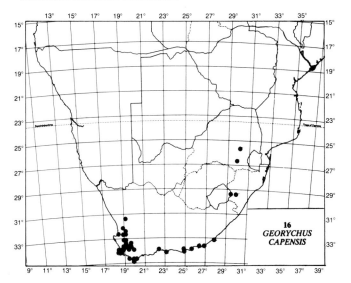

## Habitat
Cape molerats occur in sandy coastal dunes as well as in unconsolidated soils along rivers. In contrast to *Bathyergus suillus*, they occur deeper inland.

## Diet
They readily eat bulbs of *Sparaxis grandiflora* (Iridaceae), tubers of *Zantedeschia (Richardia)* spp. and a diversity of other roots and tubers, including potatoes, sweet potatoes, carrots, onions, lettuce and even carnations. Like *Cryptomys*, they often prevent these plants from sprouting by removing the buds with their incisors. Landry (1957) quotes Tullberg who pointed out that they do not masticate their food much, as pieces of roots up to 10 mm in length are found in the stomachs.

## Habits
In contrast to *Bathyergus* and *Cryptomys*, very little is known about the biology of *Georychus capensis*. It burrows rather superficially under the soil and its mounds are smaller than would be expected. Blind tunnels occur in the tunnel system, and they may store food in specially constructed chambers. Like *Cryptomys*, it uses its incisors more than its claws for digging. Echolocation may play an important role in their orientation. The voice has been described as a series of grunts.

## Predators
No information available.

## Reproduction
Nothing is known about the reproduction of the Cape Molerat. The female has two pairs of pectoral and one pair of inguinal mammae indicating that a female could probably suckle a litter of four young.

## Parasites
The parasites known to be associated with *Georychus capensis* are surprisingly few. They are susceptible to bubonic plague caused by *Yersinia pestis* (Neitz 1965).

The ciliate protozoan *Meiostoma georychi* has been identified from the caecum of *G. capensis* (Sandon 1941). Cysts of the small tapeworm *Echinococcus* sp. were obtained from the internal muscular wall of the abdominal cavity in the liver and loose in the abdominal cavity of a Blesmol specimen collected at Wynberg near Cape Town (De Graaff 1964). The nemathelminth *Trichuris* has also been taken from *Georychus capensis* collected at Wynberg (Ortlepp *pers. comm.*). The mites are represented by the Laelaptidae *(Haemolaelaps capensis, H. cryptomius* and *H. lawrencei)* and the Listrophoridae *(Listrophoroides zumpti)* (Zumpt 1961a). The only other arthropods hitherto known to be associated with *G. capensis* is the tick *Ixodes alluaudi* (Theiler 1962) and the flea *Cryptoctenopsyllus ingens* (Bedford 1932).

## Relations with man
Dreyer (1910) has stated that this genus is the most directly injurious of the molerats, because it often robs the farmer of a good percentage of his vegetable crop. Like *Cryptomys*, it spoils more than it can actually devour.

## Prehistory
Fossils identifiable as *G. capensis* have been collected at Elandsfontein, a few miles west of Hopefield where Saldanha Man *Homo sapiens rhodensiensis* was found by Jolly and Singer in 1953. These late Pleistocene remains are virtually identical to the extant species. Earlier in the Pleistocene was a giant georychid named *Gypsorhychus darti* by Broom (1930), who discovered it in breccias associated with the first ape-man *Australopithecus africanus* found at Taung in the northwestern Cape in 1925. In 1934 Broom described these fossils in greater detail and he postulated a close affinity with *Georychus capensis* based on certain features of the incisors, the structure of the nasals and the arrangement of the structure of the lateral walls of the frontals and maxillae. On the other hand, the arrangement of the molars is unique and unlike that of *Bathyergus*, *Georychus* or *Cryptomys*. Another specimen of *G. darti* was found in 1937 in Hrdlicka's Cave near Taung (Broom 1939). In 1948 Broom changed the name of this specimen to *G. minor*. Finally, a third species, *G. makapani*, was found at Makapansgat in 1946 (Broom 1948b). A similar specimen turned up in 1959 and was described by De Graaff (1965).

## Taxonomy
Roberts (1951) divided *Georychus capensis* into three subspecies *(G. c. capensis, G. c. canescens* and *G. c. yatesi)*,

based mainly on colour variations in specimens from Knysna (Cape) and Belfast (Transvaal). Ellerman *et al.* (1953), regard *G. capensis* as a monomorphic species with no subspecies. The interpretation I adhere to follows that of Ellerman *et al.* (1953) and the polytypic nature of the species as proposed by Roberts is rejected (De Graaff 1975).

Myomorpha

CRICETIDAE: White-tailed Rat, Gerbils, Giant Rat, Pouched Mouse, Climbing Mice, Fat Mice, Large-eared Mouse, Rock Mice, Vlei Rats, Karoo Rats, Whistling Rats

MURIDAE: Typical Rats and Mice

GLIRIDAE: Dormice

# Myomorpha Brandt, 1855

# Muroidea Miller & Gidley, 1918

The myomorphs (comprising rats and mice and allied forms) are characterised by a slit-like infraorbital foramen through which the medial masseter muscle passes and which is also compressed by the development of the zygomatic plate, a flattened area from which part of the lateral masseter muscle arises. The molars always number three in both maxilla and mandible.

The myomorphs are generally subdivided into three superfamilies. The superfamily Muroidea contains four families (Cricetidae, Muridae, Spalacidae and Rhizomyidae) of which only the first two are relevant to southern Africa. In addition to the Muroidea, two other superfamilies are recognised: the Gliroidea (the dormice and their relatives) include the families Gliridae, Platacanthomyidae and Seleviniidae (the latter two being extralimital); the Dipodoidea (jumping mice and jerboas) are also extralimital and include the Zapodidae and Dipodidae as families.

The myomorphs are an old and highly successful group. Romer (1971) stated that the first definitely recognisable genera date from the Oligocene, but they became common only in the Pliocene. In Recent times they have become the most widespread and abundant of rodents – indeed, of all mammals. The myomorphs include approximately two thirds of all Recent rodent species. They outnumber all other rodents in number of genera and species and their geographical range is greater than that of any other rodent group.

As indicated above, two major families are often distinguished, *viz.* the Cricetidae and Muridae. The Cricetidae include all the mouse-like rodents in the Neotropics, Nearctic and Palaeartic since the Oligocene. In contrast, the centre of evolution of the Muridae appears to have been in the eastern hemisphere (i.e. they do not occur in the Neotropics or Nearctic) and they are predominantly native to the Old World tropics.

Kingdon (1974) has discussed the relationships of the cricetids and the murids, but there is no concensus of opinion. The controversy centres on the interpretation of how the murids developed their many-cusped molar teeth. Kingdon points out that Petter (1966a) and Lavocat (1962, 1967) believe that the murids developed from the cricetids. Vandebroek (1961, .1966) interprets them as both having a common ancestor, while Misonne (1969) '… has developed the latter thesis and has suggested that the murids originated in southeast Asia with a secondary expansion and development in Africa', having no connection with the rather rat-like Dendromurinae and Cricetomyinae. According to Kingdon, this invasion of Africa probably took place in the late Miocene or Pliocene.

The Cricetidae consists of 100 genera, with approximately 567 species. Their distribution is nearly worldwide, excluding the Australian and Malayan regions. They include such well-known animals as muskrats, hamsters, voles, lemmings and gerbils. They are primarily terrestrial in habits and have remarkable cyclic fluctuations in numbers. They are prolific breeders and in the wild they have a life expectancy of less than one year. Their populations, therefore, show *r*-selection (high rate of productivity coupled to a low investment of energy in individuals which occur as the progeny or offspring) (French, Stoddart & Bobek 1975). This is associated with a short gestation period and the onset of breeding at an early age. Cricetids are important vectors of several diseases to man. They are destructive and are often in direct competition with man's agricultural ventures. They are important small mammals in terms of their effect on the environment, as well as contributing as a staple food item for many predators (Vaughan 1972).

Most cricetids show a 'standard' mouse-like form with a long tail and a generalised limb structure (Vaughan 1972). The skulls is variable in shape and this variability also applies to the occlusal surfaces of the molar teeth. The structure is based on a pattern of five crests formed by re-entrant enamel folds. The family contains a number of subfamilies of which the Cricetinae (New World rats and mice) and the Microtinae (voles, lemmings) are of importance extralimitally, (although a single representative of the Cricetinae *(Mystromys albicaudatus)* occurs in South Africa), while the Dendromurinae (climbing mice), Gerbillinae (gerbils) and Otomyinae (vlei rats) are the most important families in southern Africa.

The Muridae also include 100 genera with 457 species, ranging from Africa, Europe and Asia to the Malayan region. They are abundant in the tropics and temperate regions in practically all habitats. This family includes the well-known Old World rats and mice. Murids occupy a diversity of habitats and are often commensal with man, where they are economically important, being instrumental in the spread of disease and damage to crops and stored foods. Like the cricetids, they also breed prolifically.

They range in size from that of a small mouse to that of

a large rat. As is the case in the cricetids, the skull is variable in structure while the molars usually have crowns with cusps, tending towards a simplification of the crown pattern. The feet retain all the digits but the pollex is rudimentary.

The Spalacidae and Rhizomyidae mentioned earlier on, are extralimital to this work. The fossorial spalacids form a group of three species of one genus *(Spalax)* occurring in the eastern Mediterranean Region and southeastern Europe. The rhizomyids or bamboo rats occur in southeastern Asia and in tropical parts of eastern Africa. The family includes three genera with 18 species.

The taxonomic interpretation of the Cricetidae and Muridae in this book follows Misonne (1974). The Muri-dae include only the Murinae and the Cricetidae comprise the Cricetinae, Gerbillinae, Cricetomyinae, Dendromuri-nae, Petromyscinae and Otomyinae as southern African forms, while the Lophiomyinae (maned rats) and Micro-tinae (voles and lemmings) as well as the Nesomyinae (Malagasy rats) occur elsewhere.

According to Misonne *(op. cit.),* only the shape of the molars separates the Muridae from the Cricetidae. In the following key taken from Misonne, the Cricetomyinae (giant rats) and Petromyscinae (rock mice) have been keyed with the Dendromurinae (climbing mice). These subfamilies are very similar and distinction between them is based on small dental differences.

---

**Key to the southern African subfamilies of the Cricetidae and Muridae**

(Modified after Misonne (1974))

1 $M^3$ the largest tooth; molars laminated...Otomyinae
    Page 143

  $M^1$ the largest tooth; molars not laminated..........2
2 $M^1$ with three cusps in first row ..............Murinae
    Page 163

  $M^1$ with two cusps in first row* ........................3
3 Pattern of upper molar as in figure......... Gerbillinae
    Page 90

  Pattern of upper molar not as above....................4

4 Soles of hindfeet partly hairy ................. Cricetinae
    Page 86

  Soles of hindfeet naked** ............. Dendromurinae
    Page 119
    Cricetomyinae
    Page 111
    Petromyscinae
    Page 139

*Except *Cricetomys*, indentifiable by its large size
**Except *Malacothrix*

---

FAMILY      # Cricetidae Rochebrune, 1883

SUBFAMILY      # Cricetinae Murray, 1866

The family Cricetidae can be divided into structurally or geographically discontinuous groups (Anderson & Jones 1967) which have certain characteristics in common. The Cricetinae are an important assemblage of these rodents and is often subdivided into two major tribes, the Hespe-romyini (New World rats and mice) and the Cricetini (true hamsters). Hamsters are mouse-like with heavy-set bodies, short tails and cheekpouches (not in *Mystromys*) and occupy the Palaearctic and Afrotropical regions. In southern Africa, this subfamily is represented by only one genus south of the equator, hitherto known only from the RSA and is known as *Mystromys albicaudatus* (the White-tailed Rat).

GENUS    *Mystromys* Wagner, 1841

This attractive rodent is not frequently encountered in the wild, probably due to its secretive nature and noctur-nal mode of life.

# *Mystromys albicaudatus* (A. Smith, 1834)

## White-tailed Rat

## Witstertrot

The name is compounded from the Greek *mystron* = a spoon and *mys* = mouse. The specific name is from the Latin *albus* = white and *caudatus* = having a tail.

The White-tailed Rat *Mystromys albicaudatus* is an endemic South African rodent which, according to Dean (1978) may be in need of stricter conservation measures. It is the only African member of the subfamily Cricetinae. Present-day populations appear to be relatively local and dispersed. Dean (1978) has argued in favour of the White-tailed Rat possibly being *K*-selected, since it apparently exists at a low density throughout its range, at or near maximum population size *(K)*. Dean points out that, if this is the case, the species may be considered to be threatened.

Ellerman (1941) pointed out the uncertain relationships of this genus '. . . which seems not only isolated from the Palaearctic and Neotropical genera, but to have no marked generic characters, making it exceptionally difficult to place in the key'. Sir Andrew Smith met with this species in the vicinity of Grahamstown in 1834 as well as to the north of the Orange River (Sclater 1901). The white hands and feet prompted Wagner in 1841 to describe a second species from South Africa which he called *Mystromys albipes*. This was later shown to be a synonym of *M. albicaudatus*. In 1887, Noack described a third species *M. longicaudata* from Tanzania, but this has turned out to be a specimen of the Multimammate Mouse *Praomys (Mastomys) natalensis*.

## Outline of synonymy

1834 *Otomys albicaudatus* A. Smith, *S. Afr. Quart. Journ.* 2:148. Albany district, eastern CP. (Genotype *O. albicaudatus,* not *Otomys* F. Cuvier (1825).)

1841 *Mystromys albipes* Wagner, *Arch. f. Naturgesch.* 7 (1):133. South Africa (= Albany district). (Genotype *Mystromys albipes* Wagner = *Otomys albicaudatus* A. Smith.)

1842 *Mystromys lanuginosus* Lichtenstein, *Verzeichn. d. Doubletten, Mus. Berlin:* 10.

1843 *Malacothrix albicaudatus* Wagner. *In* Schreber's *Säugeth.* Suppl. 3:500.

1887 *Mystromys longicaudatus* Noack, *Zool. Jahrb. Jena* 2:246. Tanzania. (= *Praomys (Mastomys) natalensis.)*

1905 *M. a. fumosus* Thomas & Schwann, *Proc. zool. Soc. Lond.:* 1:137. Wakkerstroom, Transvaal.

1939 *M. a. albicaudatus* (A. Smith). *In* Allen, *Bull. Mus. comp. Zool. Harv.* 83:314.

1953 *M. albicaudatus* (A. Smith). *In* Ellerman *et al., Southern African mammals.*

1974 *M. albicaudatus* (A. Smith). Misonne, *in* Meester & Setzer (eds), *The mammals of Africa: an identification manual.*

### TABLE 15
#### Measurements of male and female *Mystromys albicaudatus*

| | Parameter | Value (mm) | N | Range (mm) | CV (%) |
|---|---|---|---|---|---|
| Males | HB | 163 | 12 | 139–184 | 8,9 |
| | T | 58 | 14 | 50–82 | 12,0 |
| | HF | 27 | 14 | 24–30 | 4,7 |
| | E | 26 | 14 | 20–28 | 7,3 |
| | Mass: | 95,7 g | 2 | 78–111 g* | — |
| Females | HB | 144 | 16 | 105–147 | 9,7 |
| | T | 63 | 16 | 53–97 | 17,0 |
| | HF | 26 | 16 | 24–28 | 4,0 |
| | E | 24 | 16 | 20–27 | 7,0 |
| | Mass: | 78 g | 2 | 75–81 g* | — |

*Rautenbach (1978b)

## Identification

The animal is thickset, mouse-like with a soft and rather woolly fur. The ears are relatively large and rounded. The body colour is buffy-grey, interspersed with black-tipped hairs. The flanks are grey, merging into white on the belly. Individual hairs are almost black at the base. The sides of the face, fore- and hindlimbs are of lighter hue than the back, while the hands, feet, tail and an area around the nose and chin are a dull white. The ears are mainly dark brown. The tail is less than half the total head-body length, unscaled, and thinly haired (plate 7). The animal has four fingers and five toes, each digit with a sharp-tipped claw. The soles of the feet are partly haired, only the digital area being naked. The female has two pairs of inguinal mammae.

## Skull and dentition

In the skull the interorbital region is constricted far back so that the braincase appears shortened. The jugal arches tend to spread widely. The zygomatic plate has a straight anterior edge. The palate extends slightly behind the last molars and the palatine foramina reach the level of the first molars. The bullae are moderately large and the infraorbital foramen has a long oval shape (fig. 32).

The cheekteeth have opposed cusps and slanting folds, resulting in a zigzag enamel pattern. The cusps are aligned in two parallel rows and the folds are without lateral cusps as in true murids. The third molar is reduced in size. $M^1$ has two inner and two outer folds. $M^2$ has two outer folds and a single inner fold (fig. 33). The incisors are smooth, pale yellow.

The mandibular condyle bends inwards, the angular portion is narrow, extending well backwards, while the coronoid is long and slender. The $M_1$ has two outer folds and three inner folds while $M_2$ shows two inner and two outer folds. The $M_3$ is small.

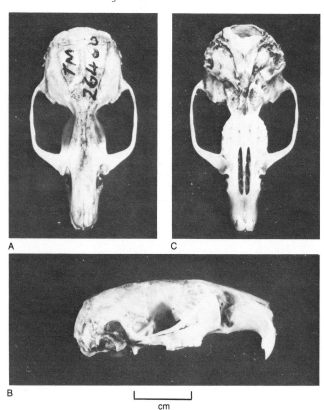

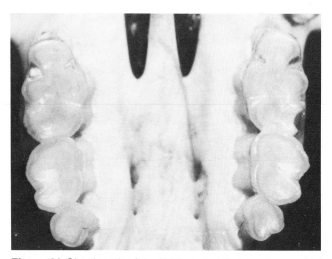

**Figure 32** Dorsal (A), lateral (B) and ventral (C) views of the skull of the White-tailed Rat *Mystromys albicaudatus*. Interorbital constriction situated far back so that braincase appears shortened. Jugal arches spread widely. Palate extends slightly beyond $M^3$ and anterior palatine foramina reach level of $M^1$.

**Figure 33** Cheekteeth of the White-tailed Rat *Mystromys albicaudatus*. The opposite cusps and slanting folds result in a zigzag enamel pattern. No lateral cusps on teeth. $M^1$ has two inner and two outer folds. $M^2$ has a single inner and two outer folds. $M^3$ reduced.

## Distribution

The White-tailed Rat occurs from the lower lying regions of the Eastern Province and ranges westwards to the western Cape, where it is regarded as a relict population (Davis 1974). It is also found in the Transvaal highveld, northwestern Natal, the OFS, Lesotho and East Griqualand (map 17). It is consequently encountered in the Southern Savanna Grassland and the South West Cape biotic zones.

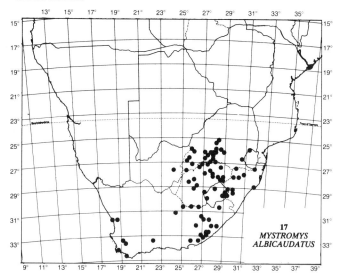

17
*MYSTROMYS ALBICAUDATUS*

## Habitat

Dean (1978) has pointed out that the White-tailed Rat is confined mainly to highveld and montane grasslands to the south of 25°S, where it occurs in several of Acocks' (1953) Pure Grassveld types as well as in Bankenveld, Southern Tall Grassveld and Natal Sour Sandveld. In the western Cape, it inhabits Succulent Karoo, Cape Macchia and Rhenosterbosveld. According to Dean, it was also recently found in an owl pellet obtained in the Reivilo district of the Cape Province extending its occurrence into the *Tarchonanthus* veld of the Vryburg Shrub Bushveld.

## Diet

No information is available about food under natural conditions. Roberts (1951) states that the White-tailed Rat feeds on seeds and vegetable matter. The total food requirements for small laboratory rodents (including *Mystromys albicaudatus*) has been discussed by Joubert (1967).

## Habits

White-tailed rats live in holes in the ground or shelter in cracks in the soil. They have also been observed to use the burrows of the Suricate *Suricata* (Walker 1975). Jameson also found them in suricate abodes and, since Grant has recorded that cats would not touch them, '... presumed that they carried some protective secretion which repelled carnivora' (Roberts 1951).

They are nocturnal and therefore often turn up in owl pellets. According to Sclater (1901) they are especially active during wet whether. They soon become docile and breed freely in captivity.

## Predators

Very little is known about this aspect of the animal's ecology. White-tailed rats occur in pellets of the Barn Owl *Tyto alba* (Vernon 1972; Davis 1973), although in consistently low proportions (Dean 1978), while they are also taken by the Marsh Owl *Asio capensis,* the Spotted Eagle Owl *Bubo africanus* (Dean 1977) and the Grass Owl *Tyto capensis* (Davis 1973).

## Reproduction

Roberts (1951) states that they breed at all times of the year, with four to five pups forming a litter. The mean litter size observed in the laboratory is 2,9 and the average birth weight of a pup is 6,5 g (Meester & Hallett 1970).

The early postnatal development of *Mystromys albicaudatus,* under laboratory conditions, has been described by Hallett & Meester (1971). After a gestation period of 37 days (Roberts 1951) the young become attached to the mother's mammae soon after birth remaining thus for about 21 days becoming detached only occasionally. Detachment is permanent after the 32nd to 35th day. The pups are pink and naked at birth and hair is developed on dorsal parts after about five days. They respond to sound after about two weeks. They can crawl about clumsily at birth, improve after five days, and gradually learn to walk and run. The incisors start erupting on the third to fifth days. Weaning is completed at about the 38th day. The eyes open between 16 and 20 days. Parental protection of the young takes the form of aggressive behaviour towards intruders, and sheltering the young by huddling.

## Parasites

It has been shown by laboratory tests that the schizomycete *Pseudomonas pseudomallei* causes melioidosis in *Mystromys albicaudatus* (Neitz 1965).

Although this organism does not occur in South Africa, it has been isolated from a patient who had been on active service in India and Malaya. In the event of unsuspected introduction it could maintain itself in *M. albicaudatus*. It is also susceptible to plague infections *(Yersinia pestis)* which occur sporadically in free living populations of this species. Neitz (1965) has also recorded high susceptibility of the White-tailed Rat to the moniliacid fungus *Histoplasma capsulatum* which often develops in man after being exposed to caves, so *M. albicaudatus* could be a useful laboratory indicator of the presence of spores causing 'cave disease'.

There seem to be no records of any endoparasitic worms associated with *Mystromys albicaudatus*. The associated mites include the laelaptid species *Haemolaelaps davisi, H. labuschagnei* and *H. mystromys* (Zumpt 1961a).

The fleas found on *M. albicaudatus* include Pulicidae *(Ctenocephalides felis, Xenopsylla eridos, X. philoxera* and *X. piriei),* Hystrichopsyllidae *(Ctenophthalmus calceatus, Dinopsyllus ellobius, Listropsylla agrippinae* and *L. fouriei)* and Chimaeropsyllidae *(Chiastopsylla numae, C. roseinnesi, C. rossi* and *C. godfreyi)* (Zumpt 1966).

## Relations with man

The White-tailed Rat has been bred in the laboratory of the Medical Ecology Centre of the South African Institute for Medical Research since 1941 for use in medical research (Hallett & Meester 1971) on poliomyelitis, benign histoplasmosis, dental caries and diseases like diabetes (Davis 1963; Hall, Persing, White & Ricketts 1967; Packer, Kraner, Rose, Stuhlman & Nelson 1970).

## Prehistory

The genus *Mystromys* has been identified as a fossil from various localities in South Africa (De Graaff 1961). Lavocat (1957) and De Graaff (1960a) reported the occurrence of *M. darti,* a small form, at Makapansgat, while Broom (1937, 1948a) recorded *M. hausleitneri (= M. hauslichtneri)* in the Schurveberg deposits near Pretoria. This species is also known from Sterkfontein (Lavocat 1957; De Graaff 1960a) and Kromdraai (De Graaff 1960a). Another fossil species, *M. antiquus,* was found at Taung, as recorded by Broom (1948a) and De Graaff (1960a).

*Mystromys darti* is evidently a rare species and has not been isolated from any deposit other than Makapansgat. In contrast, *M. hausleitneri* apparently had a wide geographical distribution. *Mystromys hausleitneri barlowi* Broom, reported from Sterkfontein, is a subspecies of uncertain status and its validity has been questioned by Lavocat (1956a).

As the Pleistocene merged into the Holocene, the cricetids (with the fossils of *Mystromys hausleitneri* as dominants) were gradually displaced in number and diversity by murids in the South African fossiliferous localities (De Graaff 1960). The predominance of fossil *Mystromys* in breccias from Kromdraai B (0–6) has been commented upon by Davis (1959): 'The subfamily Cricetinae, now represented in southern Africa by one fairly widely distributed but rather uncommon species, *M. albicaudatus,* greatly outnumbers the subfamily Murinae, which are today the dominant group, with very many species.' There has been a change in status of the subfamily between the early Pleistocene and the present day.

## Taxonomy

Ellerman (1941) was entirely at a loss to suggest relationships of this genus. Roberts (1951), Ellerman *et al.* (1953) and Meester *et al.* (1964) place them under the Cricetinae. Lavocat (1959, 1964) regards *Mystromys* as a living cricetodont (a fossil cricetid tribe known from the lower Oligocene to lower Pliocene in Europe). If Lavocat is correct, then the Cricetinae are not represented in Africa (Nel 1969).

Roberts (1951) and earlier authors recognise the sub-species *fumosus*, while Ellerman *et al*. (1953) include it in the nominate *albicaudatus*. Meester *et al*. (1964) and Misonne (1974) also interpret this species as monotypic.

# SUBFAMILY Gerbillinae Alston, 1876

The Gerbillinae as a subfamily are characterised by their morphology and habitat (Petter 1975). They all have tawny-coloured dorsal fur with white bellies. They are closely related to the cricetids, of which they are an offshoot. Their pelage is soft and the forelimbs are adapted to burrowing. They are all terrestrial, nocturnal, gregarious, vegetarian and are important carriers of fleas which spread plague.

Gerbils have large eyes and the tympanic bullae are always pronounced. The molars each have two or three more or less transverse elongate-oval sections of enamel surrounding a cement centre, not rows of cusps as in the Murinae, nor transverse layers of enamel as in the Otomyinae. The upper incisors are slender, usually grooved. The zygoma and the mandible are lightly built. The infraorbital foramen is wide and bounded by a broad zygomatic plate.

A combination of morphological, physiological and ecological adaptations of gerbils are all indicative of a steppe or desert habitat. Consequently, they are widespread in Africa and southern Asia. The hindlimbs are well developed and there is a tendency towards saltatorial locomotion.

**Key to the southern African genera of Gerbillinae**

(Modified after Roberts (1951) and Rautenbach (1978b))

1 Tail much shorter than HB length; enlarged bullae showing on dorsal surface of skull, divided into two sections, the anterior section being twice the size of the posterior section; medium-sized animals............
*Desmodillus*
(Short-tailed Gerbil)
Page 108
Not combining these characters........................ 2
2 Tail much longer than HB length; tassel of long hairs at the tip; soles of hindfeet haired; bullae large with anterior section more globular than in *Desmodillus* with the posterior section not conspicuous; cheekteeth sublaminated; small-sized animals.........
*Gerbillurus*
(Hairy-footed Gerbils)
Page 91

3 Tail equal to or longer than HB length; soles of hindfeet naked; bullae without a posterior development adjoining the occiput; molars laminated; large-sized animals .................................. *Tatera*
(Gerbils)
Page 99

(See fig. 34.)

**Figure 34** Lateral view of the skulls of the Short-tailed Gerbil *Desmodillus* (A), the Brush-tailed Hairy-footed Gerbil *Gerbillurus* (B) and the Bushveld Gerbil *Tatera* (C), showing structural differences in the auditory bulla. *Desmodillus auricularis* has a well-developed anterior (a.s.) as well as posterior section (p.s.). *Gerbillurus vallinus* has a more globular a.s. with a p.s. not so conspicuous, while *Tatera leucogaster* has virtually no development of the p.s.

# GENUS *Gerbillurus* Shortridge, 1942

The taxa included under *Gerbillurus* have the following characteristics in common: the soles of the hindfeet are partly naked to hairy; the molar cusps are less alternate in position; the alveolar pattern of the $M^1$ has three or four sockets and the $M^2$ mostly three sockets, while the $M_1$ always has four sockets (Davis 1975a).

*Gerbillurus* was proposed as a new subgenus by Shortridge (1942) with *Gerbillus (Gerbillurus) vallinus* as genotype. This specimen, collected at Berseba in Great Namaqualand, in 1923, is housed in the Kaffrarian Museum, King William's Town. *Gerbillurus* differs from *Gerbillus* in the more inflated bullae and the development of a small additional bulla beside the occiput (Roberts 1951). Its ears are short and a white patch is present behind the ear. It has a tufted tail. The soles of the hindfoot tend to be hairy '... with only a narrow central line naked' (Roberts 1951). The female has one pair of pectoral and two pairs of inguinal mammae. Furthermore, there are differences in the teeth as well. In *Gerbillus paeba* A. Smith, the last molar has two clear cusps, but in *Gerbillurus* the second cusp is only vaguely indicated. An extra anterior cusp is present on the $M^2$ (Lundholm 1955a).

Following Davis (1975a), five species of *Gerbillurus* are provisionally recognised in southern Africa: two species (*paeba* and a *Gerbillurus* sp.) can be placed in the *paeba* group and three species (*tytonis, vallinus* and *setzeri*) in the *vallinus* group.

---

**Key to the southern African species of *Gerbillurus***
(Modified after Davis (1975a))

1 Tail approximately 20% longer than HB length, the tip slightly or moderately tufted; bullae normally inflated, not extending behind the occiput and may measure up to 9 mm in length ......................... 2
  Tail approximately 40% longer than HB length (except in *setzeri*, which is distinguished by its size), the tip tassled; bullae more inflated and extending behind the level of the occiput and measures over 9 mm in length ................................................. 3

2 Pelage varying shades of brown .............. *G. paeba*
  (Hairy-footed Gerbil)
  Page 91
  Pelage grey: Cape Flats only ......... *G.* sp. aff. *paeba*
  Page 94

3 Bullae less inflated *ca* 9–10 mm; posterior palatal foramina very short; Sossosvlei and Gobabeb only ....
  *G. tytonis*
  (Dune Hairy-footed Gerbil)
  Page 94
  Bullae much inflated *ca* 10–12 mm; posterior palatal foramina long, more or less length of molar tooth row ...................................................................... 4

4 Tail longer, *ca* 40% longer than HB length .............
  *G. vallinus*
  (Brush-tailed Hairy-footed Gerbil)
  Page 95
  Tail shorter, *ca* 20% longer than HB length............
  *G. setzeri*
  (Setzer's Hairy-footed Gerbil)
  Page 97

---

# *Gerbillurus paeba* (A. Smith, 1836)

**Hairy-footed Gerbil**

**South African Pygmy Gerbil**

**Lesser Gerbil**

**Haarpoot Nagmuis**

**Klein Nagmuis**

The generic name is derived from the French *gerbille* = a small rodent. The suffix *-urus* denotes 'as belonging to'. The specific name is derived from the Greek *paidos* = child, implying small stature reminiscent of the small size of this gerbil.

This is a small species of gerbil frequenting the drier areas of southern Africa. It is characterised by its rather slender form, a slightly tufted, longish tail, well-developed ears, grooved upper incisors and the fact that the posterior portion of the bulla is not conspicuously developed, as is the case in *Desmodillus auricularis*. Furthermore, the last upper molar may be divided into two clear sections, but this is apparently not invariably the case. The pelage is various shades of brown. The claws are fairly long and adapted for burrowing.

The type specimen of this species was originally obtained by Sir Andrew Smith (1836) in country to the north of Latakoo = Litakun (= Vryburg as nominated by Roberts in 1951) in the northern Cape.

## Outline of synonymy

1836 *Gerbillus paeba* A. Smith, *App. Rpt. Exped. Explor. S. Afr.*: 43. Vicinity of Vryburg, CP, as nominated by Roberts (1951).

1842 *Gerbillus tenuis* A. Smith, *Illustr. Zool. S. Afr., Mamm.* pl. 36. Renaming of *G. paeba*. (A synonym of *G. paeba* according to Thomas, *Ann. Mag. nat. Hist.* (9) 2:64.) North of Latakoo.

1842 *Meriones (Rhombomys) caffer* Wagner, *Arch. f. Naturgesch.* 8 (1):18. South Africa.

1889 *Gerbillus tenuis* var. *schinzi* Noack. *Zool. Jahrb. Jena* 4:134. Kalahari Desert, SWA/Namibia.

1918 *Gerbillus calidus* Thomas, *Ann. Mag. nat. Hist.* (9) 2:63. Molopo, west of Morakwen, Botswana.

1918 *Gerbillus paeba broomi* Thomas, *Ann. Mag. nat. Hist.* (9) 2:64. Port Nolloth, Little Namaqualand, CP.

1925 *Gerbillus swalius* Thomas & Hinton, *Proc. zool. Soc. Lond.*: 235. Karibib, SWA/Namibia.

1927 *Gerbillus paeba leucanthus* Thomas, *Proc. zool. Soc. Lond.*: 382. Ondongwa, SWA/Namibia.

1932 *Gerbillus calidus kalaharicus* Roberts, *Ann. Transv. Mus.* 15:10. Gomodimo Pan, Botswana.

1939 *Gerbillus paeba paeba* A. Smith. *In* Allen, *Bull. Mus. comp. Zool. Harv.* 83:325.

1946 *Gerbillus paeba mulleri* Roberts, *Ann. Transv. Mus.* 20:317. Eendekuil, CP.

1951 *Gerbillus paeba swakopensis* Roberts, *The mammals of South Africa*: 404. Swakopmund, SWA/Namibia.

1953 *Gerbillus gerbillus paeba* A. Smith. *In* Ellerman *et al.*, *Southern African mammals*.

1964 *Gerbillus paeba* (A. Smith). *In* Meester *et al.*, *An interim classification of southern African mammals*.

1975 *Gerbillurus paeba* (A. Smith). Davis, *in* Meester & Setzer (eds), *The mammals of Africa: an identification manual*.

### TABLE 16
Measurements of male and female *Gerbillurus paeba*

| | Parameter | Value (mm) | N | Range (mm) | CV (%) |
|---|---|---|---|---|---|
| Males | HB | 94 | 31 | 88–99 | 3,7 |
| | T | 109 | 31 | 97–127 | 6,7 |
| | HF | 26 | 31 | 23–30 | 6,2 |
| | E | 15 | 31 | 10–19 | 11,8 |
| | Mass: | 27,7 g | 11 | 24–34 g | 8,9 |
| Females | HB | 93 | 29 | 80–101 | 4,2 |
| | T | 110 | 29 | 92–125 | 6,9 |
| | HF | 27 | 29 | 21–30 | 6,1 |
| | E | 15 | 29 | 13–17 | 7,1 |
| | Mass: | 27,0 g | 8 | 21–35 g | 14,2 |

## Identification

The Hairy-footed Gerbil is a small species with a pale reddish-orange dorsal colour. The back is pencilled with liver-brown hairs, while the lower surface, including the chin and inner surface of the limbs, is pure white (plate 4). The tail is the same colour as the back, being of slightly lighter hue and towards the tip many hairs have an amber-brown tint. The ears are a pale yellowish brown externally. Smith (1836) described the eyes as a deep reddish-brown. The distal part of the hindfoot is always haired and the claws on the digits are fairly long for digging. Geographically, there is much colour variation. It is a common species in southern Africa.

## Skull and dentition

In the bulla only the tympanic portion is considerably developed. The skull has a broad braincase with a rather long rostrum. The zygomatic arches are not wide. A strongly developed zygomatic plate is present. The infraorbital foramen is narrow. The palatine foramina are long. The occipital region is not strongly developed (fig. 35).

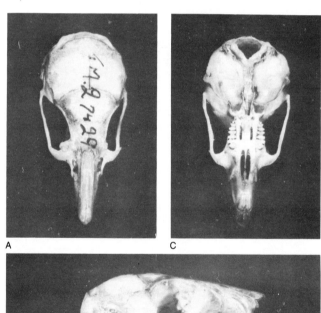

**Figure 35** Dorsal (A), lateral (B) and ventral (C) views of the skull of the Hairy-footed Gerbil *Gerbillurus paeba*. The bullae are normally inflated and do not extend beyond level of occiput. Rostrum elongated and zygomatic plate well developed. Anterior palatine foramina much bigger than posterior pair.

The upper M¹ has three laminae, the front one with a single cusp, and narrower than the rest. All the other laminae bear two cusps when unworn. The M² has two laminae. The M³ is a small tooth, showing a small posterior lamina behind the main one. The laminae of all the teeth are not usually pressed closely together (fig. 36). The grooved upper incisors are orange and the last molar is usually divided into two sections. The lower incisors are plain.

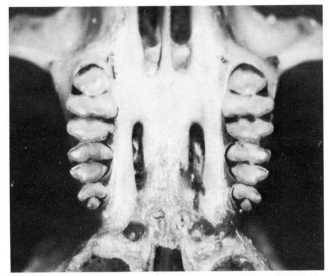

**Figure 36** Cheekteeth of the Hairy-footed Gerbil *Gerbillurus paeba*. Laminae of all three molars not closely pressed together. M³ may be divided into two clear sections, but this is not invariably the case.

## Distribution

The Hairy-footed Gerbil is widely distributed in the South West Arid and encroaches marginally into the Southern Savanna and the South West Cape biotic zones (Davis 1974). This species is consequently confined mainly to arid areas from southwestern Angola to the Cape. Isolated populations occur in the northern Namib, northern Transvaal and in sand dunes in the vicinity of Port Elizabeth. The northern Transvaal populations from Soutpansberg district occur in relict patches of Kalahari sand (map 18).

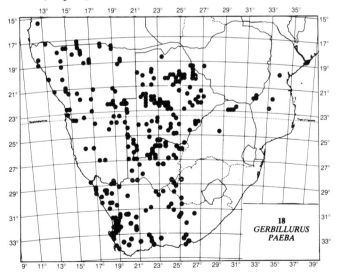

## Habitat

These gerbils prefer desert and subdesert conditions. In Botswana they occur in sandy ground or sandy alluvium with a grass, scrub or light woodland cover (Smithers 1971).

## Diet

Seeds of grasses, as well as those of bushes and trees (especially *Acacia* spp.) seem to be their main food items (Smithers 1971).

## Habits

According to Roberts (1951), hairy-footed gerbils live in burrows which are neither elaborate nor extensive. The many entrances to these burrows are usually blocked with loose soil. Those in use are characterised by ramps of loose soil at the entrance, similar to that made by *Tatera leucogaster,* but the diameters of the burrows are smaller. Like other gerbils, they are terrestrial and nocturnal and seem to be independent of water (Smithers 1971).

## Predators

No specific information exists, but they probably fall prey to smaller carnivores, certain birds of prey and snakes.

## Reproduction

The number of mammae varies from one to two pectoral pairs and two inguinal pairs. Indications are that young may be born at any time throughout the year (Smithers 1971), with an average of three to four pups per litter. In the laboratory, they only breed when given extra light hours (Keogh & Isaäcson 1978).

## Parasites

Two laelaptid mites are known: *Haemolaelaps oliffi* and *Androlaelaps marshalli* (Zumpt 1961a). The fleas are represented by four genera and 11 species. These include *Chiastopsylla mulleri, C. numae* form *rossi, Dinopsyllus ellobius, Echidnophaga gallinacaea, Xenopsylla davisi, X. h. hirsuta, X. lobengulai, X. philoxera, X. phyllomae, X. piriei* and *X. trifaria* (De Meillon, Davis & Hardy 1961). As far as ticks are concerned, immature specimens of *Rhipicephalus capensis* have been reported from *Gerbillurus paeba* (Theiler 1962).

## Relations with man

The relations of this species with man is not clear. It may act as a reservoir for fleas (e.g. *Xenopsylla* spp.) which transmit plague to man. The tick *Rhipicephalus capensis* has been proved to transmit the babesid protozoan *Theileria parva,* causing East Coast fever in domestic cattle.

## Prehistory

This species has not been recorded in any of the fossiliferous localities known in southern Africa.

## Taxonomy

*Gerbillurus paeba* was regarded as a synonym of the North African *Gerbillus gerbillus* by Ellerman *et al.* (1953). Lundholm (1955a), supported by Herold & Niethammer (1963), disagreed. Davis (1975a) is not convinced that the hairy-footed gerbils should be kept separate at generic level from the North African forms. This matter will not be elaborated upon in this text. Davis (1975a) is followed in listing the subspecies below:

*G. p. paeba:* Occurrence as described under 'Distribution' above. Synonyms include *calidus, broomi, swalius, oralis, leucanthus, kalaharicus, mulleri* and *swakopensis.* These represent described forms that appear to intergrade geographically, not forming isolated populations.

*G. p. coombsi:* Known only from the type locality in the Soutpansberg, northern Transvaal. In contrast to *G. p. paeba* it is consistently smaller in size.

*G. p. exilis:* Sand dunes along the coast from Sundays River mouth to Port Elizabeth. A remarkably pallid race.

*G. p. infernus:* Rocky Point, Skeleton Coast, SWA/Namibia.

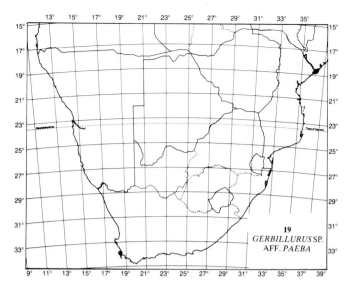

**19**
*GERBILLURUS* SP.
AFF. *PAEBA*

# *Gerbillurus* sp. aff. *paeba*

Davis (1975a) has described this species as being a '*Tatera*', grey and '. . . a trifle larger than any *G. p. paeba* so far examined'. His point of view is that it is a distinct species, though a case might be made out for its being a 'very distinct subspecies of *G. paeba*'. Matthey (1958, 1959) has investigated the karyogram of this form and has found it to be no different from a specimen of *G. paeba* from the Kalahari Gemsbok National Park in the northwestern Cape.

## Identification

Like *G. paeba,* the bullae are normally inflated, not extending behind the occiput, likewise the tail is about 20% longer than the head-body length with the tip slightly or moderately tufted. The pelage is a distinctive grey.

## Distribution

A few kilometers separate this form from the typical 'brown' *paeba* on the Cape Flats and examples are hitherto known only from the D.F. Malan Airport near Cape Town (map 19).

# *Gerbillurus tytonis* (Bauer & Niethammer, 1959)

**Dune Hairy-footed Gerbil**

**Duine Haarpoot-nagmuis**

The origin of the word *Gerbillurus* has been given under *Gerbillurus paeba*. The specific name refers to *Tyto*, the genus of the Barn Owl in whose pellets this species was first discovered.

This species was first identified and described from skulls recovered from owl pellets of the Barn Owl *Tyto alba* by Bauer & Niethammer (1960). Subsequently, live specimens have been procured and there are 13 study skins and skulls in the Transvaal Museum, Pretoria. The measurements given below were taken from the labels attached to the specimens and obviously a much bigger series is required to make the measurements more meaningful. Apart from some indications relating to the phenotype of the species (as described by Schlitter 1973) and a very limited knowledge of its distribution in SWA/Namibia, no other information on this species is available at present. It is the smallest of the *vallinus* group.

## Outline of synonymy

1959 *Gerbillus vallinus tytonis* Bauer & Niethammer, *Bonn. Zool. Beitr.* 10:236–261. Sossusvlei, SWA/Namibia.

1973 *Gerbillus (Gerbillurus) tytonis* Bauer & Niethammer. *In* Schlitter, *Bull. South. Calif. Acad. Sci.* 72(1):13–18.

1975 *Gerbillurus tytonis* (Bauer & Niethammer). Davis, *in* Meester & Setzer (eds), *The mammals of Africa: an identification manual.*

**TABLE 17**
Measurements of male and female *Gerbillurus tytonis*

|  | Parameter | Value (mm) | N | Range (mm) | CV (%) |
|---|---|---|---|---|---|
| **Males** | HB | 99 | 5 | 90–108 | 5,8 |
|  | T | 121 | 6 | 118–126 | 2,7 |
|  | HF | 31 | 7 | 27–33 | 6,3 |
|  | E | 14 | 7 | 12–15 | 9,3 |
|  | Mass: | 27,6 g | 5 | 24–30 g | 7,4 |
| **Females** | HB | 102 | 6 | 90–111 | 6,7 |
|  | T | 127 | 5 | 122–135 | 3,6 |
|  | HF | 31 | 8 | 30–33 | 5,3 |
|  | E | 14 | 7 | 13–16 | 7,2 |
|  | Mass: | 30,6 g | 5 | 29–33 g | 5,3 |

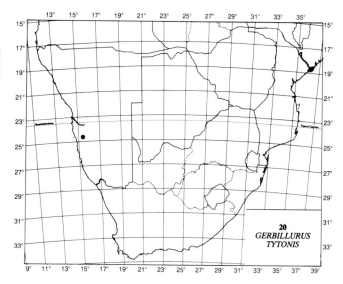

20
GERBILLURUS
TYTONIS

## Identification

The description of the animal (plate 4) can best be quoted from Schlitter (1973) who described the skins of this species based on adult specimens taken on 28 March 1966 at the type locality at Sossusvlei, SWA/Namibia:

> Upper parts near Hazel; all hairs plumbeous at base. Entire underparts, small supraorbital and well-defined postauricular spots, and dorsal surfaces of hands and feet, white; all hairs uniformly white to base. Sharp line of demarcation present between dorsal and ventral colour of pelage. Tail relatively long for subgenus and bicolored, dorsal color same as color of back, ventral color same as dorsal but with admixture of white hairs; tail with variable penicillate tip of greyish hairs. White circumorbital ring absent. Ears essentially bare and cinnamon-buff in color; narrow fringe of black pigmentation present on lateral margin of pinnae.

## Skull and dentition

In contrast to *Gerbillurus vallinus,* this species has conspicuously short posterior palatal foramina. Its bullae are intermediate in size between those of *Gerbillurus vallinus* and *Gerbillurus paeba* (with which it occurs) (Davis 1975a) – i.e. its bullae are smaller than those of *G. vallinus.*

## Distribution

Known from Sossusvlei and Gobabeb in the southern Namib in SWA/Namibia, as well as from the farm Canaan adjoining the prohibited diamond area of SWA/Namibia (map 20).

## Habitat

This species is apparently found only on the shifting red sands south of the Kuiseb River which flows from east to west through the Namib Desert.

## Diet, habits, predators, reproduction, parasites, relations with man and prehistory

No information available.

## Taxonomy

Although originally described as a subspecies of *Gerbillurus vallinus* by Bauer & Niethammer in 1959, it was raised to full species rank by Davis (1975a).

# *Gerbillurus vallinus* (Thomas, 1918)

**Brush-tailed Hairy-footed Gerbil**

**Brush-tailed Gerbil**

**Borselstert Haarpoot Nagmuis**

The origin of the word *Gerbillurus* has been explained under *Gerbillurus paeba.* The Latin word is *vallis* = valley. The species name, therefore, infers a small rodent usually frequenting valleys.

This gerbil occurs from Bushmanland through Great Namaqualand to the Namib Desert. It is a rare species in South Africa and has been listed accordingly in the *South African Red Data Book* (Meester 1976). The type specimen of *G. vallinus* was presented alive by Maj. H.A.P. Littledale to the Zoological Society of London, who transferred it on death to the National Collection (Thomas 1918). It is a well-marked species, readily distinguished by its enlarged bullae which tend to approach those of *Desmodillus auricularis* in size.

## Outline of synonymy

1918 *Gerbillus vallinus* Thomas, *Ann. Mag. nat. Hist.* (9) 2:148. Kenhardt, CP.

1942 *G. (Gerbillurus) vallinus* Shortridge, *Ann. S. Afr. Mus.* 36: 27–100. Berseba, SWA/Namibia.

1951 *Gerbillurus vallinus* (Thomas). *In* Roberts, *The mammals of South Africa.*

1953 *Gerbillus (Gerbillurus) vallinus* Thomas. *In* Ellerman et al., *Southern African mammals.*

1975 *Gerbillurus vallinus* (Thomas). Davis, *in* Meester & Setzer (eds), *The mammals of Africa: an identification manual.*

### TABLE 18
### Measurements of male and female *Gerbillurus vallinus*

| | Parameter | Value (mm) | N | Range (mm) | CV (%) |
|---|---|---|---|---|---|
| Males | HB | 81 | 8 | 58–95 | 14,5 |
| | T | 120 | 8 | 80–145 | 14,0 |
| | HF | 29 | 8 | 26–33 | 6,3 |
| | E | 14 | 8 | 12–15 | 6,0 |
| | Mass: | 39 g | 3 | 37–43 g | — |
| Females | HB | 91 | 8 | 89–109 | 11,0 |
| | T | 111 | 8 | 70–131 | 14,3 |
| | HF | 31 | 8 | 28–34 | 5,1 |
| | E | 31 | 8 | 28–34 | 1,9 |
| | Mass: | 31,5 g | 4 | 30–34 g | — |

## Identification

The general colour of the animal is sandy-buff, less inclined to russet than *Gerbillurus paeba,* with lighter post-orbital and postauricular markings scarcely perceptible (Thomas 1918) (plate 4). The forelimbs are white. A naked area is present from the heel to the middle of the sole of the level of the base of the hallux. The fur is loose and long. The tail is buffy at the base dorsally with a tassel of long hairs at the tip, and its overall length is about 40% longer than the length of the head and body. In size, *G. vallinus* approximates *G. paeba.* A grey form has also been collected in SWA/Namibia.

## Skull and dentition

The skull is remarkable for the enlarged bullae, often with the development of a small additional bulla alongside the occiput (fig. 37). The posterior breadth of the skull is, therefore, unusually great. The muzzle is slender with the zygomatic plate more projected forward than in most species of *Gerbillus.* Both anterior and posterior palatine foramina are large. The upper incisors are grooved. The upper molars tend to be laminate (fig. 38).

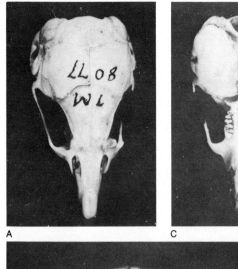

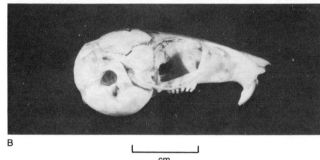

**Figure 37** Dorsal (A), lateral (B) and ventral (C) views of the skull of the Brush-tailed Hairy-footed Gerbil *Gerbillurus vallinus.* The inflated bullae extend behind the level of the occiput, and widens the skull at its posterior end. Both anterior and posterior palatine foramina are large.

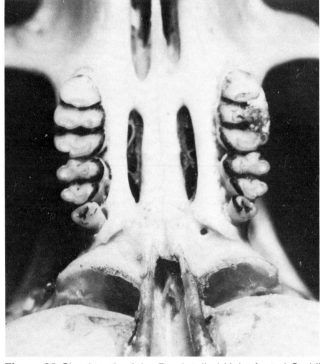

**Figure 38** Cheekteeth of the Brush-tailed Hairy-footed Gerbil *Gerbillurus vallinus.*

## Distribution

The Brush-tailed Hairy-footed Gerbil is endemic to and widespread in the South West Arid zone (Davis 1962; Meester 1976). It is found from southwestern Angola down the northern Namib to Swakopmund and in the Brukaros-Karas Mountains ranging southwards to the Orange River and Kenhardt (Davis 1975a) (map 21).

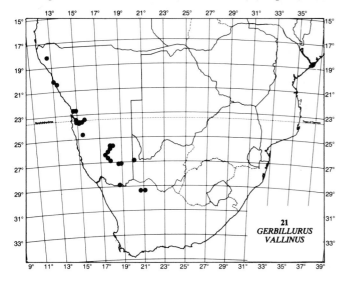

21
GERBILLURUS
VALLINUS

## Habitat

It seems to prefer the desert coastal zone along the Atlantic Ocean where there is surface sand. Mists occur frequently but the rainfall is sparse and irregular. Southwards it occurs in what Wellington (1955) calls the Cape Middleveld.

## Diet

Unknown.

## Habits

The Brush-tailed Hairy-footed Gerbil lives in burrows in sandy soils, frequently closed up at the entrance by falling sand (Roberts 1951). These burrows may penetrate to a depth of 150 cm, opening into a chamber and are often interconnected with other burrows. If the animal is dug out, it closes the burrow behind it with loose soil as it proceeds – if followed persistently it will break surface and seek refuge in another set of burrows (Roberts 1951). These gerbils are gregarious.

## Parasites

In contrast to other species of the South African Gerbillinae, no information is available on the ecto- and/or endoparasites which are associated with *Gerbillurus vallinus*.

## Reproduction

The mammae consist of one pectoral pair and two inguinal pairs. The nest is built from dry vegetable matter in a chamber. Roberts (1951) came across a female with five young in a chamber dug out of soft sand.

## Predators, prehistory and relations with man

Hitherto, no data are available pertaining to these aspects of the animal's ecology.

## Taxonomy

According to Davis (1975a) the following subspecies warrant recognition:

*G. v. vallinus* – occurs at Kenhardt and the lower Orange River.

*G. v. seeheimi* – Kuiseb River (Gobabeb area) to the Karas Mountains.

A specimen found 22,5 km southeast of Torra Bay in the northern Namib, may prove to be a third, as yet unnamed, subspecies.

# *Gerbillurus setzeri* (Schlitter, 1973)

### Setzer's Hairy-footed Gerbil

### Setzer se Haarpoot-nagmuis

The meaning of *Gerbillurus* has been given under *G. paeba*. The present species is named after Dr H. W. Setzer in honour '... of his efforts in African mammalogy and in particular for his interest in the taxonomy of desert rodents' (Schlitter 1973).

The account which follows has been gleaned from the original description of the species by Schlitter (1973) and for a more detailed account the reader is referred to that publication.

According to Schlitter there appears to be two groups of nominal species in the genus *Gerbillus* in southern Africa: a *'paeba'* group and a *vallinus* group (including *Gerbillus vallinus* Thomas, 1918 and *G. tytonis* Bauer & Niethammer 1959). The *'vallinus'* species have been included under the subgenus *Gerbillurus* Shortridge, 1942. In 1951, Roberts as well as Lundholm (1955a) elevated *Gerbillurus* to generic rank, but it was retained as a subgenus of *Gerbillus* by Ellerman *et al.* (1953). Herold & Niethammer (1963), however, felt that *Gerbillurus* was more closely related to *Tatera* when the enamel patterns of the lower first molars of juveniles and the molar alveoli of different gerbil genera are considered. They did not, as is indicated by Schlitter, state conclusively whether *Gerbillurus* is to be recognised as a distinct genus or to be retained as a subgenus of *Gerbillus*. To this array, Schlitter proposed the inclusion of a new species which he described as *Gerbillus (Gerbillurus) setzeri*.

The holotype of *Gerbillurus setzeri* is a young adult female, housed in the US National Museum of Natural History, collected 1,6 km east of the Namib Desert Research Station at Gobabeb in SWA/Namibia.

97

**Outline of synonymy**

This recently described species has as yet no synonymy.

TABLE 19

Measurements of male and female *Gerbillurus setzeri*

|  | Parameter | Value (mm) | N | Range (mm) | CV (%) |
|---|---|---|---|---|---|
| Males | HB | 105 | 1 | — | — |
|  | T | 123 | 1 | — | — |
|  | HF | 32 | 1 | — | — |
|  | E | 14 | 1 | — | — |
|  | Mass: No data available | | | | |
| Females | HB | 110 | 3 | 104–115 | — |
|  | T | 124 | 3 | 117–128 | — |
|  | HF | 32 | 3 | 32–33 | — |
|  | E | 13 | 3 | 9–17 | — |
|  | Mass: | 38,6 g | 3 | 33–48 g | — |

The figures given above are from specimens in the Transvaal Museum, Pretoria. It is too small a sample to give a true picture of the parameters of the species. In contrast, Schlitter (1973) has worked with a sample of 43 animals (males and females), including the specimens of the Transvaal Museum to give a mean total length of 233 mm, a mean tail length of 127 mm, a mean hindfoot length of 32 mm and a mean ear length of nearly 14 mm. These values lie very close to the values shown by the females housed in the Transvaal Museum.

**Identification**

The upper parts of the animal are '... near light pinkish cinnamon, with slight admixture of gray hairs ...' (Schlitter 1973). Individual hairs are plumbeous at their bases. The area around the mouth, entire underparts, supraorbital and postauricular spots, as well as the dorsal surfaces of the limbs are white, the hairs being uniformly white to base. The dorsal and ventral body colors are sharply demarcated (plate 4). The tail is mouse-grey dorsally on the distal third, otherwise coloured like the back dorsally and white ventrally. The hairs on the internal surface of the ears are white.

**Skull and dentition**

The skull is large for *Gerbillurus*. The auditory and mastoidal portions of the auditory bulla are large and inflated ventrally and posteriorly. The mastoidal section of the bulla projects beyond the occiput. The anterior palatine foramina are short and wide, while the posterior pair is long (Schlitter 1973). The upper tooth row is relatively short and strongly built.

**Distribution**

Schlitter (1973) has examined some 74 specimens of this SWA/Namibia species collected at the Namib Desert Research Station at Gobabeb and its immediate environs. It was also taken in the Swartbank Mountain (36 km west-northwest of Gobabeb), near the Tumas Mountain, at Swakopmund and vicinity, at Goanikontes, and at Hope Mine and surrounding area (map 22).

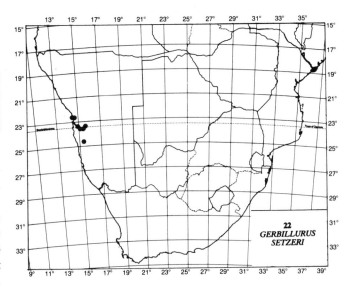

22
GERBILLURUS
SETZERI

**Habitat**

*Gerbillus (Gerbillurus) setzeri* seems to be restricted to the very pale fledspar and quartz gravel plains in the Namib Desert of SWA/Namibia. According to Schlitter (1973), it can disperse from the gravel plains across the riverbed of the Kuiseb into the adjacent red dunes where *G. (G.) tytonis* is usually found. This dispersal occurs during periods of higher population levels. In contrast, individuals of the *G (G.) paeba* group seem to be distributed ubiquitously in both habitats.

**Diet, habits, predators, reproduction, parasites, relations with man and prehistory**

No data are available on any of these facets of Setzer's Hairy-footed Gerbil.

**Taxomony**

*Gerbillurus setzeri* is a member of the *vallinus* group of gerbils, related to *G. vallinus* on the one hand and *G. tytonis* on the other.

# GENUS *Tatera* Lataste, 1882

The southern African gerbils of the genus *Tatera* fall into two groups, *viz.* an *afra* group and a *robusta* group (Davis 1975a). Davis points out that it is unfortunate that characters used to distinguish the species in one group sometimes also occur in species of the other group (examples would include colour, characteristics of the hindfoot, width of the molars, etc.). It is therefore necessary in some instances to qualify certain morphological characters by geographic locality. Davis states that in the field, in any one area, representatives of the two groups can be distinguished without much difficulty, provided enough specimens are available. This, however, does not imply that only one species of each group is present.

The two groups may be described as follows:

The *robusta* group: Colour 'brighter'; texture of fur sleek and silky; the line of demarcation between belly and flanks sharp; belly pure white; two pectoral and two inguinal pairs of mammae; tail longer than length of head and body, never white-tipped; feet narrow; upper incisors grooved, moderately to strongly opisthodont, while the molars are narrow and slightly built.

The *afra* group: Colour 'duller'; texture of fur 'fluffy' or somewhat harsh; line of demarcation between belly and flanks often indistinct; belly white, buffy or grey; one pectoral and two inguinal pairs of mammae, or as in the *robusta* group; tail longer than, equal to or shorter than length of head and body, sometimes white-tipped; feet broad; upper incisors grooved or plain, moderately opisthodont or orthodont, molars broad, heavily built.

In southern Africa, *Tatera leucogaster* is a member of the *robusta* group, while *T. afra, T. brantsii* and *T. inclusa* are representatives of the *afra* group. *Tatera robusta*, the Fringe-tailed Gerbil, is extralimital to this work and occurs from West Africa to Ethiopia and Somalia southwards to Tanzania and the Indian Ocean coast.

## Key to the southern African species of *Tatera*
(Modified after Davis (1975a))

1 Colour 'brighter'; texture of fur silky; line of demarcation between flanks and belly sharp; belly pure white; tail longer than HB length, well-haired especially towards tip, never white-tipped; upper incisors grooved, moderately to strongly opisthodont; molars narrow, lightly built........ *Tatera leucogaster* (Bushveld Gerbil) Page 99

Colour 'duller'; texture of fur 'fluffy' or somewhat harsh; line of demarcation often indistinct; belly white, buffy or grey; tail longer, equal to or shorter than HB length, sometimes white-tipped; upper incisors grooved or plain, moderately opisthodont to orthodont; molars relatively broad, more heavily built............................................2

2 Ears long (av. *ca* 24 mm); belly pure white; tail evenly coloured to tip; mammary formula (MF) 2–2 = 8; western Cape.............................. *T. afra* (Cape Gerbil) Page 102

Ears shorter, average less than 24 mm; belly white, grey or buffy, or white with greyish chest patch; tail evenly coloured above, paler below, may be white-tipped; MF 1–2 = 6 or 2–2 = 8...........................3

3 Anterior palatine foramina reaching and normally extending beyond anterior margin of molar alveoli; fair proportion of individuals in any population with white-tipped tail (southern Africa)......... *T. brantsii* (Highveld Gerbil) Page 104

Anterior palatine foramina normally not extending to anterior margin of molar alveoli; tail very rarely white-tipped (eastern Africa)................ *T. inclusa* (Gorongoza Gerbil) Page 107

# *Tatera leucogaster* (Peters, 1852)

**Bushveld Gerbil**

**Peters' Gerbil**

**Bosveldse Nagmuis**

The word *Tatera* is an invented euphonious name, without derivation or meaning. The white belly of the animal is described by the combination of the Latin words *leuco* = white and *gaster* = pertaining to belly.

A synonym of this species often encountered in the older literature, is *T. schinzi*. It is a very common gerbil, showing a wide geographical distribution.

## Outline of synonymy

1852 *Meriones leucogaster* Peters, *Monatsber. K. Preuss. Akad. Wiss. Berlin*: 274; *Reise nach Mossambique. Säugeth.*: 145. Mesuril and Boror, Mozambique.

1896 *Gerbillus leucogaster* De Winton, *Proc. zool. Soc. Lond.*: 806.

1898 *Gerbillus (Tatera) lobengulae* De Winton, *Ann. Mag. nat. Hist.* (7) 2:4. Essexvale, Zimbabwe.

1889 *Gerbillus tenuis* var. *schinzi* Noack, *Zool. Jahrb. Jena.* 4:134. Kalahari Desert.

1906 *Tatera lobengulae* Wroughton, *Ann. Mag. nat. Hist.* (7) 17:482 Bulawayo, Zimbabwe.

1906 *Tatera lobengulae bechuanae* Wroughton, *Ann. Mag. nat. Hist.* (7) 17:482. Molopo, Botswana.

1906 *T. lobengulae mashonae* Wroughton, *Ann. Mag. nat. Hist.* (7) 17:482. Mazoe, Zimbabwe.

1906 *T. lobengulae griquae* Wroughton, *Ann. Mag. nat. Hist.* (7) 17:482. Kuruman, South Africa.

*Synonyms*

1906 *Tatera miliaria stellae* Wroughton, *Ann. Mag. nat. Hist.* (7) 17:485. Kuruman, South Africa.

1906 *T. m. salsa* Wroughton, *Ann. Mag. nat. Hist.* (7) 17:485. Klein Letaba River, NE Transvaal.

1929 *T. lobengulae salsa* Roberts, *Ann. Transv. Mus.* 13:102.

1906 *Tatera panja* Wroughton, *Ann. Mag. nat. Hist.* (7) 17:486. Tete district, Mozambique.

1927 *Tatera schinzi* Thomas, *Proc. zool. Soc. Lond.*: 382. Gobabis, SWA/Namibia.

1929 *Tatera lobengulae pestis* Roberts, *Ann. Transv. Mus.* 13:102. Bothaville, OFS.

1929 *T. l. mitchelli* Roberts, *Ann. Transv. Mus.* 13:103. Wonderfontein, Transvaal.

1929 *T. l. pretoriae* Roberts, *Ann. Transv. Mus.* 13:104. Pretoria North, Transvaal.

1929 *T. l. limpopoensis* Roberts, *Ann. Transv. Mus.* 13:104. Njelele River, Transvaal.

1929 *T. l. tzaneenensis* Roberts, *Ann. Transv. Mus.* 13:105. Tzaneen Estates, NE Transvaal.

1929 *T. l. littoralis* Roberts, *Ann. Transv. Mus.* 13:105. Masiyeni, Mozambique.

1929 *T. l. beirensis* Roberts, *Ann. Transv. Mus.* 13:106. Near Beira, Mozambique.

1931 *T. l. zuluensis,* Roberts, *Ann. Transv. Mus.* 14:230. Manaba, northern Natal.

1938 *Tatera schinzi waterbergensis* Roberts, *Ann. Transv. Mus.* 19:239. Waterberg, SWA/Namibia. (In addition, see Roberts *Ann. Transv. Mus.* 20:318.)

1939 *Tatera (T.) leucogaster* (Peters). Allen, *Bull. Mus. comp. Zool. Harv.* 83:334.

1951 *Tatera schinzi schinzi* (Noack). *In* Roberts, *The mammals of South Africa.*

1953 *T. afra leucogaster* (Peters). *In* Ellerman *et al., Southern African mammals.*

1964 *T. leucogaster* (Peters). *In* Meester *et al., An interim classification of southern African mammals.*

1975 *T. leucogaster* (Peters). Davis, *in* Meester & Setzer (eds), *The mammals of Africa: an identification manual.*

**TABLE 20**
Measurements of male and female *Tatera leucogaster*

|  | Parameter | Value (mm) | N | Range (mm) | CV (%) |
|---|---|---|---|---|---|
| Males | HB | 130 | 70 | 121–146 | 5,3 |
|  | T | 150 | 69 | 120–183 | 7,5 |
|  | HF | 31 | 69 | 22–36 | 8,1 |
|  | E | 21 | 69 | 18–29 | 8,3 |
|  | Mass: | 65,2 g | 11 | 48–80 g | 12,5 |
| Females | HB | 127 | 79 | 120–144 | 3,9 |
|  | T | 150 | 79 | 135–172 | 4,3 |
|  | HF | 31 | 79 | 23–36 | 8,5 |
|  | E | 20 | 78 | 17–23 | 5,2 |
|  | Mass: | 62,4 g | 10 | 49–98 g | 16,8 |

## Identification

The Bushveld Gerbil is pale orange-buffy dorsally and pure white on the chin, throat, ventral surface, limbs, hands and feet. Lower back darkened by an admixture of hairs with black tips. Individual hairs on the dorsal surface are a slaty-grey for the basal two thirds, passing into a buffy-white which changes into a pale rufous-buff at their ends. On the ventral surface the hairs are white from the base. There is an indistinct white mark above and behind the eyes, while the sides of the muzzle is also white. The ears are dark brown. Tail, brownish above, white below, often shows a dark, terminal tuft of hair (plate 4).

The colour is geographically variable. The Kalahari harbours the palest individuals and the western Transvaal the darkest (Davis 1975a). Smithers (1971) also records colour variation in Botswana where animals from the east, northeast and the Okavango are darker and redder than those from other regions.

There are two pectoral and two inguinal pairs of mammae.

## Skull and dentition

The occipital region tends to be strongly ridged (fig. 39). The braincase as a rule does not seem to appear as enlarged and broadened posteriorly as in other gerbils. The grooved upper incisors are coloured orange to yellow, with the upper molars tending towards hypsodonty (fig. 40).

## Distribution

Bushveld gerbils range from the lower Orange River in the southwest, to KwaZulu in the southeast. There is remarkably little discontinuity in distribution, and populations intergrade evenly throughout the range of the species. They are found in Southern Savanna Woodlands and in the western margin of the Southern Savanna Grassland biotic zone. They also inhabit the South West Arid from the Orange River northwards fringing the northern Namib, but are absent from duneveld of the southwestern Kalahari and from Great Namaqualand (Davis 1974) (map 23).

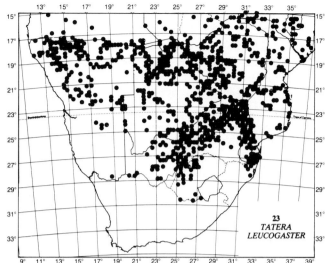

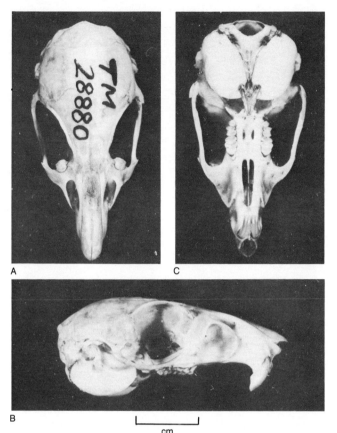

**Figure 39** Dorsal (A), lateral (B) and ventral (C) views of the skull of the Bushveld Gerbil *Tatera leucogaster*. The posterior part of the skull is not broadened to the same extent as in related gerbils.

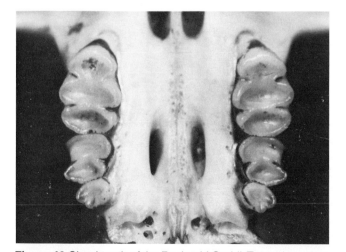

**Figure 40** Cheekteeth of the Bushveld Gerbil *Tatera leucogaster*. In contrast to related gerbils, the molars are narrow and slightly built, tending towards hypsodonty.

## Habitat

Bushveld gerbils occur in open woodland, in sandy soils, where their warrens are small and scattered, usually round bases of trees and bushes. They also occur in thornveld and bushveld and seem to shun areas with less than 250 mm mean annual rainfall. In Botswana they are encountered in sandy soils or sandy alluvium, but may also be found on hard ground. Apparently they occur in any type of vegetation in Botswana (Smithers 1971). They are usually absent from rocky hills, but occur in the valleys. *Tatera leucogaster* is very similar to *T. brantsii* in diet and habitat, although it does not tolerate conditions as waterless as those in which *T. brantsii* is found (Smithers 1971). In the Kruger National Park the present veld burning policy removes cover and litter annually and triennially from sections separated by properly constructed firebreaks. The resultant low cover and litter favours and maintains numerical domination by *T. leucogaster* over other species immediately after being burnt (Kern 1977).

## Diet

They consume seeds from trees, grasses and bushes while rhizomes and bulblets of certain grasses are also readily taken. In the Nylsvley Nature Reserve near Naboomspruit it feeds on seeds of *Peltophorum africanum* and other plant material such as the stems of *Achyranthes aspera* (Jacobsen 1977).

A study of hoarding behaviour in southern African rodent species by Pettifer & Nel (1977) revealed that *Tatera leucogaster* does not hoard its food, but that it frequently covers its food (seeds), a possible primitive form of scatter-hoarding.

## Habits

Burrows are usually dug at the base of small shrubs, but also in the open. Occasionally, the warrens are as extensive as in *T. brantsii;* in fact, the two species may occupy the same ground and may even be caught in the same

101

burrow. Ramps of loose soil are usually present near occupied burrows and indications are that the burrows are cleaned out nightly. The entrances measure 40–50 mm in diameter, leading to burrows which may extend for considerable distances (Smithers 1971). In harder soils, they do not excavate their own burrows and tend to use holes in termite mounds or shallow excavations under logs.

There is no evidence of the formation of runways among vegetation. This species is communal, entirely terrestrial and nocturnal. Populations are subject to cyclic explosions in numbers. It appears that they are entirely independent of water for drinking purposes (Smithers 1971).

## Predators
Little data available. Dean (1977) records *Tatera leucogaster* as a prey item of the Barn Owl *Tyto alba*.

## Reproduction
These are two pectoral and two inguinal pairs of mammae, making eight in total. Young are born throughout the year and average four to five per litter. Smithers (1975) states that litters number from two to nine. The diploid chromosome number is 42.

In Malaŵi, maximum production in most species occurs in the wet season or the early part of the dry season (Hanney 1965) and indications are that males of *T. leucogaster* become infertile for part of the dry season. In the Kruger National Park I have trapped males with heavily descended testes during May. They appear to breed poorly under laboratory conditions (Keogh & Isaäcson 1978).

## Parasites
Among the virus diseases which occur as zoonoses in southern Africa, it has been shown under laboratory conditions that *Tatera leucogaster* is susceptible to infection of the neurotropic strain of African horse sickness (Neitz 1965). This species is a reservoir of the plague bacillus *Yersinia pestis* and it is largely due to their presence that this disease has managed to spread so far inland from coastal ports (Smithers 1975). Periodic outbreaks of listeriosis, caused by *Listeria monocytogenes*, also occur in populations of the Bushveld Gerbil. Laboratory tests have shown that this gerbil is highly susceptible to these coryne-bacteria. It has been isolated from birds, animals and man in almost every part of the world, causing encephalitis and metritis in cattle and sheep and symptoms of the disease has given rise to the colloquial term 'circling disease'. The incidence of the disease in man appears to be increasing, but it is considered that this is due to an increasing awareness of the condition, rather than to a real increase in incidence (Andrewes & Walton 1977).

Hitherto we seem to have no indications of the nature of helminth or nematode infections which may be associated with *Tatera leucogaster*. In contrast, the arthropod parasite load is fairly well documented. The blowflies and allies known to be associated with this gerbil, includes *Cordylobia anthropophaga* (Zumpt 1961a). Fleas are represented by the families Pulicidae, Hystrichopsyllidae and Chimaeropsyllidae. The pulicid fleas are represented by some seven genera and 25 species. These include, according to Zumpt (1966), the following: *Echidnophaga bradyta, E. gallinacea, Ctenocephalides connatus, C. felis, Xenopsylla frayi, X. hipponax, X. mulleri, X. philoxera, X. phyllomae, X. piriei, X. raybouldi, X. trifaria, X. versuta, X. bechuanae, X. brasiliensis, X. scopulifer, X. syngenis, X. cryptonella, Dinopsyllus ellobius, D. lypusus, Listropsylla agrippinae, L. dorippae, L. prominens, Epirimia agarippes* and *Chiastopsylla rossi*.

## Relations with man
This gerbil is subject to impressive population explosions when its numbers rise to astronomical figures. Smithers (1975) points out that following the peaks of such explosions there is the ever-present danger (for man) '... of plague outbreaks if the populations are infested'.

## Prehistory
Although this species shows a wide geographical distribution, it has not yet been identified in any of the fossiliferous localities known in southern Africa.

## Taxonomy
On account of little discontinuity in distribution and the fact that populations appear to intergrade evenly throughout the range of the species, valid subspecies cannot readily be picked out, and Davis (1975a) makes no attempt to do so. Southern African synonyms include *schinzi, waterbergensis, bechuanae, lobengulae, mashonae, griquae, pestis, stellae (= griquae), mitchelli, pretoriae, limpopoensis, salsa, tzaneenensis, zuluensis, littoralis, beirensis* and *panja*.

# *Tatera afra* (Gray, 1830)
## Cape Gerbil
## Kaapse Nagmuis

For the meaning of *Tatera*, see *Tatera leucogaster*. The word *afra* is a declension of the Latin *afer* = African.

This species, the first to be considered in the *'afra'* group as defined by Davis (1975a), (and which also included *Tatera brantsii* and *T. inclusa*) is limited to the southwestern area of the Cape. Members of this species have so often been found to be carriers of bubonic plague through their fleas '... that special staffs have had to be employed by the Union Health Department to destroy the animals in order to prevent spread of the disease' (Roberts 1951). This is yet another species where the state of our knowledge is unsatisfactory. Meester (1976) describes it as rare with a '. . . limited distribution, but not uncommon where present' in the *South African Red Data Book*.

## Outline of synonymy

1830 *Gerbillus afra* Gray, *Spicilegia Zool.* 2:10. Cape of Good Hope.

1832 *Meriones schlegelii* Smuts, *Enum. Mamm. Cap.*: 41. Port Elizabeth. (Where the species does not occur.)

1838 *Gerbillus africanus* F. Cuvier, *Trans. Zool. Soc. Lond.* 2: pl. 26. Renaming of *afra*. No locality, but presumeably Cape of Good Hope.

1882 *Tatera* Lataste, *Le Naturaliste Paris* 4(16):126. Subgenus of *Gerbillus*. Type *G. indicus*.

1906 *Tatera afra* Wroughton, *Ann. Mag. nat. Hist.* (7)17:481.

1929 *T. a. gilli* Roberts, *Ann. Transv. Mus.* 13:100. Lamberts Bay, CP.

1939 *T. (T.) afra afra* (Gray). *In* Allen, *Bull. Mus. comp. Zool. Harv.* 83:331.

1951 *T. afra afra* (Gray). *In* Roberts, *The mammals of South Africa*.

1953 *T. (T.) afra* (Gray). *In* Ellerman *et al.*, *Southern African mammals*.

1975 *T. afra* (Gray). Davis, *in* Meester & Setzer (eds), *The mammals of Africa: an identification manual*.

### TABLE 21
Measurements of male and female *Tatera afra*

|  | Parameter | Value (mm) | N | Range (mm) | CV (%) |
|---|---|---|---|---|---|
| Males | HB | 147 | 21 | 130–157 | 3,3 |
|  | T | 148 | 21 | 141–175 | 5,1 |
|  | HF | 37 | 21 | 28–40 | 6,7 |
|  | E | 24 | 21 | 20–26 | 5,0 |
|  | Mass: | 103,2 g | 5 | 84–113 g | 10,0 |
| Females | HB | 136 | 23 | 124–152 | 3,0 |
|  | T | 156 | 23 | 133–168 | 4,9 |
|  | HF | 38 | 23 | 30–40 | 6,0 |
|  | E | 25 | 23 | 24–28 | 3,6 |
|  | Mass: | 91,0 g | 10 | 78–107 g | 9,2 |

## Identification

The back, the sides, and the outer surfaces of the limbs vary from reddish-orange to wood-brown, pencilled with umber-brown. The hairs near their roots are a dull lavender. The lower surface of the animal is white and the line of demarcation is rather clearly defined (plate 4). Internally, the ears are a light flesh colour, deep brown externally. The front limbs are short, brown outside, white inside. The hindlimbs are long.

## Skull and dentition

The skull is very much like that of *Tatera leucogaster*. The rostrum is narrow, as are the antorbital foramina. The braincase and bullae are not particularly enlarged.

The upper molar teeth tend to be simpler than in *Gerbillurus*, but show the same general arrangement. The $M^3$ is not excessively reduced, usually bilaminate (fig. 41). The anterior lamina of the $M_1$ may show a small fold in front. The upper incisors are grooved, the lower ungrooved.

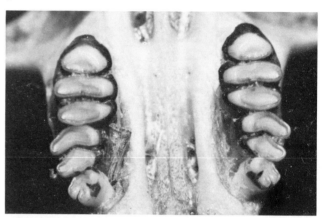

**Figure 41** Cheekteeth of the Cape Gerbil *Tatera afra*. The molars tend to be simpler in structure than found in *Gerbillurus* but the overall arrangement is the same. $M^3$ not markedly reduced in size, usually bilaminate.

## Distribution

This species is confined mainly to the sandy areas of the western Cape, northwards to Nieuwoudtville and eastwards along the coast to Herold's Bay. It is endemic to the South West Cape biotic zone (Davis 1974) (map 24).

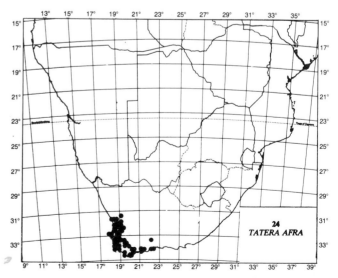

## Habitat

Cape gerbils seem to prefer loose, sandy soils and live in extensive burrows either in the open or under shelter of shrubs.

## Diet

Little information is available, but they consume grass bulbs and roots '... together with seeds and green vegetable matter' (Roberts 1951).

## Habits

Though reputed to be migratory, they are not so in the strict sense of the word. Their movements to and fro are not seasonal but are due to some local stimulus and thus sporadic in occurrence (Roberts 1951). Roberts states that they are not easily caught in baited traps. The burrows are labyrinthine with many entrances or exits leading to and from the system. These passages are often blocked with loose earth, presumably to keep out predators such as snakes. When being dug out, they follow the practice of many desert-dwelling rodents by burrowing through the sand and blocking up the passage behind them as they progress. Roberts (1951) says that when they are persistently followed in this way, they often break the surface at some unexpected point and make for other burrows in the vicinity.

Extensive burrows occur in sandy places in the open or in close proximity to shrubs and bushes. The burrows often intercommunicate leading downwards into a chamber with a nest. The latter consists of shredded grass and similar vegetation. When pursued, they often block passages with soil. Normally they do not frequent or enter houses or dwellings (Roberts 1951).

There are indications that populations tend to undertake short, local migrations. There are, however, not migrations in the true sense of the word as their movements are not back and forth.

## Predators

Their main enemies seem to be snakes.

## Reproduction

The mammae vary in arrangement and number; usually two pectoral pairs and two inguinal pairs occur. The diploid chromosome number equals 44. Longevity studies indicate that *T. afra* may have a lifespan of more than one year (Allanson 1958). Keogh & Isaäcson (1978) have found them to be poor laboratory breeders.

## Parasites

Laboratory tests have shown that this species is susceptible to infections of *Mycobacterium tuberculosis*, louse typhus caused by *Rickettsia prowazekii*, and to rat typhus caused by *R. typhi* (Neitz 1965). *Rickettsia conorii*, the cause of tick-bite fever, also affects this rodent adversely.

A fair diversity of species of mites are known to be associated with *Tatera afra*. The laelaptid mites include *Laelaps giganteus*, *L. muricola*, *L. simillimus*, *L. vansomereni*, *Haemolaelaps glasgowi*, *H. oliffi*, *H. taterae*, *Androlaelaps marshalli* and *A. theseus*. The myobiid mites are represented by *Radfordia forcipifer*. The trombiculid mites include *Schoutedenichia morosi* and *Acomatacarus gateri*, while the listrophorid mite *Listrophorus bothae* has also been recorded (Zumpt 1961a).

The fleas include pulicids (*Echidnophaga gallinacea, Xenopsylla cheopis, X. eridos, X. piriei, X. trifaria, X. davisi, X. demeilloni, X. hirsuta, X. lobengulai* and *X. sulcata*), hystrichopsyllids (*Dinopsyllus ellobius*) and chimaeropsyllids (*Chiastopsylla rossi*) (Zumpt 1966).

## Relations with man

This species is often a carrier of fleas which in turn spreads bubonic plague, so that special personnel have to be employed by the state to prevent the spread of the disease. The fleas breed in the nesting material.

Destruction of the animals is no easy matter. Gassing is not always effective, because they block the passages.

## Prehistory

Unknown.

## Taxonomy

*T. a. gilli* is a synonym of the nominate race, which was originally described as *Tatera gilli* by Roberts in 1929.

# *Tatera brantsii* (A. Smith, 1836)

**Highveld Gerbil**

**Brants' Gerbil**

**Hoëveldse Nagmuis**

For the origin of the name *Tatera*, see *Tatera leucogaster*. The species is named after A. Brants, who compiled *Het Geslacht der Muizen* in 1827. This monograph on the murids contained early descriptions of Cape rodents.

According to Sclater (1901) Sir Andrew Smith described this species on specimens obtained near the sources of the Orange and Caledon rivers in what is now Lesotho. Another interesting anecdote refers to a synonym of *T. brantsii*, *Meriones (Rhombomys) maccalinus*, which was described by Sundevall in 1846 on specimens collected by Wahlberg in the Maccali (i.e. the Magaliesberg Range extending from Pretoria in the east to Rustenburg in the west) in the Rustenburg district of the Transvaal.

*Tatera brantsii* is a second example of a species that Davis (1975a) placed in his '*afra*' group of gerbils. It has a wide geographical distribution in southern Africa and is common.

## Outline of synonymy

1836 *Gerbillus brantsii* A. Smith, *App. Rept. Exped. Explor. S. Afr.*: 43. Near the sources of the Caledon River, OFS.

1842 *Merionis montanus* A. Smith, *Illustr. Zool. S. Afr. Mamm.* pl. 36. 'Bashartoo Country' = ?Basutuland = Lesotho.

1846 *Meriones (Rhombomys) maccalinus* Sundevall, *Oefvers. K. Vet. Akad. Förh. Stockholm* 120. Magaliesberg, Transvaal.

1906 *Tatera brantsii* Wroughton, *Ann. Mag. nat. Hist.* (7) 17:480.

1906 *T. draco* Wroughton, *Ann. Mag. nat. Hist.* (7) 17:479. Wakkerstroom, SE Transvaal.

1939 *T. (T.) brantsii brantsii* (A. Smith). *In* Allen, *Bull. Mus. comp. Zool. Harv.* 83:332.

1951 *T. brantsii brantsii* (A. Smith). *In* Roberts, *The mammals of South Africa.*

1953 *T. afra brantsii* A. Smith. Ellerman *et al., Southern African mammals.*

1975 *T. brantsii* (Smith). Davis, *in* Meester & Setzer (eds), *The mammals of Africa: an identification manual.*

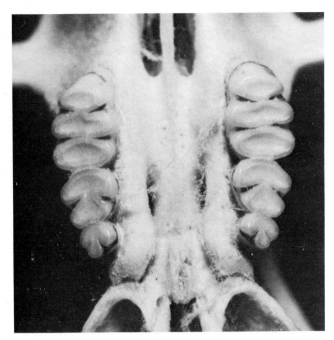

**Figure 42** Cheekteeth of the Highveld Gerbil *Tatera brantsii.* The posterior palatine foramina are small.

### TABLE 22
Measurements of male and female *Tatera brantsii*

|       | Parameter | Value (mm) | N | Range (mm) | CV (%) |
|-------|-----------|------------|---|------------|--------|
| Males | HB | 140 | 27 | 120–159 | 5,0 |
|       | T  | 148 | 27 | 125–169 | 5,6 |
|       | HF | 33  | 27 | 28–37   | 5,6 |
|       | E  | 22  | 27 | 18–25   | 6,0 |
|       | Mass: | 63 g | 8 | 37–87 g | 19,8 |
| Females | HB | 140 | 32 | 130–148 | 4,6 |
|       | T  | 154 | 32 | 132–190 | 6,7 |
|       | HF | 34  | 32 | 30–39   | 4,2 |
|       | E  | 24  | 31 | 20–26   | 4,9 |
|       | Mass: | 65 g | 11 | 40–82 g | 15,5 |

## Identification

Dorsally the animals are a light rufous-brown to reddish, freely pencilled with darker brown. The chin, throat and ventral surface are pale cream to dull white or yellowish-white (plate 4). The ears are brown and thinly covered with hair. The tail is reddish-brown above with black hair intermixed and is about as long as the head-body measurement. According to Sclater (1901), the toes of *T. brantsii* are shorter than those of *Tatera afra.*

## Skull and dentition

The incisors are larger than in *T. afra,* while the length of the diastema is smaller. The upper molars are well developed (fig. 42).

## Distribution

This species occurs in the eastern Cape, ranging northwards, skirting the Karoo, to northern KwaZulu and Transvaal. It is widely distributed throughout Botswana and spreads marginally into southwestern Zimbabwe, while it occurs fairly commonly in central and northern SWA/Namibia (map 25). It therefore inhabits the South West Arid biotic zone and the Grassland zone of the Southern Savanna and occurs marginally in the Southern Savanna Woodland biotic zone (Davis 1975a).

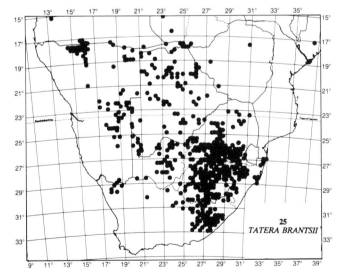

25
*TATERA BRANTSII*

## Habitat

Highveld Gerbils may be found in peaty soils around marshes and pans and may also occupy the tunnel-systems of the molerat *Cryptomys,* making entrances along

105

the course of the tunnels (Roberts 1951). They seem to shun areas where the ground is hard and prefer sandy soils, irrespective of the type of vegetation cover (Smithers 1971). They appear to be able to tolerate drier conditions than *T. leucogaster*. In Botswana they occur in areas receiving less than 250 mm rainfall annually (Smithers 1971). They are the dominant rodents in some parts of the Kalahari Gemsbok National Park.

## Diet
It takes bulbs of certain grasses and seeds of other plants. According to Pettifer & Nel (1977), *T. brantsii* does not hoard its food. De Beer (1972) found a high percentage of insect material in stomach contents and hoarding is thus not essential for survival.

## Habits
The Highveld Gerbil is an inveterate burrower in sandy ground (Roberts 1951). It occurs in small colonies in intercommunicating warrens. It is cunning and not easily trapped. It is strictly nocturnal and occupies its burrows during the day. Occupancy is readily observed from the fresh earth excavated and scattered near the entrances of the burrows.

A home range area of 0,49 ha (males) and 0,19 ha (females) for *T. brantsii* on the South African highveld is given by De Moor (1969).

## Predators
This species is taken by snakes and small carnivores. Roberts (1951) states that snakes will enter the burrows to devour the young if the adults elude them. Nocturnal raptors include the Barn Owl *Tyto alba* (Vernon 1972) and the Grass Owl *Tyto capensis* (Vernon 1972; Davis 1973).

## Reproduction
The mammae consist of one pectoral pair and two inguinal pairs. Gravid females have been taken during May, June, July, August and December in Botswana (Smithers 1971), and the number of fetuses averages 3,3. This gerbil probably breeds throughout the year. The young remain in the nests until they can run and burrow themselves. In contrast, De Moor (1969) has indicated it to be a seasonal breeder.

## Parasites
Among the schizomycetes, it has been shown that under laboratory conditions *T. brantsii* is susceptible to infections of *Pseudomonas pseudomallei,* a protophyte causing melioidosis in man. As is to be expected, sporadic afflictions of *Yersinia pestis* occur in populations of these gerbils, leading to outbreaks of sylvatic plague. Laboratory tests have also shown their susceptibility to infections of *Listeria monocytogenes* which results in bacterial listeriosis. Finally, in laboratory experiments they also react to infections of *Mycobacterium tuberculosis* (Neitz 1965).

Collins (1972) in assessing the cestodes from rodents in the RSA, reported the occurrence of *Hymenolepis microcantha, H. taterae* and *Raillietina (R.) trapezoides* from individuals of *Tatera brantsii*. In addition the nematode *Capillaria hepatica*, a relative of the whipworm, can also occur (Collins *pers. comm.*).

Fleas associated with *Tatera brantsii* include no fewer than some 29 species (Zumpt 1966). The Pulicidae include *Echidnophaga gallinacea, Pulex irritans, Ctenocephalides connatus, C. felis, Xenopsylla cheopis, X. geldenhuysi, X. hipponax, X. mulleri, X. philoxera, X. phyllomae, X. piriei, X. trifaria, X. brasiliensis, X. cryptonella* and *X. erilli.* The Hystrichopsyllidae include *Ctenophthalmus calceatus, C. natalensis, Dinopsyllus ellobius, D. lypusus, Listropsylla agrippinae, L. cerrita juliae, L. chelura, L. dorippae, L. fouriei* and *L. prominens.* The Leptopsyllidae are represented by a single species *Leptopsylla segnis.* The Ceratophyllidae are represented by *Nosopsyllus fasciatus.* Finally the Chimaeropsyllidae include *Epirimia aganippes, Chiastopsylla rossi* and *C. godfreyi.* The calliphorid fly *Cordylobia anthropophaga* is also known to occur on this gerbil (Zumpt 1966).

Theiler (1962) has listed some ticks associated with *T. brantsii.* However, she has used a broad interpretation of *Tatera afra, sensu* Ellerman, *et al.* (1953) which includes *T. leucogaster* and *T. brantsii.* Within this complex, a single species of *Ixodes* and *Haemaphysalis* has been recorded with two species of *Hyalomma* and five species of *Rhipicephalus.* In *Ixodes alluaudi* and *Haemaphysalis leachii* both adult and immature specimens occur, while only immature ticks of *Hyalomma* and *Rhipicephalus* have been taken off these gerbils (which would include *T. brantsii).*

## Relations with man
This species is one of the principal sources of epidemics of bubonic plague. They are parasitised by fleas which breed in their underground nests and, as Roberts (1951) implies, the rather sociable habits of the host has helped to establish this rodent as an important plague reservoir. During the Second World War this species was used for typhus vaccine production (Gear & Davis 1942).

## Prehistory
Hitherto, a fossil form of *Tatera cf. brantsii* has been reported from Taung and Sterkfontein (De Graaff 1960a; Lavocat 1957), as well as from Kromdraai, Makapansgat limeworks and the Makapansgat Cave of Hearths (De Graaff 1960a).

## Taxonomy
The following provisional list of subspecies follows Davis (1975a). *T. b. brantsii* is found in the eastern Cape, Lesotho and the Natal midlands, ranging westwards to the edge of the Kalahari. Synonyms include *maccalinus, draco, milliaria* and *natalensis.* The specimens tend to show buffy-grey patches on the chest (cf. *T. b. griquae)* and this

feature intergrades towards the west and the Kalahari where the latter occurs. It also has heavier molars.

*T.b. griquae* occurs in the Kalahari and its fringes from the Orange River northwards to southern Angola. Synonyms include *perpallida, breyeri, namaquensis* and *joanae*. It has a pure white belly, somewhat narrower molars and a typical pallid desert coloration. Davis (1975a) states that *namaquensis* may prove to be a valid subspecies.

*T.b. ruddi* occurs in northern KwaZulu from Richards Bay to Kosi Bay. As synonyms Davis includes *tongensis* and *maputa*. This subspecies is distinguished by its relatively long tail, whitening towards the tip, a buffy-greyish belly and a long hindfoot. Note that Davis emphasises that there is considerable geographic variation and that the interpretation of subspecies given above is provisional.

# *Tatera inclusa* Thomas & Wroughton, 1908

### Gorongoza Gerbil
### Gorongoza Nagmuis

For the origin of the name *Tatera*, see *Tatera leucogaster*. The specific name refers to the Latin *inclusus* = confined and its limited occurrence in Mozambique may have prompted Thomas & Wroughton to accentuate the confined area of its range.

This is the third of the three southern African species of *Tatera* belonging to the '*afra*' group (Davis 1975a). The other species are *T. afra* and *T. brantsii*. In southern Africa it is known from Gorongoza in Mozambique and the eastern escarpment of Zimbabwe, ranging northwards through Mozambique to northeastern Tanzania. Much remains to be learnt from this species which was described by Thomas & Wroughton in 1908 from material collected at Tambarara in the Gorongoza district of Mozambique.

## Outline of synonymy
The premises on which *T. inclusa* was described as a separate species over 70 years ago, have withstood the test of time. There is, therefore, no synonymy.

## Identification
Thomas & Wroughton (1908a) described the ground colour of the back as 'ochraceous-buff' washed with black. It is darker in colour than *T. brantsii*, especially down the back. The ventral surface is white. The individual hairs of the dorsal surface are dark slate subterminally ringed with ochraceous and tipped with black. Those on the sides are tipped with ochraceous. The hands and feet are white. The tail is darker above than below and shows

**TABLE 23**
Measurements of male and female *Tatera inclusa*

| | Parameter | Value (mm) | N | Range (mm) | CV (%) |
|---|---|---|---|---|---|
| Males | HB | 151 | 3 | 140–170 | — |
| | T | 174 | 4 | 170–185 | — |
| | HF | 38 | 4 | 32–41 | — |
| | E | 24 | 4 | 15–25 | — |
| | Mass: No data available | | | | |
| Females | HB | 146 | 3 | 143–150 | — |
| | T | 169 | 3 | 164–175 | — |
| | HF | 36 | 4 | 34–39 | — |
| | E | 22 | 4 | 20–24 | — |
| | Mass: | 58 g | 2 | 55–61 g | — |

no white towards the distal end (plate 4). This is a large gerbil of which the tail comprises more than half the total length of the animal.

## Skull and dentition
Thomas & Wroughton (1908a) described the skull of this species as being narrower than in *T. draco* (= *maccalinus* = *brantsii*). In addition, the nasals seem to be longer than in *T. brantsii*.

The molar series is longer than in *T. brantsii*, while the upper incisors are '... less deeply grooved, the groove less central' (Thomas & Wroughton 1908a).

## Distribution
The Gorongoza Gerbil occurs in the Southern Savanna Woodlands only. This species is known from the Gorongoza district in Mozambique and the eastern escarpment of Zimbabwe in southern Africa. It ranges northwards through Mozambique into northeastern Tanzania (Davis 1975a) (map 26).

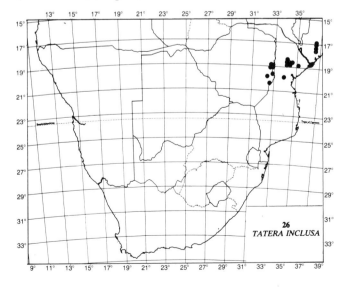

26
*TATERA INCLUSA*

107

## Habitat

According to Smithers (1975) it occurs in sandy ground or sandy alluvium, often on the fringe areas between dry and riverine woodland. In Mozambique it occurs on the same ground as *Tatera leucogaster* (Smithers & Lobão Tello 1976).

## Diet

Smithers (1975) states that Gorongoza gerbils are encountered in grain lands and that by implication they are predominantly granivorous.

## Habits

They dig their own burrows and will, under certain circumstances, '... use holes in termite mounds, cavities under rocks, fallen logs or tree roots' (Smithers 1975).

## Predators, reproduction, parasites, relations with man and prehistory

No data available.

## Taxonomy

Davis (1975a) lists three subspecies which are provisionally accepted. *Tatera inclusa coseni* and *T. i. pringlei* are both extralimital to southern Africa. The former occurs in northern Mozambique and eastern Tanzania, while the latter occurs in Tanzania. The nominate subspecies *T. i. inclusa* occurs in eastern Zimbabwe and southern Mozambique (Gorongoza) to the south of the Zambezi River.

# GENUS *Desmodillus* Thomas & Schwann, 1904

These gerbils show an array of characteristic morphological features, contrasting them clearly with *Tatera* or *Gerbillurus*. The ears are rather short, with a clear and conspicuous white spot behind each. The tympanic bullae are enormous with an additional development of the mastoids above them, reaching the levels of the parietals and projecting prominently behind the level of the occiput. The bullae can therefore be seen when the skull is viewed from above. The meatus of the ear passage is inflated and projects outwards to the same level as the slender zygomatic arch. The zygomatic plate is short and is not nearly as well developed as in *Tatera*. The upper incisors are long and slender, with a shallow groove. The upper molars are small, the $M^1$ having three sections, the $M^2$ two, while the $M^3$ has one section only. The middle section of the $M^1$ has two circular patterns of enamel.

Closely related genera to *Desmodillus* include *Desmodilliscus* from Sudan and northern Nigeria and *Pachyuromys*, from the Algerian Sahara and western lower Egypt (Ellerman 1941).

# *Desmodillus auricularis*
(A. Smith, 1834)

**Short-tailed Gerbil**

**Namaqua Gerbil**

**Kortstertnagmuis**

**Namakwa Nagmuis**

The generic name is derived from the Greek *desma* = similar to. The sufx *-illus* comes from *Gerbillus,* derived from the French *gerbille,* the name of a small rodent. *Desmodillus,* therefore, implies a similarity to the gerbil. The specific name is the Mid-Latin for *auricularis* = pertaining to the ear, no doubt calling attention to the large auditory bullae that this species characteristically shows.

This genus contains gerbils distinguished by their short tails and peculiarly shaped skulls. The original specimen described by A. Smith (1834) was collected near the Kamiesberg in Namaqualand.

## Outline of synonymy

1834 *Gerbillus auricularis* A. Smith, *S. Afr. Quart. Journ.* 2:160. Kamiesberg, Namaqualand, CP.

1838 *Gerbillus brevicaudatus* F. Cuvier, *Trans. zool. Soc. Lond.* 2:144. Cape of Good Hope.

1881 *Pachyuromys auricularis* Huet, *Le Naturaliste* 3:339.

1904 *Desmodillus* Thomas & Schwann, *Abstr. Proc. zool. Soc. Lond.*: 2:6. (Genotype *Gerbillus auricularis.*)

1910 *Desmodillus auricularis pudicus* Dollman, *Ann. Mag. nat. Hist.* (8) 6:395. Lehutiting, Botswana.

1939 *Desmodillus auricularis auricularis* (A. Smith). *In* Allen, *Bull. Mus. comp. Zool. Harv.* 83:319.

1951 *Desmodillus auricularis* (A. Smith). *In* Roberts, *The mammals of South Africa.*

1975 *Desmodillus auricularis* (A. Smith). Petter, *in* Meester & Setzer (eds), *The mammals of Africa: an identification manual*.

### TABLE 24
Measurements of male and female *Desmodillus auricularis*

|         | Parameter | Value (mm) | N | Range (mm) | CV (%) |
|---------|-----------|-----------|----|-----------|--------|
| Males   | HB        | 113       | 25 | 101–119   | 2,9    |
|         | T         | 88        | 25 | 79–103    | 6,0    |
|         | HF        | 25        | 25 | 16–27     | 8,9    |
|         | E         | 12        | 22 | 10–15     | 3,9    |
|         | Mass:     | 58 g      | 7  | 46–76 g   | 17,0   |
| Females | HB        | 110       | 23 | 105–116   | 2,0    |
|         | T         | 88        | 23 | 72–99     | 5,6    |
|         | HF        | 24        | 23 | 21–29     | 5,7    |
|         | E         | 12        | 21 | 9–16      | 4,7    |
|         | Mass:     | 54 g      | 11 | 38–78 g   | 12,3   |

## Identification

*Desmodillus auricularis* is a short and stockily built medium-sized gerbil, the untufted tail comprising approximately 78% of the head-body length (Petter 1975). The tail is pale brown above and reddish white below. The dorsal coloration is variable, predominantly ochraceous-orange, often with some black-tipped hairs. The dorsal hairs are slaty-grey at their bases, ochraceous yellow in the middle and dark brown at the tips. The ventral surface is white, with hairs pure white to the base. There is a pure white spot behind each ear (plate 4). The ears are small and oval. The digits on the hand number four and five on the foot. The palms and proximal half of the soles are naked, but the distal half of the soles and undersides of the toes are closely haired. The mammary formula of the female conforms to two pectoral and two inguinal pairs (Shortridge 1934).

## Skull and dentition

Features of the skull have been referred to above, the most conspicuous feature being the enormous development of the bullae. The posterior palatal foramina are well developed (fig. 43). The teeth have been described; the laminate molars show lamination even when much worn (fig. 44).

## Distribution

*Desmodillus auricularis* has a wide distribution in the central, western and northern Cape, the OFS, western Transvaal, Botswana and SWA/Namibia. In the latter two areas, their northernmost limit appears to be latitude 20°S, although in the western areas of SWA/Namibia they have been taken as far north as Sesfontein *ca* 19°S

(Smithers, 1971). It therefore occurs throughout the South West Arid zone with peripheral extensions into the Southern Savanna and the South West Cape biotic zones (Davis 1974) (map 27).

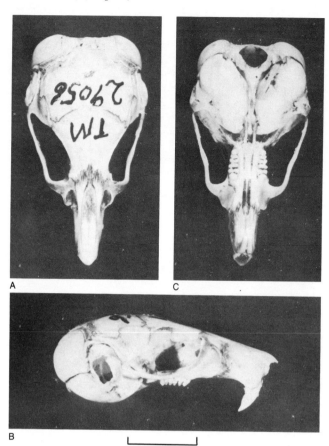

**Figure 43** Dorsal (A), lateral (B) and ventral (C) views of the skull of the Short-tailed Gerbil *Desmodillus auricularis*. Note enormous development of auditory bullae projecting prominently beyond level of occiput. The zygomatic plate not as well developed as in *Tatera*.

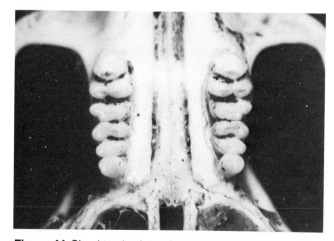

**Figure 44** Cheekteeth of the Short-tailed Gerbil *Desmodillus auricularis*. Molars are small. $M^1$ is trilaminate, $M^2$ bilaminate and $M^3$ unilaminate. Second lamina of $M^1$ shows two circular patterns of enamel.

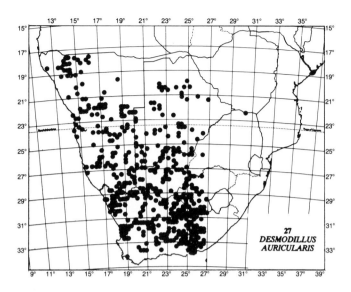

**27 DESMODILLUS AURICULARIS**

## Habitat

Their favourite sites in Botswana appear to be in and around dry pans. Smithers (1971) records them as occurring on open hard ground with some cover of grass or karroid bushes or hard sand with similar cover.

## Diet

They seem to be very partial to the prickly seeds of the 'dubbeltjie' or devilthorn *(Tribulus terrestris)*, the remains of which are found inside the entrances to the tunnels of the burrows. Furthermore, they take seeds of grasses and annuals (Smithers 1971), and possibly the pips of wild melons and locusts as well. Smith (1834) has described this species as feeding on insects.

## Habits

Burrows may be up to 2 m deep and the entrances are closed when the animal is in occupation. The burrows are not complicated, and according to Smithers (1971), they may be within short distances of each other, but do not form warrens. Food is usually eaten near the entrances of the burrow, as evidenced by scattered food remains. Hoarding behaviour by *Desmodillus auricularis* was studied under semi-natural conditions by Pettifer & Nel (1977), who found this species to be a true larder-hoarder, although it may also scatter-hoard on occasion. There appears to be another entrance to the underground shelter, nearly vertical, without any sign of excavated material or food remains. Burrow systems of *Desmodillus auricularis*, which were excavated in the Kalahari Gemsbok National Park, have been described in detail by Nel (1967). On the average these burrows lie within an area of about 7 m².

They are nocturnal and do not enter traps readily. They are inactive animals and fatten readily if plenty of food is available (Roberts 1951).

## Reproduction

Mammae consist of two pectoral and two inguinal pairs.

Young may be born throughout the year. The number of fetuses carried by females averaged 3,9 in Botswana (Smithers 1971). It seldom has litters under laboratory conditions unless given extra light hours (Keogh & Isaäcson 1978).

Nel & Stutterheim (1973) studied the early post-natal development of litters of *D. auricularis* born in captivity. Animals were live-trapped in the Kalahari Gemsbok National Park and transported to a laboratory at the University of Pretoria. They do not breed readily in captivity, probably because of animosity between the sexes. At birth the young are pink in colour and some degree of cannibalism by the mother on the pups was observed. The average mass of a newborn young in 1,85 g with a mean total length (i.e. head-body length plus tail length) of 37,7 mm. The litter size ranged from one to three with a mean value of two. The pelage proper was visible only on day 12 while the white spot behind the pinna (characteristic of adult specimens of *Desmodillus auricularis)* appears on day 14. The eyes appeared as dark slits on day 14 and in two pups observed, the eyes were completely opened on days 21 and 23. Leaving the nest was first observed on day 19. In nature, 45+ days may be the time when the young first leave the nests (and burrows) to start wandering about.

In contrast to *Tatera brantsii* (Meester & Hallett 1970), nipple-clinging is absent in *D. auricularis*. In general, compared to *Tatera*, weaning occurs rather late in *Desmodillus*, while the development of locomotory functions is also slower. Nel & Stutterheim (1973) also point to the small litter size in the mouth-carrying *Desmodillus* which may demonstrate the survival value of underground nests. This is also demonstrated by comparing the burrowing *Tatera brantsii* (2,0 per litter) and the surface-dwelling *Rhabdomys pumilio* (7,0 per litter).

## Parasites

As is the case among other southern African gerbils, the Namaqua Gerbil is also subjected to periodic zoonoses of *Yersinia pestis* outbreaks. In the laboratory, they become infected by *Listeria monocytogenes,* the coryne-bacteria responsible for listeriosis, *Rickettsia prowazekii* (louse typhus), *Rickettsia typhi* (murine or rat typhus) and *R. conorii* (tick-bite fever) (Neitz 1965).

The laelaptid mite *Androlaelaps marshalli* has been taken from this rodent (Zumpt 1961a). The fleas known to be associated with *Desmodillus auricularis* include four families, nine genera and 24 species. The pulicids include *Echidnophaga gallinacea, Pulex irritans, Procaviopsylla creusae, Xenopsylla cheopis, X. eridos, X. philoxera, X. phyllomae, X. piriei, X. trifaria, X. versuta, X. brasiliensis* and *X. demeilloni.* They hystrichopsyllids include *Dinopsyllus ellobius, Listropsylla agrippinae* and *L. dorippae. Leptopsylla ingens* represents the Leptopsyllidae. The fourth family, the Chimaeropsyllidae, is represented by *Epirimia aganippes, Chiastopsylla numae, C. quadrisetis, C. rossi, C. co-*

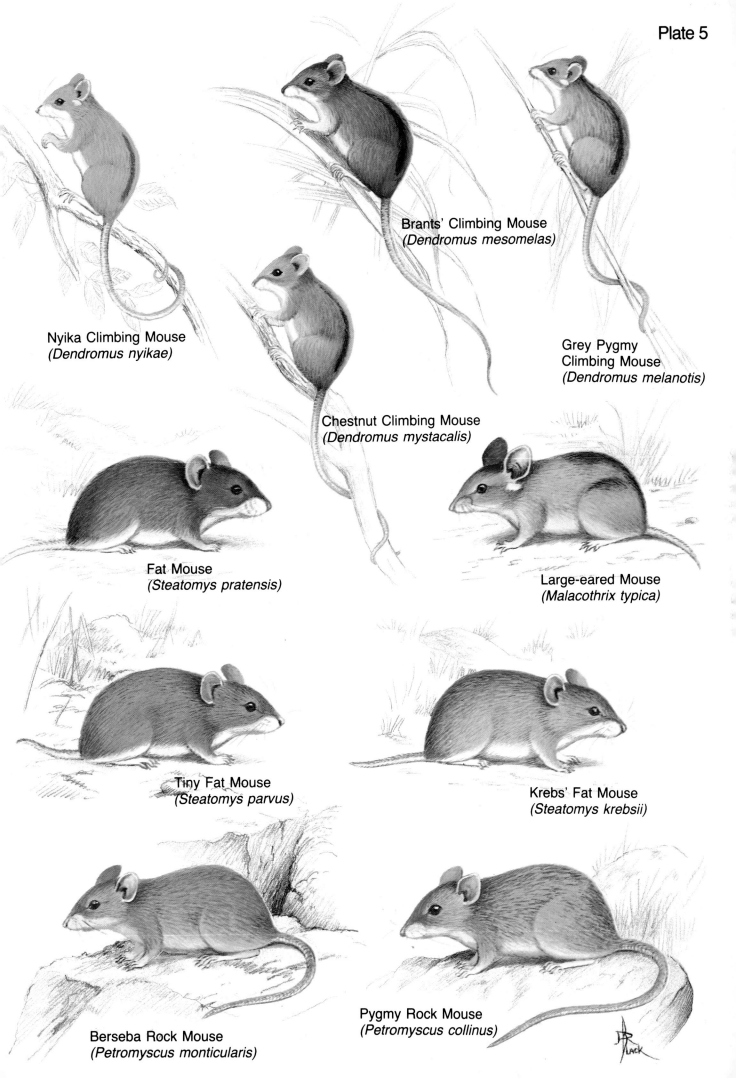

Plate 5

Nyika Climbing Mouse
(*Dendromus nyikae*)

Brants' Climbing Mouse
(*Dendromus mesomelas*)

Grey Pygmy
Climbing Mouse
(*Dendromus melanotis*)

Chestnut Climbing Mouse
(*Dendromus mystacalis*)

Fat Mouse
(*Steatomys pratensis*)

Large-eared Mouse
(*Malacothrix typica*)

Tiny Fat Mouse
(*Steatomys parvus*)

Krebs' Fat Mouse
(*Steatomys krebsii*)

Berseba Rock Mouse
(*Petromyscus monticularis*)

Pygmy Rock Mouse
(*Petromyscus collinus*)

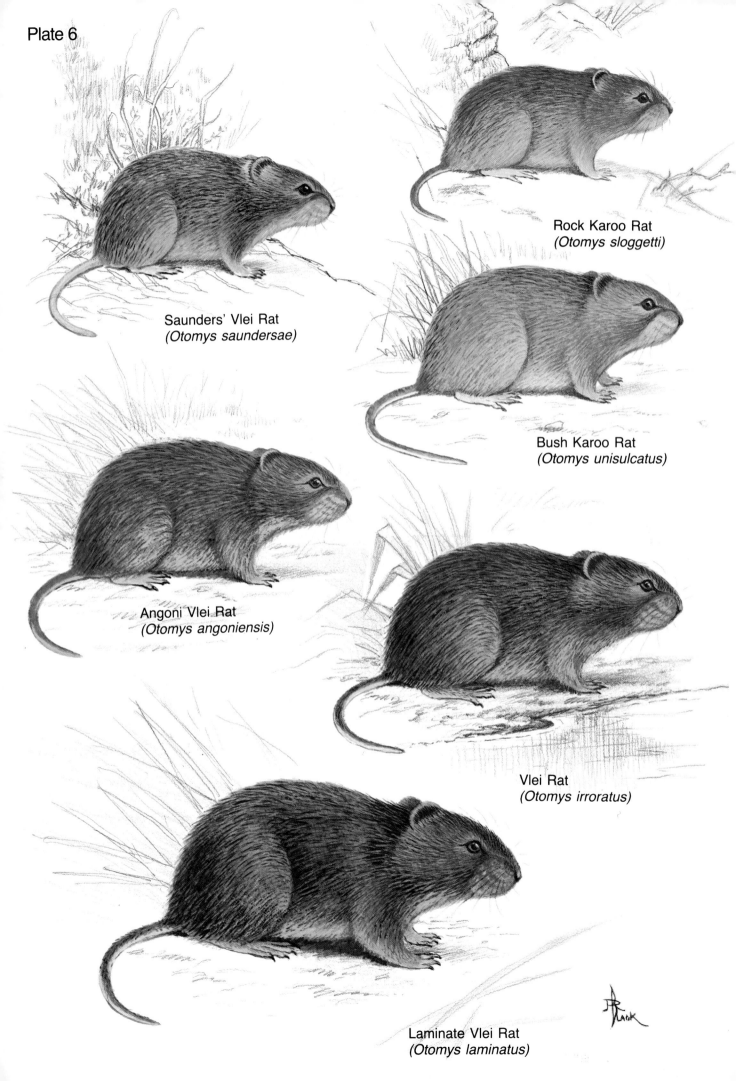

Plate 6

Saunders' Vlei Rat
(*Otomys saundersae*)

Rock Karoo Rat
(*Otomys sloggetti*)

Bush Karoo Rat
(*Otomys unisulcatus*)

Angoni Vlei Rat
(*Otomys angoniensis*)

Vlei Rat
(*Otomys irroratus*)

Laminate Vlei Rat
(*Otomys laminatus*)

*raxis, C. godfreyi, C. mulleri* and *C. pitchfordi* (Zumpt 1966).

The ticks associated with this gerbil have been recorded by Theiler (1962). They include the soft tick *Ornithodoros zumpti* (both adults and immatures) and the hard ticks *Haemaphysalis houyi* (adult), *H. leachii leachii* (immature) and *Hyalomma rufipes* (immature).

## Relations with man
A large number of flea species associated with this rodent can be involved in the transmission of plague from rodent to man. The ticks they carry also spreads a diversity of disease, e.g. *Hyalomma rufipes* has been shown to transmit *Rickettsia conorii, R. pijperii* and *Coxiella burnetii*. Under laboratory conditions they eat mainly sunflower seeds and greens, although they will take to standard mouse cubes (Keogh & Isaäcson 1978). It has been used in tick-bite fever studies and it shows partial resistance to plague *Yersinia pestis*.

## Prehistory and predators
No information is available.

## Taxonomy
Lundholm (1955b) described two new subspecies of *D. auricularis* which he named *shortridgei* and *robertsi*. The former was based on material from Port Elizabeth and the latter from Sesfontein in SWA/Namibia. Von Lehmann (1955) also described two more subspecies from SWA/Namibia naming them *hoeschi* and *wolffi* respectively. In 1960, Von Lehmann reassessed the situation and pointed to the possibility of regarding *wolffi* as a synonym of *pudicus,* a subspecies described from Lehutitung in Botswana, by Dollman in 1910 (later accepted as a synonym of *auricularis*). On working with a wide range of material from Botswana, Smithers (1971) concurs with the view that no subspecies can be accepted and that all the supposed subspecies are inseparable from *auricularis* (Petter 1975).

SUBFAMILY # Cricetomyinae Roberts, 1951

The pouched rats and mice are a small group of rodents sharing a well-defined characteristic of large cheekpouches. All genera have a grey or grey-brown colouring. They burrow, designing storage chambers for food hoarding and they move on to new abodes when the food supply is exhausted (Kingdon 1974). Three genera are included in this subfamily: *Cricetomys, Saccostomus* and the extralimital *Beamys*.

The pouched mice occurring in southern Africa have long been regarded as murine, but Petter (1964) believes that the genus *Cricetomys* should be interpreted as a separate subfamily under the Cricetidae. This also applies to the genus *Saccostomus* (Petter 1966b) and the argument hinges on the interpretation of the lingual cusps in the $M^1$ and $M^2$, which exist in these genera – i.e. whether these are relicts of a fundamentally triserial (murid) dentition or accessories to a fundamentally biseral (cricetid) pattern of cuspidation (Rosevear 1969).

*Cricetomys* and *Saccostomus* show a number of anatomical features different from other murids – especially the possession of cheekpouches which distinguishes them from the Murinae – apart from the question of cuspidation of the molars referred to above. Roberts (1951) says that *Cricetomys* is '... a very isolated genus, totally different from other members of the Murinae in many characters ...' and placed it in a new subfamily Cricetomyinae (a practice followed by Petter). He retained the subfamily in the Muridae. Similarly, *Saccostomus* was also placed in a separate subfamily, the Saccostomurinae, also under Muridae (Roberts 1951).

I follow the interpretation of Petter who argued that the teeth in *Cricetomys* and *Saccostomus* are essentially biserial, i.e. cricetid in character. Those who have regarded the teeth as biserial have '... associated the anterior cusplet with the front lamina and the second with the second lamina as in the Murinae' (Rosevear 1969). However, Petter has shown that in *Cricetomys* as well as *Saccostomus* and *Beamys* (extralimital) the foremost cusplet in fact joins the second lamina and the posterior cusplet with the third lamina. Petter argues, therefore, that the dentition is essentially cricetid rather than murid (Rosevear 1969).

GENUS # *Cricetomys* Waterhouse, 1840

This is a large rodent, resembling an oversized rat, but it possesses cheekpouches in which it carries around food.

This feature is also found in the southern African Pouched Mouse *Saccostomus campestris* and in the extralimital *Beamys hindei* occurring in East Africa. The Giant Rat, as *Cricetomys* is popularly known, is confined to the African continent and can attain a head-body length of 300 mm and more. On account of its size alone, it is a striking species. I have followed Misonne (1974) and others by placing it with the Cricetidae in this work. Only two species, spread over a large part of Africa, are generally acknowledged. *Cricetomys gambianus* has a wider distribution than *C. emini* (extralimital to this work), which is to be found in West Africa, ranging to Tanzania. The latest revision of this genus has been presented by Genest-Villard (1967). The two species are closely related and distinction between them is not easy.

# *Cricetomys gambianus* Waterhouse, 1840

**Giant Rat**

**Reuserot**

The generic name is compounded from *Cricetus,* a Palaearctic genus, and *mys,* the Greek for mouse, for it was thought to form a link between these two genera. The species is named after the Gambia River in West Africa.

Until recently, this rodent was regarded as a representative of the Murinae although there was a degree of uncertainty about such an interpretation. Waterhouse (1840) who originally described *C. gambianus* as a subgenus of *Mus* referred to it as an interesting link between *Mus* and *Cricetus,* i.e. between the Muridae and Cricetidae as understood today. Hinton (1919) has published notes on the genus. It is a timid and docile animal and shows little inclination to bite if handled (Roberts 1951). The status of this species has been described as rare, with a very limited distribution in South Africa, although widespread elsewhere in Africa, in the *South African Red Data Book* by Meester (1976). On account of its limited distribution in South Africa, it has been included in the list. Extralimitally it ranges over most of tropical and subtropical Africa from West Africa to Kenya in the east and Transvaal and northern Natal in the south.

## Outline of synonymy

1840 *Cricetomys gambianus* Waterhouse, *Proc. zool. Soc. Lond.:* 2. Gambia River, West Africa.

1842 *Mus goliath* Rüppel, *Mus. Senckenbergianus* 3:114. Senegambia.

1853 *Cricetomys gambiensis* Temminck, *Esquisses Zoologiques sur la Côte de Guine:* 168. (A *lapsus calami.*) Guinea.

1907 *Cricetomys gambianus adventor* Thomas & Wroughton, *Proc. zool. Soc. Lond.:* 295. Inhambane District, Mozambique.

1908 *Cricetomys gambianus cunctator* Thomas & Wroughton, *Proc. zool. Soc. Lond.:* 171. Tambarara, Gorongoza District, Mozambique.

1926 *Cricetomys gambianus haagneri* Roberts, *Ann. Transv. Mus.* 11:252. Soutpansberg, Transvaal.

1946 *Cricetomys gambianus selindensis* Roberts, *Ann. Transv. Mus.* 20:139. Mount Selinda, Zimbabwe.

1953 *Cricetomys gambianus* Waterhouse. *In* Ellerman *et al., Southern African mammals.*

1974 *Cricetomys gambianus* Waterhouse. Misonne, *in* Meester & Setzer (eds), *The mammals of Africa: an identification manual.*

TABLE 25
Measurements of male and female *Cricetomys gambianus*

|  | Parameter | Value (mm) | N | Range (mm) | CV (%) |
|---|---|---|---|---|---|
| Males | HB | 363 | 5 | 350–385 | 3,2 |
|  | T | 418 | 5 | 390–450 | 4,5 |
|  | HF | 72 | 5 | 68–75 | 3,0 |
|  | E | 42 | 5 | 39–45 | 4,5 |
| Mass: Up to 1,5 kg | | | | | |
| Females | HB | 343 | 6 | 330–360 | 2,6 |
|  | T | 407 | 6 | 370–445 | 5,3 |
|  | HF | 69 | 6 | 67–72 | 2,2 |
|  | E | 40 | 6 | 38–42 | 3,0 |
| Mass: Up to 1 kg | | | | | |

## Identification

The dorsal colour of these exceedingly large rats is dull buff with the flanks and face slightly lighter. The ventral surface is a dull white, while the throat is white. A dark brown ring occurs around the eyes. The tail is about as long as the length of the head and body combined, or may be even longer: it is thinly haired, brown in colour and is often white-tipped (plate 8). The tail is not scaled, as in true murids. The ears are moderately large. The hands are white, while the feet are white only on the toes. The hindfoot is powerfully built, although the claws are rather small. On the hand, the pollex is rudimentary, while the third digit is the longest. The vibrissae are numerous and long. These rats also possess cheekpouches (as in *Saccostomus*) – a distinguishing anatomical feature – and this was one of the reasons why it was originally thought that they are phylogenetically closer related to the pouched mice (family Cricetidae) than to the true Muridae. The fur is harsh to the touch, rather loose, with an element of underfur present. The fur consists chiefly of bristles which are essentially modified gutter hairs (Rosevear 1969).

The sounds they emit have been described as short and sharp. Ellerman (1941) states that when excited or pleased, the African Giant Rat makes a chirruping noise, reminiscent of a canary.

The mammae consists of two pectoral and two inguinal pairs, eight in all (Roberts 1951).

## Skull and dentition
The skull is elongated and narrow and rather flat. The postorbital skull is fairly well developed, as are the supraorbital ridges and paroccipital processes. The jugal is long, while the bullae are relatively small. The anterior palatine foramina are short, situated largely on the premaxillae. The infraorbital foramen is large (fig. 45).

The angular process of the lower jaw is not projected backwards as in true murids. The coronoid process is well developed.

Both upper and lower incisors are ungrooved. The $M^1$ has three laminae, the $M^2$ and $M^3$ two laminae each. These laminae consist basically of two rows of cusps with a variable number of supplementary and minor cusps. Some workers have regarded the molars of *Cricetomys* as triserial having associated the anterior cusplet (a supposed t1) with the first lamina and the second with the second as in the murines. However, Petter (1966a, 1966b) has shown that in *Cricetomys* (as in *Saccostomus*) the front cusplet in fact joins with the middle lamina and the back cusplet with the last lamina. This condition '. . . is not murine, or even murid, but essentially cricetid' (Rosevear 1969). The $M^2$ always shows a clear supplementary posterior cusp. The $M^3$ is nearly as large as the $M^2$ (fig. 46). In the lower molars the terminal heel of the $M_1$ and $M_2$ are well developed and there is a variable number of subsidiary cusps present on the buccal side.

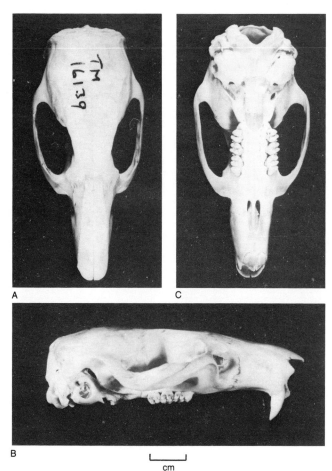

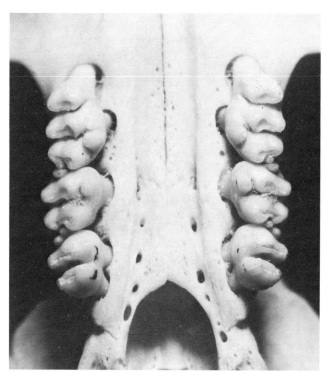

**Figure 46** Cheekteeth of the Giant Rat *Cricetomys gambianus*. The $M^1$ is trilaminate, the $M^2$ and $M^3$ both bilaminate. The laminae consist basically of two rows of cusps with a variable number of supplementary cusps. The $M^3$ is nearly as large as the $M^2$.

**Figure 45** Dorsal (A), lateral (B) and ventral (C) views of the skull of the Giant Rat *Cricetomys gambianus*. The skull is elongate, narrow and flattened. The postorbital skull is well developed with small bullae. The infraorbital foramen is large.

## Distribution
The Giant Rat occurs in southeastern Mozambique, ranging southwards to KwaZulu and westwards to the northern and central Transvaal (map 28). Extralimitally the species ranges over most of tropical and subtropical Africa from West Africa to Sudan.

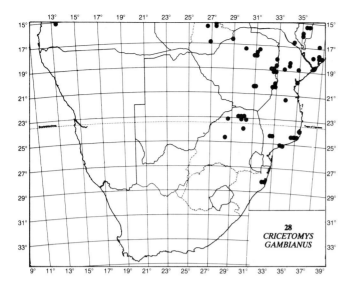

**28**
**CRICETOMYS**
**GAMBIANUS**

## Habitat

This species seems to prefer savanna conditions, but ranges into humid, evergreen forests. They shun arid and semi-arid areas. In forests in Cameroon, they live among roots of large trees or in holes in logs and under fallen tree trunks (Sanderson 1940). In West Africa, they often inhabit termite nests (Rosevear 1969).

## Diet

Giant rats feed on seeds, fruits (wild and cultivated), groundnuts and maize. Hatt (1940) reported them eating termites and there is some evidence that they consume snails (Rosevear 1969). Their main food items appear to be fruits of various kinds, wild or cultivated. They are coprophagous and are dependent on water and moist food (Kingdon 1974).

## Habits

The Giant Rat can be communal, but is usually solitary. It is terrestrial and predominantly nocturnal, although occasionally diurnal. Kingdon (1974) says that they soon die of exposure to excessive sunlight or heat, so that shelter is important to them. Its tunnel system is extensive (Morris 1963). The nest chamber is lined with plant material. Males and females occupy separate burrow systems (Ewer 1967), which consist of galleries with separate chambers for food storage, resting, defaecation and breeding (Rosevear 1969). Each system has a varying number of escape routes and they tend to change their habitation constantly. Where the ground is overgrown, they may not burrow at all (Sanderson 1940). Rosevear (1969) also states they they are infrequently found in and near human dwellings where they may steal metal objects like knives, keys, bottle-tops, rings and coins. Ewer (1967) has suggested that the impulse to carry objects to the food store is often stronger than the rats' drive to fill its pouches with food. She has shown that a succession of journeys to an abundant source of food seems to exhaust the drive to fill its pouches, while the impulse to carry strange objects may be longer-lived.

Giant rats use their buccal pouches for threat displays (Davis 1959), and for the temporary storage of food. When trapped, they soon allow themselves to be handled, without attempting to bite the handler (Roberts 1951). On account of the fairly well-developed hindlegs, they are fairly strong jumpers.

## Predators

Reports on Giant Rat predation are rare. Chapin (1932) records the Steppe Eagle *Aquila rapax nipalensis* eating a Giant Rat. Gargett (1977) records Mackinder's Eagle Owl *Bubo capensis mackinderi* as a predator in the Matopos, Zimbabwe, while Pitman & Adamson (1978) state that the Giant Eagle Owl *Bubo lacteus* also preys upon them. One of the most important predators of this species is man.

## Reproduction

The development of the Giant Rat is slower than that of murid rats like the Brown Rat. The female has a total of eight mammae: two pectoral and two inguinal pairs. According to Rosevear (1969), the gestation period is 42 days. Kingdon (1974), however, refers to 32 days. Bourlière (1948) has described the early post-natal development as follows. Two to four naked young are born with eyes and ears closed. The fur develops on the fifth day and the young animals are well covered with fur on the 17th day. The ears are also well developed at the 17th day. The eyes open partially on the 19th day (in one litter observed), while the incisors and the molars erupt on the 11th and 21st days respectively. The pups start moving about in the fourth week. Ewer (1967) has also given a detailed account of the development of the young.

Morris (1963) indicates a lifespan of 4½ years for an animal in captivity.

## Parasites

Apart from other ectoparasites, the Giant Rat carries an earwig *Hemimerus talpoides,* peculiar to itself and belonging to the order Demaptera. This is also commented upon by Ansell (1960), Roberts (1951) and Rosevear (1969). It is a blind, wingless, viviparous, furliving insect which very likely feeds on skin debris, so it is hardly a parasite but rather an epizootic animal. Initially, only the single species of *Hemimerus talpoides* was known. *H. talpoides* has now been joined by seven other species described from various parts of Africa. These have not been correlated with properly identified forms of *Cricetomys*. It probably feeds on epidermal waste of *Cricetomys,* although Morris (1963) states that it rather subsists on materials stored by the Giant Rat in its burrows (Rosevear 1969).

The mites include the following genera and species (Zumpt 1961a): Family Laelaptidae: *Andréacarus petersi, A. zumpti, Laelaps muricola* and *Haemolaelaps galagus.*

Family Trombiculidae: *Trombicula panieri, Schoengastia r. radfordi, Ascoschoengastia anomaluri, ?Euschoengastia mailloti, Schoutedenichia audyi, S. schoutedeni, S. paulus major, S. penetrans* and *Gahrliepia womersleyi*. Family Listrophoridae: *Listrophoroides aethiopicus*. Family Porocephalidae: *Armillifer armillatus*.

The earwig and fleas include the following (Zumpt 1966): Family Hemimeridae: *Hemimerus advectus, H. bouvieri, H. deceptus, H. hanseni, H. sessor, H. talpoides* and *H. vosseleri*. To these species *H. vivinus* is to be added (Rosevear 1969). A ninth species, *H. morrisi*, occurs with *Beamys*, another African rodent with cheekpouches. Family Calliphoridae: *Cordylobia anthropophaga* and *C. rodhaini*. Family Pulicidae: *Echidnophaga gallinacea, Ctenophalides felis, Procaviopsylla angolensis, Xenopsylla aequisetosa, X. nubica, X. crinita* and *X. torta*. Family Hystrichopsyllidae: *Dinopsylla semnus*. Family Ceratophyllidae: *Libyastus duratus*.

Finally, the ticks include the following (Theiler 1962): *Ornithodoros erraticus* (immatures – I), *Ixodes cumulatimpunctatus* (adults – A), *I. loveridgei* (A), *I. oldi* (I), *I. rasus* (A), *I. ugandanus* (A), *Haemaphysalis parmata* (I), *Rhipicelaphalus sanguineus (A)* and *R. simpsoni* (A). For the above list, Theiler (1962) regards the extralimital *Cricetomys emini* as a subspecies of *gambianus*.

According to Collins (1972), the following cestodes have been collected from the Giant Rat: *Catenotaenia cricetomydis, Inermicapsifer angolensis, I. guineensis* and *I. madagascariensis*.

## Relations with man

This species can become a pest in groundnut and maize fields. It will also take beans, mangoes and avocados (Kingdon 1974). In primitive societies, the animals are smoked out of their burrows and eaten as a delicacy. Because of their size, they are much sought after as food (Ajayi 1974). In some areas, (e.g. Ruwenzori) their skins are used as tobacco pouches (Kingdon 1974).

## Prehistory

These rodents have not as yet been recorded as fossils or subfossils from anywhere on the continent of Africa.

## Taxonomy

According to Ellerman (1941), 28 subspecies have been described. Roberts (1951) recognised four subspecies in southern Africa: *adventor* (near Inhambane in Mozambique), *cunctator* (from Gorongoza in Mozambique), *silindensis* (from Mount Selinda in southeastern Zimbabwe) and *haagneri* (from the Soutpansberg in northern Transvaal). Meester *et al.* (1964) consider these forms probable synonyms of *viator* Thomas, 1904. Misonne (1974) regards *Cricetomys gambianus* as monotypic and the subspecies mentioned above as conspecifics, following Genest-Villard (1967), who has divided the genus into two species only, i.e. *gambianus* and *emini*, the former embracing six races and the latter two.

GENUS *Saccostomus* Peters, 1846

# *Saccostomus campestris* Peters, 1846

**Pouched Mouse**

**Wangsakmuis**

The generic name is derived from the Greek *sakkos* = sac, *stomatos* = mouth and *mys* = mouse. A combination of these refers to the cheekpouches which open in the mouth cavity. The specific name is derived from *campestris* = of the plains (Latin).

The type specimen of this species was collected by Peters in 1846 from Tete in central Mozambique near the Zambezi River. These animals were separated from the Dendromurinae into a new subfamily, the Saccostomurinae by Roberts in 1951 because of the presence of cheekpouches and marked differences in the skull. They are robust, round, fat, mouse-like animals with short limbs and tails. By the end of the 19th century, four species had already been described which, apart from the nominate species, have been demoted with the passage of time to subspecific rank. The latest interpretations (e.g. Meester *et al.* 1964; Misonne 1974) admit no subspecies at all and advocate a monotypic species in southern Africa.

## Outline of synonymy

1846 *Saccostomus campestris* Peters, *Monatsber. K. Preuss. Akad. Wiss. Berlin*: 258. Tete, Mozambique.
1852 *Saccostomus lapidarius* Peters, *Reise nach Mossambique. Säugeth.*: 167. (Renaming of *S. campestris*.)

1852 *Saccostomus fuscus* Peters, *Reise nach Mossambique. Säugeth.*: 168. Inhambane, Mozambique.

1896 *Saccostomus mashonae* De Winton, *Proc. zool. Soc. Lond.*: 804. Mazoe, eastern Zimbabwe.

1898 *Saccostomus anderssoni* De Winton, *Ann. Mag. nat. Hist.* (7) 2:6. Damaraland, SWA/Namibia.

    *Synonyms*

    1906 *S. hildae* Schwann, *Proc. zool. Soc. Lond.*: 110. Kuruman, CP.

    1923 *S. pagei* Thomas & Hinton, *Proc. zool. Soc. Lond.*: 495. Lehutitung, Botswana.

1903 *Eosaccomys* Palmer, *Science* N.S. 17: 873. (To replace *Saccostomus* Peters, supposed to be preoccupied by *Saccostoma* Fitzinger, 1843, in reptiles.)

1914 *Saccostomus limpopoensis* Roberts, *Ann. Transv. Mus.* 4:183. Sand River, Soutpansberg district, northern Transvaal.

1914 *Saccostomus streeteri* Roberts, *Ann. Transv. Mus.* 4: 138. Hectorspruit, Transvaal.

1934 *Saccostomus anderssoni anderssoni* De Winton. *In* Shortridge, *The mammals of South West Africa*.

1934 *S. a pagei* Thomas & Hinton. *In* Shortridge, *The mammals of South West Africa*.

1939 *S. campestris* Peters. *In* Allen, *Bull. Mus. comp. Zool. Harv.* 83:357.

1951 *Saccostomus campestris campestris* Peters. *In* Roberts, *The mammals of South Africa*.

1953 *Saccostomus campestris* Peters. *In* Ellerman *et al. Southern African mammals*.

1974 *S. campestris* Peters. Misonne, *in* Meester & Setzer (eds), *The mammals of Africa: an identification manual*.

### TABLE 26
Measurements of male and female *Saccostomus campestris*

|  | Parameter | Value (mm) | N | Range (mm) | CV (%) |
|---|---|---|---|---|---|
| Males | HB | 114 | 21 | 83–145 | 11,3 |
|  | T | 50 | 21 | 32–83 | 20,9 |
|  | HF | 21 | 21 | 17–30 | 12,2 |
|  | E | 14 | 21 | 12–22 | 14,8 |
|  | Mass: | 48,5 g | 20 | 33–68 g* | — |
| Females | HB | 106 | 11 | 87–120 | 8,6 |
|  | T | 47 | 11 | 35–57 | 11,2 |
|  | HF | 19 | 11 | 17–22 | 6,5 |
|  | E | 17 | 11 | 15–22 | 10,4 |
|  | Mass: | 42,2 g | 20 | 30–54 g* | — |

*Smithers (1971)

### Identification

In relation to its overall size, *Saccostomus* is robust with a thickset head. The dorsal surface is brownish grey; the basal portion of the hairs of the fur is slaty with pale brown tips. Black-tipped hairs occur, especially along the mid-dorsal line. The entire ventral surface, as well as a tuft of fur below the ear, is white or yellowish-white without slaty bases to the fur. The demarcation in colour between the dorsal and ventral aspects is distinct. The appendages are sturdy and short, white, the forefeet small, while the hindfeet are large. The tail is short (less than half the total head-body length), thinly bristled, brown above and white below. When the animal is skinned, the skin of the tail does not slip as readily as in most small rodents (Shortridge 1934), and is not scaly. The colour is variable geographically, ranging from brown to sandy to grey dorsally (plate 7).

The upper whiskers are brown and the lower ones white. The eyes are small, set rather close together and far forward. The cheekpouches extend backwards to a level in front of the shoulders. When filled with seeds, they form two large swellings on either side of the head. The ears are short and rounded.

The fur is thick, fine and silky in texture and the skin is loosely attached to the body. There is much variation phenotypically, which may be seasonal and as a general rule specimens from the west of southern Africa are paler than those from the east.

The mammae consist of three pectoral and two inguinal pairs.

The glans penis and the caecum of *Saccostomus* resemble those of *Cricetomys* (Ellerman 1941).

### Skull and dentition

The skull is narrow and elongated with a heavy muzzle and a clear interorbital constriction. The infraorbital foramen is not greatly enlarged, while the bullae are relatively large and inflated. The zygomatic plate is rounded above. The zygomatic arches are not excessively wide. The palate is broad and the palatine foramina long, reaching the molar toothrow (fig. 47).

Both the upper and lower incisors are small, ungrooved and yellow in colour.

The cusps of the molars are basically biserial in arrangement with a variable number of additional cusps. The most important feature of the first upper molar is the absence of the antero-internal cusp (t1). In the $M^2$ both the t1 and t3 are missing. In the $M^3$ two sections occur: a posterior and an anterior one. This tooth is not greatly reduced as in the dendromurids (fig. 48).

### Distribution

*Saccostomus campestris* is found throughout southern Africa except in the southern and western Cape and along the coastal desert belt of SWA/Namibia (map 29). It also shuns the Southern Savanna Grassland biotic zone.

### Habitat

This species has a wide habitat tolerance. In SWA/Namibia it occurs in open veld, in dense bush or in forests

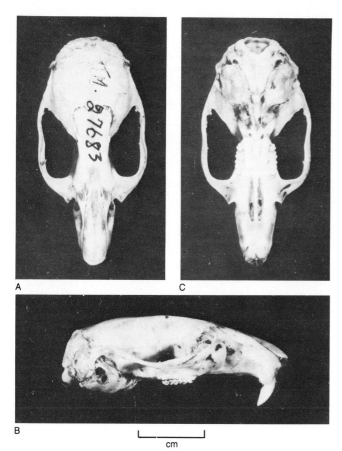

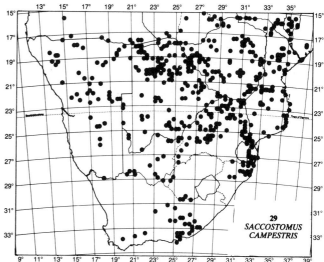

A          C

B

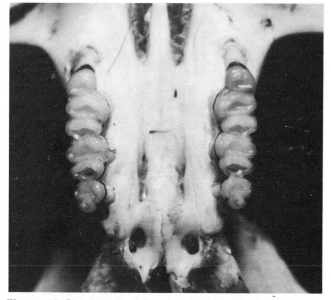

**Figure 47** Dorsal (A), lateral (B) and ventral (C) views of the skull of the Pouched Mouse *Saccostomus campestris*. The narrow, elongated skull has a clear interorbital constriction. The zygomatic plate is rounded above; bullae enlarged; zygomatic arches are not excessively wide; anterior palatine foramina do not reach toothrows.

**Figure 48** Cheekteeth of the Pouched Mouse *Saccostomus campestris*. Cusps are basically biseral in arrangement with a variable number of additional cusps. M³ bilaminate and its size is not greatly reduced as is the case in dendromurines.

(Shortridge 1934) while in Botswana it occurs in sandy areas and kopjes (Smithers 1971), as well as in open short grass near pans, in dry riverbeds and in mopane veld. Holes in termite mounds as well as other natural holes are used as homesites in northern KwaZulu (Swanepoel 1972).

Swanepoel (1972) recognised three mass classes in KwaZulu, *viz.* less than 35 g, 35–50 g and more than 50 g. These possibly correspond to juvenile, subadult and adult classes. The body mass of individuals in the Kwa-Zulu population varied seasonally: adult animals predominated in the spring sample, age classes were equally distributed in summer, but during autumn and winter subadults predominated. The number of smaller and intermediate sized animals increased progressively from spring to winter.

**Diet**

Pouched mice eat the seeds of a diversity of plant species. The seed of the Umbrella Thorn *Acacia tortilis*, the Reukpeul *A. nilotica* and the Sweet Thorn *A. karroo* have been found in their cheekpouches by Swanepoel (1972) and he also reports the storage of the seeds in their abodes of the Torchwood *Balanites maughamii*, probably for later consumption. He also observed a specimen eating the fruit of *Grewia monticola*. I have trapped this species along the Limpopo River with seeds of the Nyala Tree *anthocercis zambesiaca* in its pouches. In Botswana, they take the seeds of *Acacia giraffae*, *Grewia* sp., *Combretum* spp. and *Colophospermum mopane* (Smithers 1971). According to Pienaar *et al.* (1980) they also consume seeds of the Sekelbos *Dicrostachys cinerea*, the Raisin Bush *Grewia bicolor* as well as *G. flavescens* in the Kruger National Park. They also take termites, grasshoppers and other insects. This foraging species is a true hoarder and the transport of food is facilitated by the cheekpouches which are filled with the forefeet (Pettifer & Nel 1977). Jacobsen (1977) has reported them feeding on the seeds of *Burkea africana*, *Peltophorum africanum* and *Euclea crispa*.

117

## Habits

*Saccostomus campestris* is exclusively terrestrial and predominantly solitary and nocturnal. However, it walks about on cloudy days and during the later afternoons in search of food (FitzSimons 1919). Swanepoel (1972) has found that male and female home ranges do not differ significantly. Shortridge (1934) mentions that they may occasionally be seen wandering above ground at early dusk. They are sluggish animals and can easily be picked up, when they will bite if handled. Yet they seem to be very docile by nature and make interesting pets. Kingdon (1974) says that they cannot climb trees, shrubs or grass.

They tend to live singly or in pairs or in small family parties, while in some localities they may perhaps be gregarious. According to Roberts (1951) they make their own burrows with separate entrances and exits as well as food storage chambers, which intercommunicate (Shortridge 1934). They are adept diggers and entrances to the burrows appear to be sealed during the day (Earl 1978). Shortridge also points out that *Saccostomus* and *Desmodillus* often burrow in the same patches of ground: the former seems to throw up less soil outside the burrow, but both genera make small, perpendicular emergency outlets. On the other hand, they may use disused holes dug by antbears or springhares (Smithers 1971) and holes in termite mounds (Swanepoel 1972).

Food is accumulated during summer and stored in underground chambers for use in winter (Walker 1975) and such food is transported in the cheekpouches. The cheekpouches are elastic and distensible and in SWA/Namibia a cheekpouch can even hold a flat seedpod of the mopane '. . . which are almost as large and round as pennies', about 35 mm in diameter (Shortridge 1934). The pouches are used mostly for the transport of food for *Saccostomus* is a true larder-hoarder. Earl (1978) states that the capacity of the pouches is great and in one case 63 sunflower seeds have been transported simultaneously. Occasionally these pouches are used to carry young (Earl 1977).

According to Swanepoel (1972), females in oestrus and pro-oestrus are aggressive and have been observed to rip open the scrotal sacs of males or even kill them.

Shortridge (1934) also points out that pouched mice appear periodically in abnormal numbers and they may also be nomadic in times of drought.

## Predators

Specific information is meagre, but they are probably preyed upon by smaller carnivores, the smaller birds of prey as well as the Barn Owl *Tyto alba*. Dean (1977) and Pitman & Adamson (1978) respectively state that they are eaten by the Grass Owl *Tyto capensis* and the Giant Eagle Owl *Bubo lacteus*.

## Reproduction

The female has three pairs of pectorally situated and two pairs of inguinally situated mammae. Shortridge (1934) reported a specimen with four pectoral pairs.

Swanepoel (1972) recorded pregnant females in October and February in KwaZulu while Smithers (1971) recorded gravid females during January to April in Botswana. It thus seems likely that the young are born during the wet summer months. In a sample of eight gravid females, Smithers (1971) determined the average number of fetuses per female as 7,4 with a range of 5–10. Scrotal males have been observed by Swanepoel from August to June. Males of *S. campestris* become infertile for part of the dry season in Malaŵi. Hanney (1965) has also stated that a lifespan exceeding one year is not uncommon in this species. In the Kruger National Park the breeding season seems to be from spring to autumn (Kern 1977).

In the Kafue National Park in Zambia, Wrangham (1969) observed the post-natal development of the young, which were conceived in the wild. The litter consisted of four babies which assumed the adult appearance on the 18th day, when the eyes were still closed. From this and other evidence, Swanepoel states that it may be assumed that weaning occurs between days 19 and 25 at a mass of 11–15 g. Earl (1978) has studied the post-natal development of *Saccostomus campestris* in detail in the laboratory. Females produce their first litters at an average age of 96 days and the gestation period is 20–21 days. Litter size varies from two to eight, an average of 4,8 pups. This is lower than reported by Smithers (1971) for data from Botswana. The young are born fully haired with a mean birth weight of 2,8 g. Pups are very vocal when young and vocalisation decreases gradually as they get older.

## Parasites

The following mites are known to occur on *Saccostomus campestris* (Zumpt 1961a): Family Laelaptidae: *Laelaps giganteus, L. muricola, L vansomereni, Haemolaelaps oliffi, H. rhodesiensis, H. villosissimus, Androlaelaps marshalli* and *A. theseus*. Family Trombiculidae: *Schoutedenichia dutoiti* and *Acomatacarus theileri*. Family Listrophoridae: *Listrophorus bothae*.

Fleas and other insects include the following (Zumpt 1966): Family Muscidae: *Musca domestica*. Family Calliphoridae: *Cordylobia anthropophaga*. Family Pulicidae: *Echidnophaga gallinacea, Xenopsylla cheopis, X. hipponax, X nubica, X. philoxera, X. phyllomae, X. piriei, X. bechuanae, X. brasiliensis, X. sarodes, X. scopulifer* and *X. syngenis*, as well as *Synosternus somalicus*. Family Pygiopsyllidae: *Stivalius torvus*. Family Hystrichopsyllidae: *Ctenopthalmus evidens, Dinopsyllus lypusus* and *Listropsylla agrippinae*.

The following ticks have been described from *S. campestris* (Theiler 1962): *Ixodes* sp. (immatures), *Haemaphysalis l. leachii* (immatures), *H. l. muhsami* (adults and immatures) and *Rhipicephalus simus* (immatures).

The nematode, *Inermicapsifer madagascariensis*, commonly occurs as an endoparasite (Collins 1972).

### Relations with man

Pouched mice are attracted to some extent by cultivation. In some parts of their geographical range, they are said to be eaten by Blacks (Walker 1975). This species is used as a laboratory animal at the South African Institute for Medical Research, where it has been established as a regular breeder (Keogh & Isaäcson, 1978). It is a suitable host for the schistosome *Schistosoma haematobium* (Kingdon 1974), while it was also found to be susceptible to *S. mansoni* and *S. mattheei* (Fripp 1978).

### Prehistory

The only fossiliferous breccia in southern Africa which has yielded the Pouched Mouse as a fossil is the Cave of Hearths, Makapansgat, northern Transvaal (De Graaff 1960a).

### Taxonomy

Several subspecies are recognised by Roberts (1951). These include *streeteri, limpopoensis, hildae, anderssoni* and *pagei.* Roberts also retains *S. fuscus* as a second distinct species, as does Lundholm (1949). Meester *et al.* (1964), however, point out that '. . . it appears that *fuscus* is based on an immature animal, and that the other forms represent no more than clinal colour variation in *campestris;* hence all can be regarded as synonyms of this form'. Ellerman (1941) also thought it unlikely that there is more than one valid species of this genus. A monotypic interpretation of this species is also accepted by Misonne (1974).

## SUBFAMILY Dendromurinae Allen, 1939

This subfamily is restricted to Africa south of the Sahara and occurs predominantly on the eastern half of the continent. On account of their small size, they are referred to as mice rather than rats. This assemblage of mice varies considerably in morphology and it is well-nigh impossible to provide a generalised description of the subfamily as a whole, and the limits of the subfamily as interpreted by different authors varies a great deal. Ellerman (1941) recognises five genera and about 80 species and subspecies, while Allen (1939) went even further by listing nine genera and 99 subgeneric taxa. Nel (1969) studied the evolution and interrelationships of dendromund genera and discusses the adaptive radiation of the subfamily since early Pliocene times. He also provides a useful review of the taxonomic history and scope of the Dendromurinae, pointing out that none of the authors recognising the subfamily agree as to the genera that should be included in it.

The Dendromurinae are of uncertain origin and relationship. Like the Cricetomyinae, they have been regarded as murine but taking palaeontological evidence into account, have been placed with the cricetids by Lavocat (1959), an arrangement now generally accepted, and followed in this book.

Two characteristics serve to distinguish the dendromurids from the true rats and mice: (i) the cuspidation of the cheekteeth and (ii) the structure of the zygomatic plate. According to Rosevear (1969), the cuspidation is still open to debate. Miller & Gidley (1918) adopted the view that the cuspidation pattern is primitive and that the inner (lingual) row of cusps has never developed in the majority of teeth lamellae. On the other hand, Ellerman (1941) adopts the interpretation that such inner cusps as now exist are remnants of an original triserial arrangement,

characteristic of the murines. As indicated above, the latest opinion is at variance with both Miller & Gidley and Ellerman. Lavocat (1959) holds that the dendromurids are not murine at all and that they should be interpreted as cricetids, because of the basically biserial cusp arrangement.

The $M^1$ is always the largest tooth with the $M^2$ half the size of the first. The $M^3$ is small and usually reduced. In the $M^1$ and $M^2$ the cusps occur in two parallel, longitudinal rows and each tooth has an additional cusp lingually, adjacent to the middle lamella – this additional cusp is usually clear in the $M^1$, but rather less so in the $M^2$. Furthermore, the inner cusp row is more reduced than that of the true murids (Ellerman 1941).

The upper incisors have a single pronounced groove situated nearer to the outer margin than the inner. The incisors are opisthodont.

Historically, the subfamily under the name-form Dendromyinae was initially suggested by Alston (1876), and founded on the two genera which are still looked upon today as its central components, *Dendromus* and *Steatomys.* Tullberg (1899), Miller & Gidley (1918), Weber (1928), Allen (1939), Ellerman (1941) and Simpson (1945) all interpreted them as murine, while Miller & Gidley took them to be the most primitive group of murids (Rosevear 1969). This was also the interpretation adhered to by Roberts (1951), and Ellerman, Morrison-Scott & Hayman (1953). Meester, Davis & Coetzee (1964) do not assign them to either cricetids or murids, while Misonne (1974) places them squarely with the cricetids.

In southern Africa, three genera of the Dendromurinae are encountered, *viz. Dendromus, Steatomys* and *Malacothrix.* Of these, *Malacothrix* is confined to southern Africa as defined for the purposes of this work, while *Dendromus* and *Steatomys* show a wider distribution northwards. A fourth genus, *Petromyscus,* is sometimes placed with the Dendromurinae, but I have retained the subfamily rank accorded to them by Roberts in 1951, who referred to them as Petromyscinae. For practical purposes the Petromyscinae are included in the key differentiating between the southern African dendromurid genera. Ellerman (1941) also hinted at the possibility that this group of small mice could be another representative of the Cricetinae. The other dendromurid genera which are extralimital to this work, include *Delanymys, Dendroprionomys, Deomys, Leimacomys, Megadendromus* and *Prionomys.*

---

**Key to the southern African genera of the Dendromurinae (including the Petromyscinae)**
(Modified after Ellerman (1941) and Roberts (1951))

1 Infraorbital foramen normal: no masseter knob; hand with four, foot with five digits...... *Petromyscus*
(Rock Mice)
Page 139
Infraorbital foramen enlarged with outer border ridged; a large masseter knob present at lower border................................................................. 2
2 Hindfoot with four digits only, hallux suppressed; pterygoid fossa very broad, frontals much constricted; nares interni narrow; ears very large, tail very short..................................... *Malacothrix*
(Large-eared Mouse)
Page 136
Hindfoot with five digits; hallux retained............. 3
3 Forefoot with three digits; tail long, prehensile.........
*Dendromus*
(Climbing Mice)
Page 120
Forefoot with four digits; tail relatively short..........
*Steatomys*
(Fat Mice)
Page 129

---

# GENUS *Dendromus* A. Smith, 1829

The generic name of climbing mice is derived from the Greek *dendron* = tree and the Latin *mus* = mouse. Despite the vernacular name of tree mice (as they are often referred to), there is very little evidence that they are in fact arboreal and the usage of climbing mice is preferable. They occur predominantly in tall grass, herbs and shrubs and spend much time on the ground or in burrows.

In his review of the genus *Dendromus,* Rosevear (1969) describes the history of the taxonomy of the genus. The first attempt to rationalise the taxonomic arrangement of this genus was by Thomas (1916b), suggesting three subgeneric divisions: the subgenus *Chortomys,* a single specimen of the Three-striped Mouse from Ethiopia; the subgenus *Dendromus,* with a claw on the fifth digit; and the subgenus *Poëmys,* with a nail on the fifth digit. *Poëmys* was raised to generic rank by Shortridge (1934), but regarded as synonymous with *Dendromus* by Ellerman (1941). Bohmann (1942) subsequently reviewed the genus and abandoned Thomas' subgenera. He amalgamated the multitude of forms into three Rassenkreisen which appeared to be tantamount to species, corresponding to a large extent to the grouping as proposed by Thomas. However, he further subdivided Thomas' *Dendromus* section and also added as a fourth section, *Chortomys,* the solitary species that Thomas had also set apart.

The *Dendromus* section (= *mesomelas*) was divided into two 'species', a larger one and a smaller one, occurring sympatrically through most of Africa south of the Sahara. Bohmann was aware of the fact, however, that the differences between these two species were average rather than absolute. The smaller species was called *Dendromus pumilio* Wagner, 1841 by Bohmann, but Ellerman *et al.* (1953) regard it as not certainly identifiable and have, therefore, adopted the next available name, *mystacalis.* It appeared that the species *mesomelas* and *mystacalis* are representatives of Thomas' subdivision of *Dendromus.* However, Ellerman *et al.* (1953) state that the differences between the two groups of species (*melanotis* group and *mesomelas* group) seem to be far less than some authors (who have separated them subgenerically) would have them to believe.

In southern Africa, *Dendromus* contains the following species: *mesomelas, melanotis, mystacalis* and *nyikae.*

*Dendromus* has a wide geographical distribution. They are small mice, 60–80 mm in length, with tail slightly longer than the head-body length. The pelage is soft and full with dark-based hairs dorsally. A black mid-dorsal stripe is often present. The hair on the tail is not profuse and its scaly nature is clearly visible.

The $M^1$ is about twice as big as the $M^2$ and the $M^3$ is very small indeed. The two anterior molars have double pairs of cusps (the first with three pairs, and the second with two) and another cusp inside the middle pair of the $M^1$ and another cusp in the first row of the $M^2$ (Roberts 1951). The tail is used as a steadying organ, rather than as a prehensile appendage when climbing. The ventral pelage is lighter in colour than the dorsal. The hand has three digits, while in the hindfoot the hallux is reduced and the fifth digit large and opposable like a thumb. This is a feature unique to this genus. All the digits are long and slender. The head is rounded and the ears fairly large and rounded.

In the skull, the interorbital constriction is pronounced and the infraorbital foramen large and well developed. The muzzle is narrow. Supraoccipital crests or other ridges are absent and the zygomatic plate is narrow and vertical with a pronounced masseteric knob situated on the lower anterior corner of the plate. The anterior palatine foramina reach the level of the middle lamella of the $M^1$. The palate between the molars is broad, and terminates beyond the molar tooth row. The bullae are well developed.

All the species of *Dendromus* use specially constructed nests for shelter and breeding which have entrances on opposite sides. They may also occupy deserted nests of birds (Sclater 1901) such as weavers and some species of warblers (Roberts 1951). Three to five young form a litter (although Roberts has seen a full complement of eight embryos), and there seems to be no fixed breeding season. There are two pectoral and two inguinal mammae.

They often frequent suburban gardens and are aggressive fighters. They have been recorded killing a snake 250 mm in length (Roberts 1951). This aggressiveness is especially seen in *D. mystacalis* and *D. melanotis*, in contrast to *D. mesomelas* which seems to have a more docile character (Misonne 1963). Misonne also noticed that, in captivity, this genus dominates the equally small *Leggada* when placed in the same enclosure. *Dendromus* is also dominant to *Malacothrix* (a more stoutly built species) (Roberts 1951).

Species of *Dendromus* eat grass seeds and are also insectivorous to some extent. They are good jumpers and can climb with ease (Kingdon 1974), as well as burrow, despite the slenderness of their forelegs and digits (Roberts 1951).

## Key to the southern African species of *Dendromus*

(After Meester *et al*. (1964) and Misonne (1974))

1 Fifth hindtoe with a nail in adults*; fifth forefinger vestigeal; ears darker than back; general colour normally grey ................................. 2

  Fifth hindtoe with claw; no sign of fifth forefinger; ears about the same colour as back; general colour normally brown ........................... 3

2 Larger, adult skull normally longer than 21 mm; ears relatively small (usually less than 20% of HB length) ............................ *Dendromus nyikae*
(Nyika Climbing Mouse)
Page 128

  Smaller, adult skull normally less than 21 mm; ears relatively larger (usually 20% of HB length); variable black line on back ...................... *D. melanotis*
(Grey Pigmy Climbing Mouse)
Page 125

3 Larger, adult skull 22 mm and more; hair of underparts usually slaty-based; black line on back invariably present ................................ *D. mesomelas*
(Brants' Climbing Mouse)
Page 121

  Smaller, adult skull less than 22 mm; hairs of underparts white or ochraceous at base but not slaty .........
*D. mystacalis*
(Chestnut Climbing Mouse)
Page 124

*The characteristics of nail and claw are not always clearcut. Misonne points out that the key is unsatisfactory, but may be of use pending a much needed revision of the various species.

# *Dendromus mesomelas* (Brants, 1827)

## Brants' Climbing Mouse
## Brants se Klimmuis

The generic name is derived from the Greek *dendron* = tree and the Latin *mus* = mouse. The specific name is a combination of the Greek words *mesos* = middle and *melas* = black with reference to the black dorsal stripe.

The type specimen was collected in the vicinity of the Sondags River in the Cape Province and described by Brants in 1827 in his well-known work *Het Geslacht der Muizen*. The known distribution records are scattered and its total range is poorly defined. Rautenbach (1978b) postulates that the species may be dependent on a wooded element in its vegetational environment, while Misonne (1974) also considers it as a 'bush species'.

## Outline of synonymy

1827 *Mus mesomelas* Brants, *Het Geslacht der Muizen:* 123. Sondags River, CP.

1829 *Dendromus typus.* A. Smith, *Zool. J.* 4:439. South Africa.

1832 *Dendromys mesomelas* Smuts, *Enum. Mamm. Cap.:* 39. Substitute for *Dendromus.*

1834 *Dendromys typicus* A. Smith, *S. Afr. Quart. Journ.* 2: 158. A renaming of Smith's own *typus.*

1913 *Dendromus ayresi* Roberts, *Ann. Transv. Mus.* 4:83. Port St Johns, Transkei.

1901 *Dendromys mesomelas* (Brants). *In* Sclater, *The mammals of South Africa.*

1939 *Dendromus (D.) mesomelas mesomelas* (Brants). *In* Allen, *Bull. Mus. comp. Zool. Harv.* 83:351.

1951 *D.m. mesomelas* (Brants). *In* Roberts, *The mammals of South Africa.*

1953 *D. mesomelas* (Brants). *In* Ellerman, *et al., Southern African mammals.*

1974 *D. mesomelas* (Brants). Misonne, *in* Meester & Setzer (eds), *The mammals of Africa: an identification manual.*

### TABLE 27
Measurements of male and female *Dendromus mesomelas*

| | Parameter | Value (mm) | N | Range (mm) | CV (%) |
|---|---|---|---|---|---|
| Males | HB | 76 | 5 | 69–80 | 5,4 |
| | T | 99 | 4 | 91–105 | 5,9 |
| | HF | 20 | 4 | 19–21 | 4,2 |
| | E | 18 | 3 | 15–21 | 14,2 |
| | Mass: | 12 g | 4 | 11–13 g | 9,3 |
| Females | HB | 74 | 7 | 67–85 | 8,6 |
| | T | 103 | 7 | 94–109 | 8,1 |
| | HF | 20 | 7 | 18–22 | 6,8 |
| | E | 14 | 7 | 12–17 | 13,3 |
| | Mass: | 10,6 g | 5 | 9–14,5 g | 18,0 |

## Identification

Brants (1827) described this species as being brownish-pink in colour with a little blackish-brown admixture on the back, and a mid-dorsal black stripe beginning at the neck and reaching the root of the tail. The flanks are paler with the outside of the forelimbs and feet a clear yellowish-pink with white fingers, while the entire underside is a dirty white (plate 5). Adding to Brants' description, it may be pointed out that the hairs over the belly and chest are dark based but those of the chin, throat and anal region are whitish along their entire length. The fur is soft and thick; hairs are slaty for the basal three-quarters of their length, their tips are chestnut brown. The tail is as long as or longer than the head and body with rings formed of a series of scales concealed by bristles, dark above, lighter below.

The black stripe seems to be variable, but is generally present.

In specimens from Pondoland, in Transkei, Roberts (1913) stated that evidence pointed to females and juveniles being unstriped – though exceptions occur. Where more material became available, Roberts (1951) noted that adult males from the western Cape were not only larger, but also less heavily striped than eastern adult males.

The fifth toe of the hindfoot always bears a small claw, not a flat nail. The forelimbs are short and slender, adapted for grasping, the three middle fingers elongate and clawed, the first and fifth digit being rudimentary.

## Skull and dentition

Certain features have been discussed above in the introduction to the genus (fig. 49). The cheekteeth are narrow (fig. 50).

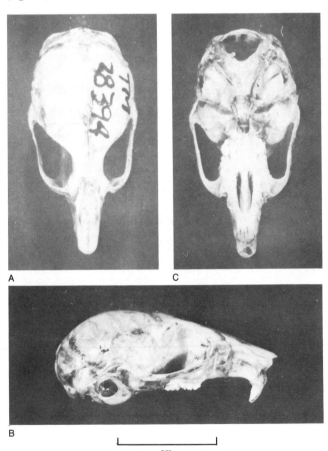

**Figure 49** Dorsal (A), lateral (B) and ventral (C) views of the skull of the Brants' Climbing Mouse *Dendromus mesomelas.* Interorbital constriction and infraorbital canal are pronounced. Crests and ridges are absent. Muzzle narrow, with masseter knob well developed. Anterior palatine foramina reach second lamella of $M^1$. The broad palate terminates beyond molars. Bullae are well developed.

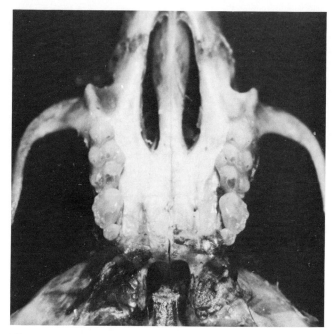

**Figure 50** Cheekteeth of the Brants' Climbing Mouse *Dendromus mesomelas*. M¹ is twice as large as M² with M³ minute. Additional inner cusps adjacent to middle lamella of M¹ and first lamella of M² occur.

## Distribution

Occurs from the southern and western districts of the Cape Province to Pondoland in the east and ranges to Woodbush in the northeastern Transvaal (map 30). The subspecies *major* occurs in the Grootfontein district of SWA/Namibia, and according to Roberts (1951) '. . . doubtless extending throughout Ngamiland'. It is a widely distributed species in central and eastern Africa as well.

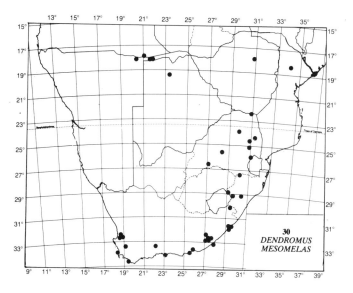

**30**
**DENDROMUS
MESOMELAS**

## Habitat

Brants' Climbing Mouse is predominantly confined to low-lying areas. Smithers (1971) reports them to occur in thick, matter vegetation growing over water 50 cm deep near the edge of swamps in the vicinity of the Gomati River in Botswana. At this locality, they also occur alongside *Dendromus melanotis*.

## Diet

Mainly seeds, but a proportion of their food consists of insects (Smithers 1971).

## Habits

Kingdon (1974) says that *D. mesomelas* is the only *Dendromus* species known to be active by day as well as by night. It lives in rank vegetation where it makes nests of shredded grass in the form of a ball, with entrances at opposite sides. It may also occupy nests of small birds, e.g. *Ploceus* sp. (Sclater 1901). It builds its nest at lower elevations than *D. mystacalis*. It burrows among boulders. It is a nocturnal animal. Not infrequently Brants' Climbing Mouse is found in shrubby growth near villages (Misonne 1963). They may be described as semi-terrestrial. Misonne also points out that this seems to be the least aggressive species of *Dendromus*. Kingdon (1974) states that in East Africa they share the runs of *Otomys* in grasses growing at higher altitudes.

## Predators

They are preyed upon by snakes and the Barn Owl *Tyto alba* (Vernon 1972; Davis 1973) as well as the Grass Owl *Tyto capensis* (Davis 1973).

## Reproduction

Loveridge *(in* Kingdon 1974) has collected many half-grown animals in February on the southern highlands of East Africa, implying that they may breed seasonally.

## Parasites

The following fleas have been taken from *D. mesomelas* (Zumpt 1966): Hystrichopsyllidae: *Ctenophthalmus verutus, C. cophurus* and *Dinopsyllus grypurus*. Ceratophyllidae: *Nosopsyllus incisus*. Furthermore, it has been demonstrated that *D. mesomelas* is susceptible to plague *(Yersinia pestis),* and sporadic outbreaks of plague do occur (Powell 1925).

## Relations with man

No data exist.

## Prehistory

Fossil fragments (maxillae, mandibulae and teeth) resembling *D. mesomelas* are known from breccias at Sterkfontein (Lavocat, 1957; De Graaff 1960a) and Makapansgat (De Graaff 1960a). A fossil species, *Dendromus antiquus*, has been identified by Broom & Schepers (1946) from Taung, but it has not been described formally and its validity is open to question (De Graaff 1961).

123

## Taxonomy

Meester *et al.* (1964) accept the following subspecies: *D. m. mesomelas* – occurring in the western, southern and eastern Cape, Natal, KwaZulu and the Transvaal; *D. m. major* – found in Ngamiland and the Caprivi Strip. A third subspecies, listed by Roberts (1951), *ayresi* (occurring at Port St Johns), is included in *mesomelas*.

# *Dendromus mystacalis* Heuglin, 1863

**Chestnut Climbing Mouse**

**Lesser Climbing Mouse**

**Roeskleurige Klimmuis**

For the derivation of *Dendromus*, see *D. mesomelas*. The specific name probably refers to the white fringe on the upper lip, reminiscent of a moustache, from the Greek *mystax*.

The Chestnut Climbing Mouse is predominantly an East African species and one subspecies reaches the extreme south of the continent. This is a common and successful species, although little is known about it in southern Africa. This may be because the species may have been misidentified in the past and consequently much remains to be learnt about its diet, predators, parasites, reproduction, prehistory and possible relations with man.

This is the smaller species of the *Dendromus mesomelas* group, referred to by Bohmann in 1942. As a type specimen, he selected *D. pumilio* erected by Wagner in 1841. However, Wagner himself stated later that this name was based on a young specimen of *Dendromus mesomelas* without an exact locality. Ellerman *et al.* (1953) said it was not being identified without doubt and proposed the adoption of the next available name which happened to be *mystacalis*.

## Outline of synonymy

1841 *Dendromys pumilio* Wagner, *Gelehrte Anzeigen* 12: 437. Cape of Good Hope. (Possibly based on a juvenile of *D. mesomelas*.)

1863 *Dendromys mystacalis* Heuglin, *Nov. Act. Acad. Leop. Dresden* 30(2):5. Ethiopia.

1909 *Dendromus jamesoni* Wroughton, *Ann. Mag. nat. Hist.* (8) 3:247. Soutpansberg, Transvaal.
  *Synonym*
    1931 *D. j. pongolensis* Roberts, *Ann. Transv. Mus.* 14:232. Pongola River, northeastern Natal.

1939 *Dendromus (Poëmys) mystacalis* Heuglin. *In* Allen, *Bull. Mus. comp. Zool. Harv.* 83:353.

1951 *Dendromus jamesoni jamesoni* Wroughton. *In* Roberts, *The mammals of South Africa*.

1953 *D. mystacalis* (Heuglin). *In* Ellerman *et al., Southern African mammals*.

1974 *D. mystacalis* (Heuglin). Misonne, *in* Meester & Setzer (eds), *The mammals of Africa: an identification manual*.

### TABLE 28
Measurements of male and female *Dendromus mystacalis*

|  | Parameter | Value (mm) | N | Range (mm) | CV (%) |
|---|---|---|---|---|---|
| Males | HB | 64 | 16 | 55–80 | 10,0 |
|  | T | 85 | 16 | 75–95 | 7,0 |
|  | HF | 18 | 15 | 16–20 | 6,3 |
|  | E | 14 | 12 | 13–15 | 5,5 |
|  | Mass: | 9 g | 8 | 8–14 g | 22,7 |
| Females | HB | 63 | 18 | 54–80 | 11,2 |
|  | T | 85 | 18 | 74–103 | 7,6 |
|  | HF | 17 | 18 | 15–18 | 6,2 |
|  | E | 14 | 15 | 12–16 | 7,5 |
|  | Mass: | 7,4 g | 9 | 6–9 g | 11,2 |

## Identification

The dorsal surface is predominantly rufous-brown, never greyish as in *D. melanotis*. A black stripe runs from the neck to the root of the tail and seems to be invariably present; the ventral surface is off-white without dark bases to the hairs. In contrast to *D. mesomelas* (whose females and young tend to be unstriped), the adult females are usually well striped (Roberts 1951), while only the young are unstriped (plate 5).

*D. mystacalis* is smaller than *D. mesomelas*. It has a normal, though small claw on the fifth toe of the hindfoot. The thumb is opposable and the tail semi-prehensile.

## Skull and dentition

As described in the introduction to the genus. The $M^1$ is an elongated tooth while the $M^3$ is minute (fig. 51).

## Distribution

Northeastern Transvaal, eastwards and southwards to Natal and Swaziland as well as into Pondoland in Transkei and further west to Knysna (map 31). Extralimitally, it ranges northwards to Central and East Africa across the width of the continent, but not into West Africa.

## Habitat

This is the only species of *Dendromus* positively recorded to be arboreal (Rosevear 1969). In Botswana it has been collected from floodplain grassland on a swamp island near Kasane (Smithers 1971). Kingdon (1974) calls it a

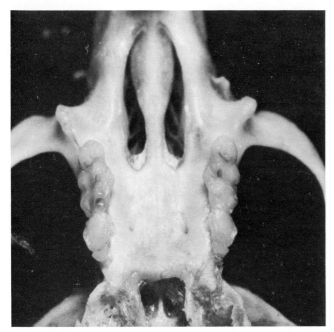

**Figure 51** Cheekteeth of the Chestnut Climbing Mouse *Dendromus mystacalis*.

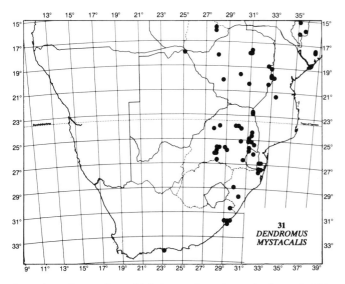

**31 DENDROMUS MYSTACALIS**

low altitude species which is not found much above 2 000 m. This species often frequents disused weaver and Bishop Bird nests (Jacobsen 1977). Smithers & Lobão Tello (1976) have described stands of *Hyparrhenia* sp. of 1–2 m high, as good habitat in Mozambique.

### Diet
Graminivorous, with a tendency towards insectivory.

### Habits
According to Misonne (1963), chestnut climbing mice are avid fighters. When on the ground at night, they move in short hops (Rosevear 1969). In contrast to the other species, they are more exclusively climbers and, in that sense '. . . may be said to be the most highly evolved' (Kingdon 1974), although they can all live on the ground.

Their nests are built in exposed situations high up in rank vegetation. Lang *(in* Kingdon 1974) noticed that nests built by solitary males are smaller than those built by females.

### Predators and prehistory
No data available.

### Reproduction
Three to four pups usually form a litter, but as many as eight have been recorded in a female collected at Barberton (eastern Transvaal) during March.

### Parasites
The trombiculid mite *Trombicula panieri* occurs as an ectoparasite on *Dendromus mystacalis* (Zumpt 1961a).

### Relations with man
Kingdon (1974) states that it is a familiar species as it has adapted to cultivation and it is not unusual to find nests in garden shrubs or even in thatched roofs.

### Taxonomy
As here understood, *D. mystacalis* includes *D. jamesoni* as listed by Roberts (1951) which in turn includes the subspecies *D. jamesoni pongolensis* (Meester *et al.* 1964). Both Allen (1939) and Roberts (1951) regard *jamesoni* as a distinct species.

*D. mystacalis* is a representative of a Rassenkreis which Bohmann (1942) called *pumilio* as originally proposed by Wagner (1841). The name *pumilio* is inadmissible, however, because it was founded on a juvenile specimen of *mesomelas* (Rosevear 1969).

# *Dendromus melanotis* A. Smith, 1834

**Grey Pygmy Climbing Mouse**
**Dark-eared Climbing Mouse**
**Grey Tree Mouse**
**Grys Dwergklimmuis**

The generic name is derived from the Greek *dendron* = tree and the Latin *mus* = mouse. When Smith erected the genus in 1834, he asserted that they are 'commonly found upon trees' – yet reliable evidence that they are arboreal is slender. The specific name is compounded from the Greek *melas, melanos* = black and *ous, otis* = ear, because the ears of the original specimen were covered with black hairs.

The specimen on which Sir Andrew Smith based this species was collected at Durban. As indicated above, this species corresponds with the subgenus *Poëmys* of Thomas' grouping (i.e. showing a nail on digit 5 of the hindfoot, the fifth digit of the hand being reduced to a mere stump and the ears conspicuously darker than the back) and Roberts (1951) retained the subgenus. They are, however, grass mice, rather than tree mice.

## Outline of synonymy

1834 *Dendromys melanotis* A. Smith, *S. Afr. Quart. Journ.* 2:158. Near Port Natal (= Durban).

1846 *Dendromys subtilis* Sundevall, *Oefvers. K. Svenska Vet. Akad. Förh. Stockholm* 3(5):120. South Africa.

1916 *Poëmys* Thomas, *Ann. Mag. nat. Hist.* (8) 18:238. As a subgenus of *Dendromus*. Type *Dendromus melanotis* A. Smith.

1916 *Dendromus (Poëmys) nigrifrons vulturnus* Thomas, *Ann. Mag. nat. Hist.* (8) 18:242. Chirinda Forest, Melsetter, Zimbabwe.

1924 *D. (P.) arenarius* Roberts, *Ann. Transv. Mus.* 10:71. Angra Pequina, Bothaville, OFS.

1924 *Dendromus (Poëmys) melanotis* Roberts, *Ann. Transv. Mus.* 10:72. Grahamstown, CP.

1926 *D. (P.) concinnus* Thomas, *Proc. zool. Soc. Lond.:* 299. Otjumbumbi, Cunene River, SWA/Namibia.

1927 *Dendromus melanotis basuticus* Roberts, *Rec. Alb. Mus.* 3:484. Thaba Putsua, Lesotho.

1929 *Dendromus (Poëmys) melanotis chiversi* Roberts, *Ann. Transv. Mus.* 13:116. Vlakfontein near Parys, OFS.

1930 *D. (P.) nigrifrons shortridgei* St Leger, *Ann. Mag. nat. Hist.* (10) 6:622. Grootfontein, SWA/Namibia.

1931 *D. melanotis thorntoni* Roberts, *Ann. Transv. Mus.* 14:231. Port Elizabeth, CP.

1931 *D. m. capensis* Roberts, *Ann. Transv. Mus.* 14:232. Wolseley, CP.

1931 *D. (P.) melanotis pretoriae* Roberts, *Ann. Transv. Mus.* 14:232. Pretoria, Transvaal.

1938 *Poëmys melanotis insignis* Shortridge & Carter, *Ann. S. Afr. Mus.* 32:287 (not of Thomas 1903). Ezelfontein, Kamiesberg, Namaqualand, CP.

1939 *Dendromus (Poëmys) melanotis melanotis* A. Smith. *In* Allen, *Bull. Mus. comp. Zool. Harv.* 83:352.

1951 *Dendromus (Poëmys) melanotis melanotis* A. Smith. *In* Roberts, *The mammals of South Africa.*

1953 *D. melanotis* A. Smith. *In* Ellerman *et al.*, *Southern African mammals.*

1974 *D. melanotis* A. Smith. Misonne, *in* Meester & Setzer (eds), *The mammals of Africa: an identification manual.*

## Identification

Instead of a claw, there is a flattened nail on digit 5 of the hindfoot with digit 1 reduced, while the fifth finger of the hand is represented by a clawless tubercle.

In addition to the dark-coloured ear, there is a white patch at the anterior lateral base of the ear, contrasting sharply with the dark dorsal surface. The dorsal stripe is very marked (from the shoulders to the root of the tail) and some specimens show a clear dark spot on the forehead (plate 5). The fur is soft and thick. The upper and lateral parts of the head, the neck, body and the upper surfaces of the extremities are ashy-grey with a rufous or rusty tint, more clearly visible on the anterior parts. The ventral surface is a greyish-white. The tail is long, not well haired, dark above, light below. There is a brownish blotch frontward of each eye.

#### TABLE 29
Measurements of male and female *Dendromus melanotis*

| | Parameter | Value (mm) | N | Range (mm) | CV (%) |
|---|---|---|---|---|---|
| Males | HB | 68 | 32 | 56–81 | 9,8 |
| | T | 90 | 26 | 76–108 | 10,0 |
| | HF | 18 | 33 | 16–21 | 7,6 |
| | E | 14 | 30 | 12–18 | 11,8 |
| | Mass: | 7,4 g | 11 | 6,0–10,0 g* | — |
| Females | HB | 70 | 17 | 60–86 | 12,1 |
| | T | 83 | 21 | 71–113 | 13,2 |
| | HF | 17 | 18 | 16–20 | 8,5 |
| | E | 16 | 20 | 13–19 | 8,4 |
| | Mass: | 7,0 g | 6 | 4,0–12,0 g* | — |

*Smithers (1971), Okavango specimens

## Skull and dentition

Characteristics of the skull have been referred to in the introduction to the genus above.

The molars are narrow. The cusp pattern (according to Ellerman 1941) is not far removed from the murine condition. In the $M^1$ two cusps are close together with occasional extra cusps in front of the first lamina. Three small cusps make up a middle one. Two cusps constitute the third lamina. In the $M^2$ the outer cusp is small. The $M^3$ is very small and reduced (fig. 52). The incisors are yellow and grooved.

## Distribution

*Dendromus melanotis* has a fairly wide distribution in southern Africa, ranging from the western and eastern Cape, Natal, Lesotho, northern OFS and Transvaal into Botswana and the Caprivi Strip (map 32). It is therefore found in the South West Arid (not in the Namib Desert), the South West Cape, as well as in the Southern Savanna Woodland and Southern Savanna Grassland biotic zones (Rautenbach 1978a). Extralimitally, it has a very extensive geographic range and is a successful and apparently dominant species (Kingdon 1974).

Plate 7

White-tailed Rat
*(Mystromys albicaudatus)*

Pouched Mouse
*(Saccostomus campestris)*

Dassie Rat
*(Petromys typicus)*

Brants' Whistling Rat
*(Parotomys brantsii)*

Littledale's Whistling Rat
*(Parotomys littledalei)*

Plate 8

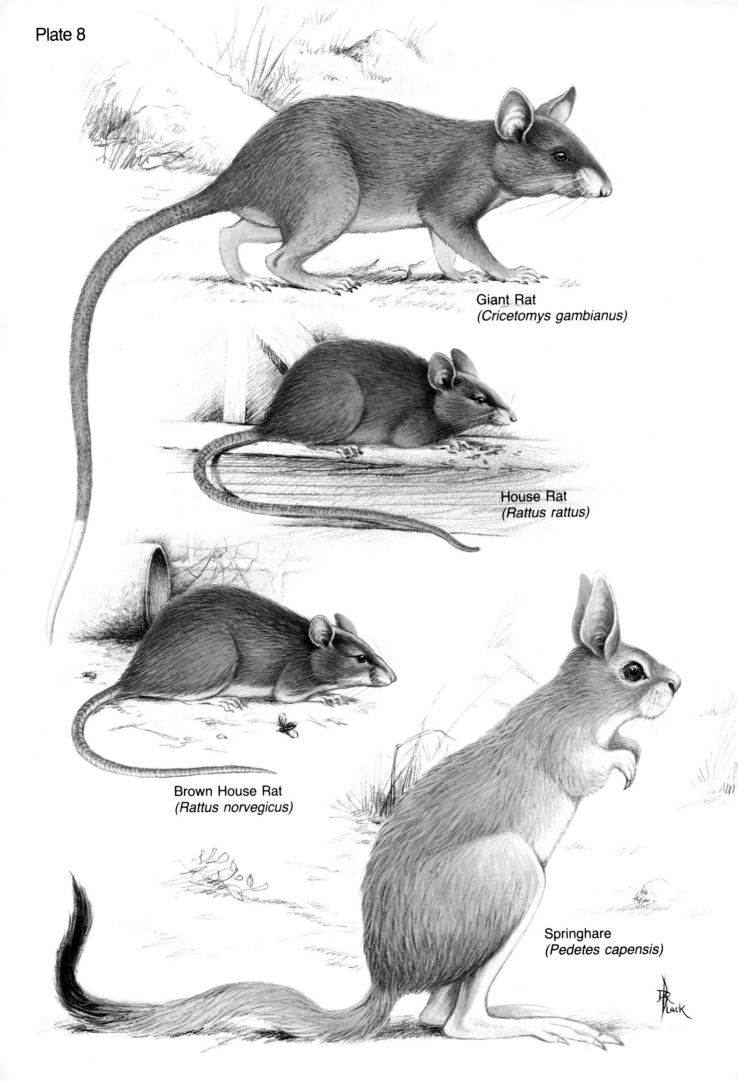

Giant Rat
(*Cricetomys gambianus*)

House Rat
(*Rattus rattus*)

Brown House Rat
(*Rattus norvegicus*)

Springhare
(*Pedetes capensis*)

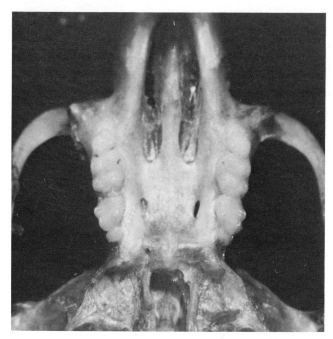

**Figure 52** Cheekteeth of the Grey Pygmy Climbing Mouse *Dendromus melanotis*. Molars are narrow; M$^1$ may show occasional extra cusps along anterior edge.

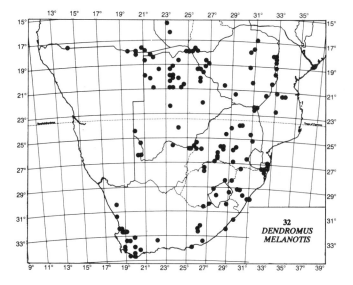

32
DENDROMUS
MELANOTIS

## Habitat

This species climbs grass in search of seeds, but tends to be more terrestrial than *Dendromus mesomelas*. It can be found at high altitudes. In Botswana it occurs in dry grassland, or on the fringes of swamps and rivers where it prefers tall grass (e.g. *Hyparrhenia* spp.), mixed with shrubs and annuals; at Kasane in Botswana it lives under carpets of dry *Salvinia* spp. on river fringes, in riverine woodland and floodplains, on swamp islands and in dry *Baikiaea* woodland adjoining rivers (Smithers 1971). Kingdon (1974) said that it '. . . seems to be the most able to cope with the exacting conditions of the African savanna'.

## Diet

The Grey Pygmy Climbing Mouse eats seeds and insects including termites (Isoptera) (Smithers 1971). Sclater (1901) says that it will devour spiders and insects and small snakes. Shortridge (1934) states that it will also eat small lizards, young birds and eggs.

## Habits

This species is largely terrestrial and nocturnal. It climbs with agility on grass stalks, where it feeds and forages for seed. In climbing and at rest, the end of the tail is lightly curled around the stalk. It appears to forage over large distances as Jacobsen (1977) has recaptured individuals 60–100 m from the original capture site.

It makes burrows up to 50 cm deep and 4–6 cm in diameter. The burrows are simple open-ended tunnels leading into a small nest chamber with an emergency exit on the opposite side, excavated to within a short distance from the surface of the soil. The burrow system may also serve as a refuge during veld fires. It may even live in holes constructed by large dungbeetles (Jacobsen 1977).

Smithers (1971) relates that it may also make nests about one metre above ground in vegetation consisting of shredded grass and fibres which are used only during the breeding season. It is not easily trapped, but may be secured during the breeding season by locating the nests, or taken at night with the aid of a dazzling light. They will fight savagely when strange individuals are introduced into the nest and in captivity will fight to the death.

This common species is not gregarious but a pair may occupy the same hole.

## Predators

Little specific information is available. Nel & Nolte (1965) recorded the Barn Owl *Tyto alba* and it seems fairly certain that they are preyed upon by snakes.

## Reproduction

A female, collected at Kasane (Botswana) contained four fetuses in the December (Smithers 1971). Gravid females with three fetuses each were taken in February in the Kruger National Park. Young are born in the nests, and in Mashonaland in Zimbabwe, they are born during December to March.

The female has two pairs of pectoral and two pairs of inguinal mammae, making eight in all. The litter size is from two to four, while Jacobsen (1977) states that as many as seven young may be born on a single occasion.

Dieterlen (1971) has bred this species in captivity. The development of the nidicolous young takes five to ten days longer than other mouse-like rodents (Kingdon 1974).

## Parasites

The following ectoparasites have been collected (Zumpt 1961a, 1966): the trombiculid mites include *Schoengastia o. oubanguiana*, while specimens of the laelaptid mite *Macronyssus bacoti* are also known from a *Dendromus* sp. The fleas include the hystrichopsyllids *Ctenophthalmus acanthurus*, *Dinopsyllus grypurus* and *Listropsylla dolosa*. These parasites have all been collected from *D. melanotis*, occurring extralimitally to southern Africa.

## Relations with man
Unknown.

## Prehistory
No fossils of *D. melanotis* from southern Africa have as yet been recorded.

## Taxonomy
As a species, *D. melanotis* corresponds exactly to Thomas' concept of the subgenus *Poëmys*. Fourteen forms are known, ranging from the Cape to Ethiopia.

Roberts (1951) recognises two species in South Africa: *D. longicaudatus* and *D. melanotis*. Ellerman *et al.* (1953) regard the former as a synonym of *D. mesomelas*, but according to Meester *et al.* (1964), this is probably incorrect. Apart from *longicaudatus*, taken as conspecific with *nyikae* in this book, nine subspecies of *melanotis* have been listed by Roberts. As the status of several is not clear, they are listed here, as is done by Misonne (1974) and the reader is referred to the synonymy given for this species. Apart from the nominate subspecies *melanotis*, *D. melanotis* may include *capensis*, *basuticus*, *chiversi*, *pretoriae*, *vulturnus*, *shortridgei*, *arenarius* and *concinnus*.

# *Dendromus nyikae* Wroughton, 1909

## Nyika Climbing Mouse
## Nyika se Klimmuis

For the meaning of *Dendromus*, see *Dendromus mesomelas*. The specific name refers to the Nyika Plateau in northern Malaŵi, where this species was first collected.

This species is very similar to *Dendromys melanotis*, but has a longer tail and a larger skull. Roberts (1913) states its colour to be a rather brighter rufous overall. In fact, it is so reddish-coloured dorsally that it might be taken for *D. mesomelas*, with which it agrees also in size but the underparts are white to the base of the hairs, while there is a white spot at the base of the dark-coloured ears. Furthermore, the fifth toe has a nail rather than a claw and the stub of the outer finger is present.

Apart from some morphological features, its geographical distribution, and a few ideas concerning its taxonomic position, no other details are available for this species. Its distribution in southern Africa is known only by a type specimen from Tzaneen in the northeastern Transvaal, another specimen from the Hans Merensky Provincial Nature Reserve, and from a few specimens from Inyanga, Zimbabwe.

## Outline of synonymy
1909 *Dendromus nyikae* Wroughton, *Ann. Mag. nat. Hist.* (8) 3:248. Nyika Plateau, Malaŵi.
1909 *Dendromus melanotis* Jameson, *Ann. Mag. nat. Hist.* (8) 4:459. Tzaneen, Transvaal.
1913 *Dendromus longicaudatus* Roberts, *Ann. Transv. Mus.* 4:83. Tzaneen Estate, NE Transvaal.
1916 *Dendromus nyasae* Thomas, *Ann. Mag. nat. Hist.* (8) 18:241. Nyika Plateau, Malaŵi.
1939 *Dendromus (Poëmys) nyikae* Wroughton. *In* Allen, *Bull. Mus. comp. Zool. Harv.* 83:354.
1951 *Dendromus (Poëmys) longicaudatus*. *In* Roberts, *The mammals of South Africa*.
1964 *Dendromus nyikae longicaudatus* (Roberts). *In* Meester *et al.*, *An interim classification of southern African mammals*.
1974 *Dendromus nyikae* Wroughton. Misonne, *in* Meester & Setzer (eds), *The Mammals of Africa: an identification manual*.

TABLE 30
Measurements of a male *Dendromus nyikae*

|  | Parameter | Value (mm) | N | Range (mm) | CV (%) |
|---|---|---|---|---|---|
| Male | HB | 79 | 1 | — | — |
|  | T | 94 | 1 | — | — |
|  | HF | 20 | 1 | — | — |
|  | E | 15 | 1 | — | — |

Mass: 13 g (a single specimen from the Hans Merensky Provincial Nature Reserve)

## Identification
In 1913 Roberts described this species as follows (plate 5): 'Very similar to *D. melanotis*, but with a longer tail and larger skull; in colour also brighter rufous in general effect.' It is so reddish-coloured dorsally that it might be mistaken for *D. mesomelas* with which it also agrees in size. The underparts are white to the base of the hairs and there is a white spot at the anterior base of the dark-coloured ears. As could be expected, the fifth toe has a nail instead of a claw and a stub as the outer finger on the hand is present.

## Skull and dentition
There are no particular features specifically attributed to

this species. The triserial arrangement in the molars are evident (fig. 53).

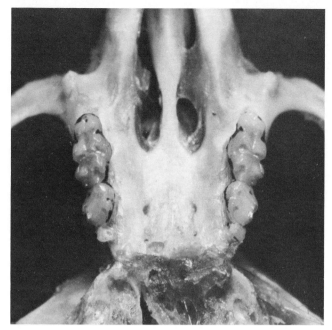

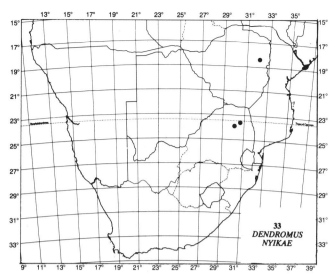

**Figure 53** Cheekteeth of the Nyika Climbing Mouse *Dendromus nyikae.*

## Distribution

It ranges along the eastern escarpment of the Transvaal and Zimbabwe (map 33). This may be a true relict species as is '. . . suggested by its sparse and scattered distribution in the southern savannas and woodlands' (Kingdon 1974).

## Habitat

It seems to be locally agumdant in 'Miombo' woodland, but seems to be rare otherwise over most of its range (Kingdon 1974). The specimen from the Hans Merensky Provincial Nature Reserve was collected in dense grass in mopane woodland (Rautenbach 1978b).

## Reproduction

In southwestern Tanzania six specimens were collected by Kingdon. A lactating female was recorded by the middle of August, while males with large testes were also present during August as well as in January (Kingdon 1974). Kingdon also refers to Loveridge who found a nest with eight young on Ukerewe Island, Lake Victoria, in June.

## Diet, habits, predators, parasites, relations with man and prehistory

No data available.

## Taxonomy

This species was first identified by Wroughton in 1909 after comparison with material in the British Museum and was subsequently named *Dendromus longicaudatus* by Roberts (1913). It has subsequently become clear that the name *nyikae* as proposed by Wroughton (1909) should be applied to this animal, as was done by Meester *et al.* (1964) and Misonne (1974) with *longicaudatus* as a subspecies or a synonym. Allen (1939) and Roberts (1951) placed *longicaudatus* in the subgenus *Poëmys*, while Ellerman *et al.* (1953) regard it as a subspecies of *mystacalis*.

The following subspecies are probably valid: *D. n. longicaudatus*, found near Tzaneen, eastern Transvaal and *D. n. bernardi*, occurring at Inyanga, Zimbabwe.

GENUS *Steatomys* Peters, 1846

This genus, widespread over Africa, comprises small to medium-sized brownish-grey mice. The body form is thick-set, the fur soft and the hands relatively large with four functional digits, each with a fairly long claw, suited to digging. The hindfoot has five clawed digits, implying that the fifth never has a flat nail as in *Dendromus*. The tail is short (about half the head-body length) and well haired. The dark, spinal stripe is never present as in *Dendromus*. The ventral surface of the body and the dorsal surfaces of the hands and feet are white. The anterior part of the sole of the hindfoot is naked, but the posterior portion is well haired.

The snout is sharp-pointed and the ears fairly conspicuous. The presence of a white spot near the base of the ear pinna is either conspicuous or obscure.

In the skull, the interorbital constriction is moderate; supraorbital and other crests are absent; the infraorbital foramen is fairly large; the zygoma curves upwards considerably towards the narrow zygomatic plate; as in *Dendromus,* a conspicuous masseter knob is present. The bullae are somewhat larger than are encountered in *Dendromus,* while the palatine foramina on the palate are as those in *Dendromus.*

The upper incisors are grooved, the lower ones plain. The other teeth are as in the rest of the subfamily, very similar to that of *Dendromus,* although the $M^3$ is not as small as in *Dendromus.*

Members of this genus live entirely on the ground and construct simple burrows. The entrances of occupied burrows are often closed, and an entrance usually leads to a living chamber lined with grass. If the animal is pursued, it often retreats into side tunnels which are sealed off behind the escaping individual and if pressed, the animal may break the surface and flee. *Steatomys* spp. occur over much of the drier and more open parts of southern Africa.

An interesting feature of this genus is the accumulation of fat beneath the skin and in the tissues. This explains the derivation of the generic name: it comes from the Greek *stear, steatos* = fat. This deposition of fat enable these animals to become inactive during unfavourable seasons. In contrast to the winter in northern hemisphere, this un-favourable season falls in the dry, arid summers and they therefore aestivate rather than hibernate. During aestivation (without parallel among other rodents in southern Africa) they are easily removed from their burrows and are eaten as delicacies by Blacks in various localities.

Ellerman (1941) has divided this genus into two groups: the *pratensis* group (smaller, head and body length less than 100 mm) and the *bocagei* group (more heavily built, head-body length more than 100 mm). Ellerman *et al.* (1953) raised these two groups to species level and many of the formerly independent forms were thereby reduced to races of these (Rosevear 1969). This arrangement was not accepted by Ansell (1958), mainly on account of the characteristics of *D. krebsii*. Rosevear (1969) is also uncertain whether the interpretations of Ellerman *et al.* (1953) are justified.

Ansell (1960) and Meester *et al.* (1964) recognise three species, which can be keyed out on the arrangement and number of mammae. Its applicability is not easy if the material is juvenile or the mammae underdeveloped and a mammary formula can only be applied to a minority of females. In the case of *pratensis* the mammae consist of five to seven pairs, while in *krebsii* and *parvus* they are arranged in two pectoral and two inguinal pairs (and in *parvus*, a pure white tail).

According to Coetzee (1977) it is impossible as yet to provide a satisfactory key to the species. He does provide a key, however, and stresses that it, as well as the species listed, should be regarded as 'provisional'.

---

**Key to the southern African species of *Steatomys***
(After Coetzee (1977))

1 Multimammate, mammae more than 2 + 2 = 8, usually 12, maximum 16, not necessarily in pairs* ...
*Steatomys pratensis*
(Fat Mouse)
Page 131
Mammae 2 + 2 = 8.................................... 2
2 Smaller, HB 63–86 mm, greatest skull length 19,5–23 mm; ear length subequal to or longer than HF (su); tail white; brownish dorsally or reddish south of 10°S latitude.......................... *S. parvus*
(Tiny Fat Mouse)
Page 133
Larger, HB 70–105 mm, greatest skull length 21–27 mm; HF (su) subequal to or longer than ear (Ansell 1969); tail brownish grey dorsally, white below**; dorsal colour brown with a greyish tinge; occurs only south of 10°S latitude ..................... *S. krebsii*
(Krebs' Fat Mouse)
Page 134

* For diagnostic purposes, the number of mammae, notwithstanding the shortcomings of this characteristic, appears to be of major importance in this genus.
** The bicoloured tail is not a consistant character either (Rautenbach 1978b).

# Steatomys pratensis Peters, 1846

**Fat Mouse**

**Vetmuis**

The name is derived from the Greek *stear, steatos* = fat and *mys* = mouse, because of the regular storage of fat in the tissues. The species name is derived from the Latin *pratum* = a meadow and *pratensis* = pertaining to a meadow.

These small, mouse-like animals are always very plump due to the storage of fat spread all over the body. The have no cheekpouches, have moderately sized ears and short limbs. This genus is restricted to the Afrotropical Region and *Steatomys pratensis* is found in all savannas of sub-Saharan Africa (Kingdon 1974). Little is known about fat mice and they are worthy of a detailed study.

## Outline of synonymy

1846 *Steatomys pratensis* Peters, *Monatsber. K. Preuss. Akad. Wiss. Berlin*: 258. Tete, on the Zambezi River, Mozambique.

1852 *Steatomys edulis* Peters, *Reise nach Mossambique. Säugeth.*: 163. Substitute for *pratensis*. Sena. Mozambique.

1929 *Steatomys natalensis* Roberts, *Ann. Transv. Mus.* 13: 117. Bergville, Natal.

1932 *Steatomys pratensis maunensis* Roberts, *Ann. Transv. Mus.* 15:11. Shorobe, Botswana.

1939 *Steatomys pratensis pratensis* Peters. *In* Allen, *Bull. Mus. comp. Zool. Harv.* 83:360.

1951 *Steatomys pratensis pratensis* Peters. *In* Roberts, *The mammals of South Africa.*

1953 *Steatomys pratensis* Peters. *In* Ellerman *et al., Southern African mammals.*

1964 *Steatomys pratensis* Peters. *In* Meester *et al., An interim classification of southern African mammals.*

1977 *Steatomys pratensis* Peters. Coetzee, *in* Meester & Setzer (eds), *The mammals of Africa: an identification manual.*

## Identification

The dorsal surface is rusty brown with the flanks of the body and outside surfaces of the limbs less brownish; the sides of the face are a buffy-rufous; the belly (clearly demarcated from the dorsal surface) and hands and feet are white; the hairs above have slaty bases, but are white to the base elsewhere (plate 5). Specimens from Botswana are more pallid in colour while those from Natal have a generally darker appearance. The fur is short and soft. The snout is short and pointed with ears which are of moderate size. Hands have four clawed digits and a flat nail to the first finger. The hindfoot has five clawed digits, shorter than those of the hand.

## TABLE 31
### Measurements of male and female *Steatomys pratensis*

|         | Parameter | Value (mm) | N  | Range (mm) | CV (%) |
|---------|-----------|-----------|----|-----------|--------|
| Males   | HB        | 95        | 25 | 83–105    | 6,0    |
|         | T         | 44        | 25 | 40–55     | 8,2    |
|         | HF        | 16        | 29 | 15–19     | 6,6    |
|         | E         | 15        | 20 | 14–18     | 7,2    |
|         | Mass:     | 26,1 g    | 13 | 22–30 g   | 9,4    |
| Females | HB        | 93        | 27 | 82–106    | 7,2    |
|         | T         | 44        | 25 | 40–51     | 8,1    |
|         | HF        | 16        | 26 | 14–18     | 5,4    |
|         | E         | 15        | 22 | 14–18     | 6,6    |
|         | Mass:     | 39,7 g    | 9  | 34–48 g   | 11,7   |

## Skull and dentition

In the synonym *natalensis,* Roberts (1951) notes a skull altogether larger than in *pratensis (sensu stricto)* and with a very large braincase (fig. 54). Furthermore the internal cusps of the $M^1$ and $M^2$ are less developed and higher (fig. 55). For the rest, the teeth of *pratensis* conform to those of the rest of the subfamily (Dendromurinae) as described above.

## Distribution

In southern Africa *Steatomys pratensis* occurs from the lower Zambezi, ranging southwards into Natal across Zimbabwe and the eastern Transvaal, below the escarpment. Towards the west, it is found in the Ngamiland and Okavango areas (map 34). It is an inhabitant of the South West Arid and Southern Savanna Woodland biotic zones (Rautenbach 1978a).

## Habitat

In Botswana, Smithers (1971) records its occurrence on sandy ground in scrub or in sandy alluvium on the fringes of swamps and rivers. It also likes open woodland and often occurs in abandoned cultivated lands.

## Diet

Graminivorous, especially grass seeds. Kingdon (1974) reminds us of the fact that the Fat Mouse is an adept digger and consequently consumes bulbs and groundnuts when available, as well as insects.

## Habits

They are nocturnal animals, terrestrial and occur singly or in pairs (Smithers 1971). According to Smithers, nests are approximately 20 cm below the surface of the soil. This has also been found to be the case in burrows which I excavated in the Kruger National Park. They make their

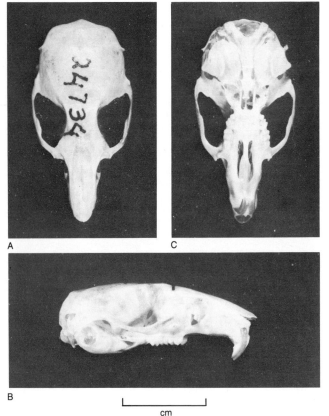

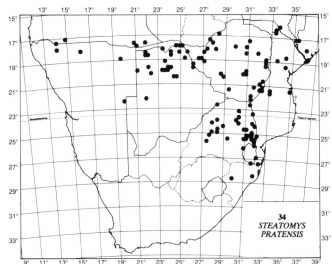

**34**
**STEATOMYS**
**PRATENSIS**

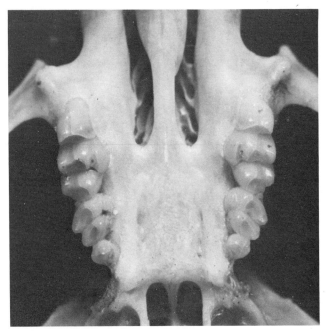

**Figure 54** Dorsal (A), lateral (B) and ventral (C) views of the skull of the Fat Mouse *Steatomys pratensis*. The interorbital constriction is moderate and no crests occur. The infraorbital foramen fairly large and zygomatic arches curve upwards considerably towards narrow zygomatic plate. The bullae are somewhat larger than found in *Dendromus*.

**Figure 55** Cheekteeth of the Fat Mouse *Steatomys pratensis*. Tooth morphology conforms to that found in dendromurines, although the M³ is not as small in *Dendromus*.

burrows in firm soils and they seem to be particularly dependent on these abodes, especially in the case of fire. The passages slant deeply down to 25 cm below soil surface and the four passages of the systems I have excavated all led to a central nesting chamber filled with shredded vegetation. Torpid individuals which were dug out from a burrow, soon recovered their normal activity.

**Predators**
Unknown.

**Reproduction**
Information is scanty and that which is available has been documented by Smithers (1971) on material from Botswana. Gravid females have been taken in February and December and lactating females from February to April, and in October and December. In Tanzania, Kingdon (1974) found a female with three small embryos in November. In Zambia, Ansell (1960) has recorded pregnancies in December and April. In the Kruger National Park, Kern (1977) reported no reproductive activity in either sex after early April until the end of November when all individuals were reproductively active.

The nest is made of coarse grass, leaves and husks.

**Parasites**
The helminth *Inermicapsifer madagascariensis* has been reported by Collins (1972). The mites include the following (after Zumpt 1961a): Family Laelaptidae: *Haemolaelaps rhodesiensis, Androlaelaps marshalli* and *A. theseus*. The following fleas are known to occur on this species (after Zumpt 1966): Family Pulicidae: *Xenopsylla nubica, X. philoxera, X. bechuanae, X. brasiliensis* and *X. scopulifer*. Family Pygiopsyllidae: *Stivalius richardi, S. torvus*. Family Hystrichopsyllidae: *Dinopsyllus ellobius* and *Listropsylla dorippae*. Ticks include the following species (after Theiler 1962): Family Ixodidae: *Haemaphysalis leachii leachii* (immatures), *Rhipicephalus evertsi* (immatures) and *R. sanguineus* (adults).

132

## Relations with man

They seem to be greatly esteemed as food by Blacks in certain areas, undoubtedly because of their accumulated fat reserves.

The mite *Haemolaelaps rhodesiensis* collected from *S. pratensis* may be regarded as an accidental host-parasite relationship (Zumpt 1961a).

The fleas of the genus *Xenopsylla* are most important from a medical point of view. Amongst them is *X. philoxera,* an important transmitter of plague among wild rodents (Zumpt 1961a).

Of the ticks, *Haemaphysalis l. leachii* has been proved to transmit *Babesia canis, Nuttalis felis, Coxiella burnetii, Rickettsia conorii* and *R. pijperii. Rhipicephalus evertsi* has been proved to transmit, *inter alia, Babesia equi, Gonderia ovis* and *Theileria parva.* Toxicosis paralysis is also transmitted by this tick in certain parts of the highveld, mostly in spring lambs (Theiler 1962).

Vesey-FitzGerald (1966) found this species common in fallow cultivation in Tanzania. Soil loosened by cultivation surely favours their spread but this '... is offset by their being a choice tit-bit in local stews' (Kingdon 1974), so that in fact they are becoming rare in certain areas. Their fat deposits fluctuate with the season. This fat is deposited all over the body, beneath the skin and around the viscera, allowing it to reduce its body temperature and overall activity surviving on a minimum of food intake.

## Prehistory

De Graaff (1961) identified *S. pratensis* as a fossil from Makapansgat.

## Taxonomy

As here understood, *S. pratensis (= S. edulus)* includes *S. p. natalensis* and *S. p. maunensis* as synonyms.

# *Steatomys parvus* Rhoads, 1896

## Tiny Fat Mouse

## Dwergvetmuis

For the meaning of *Steatomys,* see *Steatomys pratensis.* The species name is derived from the Latin *parvus* = little, small, referring to the small size of this dendromurid.

As with so many other species, this rodent has had a rough taxonomic voyage. The latest taxonomic interpretation combines previously described forms into the single *Steatomys parvus,* a name now accepted to be the correct one for this species as pointed out by Davis *(in litt.),* (Meester *et al.* 1964) and followed by Coetzee (1977). As a result, little data exist on *S. parvus* and much remains to be learnt about the significance and role this species plays in nature.

## Outline of synonymy

1896 *Steatomys parvus* Rhoads, *Proc. Acad. Nat. Sci. Philadelphia:* 529. Lake Rudolf, Kenya.

1905 *Steatomys minutus* Thomas & Wroughton, *Ann. Mag. nat. Hist.* (7)16:174. Fort Quillenges, southwestern Angola.

1926 *Steatomy swalius* Thomas, *Proc. zool. Soc. Lond.:* 300. Ondongwa, SWA/Namibia.
Synonym
1926 *S. s. umbratus* Thomas, *Proc. zool. Soc. Lond.:* 301. Cunene Falls, Cunene River, SWA/Namibia.

1939 *Steatomys minutus minutus* Thomas & Wroughton. *In* Allen, *Bull. Mus. comp. Zool. Harv.* 83:360.

1953 *Steatomys pratensis minutus* Thomas & Wroughton. *In* Ellerman *et al., Southern African mammals.*

1964 *Steatomys minutus* Thomas & Wroughton. *In* Meester *et al., An interim classification of southern African mammals.*

1977 *Steatomy parvus* Rhoads. Coetzee, *in* Meester & Setzer (eds), *The mammals of Africa: an identification manual.*

### TABLE 32
### Measurements of male *Steatomys parvus*\*

| | Parameter | Value (mm) | N | Range (mm) | CV (%) |
|---|---|---|---|---|---|
| Males | HB | 76 | 5 | 64–86 | 11,1 |
| | T | 40 | 5 | 36–50 | 12,7 |
| | HF | 15 | 5 | 14–16 | 5,9 |
| | E | 14 | 5 | 13–14 | 2,9 |
| | Mass: No data available | | | | |

\*No data could be traced on measurements of females

## Identification

The dorsal surface is a rufous grey and the ventral surface a dirty white. This white coloration extends along the limbs. The animal is well furred and individual hairs on the rump are longer than elsewhere (plate 5).

The coloration of the animal is geographically variable. Specimens from northern KwaZulu are buffy brown (forms described as *tongensis*) while they tend to be paler in northern SWA/Namibia and the central Kalahari in Botswana. Individual hairs are slaty-grey except at their distal points, which are rufous.

The tail is brown above, white below. The ears are of medium size. There is a well-marked spot at the base of the outer margin of each ear in many specimens from SWA/Namibia and Botswana (*swalius*).

The vibrissae are well developed.

### Skull and dentition

There are no particular features in the species different from those of the genus. The cheekteeth are similar to those of *Dendromus* and tend to become laminate with wear (Ellerman 1941) (fig. 56).

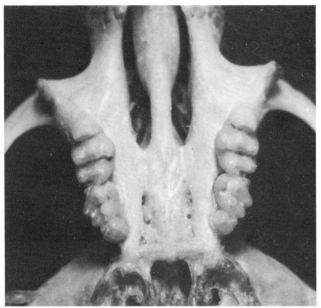

**Figure 56** Cheekteeth of the Tiny Fat Mouse *Steatomys parvus.*

### Distribution

The Tiny Fat Mouse is found in the districts of upper Natal, sandy areas of northern KwaZulu to the east of the Lebombo Mountains and ranges northwards into coastal Mozambique and westwards to the central and northern Kalahari in Botswana into the northern areas of SWA/Namibia and the pro-Namib plains as far south as 25°S (Coetzee 1977), perhaps as an isolated population (map 35). Its abodes are consequently found in the South West Arid, the Southern Savanna Woodland and the Southern Savanna Grassland biotic zones (Rautenbach 1978a).

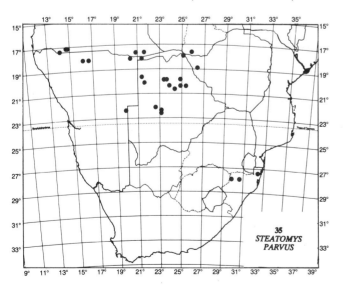

### Habitat

According to Smithers (1971) these animals have a wide habitat tolerance: they can be found in riverine flood-plains and woodland (along the Chobe and Botlele rivers), and in dry, open scrub associated with grassland (at Gutsa and Nxai). The mean annual rainfall at these localities vary considerably – more than 700 mm (at the Chobe River) and to 400 at Kaotwe Pan. According to Coetzee (1977), it prefers a drier habitat than *S. pratensis* when found on inland plateaus, but it also occurs on moister but sandy eastern coastal plains.

### Diet

Unknown.

### Habits

No information is available, except that in Botswana they are known to occur on the same ground as *S. pratensis* (Kasane and Maun) and *S. krebsii* (at Kasane) (Smithers 1971). Coetzee (1977) reports them to be nocturnal, terrestrial, graminivorous and living singly in burrows.

### Reproduction

Information is meagre. Smithers (1971) notes that a female taken in November at Gutsa Pan in Botswana was lactating.

### Parasites

Unknown.

### Predators, relations with man and prehistory

No data available.

### Taxonomy

Apart from the nominate subspecies *parvus, Steatomys parvus* also contains *kalaharicus, minutus* (including *umbratus* and *swalius*) and *tongensis (= S. chiversi tongensis)* of earlier authors as subspecies. The relationships between *kalaharicus* and *minutus* require further study (Coetzee 1977). It may also contain *loveridgei* found to the north and south of the mouth of the Zambezi River in Mozambique.

# *Steatomys krebsii* Peters, 1852

### Krebs' Fat Mouse

### Krebs se Vetmuis

For the derivation of *Steatomys,* see *Steatomys pratensis.* The species name is named after the German botanist, L. Krebs, who collected plants in South Africa during the mid-19th century.

The type locality is nominated as Graaff-Reinet by Roberts (1951), described as the 'Interior of Kaffraria' by

Peters. Coetzee *(in litt.)*, however, disagrees (Meester *et al.* 1964) and Coetzee (1977) restricted the type locality to Uitenhage. Graaff-Reinet was probably chosen as a designated type locality, because this is where Krebs lived and collected his botanical specimens. It is a bigger species than *pratensis*. This species is, however, not known from Graaff-Reinet (Ellerman *et al.* 1953).

The subspecies *Steatomys krebsii pentonyx*, occurring on the Cape Flats and ranging northwards to Tulbach, Worcester and Klaver is recorded as a rare taxon in the *South African Red Data Book* by Meester (1976). It does not appear to be endangered.

## Outline of synonymy

1852 *Steatomys krebsii* Peters, *Reise nach Mossambique. Säugeth.*: 165. Interior of Caffraria, South Africa.

1929 *S. k. transvaalensis* Roberts, *Ann. Transv. Mus.* 13: 117. Witfontein, Randfontein District, Transvaal.

1931 *Steatomys chiversi* Roberts, *Ann. Transv. Mus.* 14: 233. Blood River, Natal.

1939 *Steatomys krebsii krebsii* Peters. *In* Allen, *Bull. Mus. comp. Zool. Harv.* 83:359.

1953 *Steatomys krebsii krebsii* Peters. *In* Ellerman *et al.*, *Southern African mammals.*

1964 *Steatomys krebsii* Peters. *In* Meester *et al.*, *An interim classification of southern African mammals.*

1977 *Steatomys krebsii* Peters. Coetzee, *in* Meester & Setzer (eds), *The mammals of Africa: an identification manual.*

### TABLE 33
#### Measurements of male and female *Steatomys krebsii*\*

| | Parameter | Value (mm) | N | Range (mm) | CV (%) |
|---|---|---|---|---|---|
| Males and females | TL** | 130 | 9 | 118–146 | — |
| | T | 51 | 9 | 44–61 | — |
| | HF | 17 | 9 | 14–19 | — |
| | E | 17 | 8 | 16–18 | — |
| | Mass: | 24 g | 1 | — | — |

\*After Smithers (1971)
\*\*TL = Total length

## Identification

Peters (1852) described the dorsal surface as ochre-yellow, suffused with black towards the middle. The flanks have the same colour, while the ventral surface, inner sides of the extremities and hands are white with the feet a brownish-yellow. The tail is ochre-yellow above and on the sides, white below. All hairs are slaty-coloured at their bases (plate 5).

Roberts (1929) described the synonym *S. k. orangia* as '... A prettily coloured orange-buffy form...' while specimens from the western Cape have a general colour '...

above brown', the line of demarcation between the dorsal and ventral surfaces very clear. In these animals, the ventral hairs are without slaty bases. Generally, the dorsal colour is a colder grey than *pratensis* (Ansell 1960).

The head is pointed, ears moderately sized with four well-developed fingers on the hand. The tail is bi-coloured.

## Skull and dentition

The overall morphology has been dealt with under the introductory paragraphs to the genus. Roberts (1929) mentions that the form *transvaalensis* (a synonym to *krebsii*) has a skull '... apparently averaging larger than in *S. k. orangiae*'. The $M^3$, though small, is not as reduced as in *Dendromus* (fig. 57).

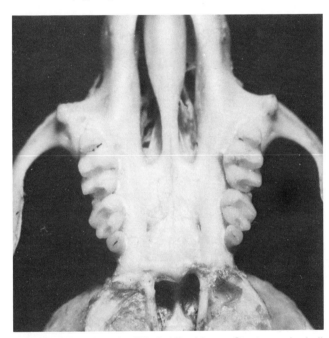

**Figure 57** Cheekteeth of Krebs' Fat Mouse *Steatomys krebsii*.

## Distribution

This species occurs in disjunct areas. It is found in the western Cape to the mouth of the Orange River, in the eastern Cape and the adjoining northern Cape, the OFS, the western Transvaal highveld and in the extreme north of Botswana, bordering the Caprivi (map 36). Additional collecting may reveal a more continuous distribution (Coetzee 1977).

## Habitat

This species is apparently restricted to open sandy plains which are covered with grass. Smithers (1971) records them in sandy alluvium on fringes of the Chobe and Kwando rivers. It seems to be more moisture tolerant than *S. pratensis* (Coetzee 1977).

## Diet

No data available, but in all probability graminivorous.

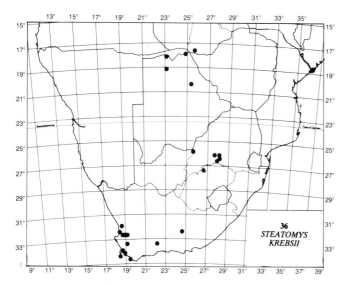

**36**
*STEATOMYS*
*KREBSII*

## Reproduction

It is said that *S. krebsii* uses finer grasses than *S. pratensis* for nest construction. The females have two pectoral and two inguinal pairs of mammae.

## Parasites

Populations of this species are subjected sporadically to outbreaks of plague, *Yersinia pestis* and this has been confirmed by laboratory tests (Powell 1925; Pirie 1927). Laboratory tests have also shown their susceptibility to infections of *Listeria monocytogenes* (Pirie 1927).

## Relations with man and prehistory

Unknown.

## Taxonomy

Four subspecies are currently accepted. These include *S. k. krebsii* (Eastern Province), *S. k. pentonyx* (western Cape) and *S. k. orangiae* (OFS and the Transvaal highveld). According to Coetzee (1977) *orangiae* includes *transvaalensis* and *S. chiversi* as synonyms. The subspecies *pentonyx* has been treated as a distinct species, both of *Steatomys* (Roberts 1951) and of *Malacothrix* (Allen 1939), and as a subspecies of *pratensis* (Ellerman *et al.* 1953). In northern Botswana *Steatomys krebsii angolensis* (including *bradleyi* and *leucorhynchus)* is a fourth subspecies.

## Habits

This species is terrestrial, nocturnal and not communal. The burrows they construct tend to run spirally downwards. Coetzee (1977) states that its general ecology may be similar to *S. pratensis* and *S. parvus*.

## Predators

Unknown.

# GENUS *Malacothrix* Wagner, 1843

This genus, endemic to South Africa, has teeth which resemble those of *Dendromus* and *Steatomys* rather closely. In contrast to these genera, however, it has only four toes on the hind foot. Furthermore, there are pronounced differences in the skull. The interorbital constriction is pronounced (to a greater extent than is found elsewhere) with a very narrow, straight-edged rostrum; the braincase is short, rounded and slants downwards posteriorly. The zygomatic arches are widely spreading. The tooth-row extends far forward in the skull with a very broad palate in between. The bullae are medium-sized and the palatine foramina large and long. The masseter knob is enlarged and the outer edge of the large infraorbital foramen is prominently ridged.

# *Malacothrix typica* (A. Smith, 1834)

**Large-eared Mouse**

**Grootoormuis**

The generic name is compounded from the Greek words *malakos* = soft, gentle and *thrix* = hair, with reference to the soft, woolly pelage of the animal. The specific name is from the Greek word *typicos* = typical, implying that the species is typical of the genus.

This dendromurid was first described by Sir Andrew Smith in 1834 from specimens obtained in the vicinity of Graaff-Reinet. The type specimen seems to be lost.

## Outline of synonymy

1834 *Otomys typicus* A. Smith, *S. Afr. Quart. Journ.* 2: 148. District east of Graaff-Reinet, CP.
1843 *Malacothrix* Wagner. *In* Schreber's *Säugeth.* Suppl. 3:496. Genotype *Otomys typicus (nec* F. Cuvier).
1898 *Malacothrix typicus* (A. Smith). De Winton, *Ann. Mag. nat. Hist.* (7)2:8.

136

1917 *Malacothrix typicus fryi* Roberts, *Ann. Transv. Mus.* 5:268. Krugersdorp district, Transvaal.

1926 *Malacothrix egeria* Thomas, *Proc. zool. Soc. Lond.:* 301. Ondongwa, SWA/Namibia.

1932 *Malacothrix typicus kalaharicus* Roberts, *Ann. Transv. Mus.* 15:10. Kuke Pan, Botswana.

1932 *M. t. damarensis* Roberts, *Ann. Transv. Mus.* 15:10. Gobabis, SWA/Namibia.

1933 *M. t. molopoensis* Roberts, *Ann. Transv. Mus.* 15:266. Pitsani, Botswana.

1951 *M. t. harveyi* Roberts, *The mammals of South Africa:* 455. (Probably a *lapsus* for *fryi*.)

1951 *M. typicus typicus* (A. Smith). *In* Roberts, *The mammals of South Africa.*

1953 *M. typica* (A. Smith). *In* Ellerman *et al., Southern African mammals.*

1964 *M. typica* (A. Smith). *In* Meester *et al., An interim classification of southern African mammals.*

1974 *M. typica* (A. Smith). Misonne, *in* Meester & Setzer (eds), *The mammals of Africa: an identification manual.*

**TABLE 34**
Measurements of male and female *Malacothrix typica*

| | Parameter | Value (mm) | N | Range (mm) | CV (%) |
|---|---|---|---|---|---|
| Males | HB | 77 | 36 | 67–86 | 3,9 |
| | T | 34 | 36 | 28–42 | 6,7 |
| | HF | 18 | 36 | 16–21 | 5,2 |
| | E | 19 | 36 | 14–21 | 6,4 |
| | Mass: | 16 g | 4 | 12–23 g | — |
| Females | HB | 79 | 21 | 71–92 | 4,2 |
| | T | 34 | 21 | 20–40 | 11,4 |
| | HF | 18 | 21 | 16–21 | 5,6 |
| | E | 20 | 21 | 18–22 | 4,4 |
| | Mass: | 15 g | 5 | 15–20 g | 17,1 |

**Identification**

Dorsally a pale brown, lightly brindled with black hairs: areas around the mouth, chin, throat, belly and legs a dull white. Individual hairs are a dark slate near their bases. The relatively short tail is fairly well covered with short white hair. The ears are very large for the size of this animal, brownish black on the outer surfaces. Whiskers are well developed.

An ill-defined dark patch occurs between the ears, while a black stripe occurs along the mid-dorsal line, and a black patch on the top of each hip (plate 5).

The intensity of the dorsal pelage colour gets paler as one moves westwards from Botswana into SWA/Namibia. Western individuals have pure white hairs on chests and bellies (Smithers 1971). The fur is of peculiar quality and very soft to the touch.

The hand has four functional digits, and this also applies to the hindfoot which is very narrow. The sole of the hindfoot is hairy. The hallux is lost while the fifth digit is not much shorter than the second, third and fourth digits.

**Skull and dentition**

The characteristic features of the skull have been dealt with in the introductory paragraph to the genus (see above) (fig. 58).

The upper incisors are very narrow, the $M^1$ with an additional cusp adjacent to the two cusps of the middle lamina which are more or less equal in size. The second lamina consists of two cusps; the third lamina is also made up of two cusps often with an additional, small outer posterior dusp. The $M^2$ has a well-developed additional anterior cusp, while the second and third laminae are like those in the $M^1$. The $M^3$ is very small (fig. 59).

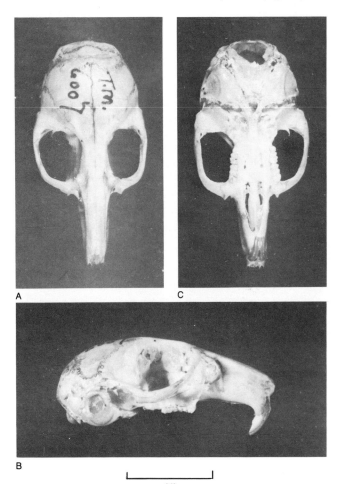

**Figure 58** Dorsal (A), lateral (B) and ventral (C) views of the skull of the Large-eared Mouse *Malacothrix typica*. Interorbital constriction pronounced with muzzle narrow and straight-edged; the short, rounded braincase slants posteriorly; zygomatic arches spread widely; a prominent masseter knob present; anterior palatine foramina lengthened. The outer edge of large infraorbital foramen markedly ridged.

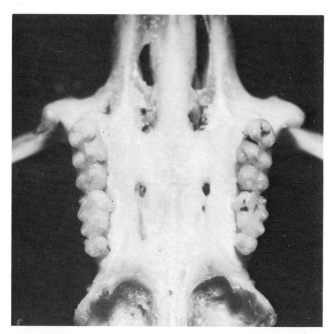

**Figure 59** Cheekteeth of the Large-eared Mouse *Malacothrix typica*. The tooth rows are situated far forward in the skull with a broad palate between them. An additional posterior cusp may be present on the posterior edge of the M$^1$ while the M$^2$ often shows an additional anterior cusp. The M$^3$ is very small.

## Distribution

The species ranges from Van Rhynsdorp in the western Cape into Ovambo in the north, as well as into the eastern Cape, OFS, southern and western Transvaal and Botswana (map 37). They consequently occur in the following biotic zones: the South West Arid and the Southern Savanna Grassland (Rautenbach 1978a).

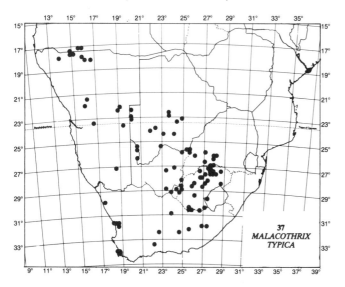

37
MALACOTHRIX
TYPICA

## Habitat

In Botswana it is associated with short grass on hard ground, on fringes of pans, or on pans with a cover of karroid vegetation, or on hard calcareous ground (Smithers 1971).

## Diet

Smithers (1971) has found that it subsists predominately on green vegetable matter. In captivity it readily takes lettuce, fresh lucerne and finger millet. Roberts (1951) reports that it is very partial to cooked farinaceous food, discarded by Blacks near their huts.

## Habits

Roberts (1951) gives the following information: large-eared mice often travel long distances at night along cattle tracks or human footpaths to particular feeding grounds. Roberts also described the burrow as follows: a passage slants deeply downwards at the bottom of which a chamber is made. Within this chamber, a nest is constructed of soft straws and other material (often including feathers) in which the young are born. From this nest chamber a new passage is contructed upwards to the surface, the excavated soil being used to fill up the original burrow beyond the chamber; on reaching the surface, no soil is thrown out and all that can be seen is a tiny open hole some 2–3 m from the original entrance, which may be present as a mound of earth without a burrow below.

The burrows are 20–25 mm in diameter. Smithers (1971) states that the Large-eared Mouse is difficult to trap. Once trapped, the animals are docile, easily handled and not apt to bite. Roberts (1951) also pointed out that they are not aggressive, easily tamed, and make delightful pets. In captivity they build domed nests constructed with variable materials (e.g. grass) cut into small pieces for the purpose.

They may attain an age of 2,5 years in captivity (Smithers 1971).

## Predators

It is likely that they are preyed upon by smaller carnivores, snakes and certain birds of prey, particularly the Barn Owl *Tyto alba* (Nel & Nolte 1965; Dean 1977).

## Reproduction

Four pairs of mammae occur; one pectoral, one pectolateral, one lateral and one inguinal.

Smithers (1971) states that young may be born in the wetter months from about August to March (Botswana). Under captive conditions, no breeding seems to occur until the onset of warmer weather during August. The gestation period varies between 22 and 26 days.

Birthweights of two pups were 1,1 and 1,15 g. The pups were born pink and naked.

Indications are that the species is apparently capable of producing more than one litter during the breeding season and of breeding at an age of 51 days (Smithers 1971).

## Parasites

It has been shown that this species is susceptible to infections of *Yersinia pestis* (Powell 1925; Pirie 1927) and to *Lysteria monocytogenes* (Pirie 1927).

The only other group hitherto known to be associated with *Malacothrix typica,* are fleas. According to Zumpt (1966), three families are represented. The Pulicidae include *Xenopsylla philoxera, X. piriei* and *X. brasiliensis.* The Hystrichopsyllidae have *Dinopsyllus ellobius* and *Listropsylla dorippae* as representatives. The Chimaeropsyllidae are represented by *Chiastopsylla rossi.*

### Relations with man

Unknown, but this species may be a reservoir of plague if the associated fleas are taken into consideration. They have been successfully bred in the South African Institute for Medical Research's animal house (Keogh & Isaäcson 1978).

### Prehistory

*M. typica* has been recovered from only one fossiliferous locality in South Africa – at the Pleistocene site Taung in the northeastern Cape (De Graaff 1961).

### Taxonomy

*Malacothrix* consists of one species only, and some subspecies have been described. These are *fryi (= 'harveyi'* of Roberts 1951, probably a *lapsus* for *fryi), egeria, kalaharicus, molopoensis* and *damarensis,* but all or most appear to be doubtfully valid (Meester *et al.* 1964).

---

SUBFAMILY ## Petromyscinae Roberts, 1951

Ellerman (1941) followed Hinton *(in* Thomas 1926a) in retaining the genus *Petromyscus* under the Dendromurinae, but Ellerman also pointed out that he was not altogether convinced that this group of small mice is not yet another South African representative of the Cricetinae. It will be recalled that the only member of the Cricetinae occurring in southern Africa (as interpreted for the purposes of this book), is the White-tailed Rat *Mystromys albicaudatus.* There is a fair degree of correspondence in the molars of *Petromyscus* with the Cricetinae, while the molars are rather different in structure from those of the Dendromurinae as discussed earlier on in this book.

This uncertainty pertaining to divergent taxonomic interpretations probably prompted Roberts in 1951 to place *Petromyscus* in a subfamily of its own. He referred to *Petromyscus* as '… a relict genus of Murines…' and gives a good account of his reasoning.

The first specimen (collected by Shortridge) was originally placed in the murine genus *Praomys,* but was subsequently transferred to a new genus, *Petromyscus,* by Thomas in 1926.

It differs from the Dendromurinae in having ungrooved incisors, while the molars have a zig-zag pattern of enamel on the occlusal surfaces (cf. *Mystromys).* In addition, it shows greater correspondence with the murines in having a long, ringed and scaly tail, thinly bristled. On the other hand, specimens mainly show two pairs of inguinal mammae, as in the case of *Mystromys,* although in some instances a pair of pectoral mammae may also be present. In the South African Dendromurinae *(Dendromus, Steatomys* and *Malacothrix)* two pectoral and two inguinal pairs of mammae are usually encountered. Otomyinae always have only two inguinal pairs of mammae, while the Gerbillinae have two pairs of inguinal and one or two pairs of pectoral mammae. In the Murinae there are from two pairs inguinal and one pair pectoral to a total of 10 to 12 pairs in the Multimammate Mouse *Praomys (Mastomys) natalensis.* Roberts suggests that the tendency seems to be '… from a start with two pairs of inguinal and no pectoral mammae, to an advance with one or more pairs of pectoral mammae to the development of lateral pairs and a general climax in a large number'.

Because *Petromyscus* shares features with both cricetids and murids, it seems to be sufficiently isolated to warrant its being grouped in a subfamily of its own.

---

GENUS ## *Petromyscus* Thomas, 1926

Members of this genus are small mouse-like animals with a soft pelage and a moderately haired tail. The hindfoot has the fifth digit lengthened, while the hallux is not reduced. The mammae number either two inguinal pairs only or one pectoral pair with two inguinally situated glands.

In the skull, the braincase is broad and flattened; the interorbital constriction is moderate; the infraorbital foramen is not enlarged and the palate relatively broad.

The dentition shows compressed upper incisors, not pro-odont. The molars differ markedly from those of the dendromurids and the re-entrant folds or spaces between the laminae play an important role in the resulting occlusal pattern. The posterior lamina of the $M^1$ and $M^2$ appears more or less doubled or twisted round itself. The $M^3$ is small and reduced. The structure of the lower molars are reminiscent of those of *Mystromys*, of which Ellerman (1941) states '... it occurs to me that this genus may belong to that subfamily' (i.e. Cricetinae).

---

**Key to the southern African species of *Petromyscus***
(After Ellerman *et al.* (1953))

1 Tail shorter than HB length; ear about 11–12 mm .....
*Petromyscus monticularis*
(Berseba Rock Mouse)
Page 142

Tail averages longer than HB length; ear 13 mm or more................................................. *P. collinus*
(Pygmy Rock Mouse)
Page 140

---

# *Petromyscus collinus* (Thomas & Hinton, 1925)

**Pygmy Rock Mouse**

**Damara Pygmy Rock Mouse**

**Dwerg Klipmuis**

The generic name is derived from the Greek *petra* = rock and *mys, myskos* = diminutive of mouse. The specific name is derived from the Latin *collina* = hilly. This little rodent is, therefore, a small rock mouse frequenting the hills.

The type specimen of *Petromyscus collinus* is housed in the British Museum (Natural History), London. It was collected by Capt. G.C. Shortridge and placed in a new genus *Petromyscus* by Thomas in 1926 using *Praomys collinus* Thomas & Hinton, 1925 as genotype. It appears to be a relict genus and its molar structure is very similar to that of *Delanymys* (extralimital). Lavocat (1964) described the phylogenetic importance of *Petromyscus* and *Delanymys* and stated that they form perfect structural links between *Mystromys* on the one hand and the Dendromurinae on the other. The presence of the characteristic third internal cusp (next to the two cusps of the second lamella) in *Petromyscus* indicate that this tubercle is not a remnant portion of a murid tooth, but is a newly elaborated structure found in the teeth of cricetids. Kingdon (1974) provides excellent illustrations of this comparison from the cricetic condition to the dendromurid condition *via* the situation found in the petromyscine mice.

## Outline of synonymy

1925 *Praomys collinus* Thomas & Hinton, *Proc. zool. Soc. Lond.*: 237. Karibib, SWA/Namibia.
1925 *P. c. bruchus* Thomas & Hinton, *Proc. zool. Soc. Lond.*: 302. Great Brukaros Mountain, SWA/Namibia.

1926 *Petromyscus shortridgei* Thomas, *Proc. zool. Soc. Lond.*: 302. Ruacana Falls, Cunene River, SWA/Namibia.
1938 *P. s. kaokoensis* Roberts, *Ann. Transv. Mus.* 19:239. Kamanjab, SWA/Namibia.
1938 *Petromyscus barbouri* Shortridge & Carter, *Ann. S. Afr. Mus.* 32:288. Witwater, Kamiesberg, CP.
1938 *Petromyscus collinus capensis* Shortridge & Carter, *Ann. S. Afr. Mus.* 32:289. Goodhouse, CP.
1948 *P. c. namibensis* Roberts, *Ann. Transv. Mus.* 21:65. Okombahe, SWA/Namibia.
1955 *P. c. rufus* Lundholm, *Ann. Transv. Mus.* 22:299. Sesfontein, SWA/Namibia.
1955 *P. c. variabilis* Lundholm, *Ann. Transv. Mus.* 22:299. Seeheim, SWA/Namibia.

**TABLE 35**
Measurements of male and female *Petromyscus collinus*

| | Parameter | Value (mm) | N | Range (mm) | CV (%) |
|---|---|---|---|---|---|
| Males | HB | 103 | 5 | 92–112 | 7,0 |
| | T | 95 | 14 | 88–103 | 3,8 |
| | HF | 17 | 12 | 16–20 | 6,3 |
| | E | 14 | 19 | 12–16 | 7,2 |
| | Mass: | 20,2 g | 11 | 3 17–24 g | 9,69 |
| Females | HB | 82 | 7 | 76–87 | 4,1 |
| | T | 86 | 12 | 78–91 | 5,8 |
| | HF | 16 | 14 | 14–20 | 9,1 |
| | E | 15 | 13 | 13–17 | 8,7 |
| | Mass: | 22 g | 8 | 19–24 g | 8,8 |

## Identification

The fur of this species is buffy-yellow, reminiscent of the

Gerbillidae. The fur is fine and soft and silky to the touch. The ventral surface is a greyish-white, as are the hands and feet (plate 5). There are no guard hairs and individual hairs of both upper and lower surfaces have dark, slaty bases extending along the shaft of the hair with the last quarter of the shaft tipped brown above and white below.

The tail is longer than the combined head-body length, its scaly nature clearly visible through the rather sparse hairs. In the hindfoot, the fifth digit is long, the hallux well developed. The feet and hands are relatively short. The ears are large and conspicuous.

The number of mammary pairs may vary: the pectoral mammae are often closely surrounded by hair and their presence is easily overlooked; consequently the mammary formula varies between one pectoral pair and two inguinal pairs to two inguinal pairs only.

The juveniles are smoky-grey in colour (Shortridge 1934).

Individual from Kamiesberg are a drab grey with a bicoloured tail (drab grey above and white below).

## Skull and dentition

The skull is lightly constructed. The braincase is broad and flattened, the infraorbital foramen normal. The palatal foramina are of medium length, reaching the level of the $M^1$. The palate is broad, while the bullae are of medium size (fig. 60).

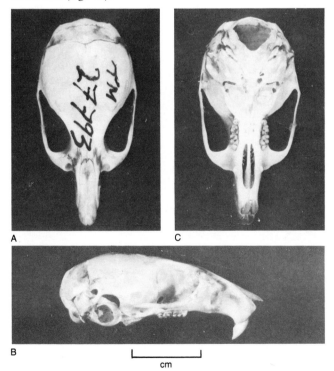

**Figure 60** Dorsal (A) lateral (B) and ventral (C) views of the skull of the Pygmy Rock Mouse *Petromyscus collinus*. The skull is flattened, with a broad braincase and a moderate interorbital constriction. The infraorbital foramen is small. The palate is relatively broad and anterior palatine foramina long, while the bullae are not enlarged.

The incisors are ungrooved and compressed laterally. The molars are small and delicate. Only two rows of cusps occur, the internal cusps occurring in the murids being absent. The re-entrant folds between the laminae form the pattern of the occlusal surface. The posterior laminae of the $M^1$ and the $M^2$ appear more or less doubled on themselves. The $M^3$ is reduced, while the occlusal pattern of the molars is easily obscured by wear (fig. 61).

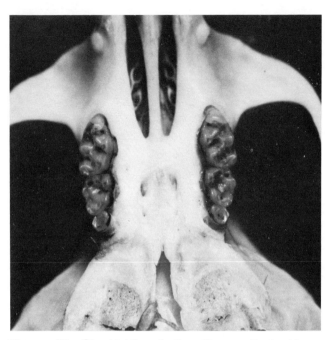

**Figure 61** Cheekteeth of the Pygmy Rock Mouse *Petromyscus collinus*. The re-entrant folds between the laminae determine the resultant occlusal pattern. The presence of a third interior cusp adjacent to the two cusps of the second lamella of the $M^1$ is a newly evolved structure. The posterior lamina of both the $M^1$ and $M^2$ appears twisted round itself. The $M^3$ is small.

## Distribution

This species occurs in the southwestern Cape (Kamiesberg) ranging northwards to the dry hills about Karibib in SWA/Namibia and beyond to the Erongo Mountains and across the Cunene River into southern Angola (map 38). It is therefore an inhabitant of the South West Arid biotic zone (as well as the Namib Desert subzone, as interpreted by Davis (1974)). The latter has been raised to full biotic zone rank by Rautenbach (1978a).

## Habitat

It occurs in the arid areas in the western half of southern Africa in areas where large boulders and rocky outcrops predominate.

## Diet

Apparently omnivorous, as it takes nearly every form of bait (Roberts 1951).

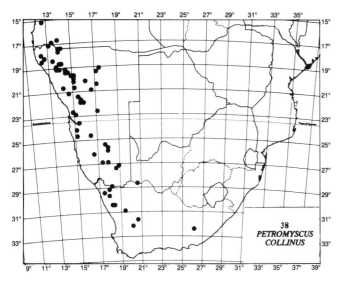

## Habits

The Pygmy Rock Mouse is predominantly nocturnal, and the nominate race *P. c. collinus* seems to be fairly abundant where it occurs.

## Predators

Unknown.

## Reproduction

Information is meagre. Shortridge (1934) records taking a pregnant female with two embryos in September (Great Brukaros Mountains, SWA/Namibia) and that at least one dozen females with two to three embryos each were taken at Otjitundua in May.

## Parasites

Information is meagre. Some data are available from Zumpt (1961a, 1966) and Theiler (1962) for the arthropods and Mönnig (1931) for the nematodes. The mites are known by a single family only, the laelaptids *Laelaps brandbergensis* and *Androlaelaps zuluensis*. The fleas occur in two families. The Pulicidae are represented by *Xenopsylla philoxera,* while the Chimaeropsyllidae are known by *Epirimia aganippes* and *Chiastopsylla nama*. The tick *Haemaphysalis leachii* (adults) has been recorded also. Finally, the occurrence of the nematode *Acanthoxyuris shortridgei* has been noted by Mönnig.

## Relations with man and prehistory

Unknown.

## Taxonomy

Ellerman (1941) divided the genus into two species groups: the *collinus* group and the *monticularis* group. Under the former he placed *bruchus* and *shortridgei* as subspecies in addition to the nominate subspecies *P. c. collinus.*

Roberts (1951) increased the number of subspecies by describing and listing *kaokoensis* and *namibensis* and also recognised *P. barbouri* and *P. capensis* as described by Shortridge & Carter (1938) as good species, while raising *P. c. bruchus* to species level (with *bruchus* and *shortridgei* as subspecies).

Ellerman *et al.* (1953) again demoted *P. barbouri* and *P. capensis* to subspecific rank. In addition, the subspecies *shortridgei, rufus* and *variabilis* were described by Lundholm in 1955.

Meester *et al.* (1964) state that most, if not all, of the named species and subspecies are probably synonyms of *P. collinus collinus,* the interpretation followed in this book.

# *Petromyscus monticularis* (Thomas & Hinton, 1925)

**Berseba Rock Mouse**

**Short-eared Pygmy Rock Mouse**

**Berseba Klipmuis**

For the explanation of *Petromyscus,* see *Petromyscus collinus*. The species name is derived from the Latin *montis* = mountain and *colere* = to dwell.

The type specimen, collected on the Great Brukaros Mountains near Berseba in SWA/Namibia, is housed in the British Museum (Natural History), London. Ellerman (1941) separated this species into the *monticularis* group as it is distinct from the other described forms (the *collinus* group), having much narrowed internal nares and shorter ears. It is also smaller than *P. collinus*. To what extent these differences are relative rather than absolute, is not clear.

This species is poorly represented in southern African museums and I was unable to compile statistical parameters.

## Outline of synonymy

1925 *Praomys monticularis* Thomas & Hinton, *Proc. zool. Soc. Lond.*: 238. Great Brukaros Mountains, SWA/Namibia.

1934 *Petromyscus monticularis* (Thomas & Hinton). *In* Shortridge, *The mammals of South West Africa.*

1951 *Petromyscus monticularis* (Thomas & Hinton). *In* Roberts, *The mammals of South Africa.*

## Identification

This is a very small species, smaller than *P. collinus*. Its colour is a brownish-buff, while the ventral surface is grey. As in *P. collinus,* the individual hairs have slaty-grey bases. The ears are unusually short and are dark brown. The hands and feet are white. The tail is short, slightly thickened and shorter than the head-body length (plate

5). There are one pectoral and two inguinal pairs of mammae.

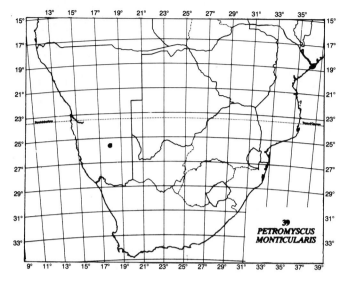

39
PETROMYSCUS
MONTICULARIS

### Skull and dentition
The skull of *P. monticularis* is smaller than that of *P. collinus*. The posterior nares are narrowed (wider in *collinus*) and the mesopterygoid fossa is also narrower. The bullae are smaller than in *collinus*.

### Distribution
This species seems to be very localised and is apparently confined to the Great Brukaros Mountains in SWA/Namibia. It may even be described as rare (map 39).

### Habitat, diet, habits, predators, reproduction, parasites, relations with man and prehistory
No data available.

### Taxonomy
*P. monticularis* has been placed in the *monticularis* group in Ellerman (1941) '... which is very distinct' from the other named forms in the *collinus* group. No subspecies have been described.

# SUBFAMILY Otomyinae Thomas, 1897

Generally, the Otomyinae are a small subfamily of specialised rats, confined to Africa south of the Sahara (Rosevear 1969). They have on occasion been included in the Muridae, but Davis (1965) and Dieterlen (1969) both suggested that they are to be placed under the cricetids. Misonne (1969) also referred to the lamellate nature of the molars and the trend towards an enlarged $M^3$. He stated this to be foreign to the Muridae and they they should therefore also be included in the Cricetidae. They are compact, stockily built animals with blunt faces, large ears and a shaggy pelage. Their tails are never much more than half the length of the head and body. Members of this subfamily have short limbs and tails, are well furred and predominantly terrestrial. According to the latest taxonomic interpretation, two genera are encountered in southern Africa, *viz. Otomys* and *Parotomys*. They are characterised by laminate molars: the $M^1$ with three, the $M^2$ with two and the $M^3$ with three to nine laminae, while the $M_1$ has four to seven, the $M_2$ two and the $M_3$ two laminae each. This implies that the $M^3$ and $M_1$ are always the largest teeth in the maxilla and mandible respectively. The cheekteeth have no cusps and the individual teeth lie closely together, creating an overall impression of a single, uniform tooth. Each tooth has a different and constant number of laminae and identification of genera is based largely on this character. In *Otomys* the upper incisors are deeply grooved, while the lower incisors show a shallow groove. The lower incisors in *Parotomys* are smooth.

They subsist on green vegetation and construct characteristic runways through the vegetation. They are diurnal, often living in the vicinity of marshes, and are competent swimmers.

Roberts (1951) states that all the species appear to have only two pairs of inguinal mammae.

The scientific name refers to their large, rounded ears, meaning 'ear-mice', (derived from the Greek *otis, otis* = ear, in reference to the large conspicuous ears) while they are also referred to as 'swamp rats' or 'vlei rats'. The latter two vernacular names are misleading, however, for they often occur far away from swamps or other open surfaces of water. They are on occasion also called 'groove-toothed rats'.

The genus *Otomys* is rather difficult to diagnose and as Kingdon (1974) has pointed out, the relationships are complex. The taxonomy is still far from clear and a great deal of additional research is needed on this subfamily (Pocock 1976). In southern Africa, six species of *Otomys* occur, including *irroratus, laminatus, angoniensis, saundersae, unisulcatus* and *sloggetti*.

The genus *Parotomys* is represented in southern Africa by two species, known as *brantsii* and *littledalei*.

GENUS *Otomys* Cuvier, 1823

The main features of this genus have been mentioned above when the subfamily was described.

In Africa ten species of *Otomys* occur, of which two thirds occur in southern Africa. The extralimital species are *O. typus*, *O. anchietae*, *O. denti* and *O. tropicalis*. *O. typus* can be found mainly on African mountains over 2 360 m above sea level and are consequently found on Kilimandjaro, Mount Kenya, the Ruwenzori Mountains and the mountains of Ethiopia. *Otomys anchietae* is a little-known species, distributed in Angola (not in the south), southern Tanzania and possibly near Mount Elgon in Uganda. *Otomys denti* is also restricted to a few mountainous areas including the Ruwenzori Mountains, certain forests in Uganda, the Nyika Plateau in Malaŵi and Zambia, as well as a number of localities in Tanzania (Misonne 1974). *Otomys tropicalis* is found in Zaïre, Kenya and Uganda.

The key to the species of *Otomys* in southern Africa given below, is taken from Misonne (1974). The extralimital species have been included as well.

# *Otomys irroratus* (Brants, 1827)

**Vlei Rat**

**Vleirot**

The word *Otomys* is derived from the Greek *otos* = ear and *mys* = mouse, with reference to its rather large and conspicuous ears. The species name is derived from the Latin *irroro* = to sprinkle with dew. This evidently refers to a sprinkling of small, light-coloured spots which is to be seen on the pelage of this species.

Sclater (1901) records that the first collector of this animal was Delalande, whose specimens were subsequently described by F. Cuvier in his work *Dents des mammifères* (1821–1823) under the name of 'Otomie Namaquois' (vernacular 1823). He was under the impression that they were collected in or near Namaqualand. The animal is, therefore, described without mention of any species by name (Ellerman *et al.* 1953). In 1827 the species was officially described as *Euryotis irrorata* by Brants in his well-known work *Het Geslacht der Muizen*. No type locality was designated and Roberts (1929) elected Uitenhage, Cape, as the type locality.

The ecology and life history of the Vlei Rat *Otomys irroratus* occurring on the Maria Van Riebeeck Nature Reserve near Pretoria, has been treated in depth by Davis (1973) and it is a useful contribution to our knowledge of the biology of this species.

## Outline of synonymy

1827 *Euryotis irrorata* Brants, *Het Geslacht der Muizen*: 94. No locality given. Uitenhage nominated by Roberts 1929, *Ann. Transv. Mus.* 13:111.

1829 *Otomys bisulcatus* F. Cuvier. *In* Geoffroy & Cuvier, *Hist. Nat. Mamm.* pt. 63.

1829 *Otomys capensis* G. Cuvier, *Regne Animal*. ed. 2. 1:208 (synonym of *irrorata*).

1832 *Otomys irroratus* Smuts, *Enum. Mamm. Cap.*: 45.

1834 *Euryotis typicus* A. Smith, *S. Afr. Quart. Journ.* 2: 149. Constantia, Cape Town, CP.

1842 *Euryotis obscura* Lichtenstein, *Verzeichn. Samml. d. Säugeth. u. Vögeln Kaffernlande*: 10. South Africa.

1906 *Otomys irroratus cupreus* Wroughton, *Ann. Mag. nat. Hist.* (7) 18:273. Soutpansberg, Transvaal.

1906 *Otomys irroratus auratus* Wroughton, 1906. *Ann. Mag. nat. Hist.* (7) 18:727. Vredefort, OFS.

1918 *O. i. coenosus* Thomas, *Ann. Mag. nat. Hist.* (9) 2: 108. Kuruman, CP.

1929 *O. i. randensis* Roberts, *Ann. Transv. Mus.* 13:112. Fontainebleau, Johannesburg, Transvaal.

1929 *Otomys irroratus natalensis* Roberts, *Ann. Transv. Mus.* 13:111. Kilgobbin, Dargle district, Natal.

1939 *Otomys (O.) irroratus irroratus* (Brants). *In* Allen, *Bull. Mus. comp. Zool. Harv.* 83:344.

1946 *Otomys irroratus orientalis* Roberts, *Ann. Transv. Mus.* 20:318. Umzimkulu, Natal.

1946 *Otomys cupreus cupreoides* Roberts, *Ann. Transv. Mus.* 20:318. Newgate, Soutpansberg, Transvaal.

1951 *O. irroratus irroratus* (Brants). *In* Roberts, *The mammals of South Africa*.

1953 *O. irroratus* (Brants). *In* Ellerman *et al.*, *Southern African mammals*.

1974 *O. irroratus* (Brants). Misonne, *in* Meester & Setzer (eds), *The mammals of Africa: an identification manual*.

### TABLE 36
### Measurements of male and female *Otomys irroratus*

| | Parameter | Value (mm) | N | Range (mm) | CV (%) |
|---|---|---|---|---|---|
| Males | HB | 168 | 47 | 153–202 | 6,1 |
| | T | 103 | 47 | 89–128 | 11,0 |
| | HF | 29 | 47 | 23–36 | 7,8 |
| | E | 22 | 45 | 19–26 | 5,8 |
| | Mass: | 146 g | 8 | 100–173 g | 13,0 |
| Females | HB | 164 | 39 | 150–175 | 4,3 |
| | T | 97 | 39 | 75–125 | 7,3 |
| | HF | 28 | 39 | 24–35 | 6,4 |
| | E | 22 | 38 | 17–26 | 6,7 |
| | Mass: | 140 g | 10 | 96–178 g | 7,0 |

## Identification

The dorsal colour is a speckled buffy-brown. The individual hairs are a dark slaty colour for four fifths of their length, buffy subterminally with dark tips. These hairs are interspersed with long, wholly black hairs. The sides and ventral parts of the pelage are paler, with the tail buffy below and on the sides and dark brown on its upper surface. The feet are dull greyish.

The ears are large, conspicuous and rounded, fairly well covered with hair. The forelimbs are short with second, third, fourth and fifth digits well developed, the first digit being rudimentary.

The sides of the muzzle are buffy to rust-coloured and the cheeks are paler, like the throat (plate 6).

Cases of pronounced melanism occur frequently in this species and virtually pitch-black individuals are often encountered. Furthermore, coloration is geographically variable. Individuals from the drier west are, however, less variable that those from other parts in its range of distribution. The baculum can also be used as a character in differentiating between species of *Otomys*. The base of the baculum is obovate in *O. irroratus,* but spatulate in *O. angoniensis*. According to Davis (1973), the base of the baculum in *O. irroratus* changes significantly with age, developing processes on the side.

## Skull and dentition

The skull is sturdily built, with a narrow interorbital constriction. The supraorbital ridges are prominent. A typical feature of this genus can be seen in the nasals: there is a pronounced angle at the anterior portion of the nasal, an expansion deflected downwards resulting in a broad rostrum. The bullae are small to moderate in size and the zygomatic plate is typically murine in form. The posterior petrotympanic foramen is a relatively large, round hole. The palate is very narrow, with deep pterygoid fossae. Palatal foramina are long, nearly reaching the $M^1$. The basioccipital is not as narrowed down as in *Parotomys* (fig. 62).

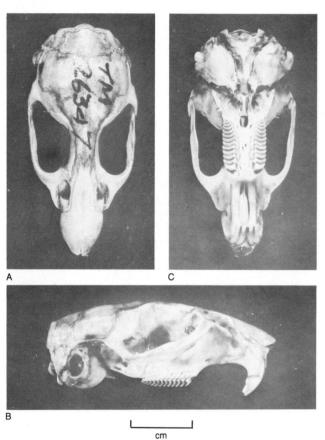

cm

**Figure 62** Dorsal (A), lateral (B) and ventral (C) views of the skull of the Vlei Rat *Otomys irroratus*. The skull is sturdily built with a marked interorbital constriction; nasals show a characteristic widening anteriorly and this expansion deflects downwards resulting in a broad muzzle; zygomatic plate is murine in shape; bullae small to moderate in size; palate narrow and pterygoid fossae deep; anterior palatine foramina long; basioccipital is much broader than in *Parotomys*.

The mandible is powerfully built, short and deep, heavily ridged.

The upper incisor has a deep groove externally. The lower incisor usually has one deep outer and one shallow inner groove. The cheekteeth are lophodont, i.e. showing a series of straight, transverse plates. The $M^1$ always has three laminae, the $M^2$ two and the number of laminae

on the $M^3$ may vary (fig. 63). In *O. irroratus* there are usually six laminae present. In the lower molars, the $M_1$ is the largest tooth with a variable number of laminae (usually four), followed by the $M_2$ and $M_3$ with two laminae each.

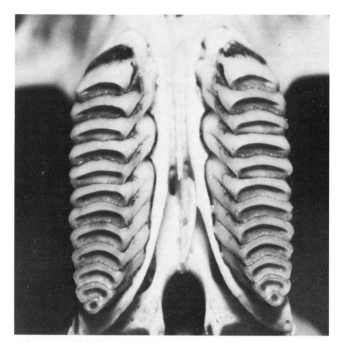

**Figure 63** Cheekteeth of the Vlei Rat *Otomys irroratus*. The lophodont teeth consist of a $M^1$ with three laminae, a $M^2$ with two laminae and a $M^3$ with a varying number of laminae (usually six, but seven in this case).

## Distribution

The geographical distribution of this species is often markedly discontinuous, but it occurs throughout the RSA. In Zimbabwe it seems to be confined to the eastern escarpment. It occurs mainly in the highveld and coastal, montane and submontane grasslands of the Southern Savanna biotic zone, but with a strong distribution in the Forest and South West Cape zones (map 40). It does not occur in the Namib biotic zone.

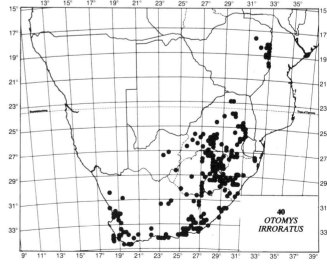

40
*OTOMYS
IRRORATUS*

## Habitat

They prefer grass-covered ground in proximity to streams and marshes (although this is not invariably so). Definite tunnels and runways occur in the vegetation. Davis (1973) was able to show a highly significant positive association of O. irroratus with the following plant species growing on the experimental grid in the Maria van Riebeeck Nature Reserve near Pretoria: *Mariscus congestus, Eleocharis dregeana, Berula erecta, Agrostis lachnantha, Juncus punctorius, Pennisetum thunbergii* and *Circium vulgare*.

## Diet

It is primarily vegetarian and eats grass stems, grass leaves and herbs.

## Habits

This species is terrestrial and depending on the surrounding habitat, may become partly amphibious. It is an able swimmer. It is semi-gregarious and diurnal, although some nocturnal activity has been observed. Where it occurs, it is usually plentiful. Its presence can be inferred immediately by the communal runs constructed through matted vegetation in which small heaps of cut grass stems and stalks occur at irregular intervals. The Vlei Rat lives primarily above ground and seldom burrows and when it occupies a burrow, it often utilises abandoned ones of other animals, doing little digging of its own (Davis 1973). Davis has also found that O. irroratus specimens are not easily captured with snap-traps and he suggests the use of live-traps to capture O. irroratus, even for removal trapping. They are anti-social animals and tend towards adult isolation which is demonstrated by the existence of marking behaviour and the probability of territoriality.

Perrin (in press a) has found that O. irroratus produces few offspring per litter, but breeds throughout the year by cropping abundant but poor quality herbage. It can therefore be termed a *K*-strategist. The iteroparity of the *K*-strategist may be attributable to the energy liberated from caecal microbial fermentation.

## Predators

The Vlei Rat is potentially preyed upon by the smaller carnivors, snakes, diurnal birds of prey and the Barn Owl *Tyto alba,* as well as the Grass Owl *Tyto capensis* and the Marsh Owl *Asio capensis*. This implies that they may be partially nocturnal as well. FitzSimons (1920) listed the following snakes as predators of O. irroratus: *Python sebae* (Python), *Pseudaspis cana* (Mole Snake), *Hemachatus haemachatus* (Rinkals), *Naja nivea* (Cape Cobra) and *Bitis arietans* (Puffadder). Likewise, FitzSimons (1920) reported the following carnivores of O. irroratus: mongoose (species unrecorded), African Polecat *(Ictonyx striatus),* Genet *(Genetta* sp.), Serval *(Felis serval),* Wild Cat *(F. lybica),* Jakal *(Canis* sp.), fox (species unrecorded) and the Honey Badger *(Mellivora capensis)*. Mendelsohn has found that the Black-shouldered Kite *Elanus caeruleus* eats *Otomys*

irroratus (Maclean *in litt.*). The Long-crested Eagle *Lophaetus occipitalis* is also an important predator as was found by Dorothy Hall (Maclean *in litt.*).

## Reproduction

There seems to be no fixed breeding season. The young number two to three per litter and are born with the incisors slightly erupted, though apparently not so as to cause discomfort to the female while suckling. The female has two pairs of inguinally situated mammae. The young are born in nests made of grass in dense vegetation. According to Roberts (1951), the young do not exceed four in number. Davis (1973) found the litter size to vary from one to four pups with an average of 2,33 per litter. The young are born precocial and are of a large size. Gestation lasts at least 35 days but 40 days or longer is more probable. Newly born pups have the head and body well furred, eyes and ears sometimes functional, incisors erupted while being ambulant with clawed toes. Davis' animals grew rapidly during the first ten weeks of life, gaining some 1,29 g per day in weight. Many adult behaviour patterns become manifest after 14 days. Weaning was nearly complete by day 13. Nipple-clinging is a prominent feature of the young Vlei Rat, especially during the first two weeks of life. Females reach sexual maturity at about nine to ten weeks of age, the males at about 13 weeks of age.

## Parasites

The following diseases and ecto- and endoparasites have been recorded from O. irroratus: Field tests in the Transvaal have revealed the presence in natural vlei rat populations of the bluetongue virus (Du Toit & Goosen 1949), while a sporadic natural incidence of bubonic plague *(Yersinia pestis)* has been demonstrated in all four provinces of South Africa, substantiated by laboratory tests (Powell 1925; Pirie 1927). Laboratory tests have also revealed their susceptibility to *Rickettsia conorii*, the causative agent of tick-bite fever. Apparently these organisms are endemic in O. irroratus (Wolstenholme & Harwin 1951, 1952). Similarly, *Coxiella burnetii,* causing Q-fever, has been found endemically and their susceptibility to this zoonosis is substantiated by laboratory tests (see Neitz 1965).

The arthropod parasites are well represented and rather well known. As recorded by Zumpt (1961a), 22 species of mites have been collected from the Vlei Rat. These include (i) the Laelaptidae: *Laelaps giganteus, L. muricola, L. parvulus, L. transvaalensis, Haemolaelops glasgowi, H. labuschagnei, H. murinus* and *H. taterae;* (ii) the Myobiidae: *Myobia otomyia;* (iii) the Trombiculidae: *Trombicula panieri, Schoengastia gigantica, S. lavoipierrei, S. oubanguiana, S. radfordi radfordi, Euschoengastia africana, E. otomyia, Schoutedenichia andrei, S. panai bukavuensis, S. p. paradoxa, S. penetrans, S. schoutedeni* and *Gahrliepia longiscutullata;* and (iv) the Listrophoridae: *Listrophoroides womersleyi*.

147

Four families of fleas are known from *Otomys irroratus:* (i) the Pulicidae: *Echidnophaga gallinacea, Pulex irritans, Xenopsylla cheopis. X. eridos, X. brasiliensis* and *X. hirsuta;* (ii) the Hystrichopsyllidae include *Ctenophthalmus calceatus, Dinopsyllus abaris, D. ellobius, D. lypusus, D. tenax, Listropsylla agrippinae, L. prominens, L. chelura* and *L. fouriei;* (iii) the Chimaeropsyllidae are represented by *Hypsophthalmus montivagans, H. temporis, Epirimia aganippes, Chiastopsylla roseinnesi, C. rossi* and *C. carus;* (iv) finally, the Leptopsyllidae is represented by *Leptopsylla segnis* (Zumpt 1966).

Further ectoparasites include the following ticks: *Ornithodoros zumpti, Ixodes alluaudi, Haemaphysalis leachii leachii, H. l. muhsami, Rhipicephalus appendiculatus, R. capensis, R. sanguineus* and *R. simus* (Theiler 1962). Only one species of louse, *Polyplax otomydis* is known from *O. irroratus* (Johnson 1960).

The following cestodes have been recorded: *Paranoplocephala omphalodes, P. otomyos, Inermicapsifer congolensis, I. madagascariensis* and *Raillientina (R.) thryonomysi* (Collins 1972). A single trematode, *Fasciola hepatica* is known to occur occasionally (Onderstepoort records). The nematodes found in *O. irroratus* to date include *Longistriata (L.) capensis* (Onderstepoort records), *Paralibyostrongylus* sp. (Onderstepoort records) and *Physaloptera africana* (Ortlepp 1937). Davis (1973) records *Capillaria hepatica* as yet another nematode in *O. irroratus.*

## Relations with man

The Vlei Rat is of economic importance in the sense that it is a probable reservoir for various zoonoses, including bluetongue, bubonic plague, tick-bite fever *(Rickettsia conorii)* and Q-fever *(Rickettsia burnetii).* The trombiculids are known to be of some medical and veterinary importance, and have been suspected of transmitting rickettsiae and toxoplasma in the Afrotropical Region, but according to Zumpt (1961a) this requires confirmation. They inflect considerable damage to young trees in pine plantations (Davis 1973), as well as destruction of agricultural products.

The pulicid fleas are the principal vectors of plague and of the most important pests of man and domestic animals. These abound on *O. irroratus.* On the other hand, it is thought that vlei rats do not play a primary role in plague transmission, owing to their patchy distribution (Davis 1973). The ticks *Haemaphysalis leachii, Rhipicephalus appendiculatus* and *R. simus* have been incriminated as transmitters of tick-bite fever (Gear 1954).

## Prehistory

The genus *Otomys* has been identified in many Pleistocene deposits in South Africa, where it appears as one of the predominant rodents. *O. irroratus* has also been reported by Dreyer & Lyle (1931) from Florisbad.

Two closely related fossil species are also known to science: *Palaeotomys gracilis* is known from Schurveberg

near Pretoria (Broom 1937) and also from Taung, Sterkfontein, Kromdraai and Makapansgat (Lavocat 1957; De Graaff 1960a). *Prototomys campbelli* has been recorded from Taung only (Broom & Schepers 1946).

## Taxonomy

As here understood, *O. irroratus* may be conspecific with the extralimital *O. tropicalis* found in the Congo, Kenya and Uganda. Roberts (1951) recognised the following subspecies: *irroratus, orientalis, natalensis, auratus, randensis* and *coenosus.* I accept these as synonyms, as is the species *cupreus* (with subspecies *cupreus* and *cupreoides).* This monotypic interpretation of *Otomys irroratus* follows Misonne (1974).

# *Otomys angoniensis* Wroughton, 1906

## Angoni Vlei Rat
## Angoni Vleirot

For the meaning of *Otomys,* see *Otomys irroratus.* This species is named after Angoniland *(ca* 14°30′S, 35°00′E) in Malaŵi, where it was first collected in 1906.

This vlei rat is rather similar to *Otomys irroratus* and is often referred to as a sibling species of the latter, but it shows some seven laminae in the $M^3$ in contrast to six laminae in *irroratus.* This species was first described as a subspecies in 1906 by Wroughton from material collected in Malaŵi. It has subsequently been shown to occur southwards as well and is now known from Zimbabwe, Mozambique, the Okavango area in Botswana, the northern Cape, Transvaal and Natal. Ellerman *et al.* (1953) have included many described forms of *angoniensis* under *O. irroratus* but Meester *et al.* (1964) agree with Davis (1962) and Ansell (1964) that *O. irroratus* and *O. angoniensis* are distinct species.

## Outline of synonymy

1906 *Otomys irroratus angoniensis* Wroughton, *Ann. Mag. nat. Hist.* (7) 18:274. M'Kombhuie, Angoniland, Malaŵi.

1918 *Otomys rowleyi* Thomas, *Ann. Mag. nat. Hist.* (9) 2:209. Inhambane District, Mozambique.

1918 *Otomys mashona* Thomas, *Ann. Mag. nat. Hist.* (9) 2:210. Mazoe, Zimbabwe.

1924 *Otomys irroratus maximus* Roberts, *Ann. Transv. Mus.* 10:70. Machile River, SW Zambia.

1929 *Otomys tugelensis* Roberts, *Ann. Transv. Mus.* 13:113. Utrecht, Natal.

1929 *Otomys tugelensis sabiensis* Roberts, *Ann. Transv. Mus.* 13:114. Mariepskop, Transvaal.

1929 *O.t. pretoriae* Roberts, *Ann. Transv. Mus.* 13:114. Fountains Valley, Pretoria, Transvaal.

1939 *O. (O.) angoniensis angoniensis* Wroughton. *In* Allen, *Bull. Mus. comp. Zool. Harv.* 83:343.

1951 *Otomys maximus* Roberts. *In* Roberts, *The mammals of South Africa.*

1955 *Otomys angoniensis davisi* Lundholm, *Ann. Transv. Mus.* 22:279–303.

1964 *O. angoniensis* Wroughton. *In* Meester et al., *An interim classification of southern African mammals.*

1974 *O. angoniensis* Wroughton. Misonne, *in* Meester & Setzer (eds), *The mammals of Africa: an identification manual.*

TABLE 37
Measurements of male and female *Otomys angoniensis*

|  | Parameter | Value (mm) | N | Range (mm) | CV (%) |
|---|---|---|---|---|---|
| Males | HB | 159 | 14 | 145–182 | 5,8 |
|  | T | 93 | 15 | 77–112 | 8,7 |
|  | HF | 27 | 15 | 20–36 | 14,0 |
|  | E | 20 | 15 | 17–23 | 6,6 |
|  | Mass: | 99 g | 3 | 89–115 g | — |
| Females | HB | 157 | 18 | 133–184 | 6,5 |
|  | T | 88 | 19 | 75–110 | 7,6 |
|  | HF | 25 | 19 | 23–30 | 6,0 |
|  | E | 21 | 18 | 19–24 | 5,4 |
|  | Mass: | 104 g | 9 | 87–126 g | — |

## Identification

The dorsal pelage is a pale buff, slightly washed with black, but not to such an extent as in *Otomys irroratus*. The individual hairs are a slaty black, each with a conspicuous broad band of buffy yellow distally which may be continued to the tip of the hair, or the tip may be dark. The nose is a uniform buffish yellow. Yellow patches surround the throat and eyes. The tail is darkly coloured above with the sides and under-surface a buffy white. The ventral coloration is a dull, dark grey (plate 6). The length of the hindfoot is shorter in *O. angoniensis* than in *O. irroratus*. Davis (1973) reports a range of 25–28 mm in the former and 29–34 mm in the latter (N = 15 and 153 respectively). The species varies markedly in size and colour in different localities. Furthermore, in contrast to *O. irroratus*, *O. angoniensis* shows a ring of orange hair approximately 2 mm in width around the eyes.

## Skull and dentition

The skull is smaller than in *O. irroratus* and tends to be more arched. The posterior petro-tympanic foramen on the bulla is smaller in *O. angoniensis* than in *O. irroratus*, forming a slit-like orifice. Furthermore, the degree of angular transition between the broad and the narrow portions of the nasals is not so sudden in *O. angoniensis* as in *O. irroratus*.

Often there are seven laminae to the third upper molar (fig. 64).

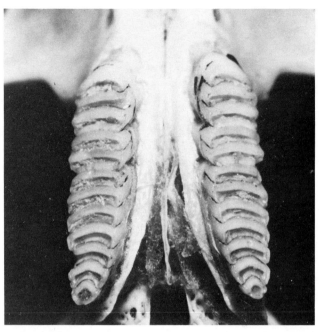

**Figure 64** Cheekteeth of the Angoni Vlei Rat *Otomys angoniensis.*

## Distribution

This species occurs in the central parts of the Transvaal and the eastern escarpment of the RSA, ranging into Swaziland, Natal, Mozambique and Zimbabwe. It is also found in the Caprivi and along the Okavango River in SWA/Namibia (Coetzee 1972), and in Angola and Kenya (Misonne 1974). Its geographical distribution could be said to be virtually entirely confined to the Woodland and Grassland regions of the Southern Savanna as well as the South West Arid biotic zones (map 41).

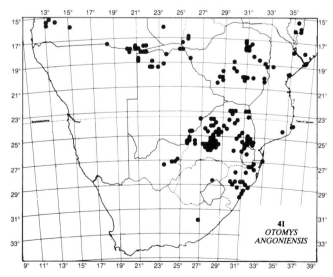

41
*OTOMYS ANGONIENSIS*

## Habitat

It seems to prefer savanna woodlands and grasslands lower than 1 000 m above sea level. In Botswana it occurs on the fringes of rivers with reed beds, sedges and semi-aquatic grasses (Smithers 1971). It may be that it prefers drier areas than *O. irroratus* or the extralimital *O. tropicalis,* although it has been trapped in the same runways as the former. Apart from being sympatric with *irroratus,* they also occur sympatrically with *O. laminatus* (Coetzee 1972). In the Maria van Riebeeck Nature Reserve near Pretoria, Davis (1973) did a Chi-square association test between the distributions of 29 most common plant species and *O. angoniensis* on the study grid. There was a very strong positive association with *Helictotrichon turgidulum, Cynodon dactylon, Melilotis alba, Hyparrhenia dregeana, Berkheya radula* and *Chironia palustris.*

## Diet

They subsist on an array of grass species, grains and seeds.

## Habits

They are predominantly nocturnal, but sometimes partly diurnal. Individuals tend to occur singly, in pairs or small family parties. Shortridge (1934) reports them as burrowers in some areas in SWA/Namibia. Their nests are made of grass, usually near permanent water.

## Predators

Smithers (1971) states that the flesh of *O. angoniensis* is attractive to other rodents and shrews, because individuals caught in traps are often eaten if the traps are not emptied regularly at short intervals. Davis (1973) records them being eaten by puffadders *(Bitis arientans),* while Dixon (1966) reported finding an *Otomys* (probably *O. angoniensis)* in the stomach of a Black Mamba *(Dendroaspis polylepis)* in northern Natal.

Black-shouldered kites have been observed taking adult *O. angoniensis* in the Caprivi (J. Taylor *pers. comm., in* Davis 1973). Owls appear to be prominant predators of *O. angoniensis* and include the Barn Owl *Tyto alba* (Ranger 1927; De Graaff 1960b; Dean 1977), the Grass Owl *T. capensis* (Benson 1965; Vernon 1972; Dean 1977) the Marsh Owl *Asio capensis* and the Spotted Eagle Owl *Bubo africanus* (Dean 1977). Smithers *(pers. comm., in* Davis 1973) states that *O. angoniensis* is a common prey of the Serval *Felis serval* in Zimbabwe. In Botswana this rodent was found in the stomach contents of the Side-striped Jackal *Canis adustus* as well as in the Small-spotted Genet *Genetta genetta* (Smithers 1971), while an analysis of the stomach contents of the Large Grey Mongoose *Herpestes ichneumon* and the Water Mongoose *Atilax paludinosus* in Zimbabwe also revealed the remains of *O. angoniensis* (Smithers, *in* Davis 1973; Pienaar *et al.* 1980).

## Reproduction

Information is rather scanty. Smithers (1971) took three gravid females during October, December and March from Kasane in Botswana, and a number of post-natal young in September, November, January and February. This indicates the dropping of young during the warm, wet months from August to March. Further data are required, however. The young referred to above weighed between 25 and 65 g. A female with three embryos was taken in November near Vryburg according to records of the Transvaal Museum. A female with three well-advanced fetuses was taken in August in the Kruger National Park (Pienaar *et al.* 1980). Taylor & Green (1976) have studied the proximate factors – particularly diet – that influence reproduction in four rodent species common in cereal-growing areas of the Kenya Highlands and it became clear that *O. angoniensis* breed mostly in the wetter months of the year (April–September), although it could occur throughout the year.

According to Matthey (1964) *O. angoniensis* has a 2n = 56 chromosome complement.

## Parasites

The schizomycete *Pseudomonas pseudomallei* has been recorded as associated with this rodent (Neitz 1965). The fleas include the following species: (i) Pulicidae: *Xenopsylla hipponax, X. versuta* and *X. brasiliensis;* (ii) Hystrichopsyllidae: *Ctenophthalmus acanthurus, C. evidens, C. calceatus, C. eumeces, Dinopsyllus ellobius, D. longifrons, D. lypusus, Listropsylla fouriei* and *L. prominens;* (iii) Chimaeropsyllidae: *Chiastopsylla rossi* (Zumpt 1966). The nematode *Longistriata capensis* has been recorded by Davis (1973) as is the case of the tapeworm *Paranoplocephala.*

It is interesting to note that such well-known parasite groups as the mites, chiggers, flatworms and roundworms are poorly known from this species – in contrast to the fleas, as listed above. The mites include *Laelaps transvaalensis* and *Androlaelaps dasymys* (Davis 1973). The ticks are also poorly known. These include specimens of the *Haemaphysalis leachii* group (immatures), *Ixodes thomasae* (adults) and *Rhipicephalus hurti* (immatures) (Walker *pers. comm.*) The louse *Polyplax otomydis* also occurs on this rodent (Davis 1973).

Bilharziasis *(Schistosoma haematobium, S. mansoni* and *S. mattheei)* have also been isolated from *O. angoniensis* (Pitchford 1959) while two species of filariosis are known from Kenya (Nelson, Teesdale & Highton 1962).

## Relations with man

Shortridge (1934) states that in the Okavango area these rats are eaten by the indigenous people. The reed beds are burnt during the dry season and, with the aid of dogs, enormous numbers of *O. angoniensis* are captured.

The Angoni Vlei Rat harbours *Pseudomonas pseudomallei* which causes melioidosis, a disease that does not normally occur in South Africa. It was isolated from a patient who had been on active service in India and Malaya. Laboratory tests have revealed that several rodents (e.g. *Tatera brantsii, Praomys natalensis, Mystromys albicaudatus, Rhabdomys pumilio*, as well as *O. angoniensis*) are susceptible to infection. In the case of unsuspected introduction of the organism it could maintain itself in these species (Neitz 1965).

The pulicid fleas found on *O. angoniensis* are the principal vectors of plague and one of the most important pests of man and domestic animals, giving it potential economic and medical importance (Zumpt 1966).

## Prehistory
Unknown.

## Taxonomy
*Otomys angoniensis* is a sibling species of *O. irroratus*. The slit-like small petro-tympanic foramen is, however, a fairly constant character.

The following southern African subspecies are considered as synonyms: *maximus, pretoriae, rowleyi, mashona, sabiensis* and *tugelensis*. This arrangement follows Misonne (1974). It is possible that *mashona, tugelensis, pretoriae* and *sabiensis* show a closer relationship to *rowleyi*, while *maximus* includes *davisi*. In an earlier classification Meester *et al.* (1964) consequently accepted *O. angoniensis rowleyi* and *O. angoniensis maximus* as subspecies.

# *Otomys saundersae* Roberts, 1929

**Saunders' Vlei Rat**

**Saunders se Vleirot**

For the meaning of *Otomys*, see *Otomys irroratus*. This species has been named after Miss Saunders, who made a study of the rodents in the Grahamstown district and first brought it to the attention of Austin Roberts.

This vlei rat was first described by Roberts in 1929 on material handed to him by Miss Saunders who collected the specimens in Grahamstown. He originally described it as a subspecies of *O. tugelensis* (= *O. angoniensis*), but by 1951 Roberts gave it separate species status. This is evidently a relict species of the eastern and western Cape, sharing morphological characteristics and habitat with *O. irroratus* and *O. sloggetti* (Misonne 1974).

## Outline of synonymy
1929 *Otomys tugelensis saundersae* Roberts, *Ann. Transv. Mus.* 13:115. Grahamstown. CP.
1931 *Otomys karoensis* Roberts, *Ann. Transv. Mus.* 14:231. Wolseley, CP.

1939 *Otomys (Otomys) tugelensis saundersae* Roberts *In* Allen, *Bull. Mus. comp. Zool. Harv.* 83:348.
1951 *Otomys saundersiae* Roberts. *In* Roberts, *The mammals of South Africa.*
1953 *O. saundersae* Roberts. *In* Ellerman *et al., Southern African mammals.*
1974 *O. saundersae* Roberts. Misonne, *in* Meester & Setzer (eds), *The mammals of Africa: an identification manual.*

TABLE 38
Measurements of male and female *Otomys saundersae*

| | Parameter | Value (mm) | N | Range (mm) | CV (%) |
|---|---|---|---|---|---|
| | HB | 151 | 13 | 133–172 | 8,9 |
| | T | 95 | 12 | 79–111 | 12,0 |
| Males | HF | 25 | 13 | 22–29 | 10,7 |
| | E | 21 | 12 | 19–23 | 5,8 |
| | Mass: | 111 g | 7 | 100–134 g | 10,3 |
| | HB | 140 | 14 | 124–165 | 7,9 |
| | T | 82 | 11 | 74–92 | 9,2 |
| Females | HF | 26 | 10 | 24–29 | 5,6 |
| | E | 20 | 14 | 19–24 | 8,1 |
| | Mass: | 95 g | 4 | 84–107 g | — |

## Identification
The species has been described as follows by Roberts (1929): 'Colour above light buffy speckled with dark-brown tipped hairs so that the flanks become pale buffy on a pale ground, the base of all the hairs being ashy. Tip of muzzle, ring round the eye and tuft of hair in front of the ears buffy-yellow; hair immediately behind the ears buffy to white.' The ventral surface of body and tail is grey with throat and distal appendages whiter in colour. The tail is buffy-white dorsally (plate 6).

Generally, the species shares many morphological characters with *O. irroratus* and *O. sloggetti*.

## Skull and dentition
The skull resembles that of *Otomys unisulcatus*, but with the nasals sharply expanded and the zygomatic arches rather broader towards the middle portion. In addition, the beading of the interorbital edges are sharply ridged, while the palatal foramina are long and narrow extending nearly to the incisors '…and, therefore, more advanced than in other species' (Roberts 1929).

Both upper and lower incisors are deeply grooved with the same laminal formula as in *O. irroratus*, but the posterior lamina of the $M_1$ more curved, as in *O. unisulcatus*. The laminal formula corresponds to that of *O. irroratus* (fig. 65).

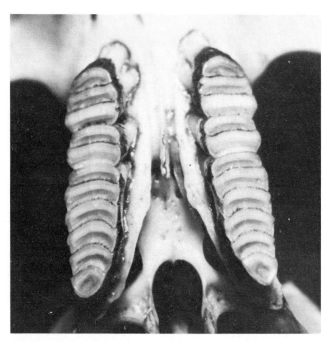

**Figure 65** Cheekteeth of Saunders' Vlei Rat *Otomys saundersae*. It has the same laminal formula as *Otomys irroratus*.

## Distribution

*Otomys saundersae* has a wide distribution, from the Eastern Province to the Western Province and Roberts (1951) states that it must be an old established species to have such an extensive distribution over a distance of 1 000 km or more. It is resident in the South West Cape and Southern Savanna Grassland biotic zones (Rautenbach 1978a) (map 42).

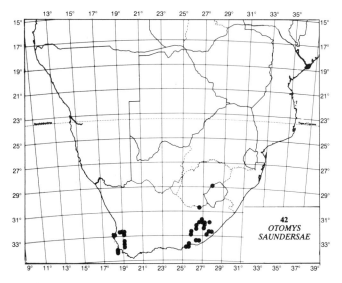

## Habitat

This species prefers a mountainous habitat and inhabits belts of dry rushes in heath country on high mountain slopes (Shortridge 1934).

## Diet

In all probability, entirely vegetarian.

## Habits

Available information virtually non-existent. Shortridge (1934) states it to be diurnal.

## Predators and reproduction

Unknown.

## Parasites

Some information on the arthropod ectoparasites found on *O. saundersae* is available in the literature. The mites are represented by three laelaptid species, which include *Laelaps muricola*, *Haemolaelaps glasgowi* and *H. labuschagnei* (Zumpt 1961a).

The fleas are represented by two families: the Hystrichopsyllidae with *Dinopsyllus ellobius* and *Listropsylla agrippinae* as species and the Chimaeropsyllidae with *Chiastopsylla rossi* and *C. godfreyi* (Zumpt 1966).

The ticks, as recorded by Theiler (1962), consist of the soft tick *Ornithodoros zumpti* (adults) and (immatures) and the hard ticks *Haemaphysalis leachii leachii* (immatures), *H. leachii* subsp. (immatures), an *Ixodes* sp. (adults and immatures) and *Rhipicephalus simus* (adults and immatures).

## Relations with man

The presence of the soft tick *Ornithodoros zumpti* on *O. saundersae* could lead to distribution of a neurotropic strain of spirochaete *Borrelia tillae*. The ticks mainly transmit various forms of rickettsiae.

## Prehistory

Unknown.

## Taxonomy

This species was first described as a subspecies of *O. tugelensis* (now *O. angoniensis*) by Roberts in 1929. It proved to be a senior homonym of *O. karoensis* described by Roberts in 1931 and was also referred to as *Myotomys karoensis* (Broom 1937). It was raised to species rank by Roberts (1951) and maintained as such by Misonne (1974). Davis (1962) recognised *karoensis* subspecifically, while Meester *et al.* (1964) follow Davis by listing *karoensis* as a subspecies.

# *Otomys laminatus* Thomas & Schwann, 1905

**Laminate Vlei Rat**

**Bergvleirot**

For the derivation of *Otomys*, see *Otomys irroratus*. The Greek *lamina* = laminate and refers to the lamination of the teeth in this species.

152

This species was first described from material collected at Sibudeni in KwaZulu, by Thomas & Schwann in 1905. They form what Ellerman (1941) calls the *laminatus* group of *Otomys*. This implies that the M³ has nine or ten laminae, while the M₁ shows seven laminae. In the lower incisor, a deep outer and a shallow inner groove can be seen. In 1918 Thomas created the subgenus *Lamotomys* for this group. Although the difference in laminae is small, Roberts (1951) considered that the difference from *Otomys* must be deep-seated in view of the fact that *Otomys* and *Lamotomys* constantly occur side by side without interbreeding. Roberts (1951) consequently raised *Lamotomys* to full generic rank. Ellerman *et al.* (1953), however, treat it as a synonym of *Otomys*, a procedure which has been followed by Meester *et al.* (1964) and Misonne (1974).

The geographical distribution of this species indicates that it occurs near mountains or at least in hilly and undulating country, explaining the Afrikaans vernacular name of Bergvleirot.

**Outline of synonymy**

1905 *Otomys laminatus* Thomas & Schwann, *Abstr. Proc. zool. Soc. Lond.*: 18:23. Sibudeni, Natal.

1918 *Lamotomys* Thomas, *Ann. Mag. nat. Hist.* (9) 2:208. As a subgenus of *Otomys*. (*Otomys laminatus* Thomas & Schwann taken as genotype.)

1919 *Otomys silberbaueri* Roberts, *Ann. Transv. Mus.* 6:114. Lormarins, Paarl, CP.

1924 *Otomys (Lamotomys) laminatus pondoensis* Roberts, *Ann. Transv. Mus.* 10:71. Ngqueleni, Pondoland, Transkei.

1929 *O. (L.) l. mariepsi* Roberts, *Ann. Transv. Mus.* 13:110. Mariepskop, Transvaal.

1939 *O. (L.) laminatus laminatus* Thomas & Schwann. *In* Allen, *Bull. Mus. comp. Zool. Harv.* 83:348.

1951 *Otomys laminatus fannini* Roberts. *In* Roberts, *The mammals of South Africa*: 426. Kilgobbin Farm, Dargle, Natal.

1953 *O. laminatus* Thomas & Schwann. *In* Ellerman *et al.*, *Southern African mammals*.

1974 *O. laminatus* Thomas & Schwann. Misonne, *in* Meester & Setzer (eds), *The mammals of Africa: an identification manual*.

**Identification**

The colour of the animal has been described as 'raw umber' (Thomas & Schwann 1905). The fur is soft, fine and thick, approximately 20 mm in length. Individual hairs are coloured blackish basally, followed by a subterminal rufous ring, with the extreme tip black. The ventral surface is dull yellowish, with the base of the hair grey. The throat is dull yellowish white. The ears are naked externally and the hands and feet are a blackish-grey on their dorsal surfaces. The tail is thickly haired, blackish above and a dull buffy below (plate 6). The

colour tends to vary geographically on the basis of which supposed subspecies were described in the past.

TABLE 39
Measurements of male and female *Otomys laminatus*

| | Parameter | Value (mm) | N | Range (mm) | CV (%) |
|---|---|---|---|---|---|
| Males | HB | 199 | 5 | 188–213 | 4,2 |
| | T | 109 | 5 | 97–115 | 5,4 |
| | HF | 30 | 5 | 22–34 | 13,0 |
| | E | 21 | 5 | 17–26 | 13,6 |
| | Mass: | 190 g | 1 | — | — |
| Females | HB | 178 | 3 | 158–197 | — |
| | T | 107 | 3 | 105–111 | — |
| | HF | 30 | 3 | 30–31 | — |
| | E | 24 | 3 | 22–26 | — |
| | Mass: | 140 g | 1 | — | — |

**Skull and dentition**

The skull resembles that of *O. irroratus*, but the M³ always has nine or more laminae (in contrast to six laminae in *O. irroratus*) (fig. 66). The laminal formula reads 3.2.9.

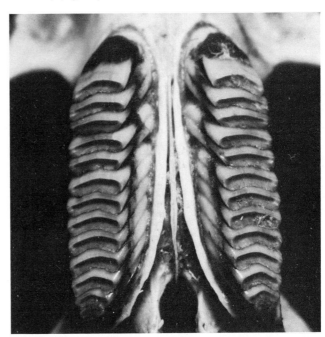

**Figure 66** Cheekteeth of the Laminate Vlei Rat *Otomys laminatus*. The number of laminae in the M³ can vary from eight to ten, nine being average. In this case there are eight only.

**Distribution**

This species has a discontinuous distribution on the low-veld side of the Eastern Escarpment/Drakensberg Range (Coetzee 1972) as well as in central Natal and Pondoland in Transkei. A relict population is known from the Paarl Valley in the western Cape. It occurs sympatrically with *O. irroratus* and *O. angoniensis* (Coetzee 1972). An example of sympatry is found *inter alia* at Kilgobbin Farm in the Dargle district, where it occurs with *O. irroratus* (Roberts 1951) (map 43). It is therefore encountered in the Southern Savanna Woodland, Southern Savanna Grassland as well as in the Forest biotic zones (Rautenbach 1978a).

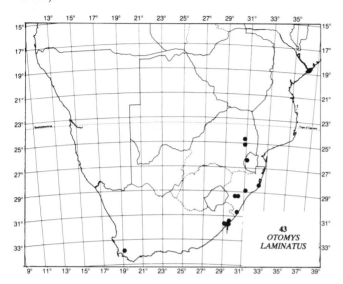

43
OTOMYS
LAMINATUS

## Habitat
It prefers grassland, submontane and coastal zones in South Africa.

## Diet
As far as is known, it is entirely vegetarian, taking shoots and stems of grasses and small shrubs.

## Habits
Very little information is available. As indicated above, it has been captured together with *O. irroratus* and *O. angoniensis,* apparently not interbreeding with them.

## Predators, reproduction, parasites and prehistory
Unknown.

## Taxonomy
*Otomys laminatus* was given subgeneric rank in 1918 by Thomas, as *Lamotomys,* a subgenus of *Otomys.* Roberts (1951) raised it to generic rank, recognising *silberbaueri* and *laminatus* as two species, the latter with four subspecies, *viz. mariepsi, laminatus, fannini* (probably synonymous with *irroratus*) and *pondoensis.* This was not accepted by Ellerman *et al.* (1953), who synonymised *Lamotomys* with *Otomys.*

The monotypic interpretation, which subsequently evolved and adhered to in this book, follows Misonne (1974).

# *Otomys unisulcatus* F. Cuvier, 1829

**Bush Karoo Rat**

**Bos Karoorot**

For the derivation of *Otomys,* see *Otomys irroratus.* The specific name is derived from the Latin *uni* = single and *sulcus* = groove with reference to the single faint groove which may persist on the lower incisor.

As is the case in *Otomys irroratus,* the original specimens of *O. unisulcatus* were also collected by Delalande and described by F. Cuvier in 1829 (Sclater 1901). A characteristic feature is the ungrooved lower incisors which is also the case in *O. sloggetti* (to be described below). Its presence in an area (especially the Karoo) is readily ascertained by the large spherical or irregular masses of dry twigs surrounding shrubs and their branches. These twigs are amassed by the animal and traversed in many directions by its burrows which extend into the ground below them. On the eastern side of their range, they live among piles of rocks in which nests are built of weeds and grasses (Roberts 1951). Ellerman (1941) has placed them in his *unisulcatus* group (together with the related *Otomys sloggetti*) because of $M^3$ shows four laminae only, (occasionally five).

## Outline of synonymy
1829 *Otomys unisulcatus* F. Cuvier, *Hist. Nat. des Mamm.* pt. 60. South Africa. Type locality nominated as Matjiesfontein by Roberts (1951).
1834 *Euryotis unisulcatus* A. Smith, *S. Afr. Quart. Journ.* 2:149.
1902 *Otomys broomi* Thomas, *Ann. Mag. nat. Hist.* (7) 10:313. Port Nolloth, CP.
1918 *Myotomys unisulcatus* Thomas, *Ann. Mag. nat. Hist.* (9) 2:204. South Africa.
1929 *Otomys unisulcatus brantii* Thomas, *Ann. Mag. nat. Hist.* (7) 10:312. Deelfontein, north of Richmond, CP.
1929 *Otomys unisulcatus bergensis* Roberts, *Ann. Transv. Mus.* 13:108. Lamberts Bay, CP.
1931 *Otomys (Myotomys) unisulcatus grantii* Thomas. *In* Hewitt, *Guide to the Albany Museum.* Grootfontein, CP.
1939 *Myotomys unisulcatus unisulcatus* (F. Cuvier). *In* Allen, *Bull. Mus. comp. Zool. Harv.* 83:342.
1946 *Myotomys unisulcatus albaniensis* Roberts, *Ann. Transv. Mus.* 20:318. Near Committees Drift, Albany district, CP.
1951 *Myotomys unisulcatus unisulcatus* (F. Cuvier). *In* Roberts, *The mammals of South Africa.*
1953 *Otomys unisulcatus* F. Cuvier. *In* Ellerman *et al.,* *Southern African mammals.*

1974 *Otomys unisulcatus* F. Cuvier. Misonne, *in* Meester & Setzer (eds), *The mammals of Africa: an identification manual.*

shrub- and Karoo-like vegetation, usually interspersed with rocks and stone.

### TABLE 40
Measurements of male and female *Otomys unisulcatus*

|         | Parameter | Value (mm) | N  | Range (mm) | CV (%) |
|---------|-----------|------------|----|------------|--------|
| Males   | HB        | 148        | 13 | 129–160    | 6,2    |
|         | T         | 96         | 13 | 77–115     | 10,2   |
|         | HF        | 26         | 13 | 23–28      | 4,6    |
|         | E         | 24         | 14 | 17–29      | 11,6   |
|         | Mass:     | 139 g      | 11 | 125–156 g  | 7,2    |
| Females | HB        | 143        | 18 | 131–150    | 3,3    |
|         | T         | 96         | 18 | 88–110     | 4,6    |
|         | HF        | 26         | 18 | 24–29      | 4,0    |
|         | E         | 23         | 17 | 20–28      | 6,5    |
|         | Mass:     | 110 g      | 12 | 101–135 g  | 8,4    |

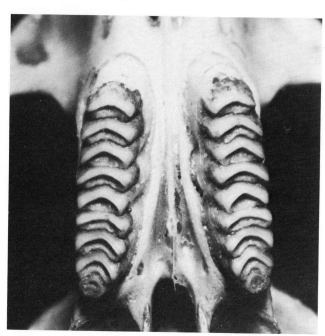

**Figure 67** Cheekteeth of the Bush Karoo Rat *Otomys unisulcatus*. The $M^3$ has four distinct laminae (the fourth as a circular portion).

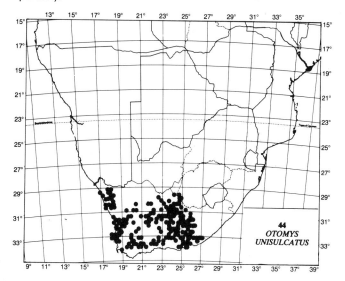

## Identification
The pelage of the upper- and underparts is ashy grey, because individual hairs are that colour for two thirds from the base, the remaining third is light buffy yellowish (on dorsal part of head and body). Interspersed are completely black hairs, longer than the others, but absent on the sides of the greyer flanks. The ventral surface is a buffy white (plate 6). The feet and hands are a uniform pale buffy. The tail is dull white below with a distinct black line along the top.

## Skull and dentition
The third upper molar has four distinct laminae and a small posterior circular portion (often worn in adults) (fig. 67). The first lower molar has two laminae with a distinct kidney-shaped anterior circular portion.

The upper incisors are shallow-grooved towards the outer edge. The lower incisors are not grooved. Some specimens may, however, show a slight, shallow groove in the lower incisors.

## Distribution
It is distributed from Little Namaqualand in the west to the Albany district in the east. It is also found in the Karoo and Little Karoo extending along the eastern escarpment (map 44). Rautenbach (1978a) lists its occurrence as the South West Arid biotic zone only, but it may also occur marginally in the South West Cape biotic zone as was indicated by Davis (1974).

## Habitat
*Otomys unisulcatus* prefers drier habitats and tends to shun damp situations in contrast to *O. irroratus*. It occurs in

## Diet
Feeds on karroid vegetation and other green vegetable matter.

## Habits
It builds rather large shelters of sticks and twigs in dry water courses and often shallow holes are made in the ground beneath the shelters. This species is diurnal and runways often extend from one shelter to the next.

## Predators, reproduction and prehistory
Unknown.

## Parasites

An impressive array of associated inverbrates is known to occur on *Otomys unisulcatus* or in its abodes. Pirie (1927) has commented on the presence of *Yersinia pestis*, the micro-organism responsible for sylvatic plague in natural populations of rodents and bubonic plague in man.

The cestodes include *Inermicapsifer congolensis* and *Paranoplocephala omphalodes* (Collins 1972).

The chelicerid arthropods include adults (A) and/or immatures (I) of both the soft ticks and the hard ticks. The former are represented by *Ornithodoros zumpti* (A, I), while seven species of the latter are known to occur with *O. unisulcatus*. These are *Amblyomma hebraeum* (I), *Haemaphysalis leachii* subsp. (A, I), *Rhipicephalus* sp. (I), *R. capensis* (I), *R. oculatus* (I), *R. simus* (I) and *Magaropus winthemi* (I) (Theiler 1962).

The mandibulate arthropods are well represented by the fleas. The pulicid family includes *Echidnophaga gallinacea*, *Pulex irritans*, *Xenopsylla cheopis*, *X. eridos*, *X. piriei*, *X. trifaria*, *X. versuta*, *X. brasiliensis* and *X. hirsuta*. The following hystrichopsyllid fleas also occur: *Dinopsyllus ellobius*. *D. tenax*, *Listropsyllus agrippinae*, *L. cerrita* and *L. chelura*. The chimaeropsyllid fleas are well represented by *Epirimia aganippes*, *Chiastopsylla numae*, *C. quadrisetis*, *C. rossi*, *C. capensis*, *C. carus*, *C. coraxis*, *C. godfreyi*, *C. mulleri*, *C. octavii* and *C. pitchfordi*. Finally there is a single representative of the Leptopsyllidae, *viz. Leptopsylla segnis*, known to be associated with *Otomys unisulcatus* (Zumpt 1966).

## Relations with man

On account of the variety of flea species found associated with this rodent, especially those transmitting bubonic plague, *O. unisulcatus* is an important vector in the spread of plague. The ticks they harbour transmit babesiases, rickettsiases, theileriosis and other micro-organisms which are of great veterinary importance.

## Taxonomy

The genotype *Otomys unisulcatus* was placed in the genus *Myotomys* by Thomas in 1918. This interpretation was also followed by Roberts (1951) who placed both *unisulcatus* and *sloggetti* under *Myotomys*. This procedure was not accepted by Ellerman *et al.* (1953) who synonymised *Myotomys* with *Otomys* and interpreted *unisulcatus* as a good species of *Otomys*. Meester *et al.* (1964) provisionally accepted *Otomys unisulcatus unisulcatus* (including *grantii*, *bergensis* and *albaniensis*) as well as *O. u. broomi* as subspecies, the former group occurring in the southeastern, central and southwestern Cape, as well as the southern OFS, while *broomi* occurs in Namaqualand. Misonne (1974) treated all the described forms as conspecific, an interpretation followed and accepted in this work.

# *Otomys sloggetti* Thomas, 1902

**Rock Karoo Rat**

**Klip Karoorot**

For the meaning of *Otomys*, see *Otomys irroratus*. The species has been named after Col. A.T. Sloggett RAMC, who presented a first collection of mammals from Deelfontein to the British National Museum in 1902.

This *Otomys* has the distinction of having chosen as a favourite habitat one of the highest points in southern Africa – the summit of Mont-aux-Sources (3 282 m above sea level) which forms part of the Drakensberg Range bordering Lesotho, Natal and the OFS.

## Outline of synonymy

1902 *Otomys sloggetti* Thomas, *Ann. Mag. nat. Hist.* (7) 10:311. Deelfontein, CP.

1907 *O. turneri* Wroughton, *Ann. Mag. nat. Hist.* (7) 20:31. Aberfeldy, near Harrismith, OFS.

1918 *Myotomys sloggetti* Thomas, *Ann. Mag. nat. Hist.* (9) 2:204. Genotype *O. unisulcatus* Brants.

1918 *M. turneri* Thomas, *Ann. Mag. nat. Hist.* (9) 2:204.

1927 *M. robertsi* Hewitt, *Rec. Albany Mus.* 3:430. Summit of Mont-aux-Sources, OFS/Lesotho border.

1929 *M. sloggetti jeppei* Roberts, *Ann. Transv. Mus.* 13:109. Jamestown, CP.

1929 *M. s. basuticus* Roberts, *Ann. Transv. Mus.* 13:110. Bolopeletsa, Lesotho.

1937 *Myotomys (Metotomys) turneri* Broom, *S. Afr. J. Sci.* 33:763.

1939 *Myotomys sloggetti sloggetti* (Thomas). *In* Allen, *Bull. Mus. comp. Zool. Harv.* 83:342.

1951 *M.s. sloggetti* (Thomas). *In* Roberts, *The mammals of South Africa*.

1953 *O. sloggetti* Thomas. *In* Ellerman *et al.*, *Southern African mammals*.

1974 *O. sloggetti* Thomas. *In* Meester & Setzer (eds), *The mammals of Africa: an identification manual*.

## Identification

According to Thomas (1902), the dorsal surface of the animal is 'vinaceous-brown', the flanks are a dull buff, the ventral surface is a soiled buffy, while the individual hairs from chin to anus have slaty-coloured bases. The head is a greyish-brown, contrasting with the rest of the body. The ears, which are not very large, have a dark edging on the back of the rim. Upper surfaces of hands and feet are buff-coloured. The tail is thin and short, with a narrow blackish line along the upper surface. The pelage of the body is soft, fine and thick (plate 6).

## Skull and dentition

The skull is somewhat smaller than that of *O. unisulcatus*. The palatal foramina are not widely open. The upper

## TABLE 41
Measurements of male and female *Otomys sloggetti*

|         | Parameter | Value (mm) | N | Range (mm) | CV (%) |
|---------|-----------|------------|---|------------|--------|
| Males   | HB        | 144        | 6 | 130–153    | 5,5    |
|         | T         | 63         | 6 | 55–71      | 7,3    |
|         | HF        | 23         | 6 | 22–25      | 3,9    |
|         | E         | 18         | 6 | 14–21      | 11,3   |
|         | Mass:     | 137 g      | 3 | 125–146 g  | —      |
| Females | HB        | 143        | 6 | 135–150    | 3,8    |
|         | T         | 63         | 6 | 58–70      | 5,5    |
|         | HF        | 23         | 6 | 20–24      | 5,4    |
|         | E         | 19         | 6 | 16–22      | 9,1    |
|         | Mass:     | 121 g      | 4 | 113–128 g  | —      |

at higher elevations of the Southern Savanna Grassland zone.

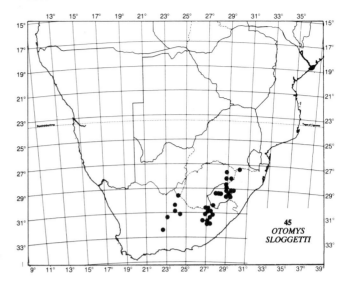

incisors have one groove, not as closely placed to the margin of the tooth as in *unisulcatus*. The single groove in the lower incisor is very indistinct. The M³ has four laminae (fig. 68) and a posterior portion (i.e. five), while the M₁ has three laminae with a rounded anterior portion (i.e. four).

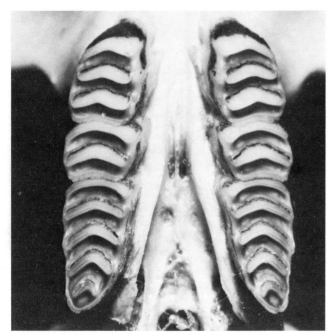

**Figure 68** Cheekteeth of the Rock Karoo Rat *Otomys sloggetti*. The M³ usually consists of four laminae plus an additional posterior one (i.e. a total of five).

## Distribution

These animals occur on the highveld and in montane grassland south of 27°S in the southeastern Transvaal, OFS, Lesotho and the eastern Cape and westwards to Deelfontein in the Karoo (map 45). It therefore occurs in the eastern Karoo of the South West Arid biotic zone and

## Habitat

In Lesotho they occur in rocky habitats up to 1 800 m above sea level. Roberts (1951) quotes the Rev. Ellenberger, who states that this animal is known as the 'Ice Rat' to the Basotho, on account '...of its coming out of its retreats in the rocks and sunning itself, when snow is on the ground'.

## Diet

Unknown.

## Habits

It inhabits crevices amongst rocks and boulders.

## Predators, reproduction and prehistory

Unknown.

## Parasites

The only parasites hitherto known are mites, fleas and ticks. The mites are represented by two families, *viz. Haemolaelaps labuschagnei* (Laelaptidae) and *Listrophoroides womersleyi* (Listrophoridae) (Zumpt 1961a). The fleas of the following families have been recorded (Zumpt 1966): Pulicidae: *Xenopsylla philoxera* and *X. brasiliensis*. Hystrichopsyllidae: *Ctenophthalmus calceatus*, *Dinopsyllus ellobius* and *Listropsylla agrippinae*. Leptopsyllidae: *Leptopsylla segnis*. Chimaeropsyllidae: *Epirimia aganippes*, *Chiastopsylla rossi*, *C. godfreyi*, *C. monticola* and *C. roseinnesi*.

The ticks include *Haemaphysalis leachii leachii* (adults and immatures), *H. leachii* subspecies (immatures) and *Rhipicephalus simus* (immatures) (Theiler 1962).

## Predators, reproduction and prehistory

Unknown.

157

## Taxonomy

In the past, this species has been considerably divided into subspecies by taxonomists. Roberts (1951) refers to the following as subspecies: *sloggetti, jeppei, basuticus, turneri* and *robertsi*. Clearly, this species is in need of taxonomic revision. Meester *et al.* (1964) listed the following arrangement: *O. s. sloggetti* (including *jeppei* and *basuticus* of earlier authors) – eastern and central Cape; *O. s. turneri* – northeastern OFS and southeastern Transvaal; *O. s. robertsi* – northern Lesotho. Misonne (1974), however, interprets all these forms as conspecific.

# GENUS *Parotomys* Thomas, 1918

This otomyid genus is clearly distinct from *Otomys* by the conspicuously inflated auditory bullae, while the auditory meatus has a strongly projecting thickened process on its anterior edge. This genus prefers arid habitat which is to be found predominantly in the westerly areas of southern Africa. Two species are acknowledged: *Parotomys brantsii* (with grooved upper incisors) and *P. littledalei* (with plain upper incisors).

---

**Key to the southern African species of *Parotomys***

(Modified after Ellerman *et al.* (1953) and Misonne (1974))

1 Upper incisors grooved; tail in British Museum material averages 59–64% of HB length ... *P. brantsii* (Brants' Whistling Rat) Page 158

Upper incisors plain; tail in British Museum material averages 77% of HB length ..... *P. littledalei* (Littledale's Whistling Rat) Page 161

---

# *Parotomys brantsii* (A. Smith, 1834)

**Brants' Whistling Rat**

**Brants se Fluitrot**

The generic name is compounded from the Latin *par* = pertaining to, and the Greek *ous, otos* = ear and *mys* = mouse. It is, therefore, an animal close to *Otomys*. It was named by Smith after Brants, who described many South African rodents during the early years of the 19th century.

According to Sclater (1901), Sir Andrew Smith's specimen originally came from the neighbourhood of the Orange River.

## Outline of synonymy

1834 *Euryotis brantsii* A. Smith, *S. Afr. Quart. Journ.* 2:150. Namaqualand. Type locality fixed as Port Nolloth, Thomas & Schwann (1904), *Proc. zool. Soc. Lond.*: 178.

1841 *Euryotis pallida* Wagner, *Arch. f. Naturgesch.* 7, 1:134. South Africa. Type locality fixed as Van Rhynsdorp by Roberts (1929), *Ann. Transv. Mus.* 13:108.

1842 *Otomys rufifrons* Rüppel, *Mus. Senckenbergianus.* 1:28. *Nomen nudem.*

1843 *Euryotis rufifrons* Wagner. *In* Schreber's *Säugeth.* Suppl. 3:507. Cape of Good Hope. Cradock nominated as type locality by Roberts (1929).

1898 *Otomys brantsii* De Winton, *Ann. Mag. nat. Hist.* (7) 2:6.

1904 *Otomys brantsii luteolus* Thomas & Schwann, *Proc. zool. Soc. Lond.* 1:178. Deelfontein, CP.

1918 *Parotomys* Thomas, *Ann. Mag. nat. Hist.* (9) 2:204. (Genotype *Euryotis brantsii*, A. Smith.)

1929 *Parotomys brantsii rufifrons* (Wagner). Roberts, *Ann. Transv. Mus.* 13:108.

1933 *Parotomys brantsii deserti* Roberts, *Ann. Transv. Mus.* 15:267. Bushman Pits, Kuruman River (22°E), CP.

1939 *Parotomys (Parotomys) brantsii brantsii* (A. Smith). *In* Allen, *Bull. Mus. comp. Zool. Harv.* 83:349.

1951 *Parotomys brantsii brantsii* (A. Smith). *In* Roberts, *The mammals of South Africa.*

1953 *Parotomys brantsi* (A. Smith). *In* Ellerman *et al., Southern African mammals.*

1974 *Parotomys brantsi* (A. Smith). Misonne, *in* Meester & Setzer (eds), *The mammals of Africa: an identification manual.*

## TABLE 42
### Measurements of male and female *Parotomys brantsii*

|         | Parameter | Value (mm) | N  | Range (mm) | CV (%) |
|---------|-----------|-----------|----|-----------|--------|
| Males   | HB        | 148       | 15 | 123–165   | 6,5    |
|         | T         | 91        | 15 | 77–103    | 6,0    |
|         | HF        | 28        | 15 | 25–32     | 5,6    |
|         | E         | 18        | 15 | 14–20     | 8,2    |
|         | Mass:     | 85 g      | 2  | 84–86 g   | —      |
| Females | HB        | 148       | 31 | 127–156   | 4,2    |
|         | T         | 89        | 31 | 73–104    | 6,3    |
|         | HF        | 28        | 31 | 23–30     | 4,4    |
|         | E         | 17        | 31 | 14–25     | 11,2   |
|         | Mass:     | 107 g     | 10 | 92–122 g  | 7,5    |

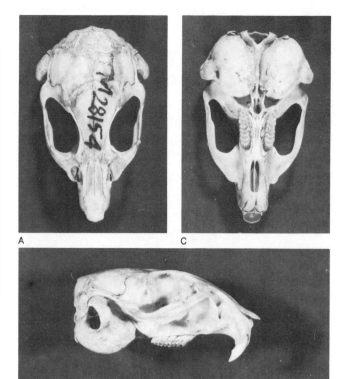

cm

**Figure 69** The dorsal (A), lateral (B) and ventral (C) views of the skull of Brants' Whistling Rat *Parotomys brantsii*. The interorbital constriction is not pronounced. In contrast to *Otomys* the nasals are only slightly broadened anteriorly. Bullae are inflated and consequently the basioccipital region is narrowed. The external auditory meatus has a thickened process on anterior edge.

## Identification

The upper and lateral parts of the head and body are a pale sienna-yellow, variegated with blackish or umber brown, while the sides of the neck and head tend towards a greyish-white, like the ventral surface of the body (plate 7). The fur is soft. The nose and the first half of the tail are a reddish-orange, while the last half of the tail is brownish-red. The ears are yellowish with pale cream yellow extremities. Four inguinally situated mammae are present.

## Skull and dentition

The bullae are very large and inflated. The thickened anterior process of the auditory meatus is visible from above. The interorbital constriction is not pronounced and the nasals are but slightly broadened anteriorly. The basioccipital region is narrowed (fig. 69).

In the cheekteeth, the $M^1$ has three laminae, the $M^2$ two and there are two complete laminae in the $M^3$. This tooth usually has a long, backwardly pointing heel in which two laminal elements occur (fig. 70). In the lower jaw, the $M_2$ and the $M_3$ have two laminae each as usual; the $M_1$ shows four laminae, the two anterior ones not clearly separated from each other (Ellerman 1941). The upper incisors have a deep outer and a shallow inner groove. The lower incisors are not grooved.

## Distribution

*Parotomys brantsii* is found in the central, northern and northwestern Cape as well as in the southern parts of SWA/Namibia and Botswana (map 46). It is a South West Arid form, extending marginally into the South West Cape (Davis 1974).

## Habitat

An arid habitat is preferred by this species. It burrows in

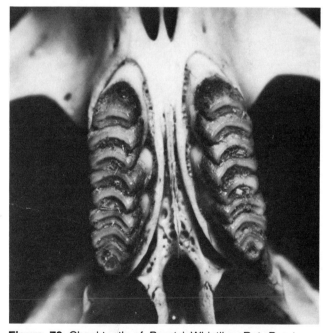

**Figure 70** Cheekteeth of Brants' Whistling Rat *Parotomys brantsii*. The $M^1$ has three laminae, the $M^2$ two and the $M^3$ has two complete laminae followed by two laminate elements on the backwardly pointing heel.

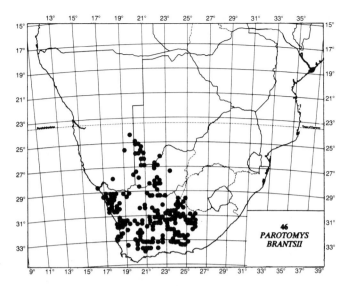

46
PAROTOMYS
BRANTSII

hard, sandy soils. In Botswana it occurs within the limits of the 300mm isohyet (Smithers 1971). In the Kalahari Gemsbok National Park, Nel & Rautenbach (1974) found the distribution of *P. brantsii* largely correlated with the occurrence of the Driedoring *Rhigozum trichotomum,* shrubs on dune slopes, dune troughs and the banks of the normally dry riverbeds.

## Diet

Entirely vegetarian. It subsists on grass and grass seeds and shoots of low-growing shrubs. In the Kalahari Gemsbok National Park, Nel & Rautenbach (1974) observed the following plants eaten by *P. brantsii:* Gramineae: *Asthenatherum glaucum* (Ghagrass), *Schmidtia kalahariensis* (Kalahari Sour Grass), *Stipagrostis uniplumis* (Silky Bushman Grass), *Brachiaria glomerata;* Aizoaceae: *Galenia* sp.; Cyperaceae: *Fimbrystylis hispidusa;* Labiatae: *Acrotome angustifolia;* Leguminosae: *Requienia sphaerosperma, Indigofera alternans, I. flavicans;* Phytolaccaceae: *Limeum viscosum;* Acanthaceae: *Monechma incanum* (Bloubos); Zygophyllaceae: *Tribulus zeyheri* (Dubbeltjie); and Bignoniaceae: *Rhigozum trichotomum* (Driedoring), *Cataphractus alexandri* (Gannabos).

## Habits

*Parotomys brantsii* lives in small colonies inhabiting intricate and extensive tunnel systems. The number of entrances to a system varies, but only some openings are constantly in use.

A more detailed description of the tunnel system made by *P. brantsii* can be found in De Graaff & Nel (1965). In the Kalahari Gemsbok National Park where their systems were followed, they occurred in the soft, deep, sandy soil of the dunes. The entrances (varying between 13 and 21 in three systems analysed), occur either on open ground or frequently under shrubs such as the Driedoring *Rhigozum trichotomum.* The systems were all longer than their widths with tunnels averaging 80 mm in diameter. The deepest tunnel was 75 cm below the surface of the soil. A

nest was found in one of these systems situated in a rounded nest chamber (diameter 25 cm) with a number of tunnels radiating from it. The nest itself weighed 190 g and was loosely constructed of *Schmidtia bulbosa, Aristida uniplumis* (Gramineae), *Rhigozum trichotomum* (Bignoniaceae), *Monechma incanum* (Acanthaceae) and *Helichrysum* sp. as well as *Pteronia mucronata* (Compositae).

The temperature prevailing in the tunnels range between 28°C to 31°C in summer (Bolwig 1958). Bolwig recorded up to 95% relative humidity (RH) in *Parotomys* warrens in May, although he points out that the true figure may be slightly lower, but definitely well above 50% RH.

The associated fauna found in the tunnels contains both vertebrates and invertebrates. The lizard *Mabuya striata,* the scorpions *Opisthopthalmus pictus, O. wahlbergi, O. carinatus, Parabuthus granulatus* and a number of beetles (especially the predaceous ground beetle *Anthia* sp.) proved to be the most common co-inhabitants (De Graaff & Nel 1965).

This species is diurnal and very wary, never foraging far from its burrows. It often squats near the burrows in an upright position and, when alarmed, emits an audible whistle. The species exhibits marked fluctuations in numbers every three to four years (Nel & Rautenbach 1974).

## Predators

Predators include the Barn Owl *Tyto alba* (Nel & Nolte 1965) as well as the Black-backed Jackal *Canis mesomelas* in the Kalahari Gemsbok National Park *(pers. obs.).*

## Reproduction

The female has two pairs of inguinally situated mammae. Nests are built in excavated chambers in the tunnel system with a number of tunnels radiating from it. Often one of these tunnels obliquely leads directly to the surface of the soil in a straight line. The average diameter of a nest chamber is 250 mm with the nesting material filling the entire space. A single nest excavated by De Graaff & Nel (1965) in the Kalahari Gemsbok National Park weighed 190 g and was loosely constructed from the following plant material: Gramineae: *Schmidtia bulbosa* (Kalahari Sandkweek) and *Aristida uniplumis* (Blinkblaargras); Bignoniaceae: *Rhigozum trichotomum* (Driedoring); Acanthaceae: *Monechma incanum* (Bloubos); and Compositae: *Helichrysum* sp., *Pteronia mucronata.*

The nesting material is finely shredded.

A litter of young seems to average four. Smithers (1971) reports a pregnant female during February in Botswana.

## Parasites

*Parotomys brantsii* is susceptible to sporadic natural outbreaks of plague *Yersinia pestis.* Laboratory tests also indicate susceptibility to *Listeria monocytogenes,* a form of listeriosis (Neitz 1965).

The only other endoparasite known is the cestode *Raillientina (R.) trapeziodes* (Collins 1972).

Ectoparasites include the following fleas: Pulicidae: *Echidnophaga gallinacea, Xenopsylla eridos, X. philoxera, X. piriei, X. erilli* and *X. hirsuta*. Hystrichopsyllidae: *Dinopsyllus ellobius* and *Listropsylla agrippinae*, while *L. dorippae* and *L. cerrita* have been recorded from *Parotomys* sp. Chimaeropsyllidae: *Epirimia aganippes, Chiastopsylla numae, C. quadrisetis, C. rossi, C. capensis, C. coraxis, C. mulleri, C. octavii* and *C. pitchfordi* (Zumpt 1966).

The ticks include adults and immatures of *Nuttalliella namaqua*, while *Rhipicephalus pravus* (adults) and *R. simus* (immatures) have been collected from *Parotomys* sp. (Theiler 1962).

The occurrence of the louse *Polyplax otomydis* has been listed by Ledger *(in* Davis 1973). According to Davis, this record is doubtful and the specimen is probably referable to *Polyplax myotomydis* recorded from *P. brantsii* and *Otomys unisulcatus*, which both occur in an arid habitat.

### Relations with man

On account of the diversified fleas they harbour, this species may be an important vector in the spread of bubonic plague.

### Prehistory

Unknown.

### Taxonomy

Roberts (1951) recognises four other subspecies apart from the nominate race *brantsii*. These are *deserti, pallida, luteolus* and *rufifrons*. These were listed by Meester *et al.* (1964) as well, but it was understood that *rufifrons* included *luteolus*. All the described forms are listed as synonymous by Misonne (1974).

# *Parotomys littledalei* Thomas, 1918

### Littledale's Whistling Rat
### Littledale se Fluitrot

For the meaning of *Parotomys*, see *Parotomys brantsii*. The species was named after Maj. H.A.P. Littledale, who had collected the type specimen at Kenhardt on 16 July 1911.

This rodent was described as a new subgenus, *Liotomys*, by Thomas in 1918, based on material collected by Littledale near Kenhardt in the northwestern Cape. *Parotomys (Liotomys) littledalei* was introduced to the scientific world as an otomyid without grooves on either the upper or lower incisors. The molars, however, are the same as those found in *Parotomys*: the $M^3$ with three distinct laminae and a small posterior portion and a $M_1$ with two

distinct laminae and an anterior kidney-shaped portion that wears down to another distinct lamina and a small anterior portion. It may, however, retain its kidney shape, even when well worn (Roberts 1951).

For the rest, its size and appearance are very much like that of *Parotomys brantsii*. The colour is variable geographically.

### Outline of synonymy

1918 *Liotomys* Thomas, *Ann. Mag. nat. Hist.* (9) 2:205. *Parotomys (Liotomys) littledalei* Thomas. Kenhardt, CP. Subgenus of *Parotomys*. Type *P. littledalei* Thomas.

1933 *Parotomys (Liotomys) littledalei molopoensis* Roberts, *Ann. Transv. Mus.* 15:267. Hakskeenpan, northern CP.

1933 *P. (L.) l. namibensis* Roberts, *Ann. Transv. Mus.* 15: 268. Swakopmund, SWA/Namibia.

1939 *Parotomys littledalei littledalei,* Thomas. *In* Allen, *Bull. Mus. comp. Zool. Harv.* 83:349.

1951 *Liotomys littledalei littledalei* Thomas. *In* Roberts, *The mammals of South Africa.*

1964 *Parotomys littledalei* Thomas. *In* Meester *et al., An interim classification of southern African mammals.*

1974 *Parotomys littledalei,* Thomas. Misonne, *in* Meester & Setzer (eds), *The mammals of Africa: an identification manual.*

### TABLE 43
**Measurements of male and female *Parotomys littledalei***

| | Parameter | Value (mm) | N | Range (mm) | CV (%) |
|---|---|---|---|---|---|
| Males | HB | 150 | 6 | 135–170 | 7,0 |
| | T | 103 | 5 | 93–122 | 9,9 |
| | HF | 27 | 6 | 26,5–27,5 | 1,5 |
| | E | 19 | 6 | 18–20 | 3,8 |
| | Mass: No data available | | | | |
| Females | HB | 146 | 9 | 130–161 | 7,3 |
| | T | 105 | 9 | 92–117 | 8,4 |
| | HF | 26 | 9 | 24–28 | 4,8 |
| | E | 18 | 9 | 17–20 | 6,1 |
| | Mass: | 127 g | 2 | 110–145 g | — |

### Identification

The size and overall phenotypic appearance of this species are very much like those of *Parotomys brantsii*. It does, however, tend to be somewhat darker in colour. The back has been described by Thomas as being rather darker than 'cinnamon-buff', the sides and belly pale buff with the individual hairs a slate colour basally. The hands and feet are a buffy white. *P. littledalei* probably has a slightly longer tail than *P. brantsii*, a remark which is upheld if the tables given in this book are compared. The tail is well haired, dark buffy above and paler below. A variable portion of the distal tail (on the upper side) may be brownish or blackish (plate 7).

### Skull and dentition
Both the upper and lower incisors are smooth. The third upper molar has three laminae as well as a small posterior portion (fig. 71), while the first lower molar has two laminae and an anterior, kidney-shaped portion.

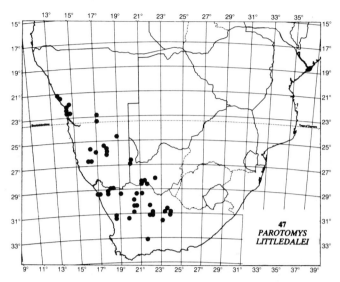

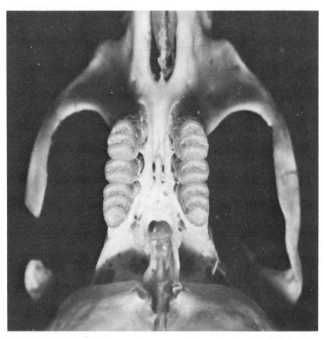

**Figure 71** Cheekteeth of Littledale's Whistling Rat *Parotomys littledalei.*

### Distribution
*P. littledalei* is found in the northern and northwestern Cape Province as well as in SWA/Namibia (map 47). Its distribution is often discontinuous.

### Habitat
It prefers dry, arid areas.

### Diet
Unknown.

### Habits
This species constructs nests at the base of dense bushes with burrows below the nests into which it retreats when disturbed. It is colonial and utters a melancholy note from its retreat when an intruder approaches the nest (Roberts 1951). This is in contrast to *Parotomys brantsii*, which lives in burrows which are made in open ground among scattered vegetation.

### Parasites
The only ectoparasites known are the mite *Laelaps giganteus* (Laelaptidae) (Zumpt 1961a) and a number of fleas. These include the following: Pulicidae: *Xenopsylla eridos, X. occidentalis* (on subspecies *namibensis*), *X. piriei* and *X. versuta.* Hystrichopsyllidae: *Listropsylla agrippinae* and *L. dorippae.* Chimaeropsyllidae: *Chiastopsylla numae* and *C. mulleri* (Zumpt 1966).

### Predators, reproduction, relations with man and prehistory
Unknown.

### Taxonomy
Roberts (1951) lists three subspecies, *viz. littledalei, namibensis* and *molopoensis,* following Ellerman (1941) in this respect. This arrangement was repeated by Ellerman *et al.* (1953) and Meester *et al.* (1964). Misonne (1974) has taken *molopoensis* and *namibensis* as synonymous with *littledalei,* an interpretation followed in this book.

FAMILY **Muridae** Gray, 1821

SUBFAMILY **Murinae** Murray, 1866

**Key to the southern African murine genera**
(Modified after Misonne (1974))

GENUS *Zelotomys* Osgood, 1910

*Zelotomys* can be readily confused with the Multimammate Mouse *Praomys (Mastomys) natalensis* on external characteristics. However, the tail is usually, although not invariably, white or whitish, the molars tend towards hypsodonty and the incisors are pro-odont. Furthermore, there is a conspicuous difference in the number of mammae.

163

# Zelotomys
## woosnami (Schwann, 1906)

**Woosnam's Desert Rat**

**Woosnam se Woestynrot**

The generic name is derived from the Greek word *zelos* = emulation, referring to the external resemblance of *Zelotomys* to the *Rattus* group, especially *Praomys (Mastomys) natalensis*. The species name refers to R. B. Woosnam, who collected it near Kuruman on 13 July 1904.

This rodent is a little-known species, first described as *Mus woosnami* by Schwann (1906) on material from the Molopo River between the RSA and Botswana. Its generic name was changed to *Epimys* by Dollman (1910), while Thomas (1920) placed it in the genus *Ochromys*. Thomas based the description of *Ochromys* upon the paler overall colour, the short tail, the absence of supraorbital ridges and unspecified differences in the M¹ (Roberts 1951). Both Allen (1939) and Roberts (1951) retained *woosnami* in *Ochromys* which Ellerman *et al.* (1953) regard as a synonym of *Rattus*. Davis (1962) has taken the situation one step further by regarding *Ochromys* as a synonym of *Zelotomys* and this interpretation was followed by Meester *et al.* (1964) and Misonne (1974).

## Outline of synonymy

1906 *Mus woosnami* Schwann, *Proc. zool. Soc. Lond.* 1:108. Molopo River, northern CP.

1910 *Epimys woosnami* Dollman, *Ann. Mag. nat. Hist.* (8) 6:398. Tamalakane River, Botswana.

1920 *Ochromys* Thomas, *Ann. Mag. nat. Hist.* (9) 5:142. Genotype *Mus woosnami* Schwann.

1941 *Rattus (Ochromys) woosnami* (Schwann). *In* Ellerman, *Families and genera of living rodents*.

1953 *Rattus woosnami* (Schwann). *In* Ellerman *et al.*, *Southern African mammals*.

1964 *Zelotomys woosnami* (Schwann). *In* Meester *et al.*, *An interim classification of southern African mammals*.

1974 *Zelotomys woosnami* (Schwann). Misonne, *in* Meester & Setzer (eds), *The mammals of Africa: an identification manual*.

## Identification

Paraphrasing Schwann (1906), the dorsal surface of the animal is between 'smoky-grey' and 'drab-grey', pencilled with black, while the flanks are paler, without pencilling. The individual hairs are about 15 mm in length, coloured a slate-grey for about two thirds of their length and termination in a lighter grey for their last third. Some hairs have black distal ends, giving a pencilled effect. The ventral surface is a creamy-white with grey bases of hairs variable. A white line runs from the muzzle to the inner sides of the forelimbs. The dorsal surfaces of the hands

and feet are white, the fur not extending over the claws. Likewise, both dorsal and ventral surfaces of the tail are white except for 10–15 mm near the distal end, where it darkens slightly (plate 11).

The overall impression gained is that it is not a heavily haired species. The fur is fine and silky. Smithers (1971) states that variation in colour can be seen if a series of skins are seen simultaneously and also remarks that juveniles are slightly darker than adults, lacking the pencilling in the upper parts.

The whiskers are well developed and the ears can be described as being medium-sized and sparsely covered with hair. The scale rings on the tail are about 33 to the centimetre (Schwann 1906). The tail is shorter than the head-body length. Ellerman *et al.* (1953) state it to average 82% or less of the head-body length. The hindfoot attains a length of 26 mm (Ellerman *et al.* 1953). In the small sample available in the Transvaal Museum, this value is exceeded in some cases and this was also found to be the case by Smithers (1971) in males from Botswana.

The mammae consist of three pectoral pairs and two inguinal pairs.

## TABLE 44
Measurements of male and female *Zelotomys woosnami*

| | Parameter | Value (mm) | N | Range (mm) | CV (%) |
|---|---|---|---|---|---|
| **Male** | HB | 138 | 1 | — | — |
| | T | 115 | 1 | — | — |
| | HF | 29 | 1 | — | — |
| | E | 20 | 1 | — | — |
| | Mass: | 75 g | 1 | — | — |
| **Females** | HB | 151 | 4 | 142–162 | 5,3 |
| | T | 98 | 4 | 83–113 | 11,5 |
| | HF | 27 | 4 | 26–29 | 4,5 |
| | E | 20 | 4 | 19–22 | 5,4 |
| | Mass: | 51 g | 1 | — | — |

## Skull and dentition

The skull is smooth, i.e. not ridged, somewhat rounded horizontally and shows a relatively narrow, short braincase, a short rostrum and a slight interorbital constriction. The palatal foramina are of medium length penetrating between the front molars. The pterygoids tend to be thickened. The anterior edge of the zygomatic plate (antorbital plate) shows considerable variation (Schwann 1906) '. . . from convexity to being nearly straight', but its border is not concave (Ellerman *et al.* 1953) (fig. 72).

According to Ellerman (1941) the skull of *Z. woosnami* is essentially like *Zelotomys hildegardeae* (an extralimital species occurring in Angola, Zambia, Malaŵi and further north in Africa) in cranial and dental characteristics, but the pro-odonty is not very pronounced.

The upper toothrow is 18–19% of the condylobasal length (Ellerman 1941). The molars are strongly cuspidate with the labial row of cusps well developed and these cusps tend to project outwards. The individual laminae are broad. The $M^1$ is longer than the combined length of the $M^2$ and $M^3$. The $M^2$ is about as broad as or broader than long, while the $M^3$ is strongly reduced and narrower than the $M^2$ (fig. 73).

There is very little inward distortion of the t1 in the three-rooted $M^1$ and this tooth has a total of eight cusps. The t2 is larger than the t5 or t8 while the t1 tends to fuse with the t2 after wear. The $M^3$ is a simple circular tooth usually with only a single large median cusp (?t5).

The upper incisors are slender, slightly protruding, orange to light yellow on their anterior surfaces and ungrooved (fig. 72).

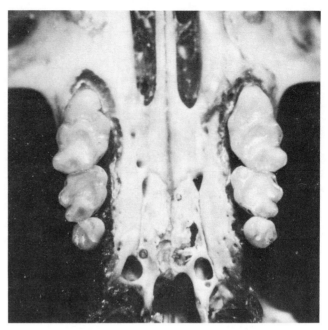

**Figure 73** Cheekteeth of Woosnam's Desert Rat *Zelotomys woosnami.* The labial row of cusps are well developed and tend to project outwards. The eight-cusped three-rooted $M^1$ is longer than the combined length of $M^2$ and $M^3$.

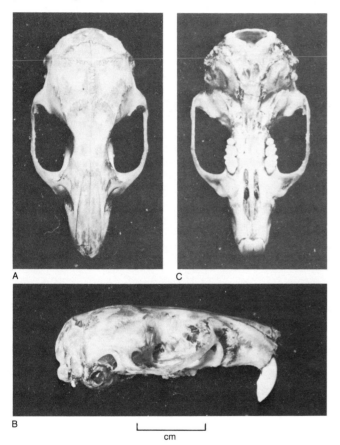

**Figure 72** Dorsal (A), lateral (B) and ventral (C) views of the skull of Woosnam's Desert Rat *Zelotomys woosnami.* Note: interorbital constriction not pronounced; absence of crests; small interparietals; anterior palatal foramina of medium length; thickened pterygoids; short rostrum; orthodont to opisthodont incisors. Anterior edge of zygomatic plate variable from convex to straight.

## Distribution

Woosnam's Desert Rat is known from the Kalahari

thornveld (South West Arid) along the Molopo River and ranges into the margin of the Southern Savanna Woodland in Ovambo and Ngamiland, but it does not extend across the Cunene–Zambezi boundary (Davis 1974) (map 48).

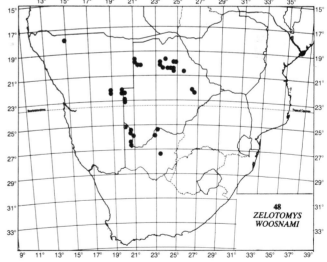

## Habitat

The species seems to prefer the semi-arid environs of the Kalahari. In Botswana, Smithers (1971) records them as occurring on sandy ground covered with sparse grass and thin, open shrub consisting of *Acacia, Grewia* and *Terminalia.* It has also been recorded from riparian *Acacia.* Smithers *(op. cit.)* points out that all the localities where this species has been recorded in Botswana, fall within a mean annual rainfall range of 200–500 mm.

Likewise, specimens collected by J. A. J. Nel at Nossob Camp and near Dankbaar in the Kalahari Gemsbok National Park (by myself), have also been taken in the adjoining dry riverbed and in open *Acacia* woodland respectively.

## Diet

Smithers (1971) and Birkenstock & Nel (1977) have contributed significantly to our knowledge of this species. Smithers states it to be graminivorous. He also refers to Liversidge *(pers. comm.)* who found hair of other rodents in stomachs of *Z. woosnami,* suggesting carnivory. In addition to the fact that they may be insectivorous as well, Smithers also refers to Ansell, who noted a captured specimen eating the carcasses of other rats. This was also noticed by Birkenstock & Nel (1977). On three out of five occasions one individual was killed, the winner eating part of the victim, starting at the rectum and pulling out the intestines. Stomach contents of a wild-caught specimen contained 60% white plant matter (seeds), 20% insects (Coleoptera) and 20% nematodes *(Physaloptera aduensis* females and larvae and *Spirura* sp. female – Verster, *in litt.).* Birkenstock & Nel (1977) also observed coprophagy in adults.

## Habits

Woosnam's Desert Rat lives in holes dug under shrubs and trees and in the Kalahari Gemsbok National Park where I found them in association with *Acacia reficiens* (False Umbrella Thorn).

Shortridge (1934) stated that they may habitually live in deserted or still occupied burrows of other rats, suggesting *Tatera* species. It does not seem to be a particularly active species, although this statement is difficult to substantiate, for they are primarily nocturnal and terrestrial. Smithers (1971) notes that they occur singly or in family parties. Birkenstock & Nel (1977) state that wild-caught individuals, when placed together in cages, are mutually intolerant and fight, usually resulting in death of one of a pair. Even where such a pair consists of both sexes, or consists of adults or juveniles of the same or different sexes, fighting is elicited. Trapping results indicate that individuals are widely spaced and it seems, therefore, that *Zelotomys* is asocial and usually solitary.

Although Shortridge has reported successful trapping among scattered bushes in or near dry water courses at no great length from water holes, specimens which I have trapped near Dankbaar in the Kalahari Gemsbok National Park were at least 5 km from the nearest open and available water. Another interesting feature mentioned by Shortridge *(op. cit.)* is that specimens tend to have a musky smell, not unlike that of *Mus musculus* or that found in certain *Crocidura* species.

A burrow system excavated in the Kalahari Gemsbok National Park contained two nests. These were respectively 55 cm and 40 cm below the soil surface with nest chambers dimensions of 14x10x10 cm and 14x16x15 cm respectively. Birkenstock & Nel (1977) state that both nests were rather shapeless and consisted of finely shredded 'Kortbeenboesmangras' *Stipagrostis obtusa.*

## Predators

Unknown.

## Reproduction

Information on the reproduction of *Z. woosnami* is meagre and existing data have been compiled by Smithers (1971). In Botswana and the eastern part of SWA/Namibia, the evidence suggests the dropping of pups from December to March, i.e. during the warm, wet months. The number of fetuses varies between five and 11. Shortridge (1934) refers to a female collected at Sandfontein in SWA/Namibia containing two fetuses on 7 November.

Birkenstock & Nel (1977) have described the post-natal development of *Z. woosnami* under experimental conditions. At birth (day 0) the pups are pink and show a dorsal grey tinge due to the start of hair proliferation. Hair eruption takes place by day 6. The auditory meatus opens on day 14. The incisors erupt between days 10 and 12, with the lower incisors slightly in advance to the upper ones. At day 0 the lips of the mouth are almost totally fused with only a small orifice in the centre (1–1,5 mm), permitting suckling. By day 10 the mouth can be almost fully opened, but the corners are still fused. The eyes open on days 16 or 17 and marks the beginning of a high level of activity in the pup. Birkenstock & Nel (1977) found that their data on numbers and developmental time of *Zelotomys woosnami* could be interpreted in the framework of the *r*- and *K*-continuum (Pianka 1970). This species shows *r*-correlates of inhabiting an area with a variable and unpredictable climate (expecially rainfall), showing a low population density, having relatively large litters and a fast development rate. To these correlates, a wide habitat utilisation and a diversity of food niches could perhaps be added. In contrast, other attributes, e.g. large body size, the possibility of multiple breeding (i.e. having a postpartum oestrus) as well as a long lifespan, do not fit the theory of *Zelotomys* being *r*-selected (Birkenstock & Nel 1977).

## Parasites

Two species of fleas, representing two families, have been taken from *Z. woosnami.* These include the important plague vector *Xenopsylla philoxera* (Pulicidae) and *Listropsylla dorippae* (Hystrichopsyllidae) (De Meillon *et al.* 1961).

The only other ectoparasites known to occur on *Z. woosnami* are immatures of the tick *Hyalomma rufipes* (Theiler 1962).

The two specimens collected by Woosnam on the east

bank of the Tamalakane River in Botswana in 1904 had been attacked by some parasitic fly, probably *Cordylobia anthropophagus* '... or some near species: the maggots were in both cases in the scrotum, which was much enlarged and full of pus; in one case, the testes were diseased and one almost entirely sloughed away. The mice were otherwise in good condition and fat' (Dollman 1910). It is interesting to note that Woosnam observed that it '... is possible that, owing to some habit, this mouse is peculiarly subject to these parasitic flies which, destroying the reproductive organs, are exterminating the species in the locality'. Today, as we know, this species is listed in the *South African Red Data Book* on rare and endangered mammals (Meester 1976).

### Relations with man
Uncertain and unlikely to be of any magnitude, chiefly on account of the rarity of the species. How important it is in the transmission of plague is not clear, but it should be borne in mind that the important plague vector, the flea *Xenopsylla philoxera*, has been collected from *Z. woosnami*.

### Prehistory
This genus has been provisionally recorded as a fossil from the well-known fossiliferous locality at Makapansgat near Potgietersrust in the northern Transvaal (Lavocat 1957).

### Taxonomy
Apart from being reshuffled and placed under various genera *(Mus, Epimys, Ochromys, Rattus* and presently *Zelotomys)*, *Zelotomys woosnami* has not been divided subspecifically and is therefore interpreted as a monotypic species.

A second species, *Zelotomys hildegardeae*, referred to as a '... rare species' by Misonne (1974), occurs extralimitally in Angola, Zambia, Malaŵi and northwards to Tanzania, southern Kenya and western Uganda.

GENUS *Thamnomys* Thomas, 1907
SUBGENUS *Grammomys* Thomas, 1915

In 1907 Thomas erected the genus *Thamnomys* (separating it from *Mus*) for those African species of rats in which the postero-internal cusp (t7) of the upper molars is retained. The type specimen was described as *Thamnomys venustus* (Thomas 1907) encountered chiefly in Uganda and Ruwenzori (Ellerman 1941). He also pointed out that the new genus could really be divided into two sections according to differences in molar pattern. During a subsequent revision of the genus (1915a), Thomas erected the genus *Grammomys* for those individuals formerly placed under *Thamnomys,* where the postero-internal cusps (t7) of the M¹ and M² are lacking or becoming reduced to mere ridges and in doing so, he used *Mus dolichurus* Smuts, 1832 as the type. In 1919, Hollister remarked that the characteristics proposed by Thomas for *Grammomys* were '... too vague even for subgeneric recognition, and synonymised it with *Thamnomys,* from which it was split' (Ellerman 1941). On the other hand,

Roberts (1951) states that *Grammomys* consists of individuals of which the overall size is smaller and the upper molar series less than 5 mm in length, in contrast to the *Thamnomys venustus* group, and consequently thought that *Grammomys* was entitled to subgeneric rank and that it be accepted as such. These sentiments were also expressed by Ellerman (1941) and Ellerman *et al.* (1953) who refer to '... an unquestionable difference in the dentition of the two genera, particularly the development of t7...'. In 1908, Thomas & Wroughton described *Thamnomys cometes* from Inhambane, Mozambique (1908b). It was eventually interpreted as *Grammomys cometes* by Ellerman (1941) and as *G. dolichurus cometes* by Ellerman *et al.* (1953). I regard *Thamnomys (Grammomys) cometes* and *T. (G.) dolichurus* as two species, even though they are not easily separable from one another and they seem to coexist in the same areas. I am following Misonne (1974) in this regard.

---

**Key to the southern African species of *Thamnomys (Grammomys)***
(After Misonne (1974))

1 Postero-internal cusp (t7) of M¹ small but visible.....
　　　　　　　　　　　　　　　*T. (G.) rutilans*
　　　　　　　　　　　　　　　(extralimital)
　Postero-internal cusp (t7) of M¹ reduced to a clear ridge connecting t4 to t8 ............................... 2

2 Ears with subauricular tuft of white hairs, although not always present; larger and greyer *T. (G.) cometes*
　　　　　　　　　(Mozambique Woodland Mouse)
　　　　　　　　　　　　　　　　　　Page 170
　No such tuft or white hairs; smaller, less grey ..........
　　　　　　　　　　　　　　　*T. (G.) dolichurus*
　　　　　　　　　　　　　　　(Woodland Mouse)
　　　　　　　　　　　　　　　　　　Page 168

# Thamnomys (Grammomys) dolichurus (Smuts, 1832)

**Woodland Mouse**

**Woudmuis**

The generic name is derived from the Greek *thamnos* = thicket or shrub, and *mys* = mouse, while the specific name is derived from the Greek *dolich* = long, lengthy and *oura* = a tail. Smuts obviously referred to the conspicuous length of the tail of the species.

According to Ellerman *et al.* (1953), the type locality of this species is usually quoted as 'near Cape Town', but 'where we do not think the animal occurs'. This remark is certainly borne out by distribution maps which we have at our disposal (see Davis 1974 and map 49 in this book). A. Smith (1834) gave Uitenhage as the area to which the type locality of *T. (G.) dolichurus* is now restricted.

## Outline of synonymy

1832 *Mus dolichurus* Smuts, *Enum. Mamm. Cap.*: 38. Near Cape Town. (Uitenhage nominated by A. Smith, 1834, *S. Afr. J.* 2:156.)

1852 *Mus arborarius* Peters, *Reise nach Mossambique. Säugeth.*: 152.

1907 *Thamnomys venustus* Thomas, *Ann. Mag. nat. Hist.* (7) 19:122. Ruwenzori East, Uganda.

1909 *Thamnomys dolichurus* (Smuts). Jameson, *Ann. Mag. nat. Hist.* (8) 4:464. Malvern, Natal.

131 *Rattus (Grammomys) dolichurus* (Smuts). *In* Hewitt, *Guide to the Albany Museum.*

1931 *Grammomys dolichurus* (Smuts). St Leger, *Proc. zool. Soc. Lond.*: 957–997.

1939 *Thamnomys (Grammomys) dolichurus dolichurus* (Smuts). *In* Allen, *Bull. Mus. comp. Zool. Harv.* 83:419.

1953 *Grammomys dolichurus* (Smuts). *In* Ellerman *et al.*, *Southern African mammals.*

1964 *Thamnomys dolichurus* (Smuts). *In* Meester *et al.*, *An interim classification of southern African mammals.*

1974 *T. (G.) dolichurus* (Smuts). Misonne, *in* Meester & Setzer (eds), *The mammals of Africa: an identification manual.*

## Identification

The upper parts of these slender, long-tailed mice are brown to dull tawny with a reddish fringe particularly noticeable on their rumps (Smithers 1975). If a large number of specimens are seen simultaneously, their colour is not as grey as *T. (G.) cometes,* while the overall coloration differs geographically. There is no tuft of white hairs subauricularly. The individual hairs are slaty coloured to the base.

Ventral surface is pure white with individual hairs white to the base and sharply defined from the darker upper coloration. The length of the head and body is usually 120 mm.

The large and conspicuous ears are more brownish than the colour of the back, while the hands and feet are whitish. The tail (about 60% of the combined head-body and tail length) is a uniform brown above and below and usually less than 170 mm in length (plate 9).

Meester *et al.* (1964) and Ansell (1960) state that the mammae consist of one pectoral pair and two inguinal pairs, while Rosevear (1969) gives two inguinal pairs only as another possibility.

### TABLE 45
#### Measurements of male and female Thamnomys (Grammomys) dolichurus

|  | Parameter | Value (mm) | N | Range (mm) | CV (%) |
|---|---|---|---|---|---|
| Males | HB | 114 | 4 | 105–124 | — |
|  | T | 162 | 4 | 154–168 | — |
|  | HF | 23 | 4 | 21–24 | — |
|  | E | 20 | 4 | 18–23 | — |
|  | Mass: Three weights are known: 30 g, 34 g and 44 g respectively | | | | |
| Females | HB | 113 | 11 | 102–136 | 7,8 |
|  | T | 173 | 11 | 158–190 | 4,7 |
|  | HF | 23 | 11 | 19–25 | 7,5 |
|  | E | 19 | 10 | 17–25 | 10,5 |
|  | Mass: | 50,3 g | 6 | 41–58 g | 11,1 |

## Skull and dentition

The skull of *T. (G.) dolichurus* tends to be smaller than that of *T. (G.) cometes* with a condylobasal length of 27 mm or less (Meester *et al.* 1964). Ellerman *et al.* (1953) state that the bullae are rarely more than 6 mm (about 17% of the occipitonasal length) while the palatal foramina average more than 22% of the occipitonasal length in South African *dolichurus* (fig. 74).

The dentition corresponds rather closely to that of *T. (G.) cometes.* The upper molar row is usually more than 5 mm in length in *cometes* and less than 5 mm in *dolichurus* (Meester *et al.* 1964) (fig. 75).

## Distribution

This species occurs in the forested or well-wooded areas in the Southern Savanna Woodlands from the Great Escarpment to the coast as far as Van Staden's Pass, west of Port Elizabeth (Davis 1974). It also occurs in KwaZulu,

Mozambique and the eastern parts of Zimbabwe (map 49).

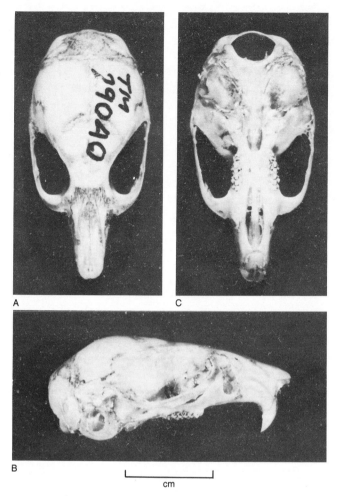

A

B

C

cm

**Figure 74** Dorsal (A), lateral (B) and ventral (C) views of the skull of the Woodland Mouse *Thamnomys (Grammomys) dolichurus*. Supraoccipital ridges weak to absent; braincase relatively wide; bullae small, about equal in length to the toothrow; condylobasal length less than 27 mm (making the skull slightly smaller than that of the closely related *T. (G.) cometes*). Anterior edge of zygomatic plate vertical.

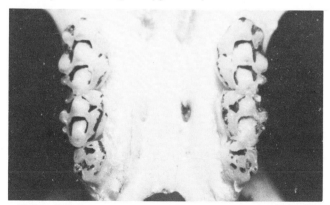

**Figure 75** Cheekteeth of the Woodland Mouse *Thamnomys (Grammomys) dolichurus*. The toothrow is usually less than 5 mm in length. The centre row of cusps tends to be enlarged in this genus.

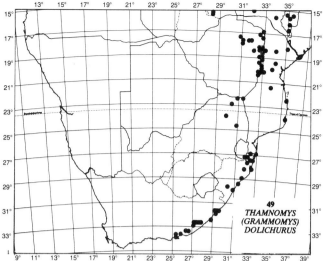

49
THAMNOMYS
(GRAMMOMYS)
DOLICHURUS

## Habitat

In South Africa, the species seems to prefer forests and thickets while extralimitally in Zambia, it occurs in gallery and montane forests and it also uses banana trees where it sometimes nests (Ansell 1969). In Zimbabwe it is generally confined to forests (Smithers 1975). A specimen in the Transvaal Museum was trapped in a large tree in riverine forest on the Limpopo River, while another was trapped in the thorny top of a *Euphorbia* species.

According to Rosevear (1969), the Woodland Mouse is moderately common in the right localities, but it is not abundantly represented in collections. This is certainly the case as far as the Transvaal Museum material is concerned. Rosevear speculates that this might be attributed to the fact '. . . of their essentially being inhabitants of fairly thick tangle in shrubs and small trees and thus not so readily trapped as purely ground dwellers'.

## Diet

Smithers (1975) states that they are predominantly arboreal, climbing around in the canopies of trees '. . . in search of the kernels of nuts or of wild fruits, although they will descend to the ground to search for those that have fallen or to eat grass and shrub seeds'.

## Habits

It is known that they are arboreal, communal and nocturnal. The nests are placed 30 cm to 4 m above the ground and are made of leaves and grass (Misonne 1974). Specimen 7222 in the Transvaal Museum is a female caught in the nest of a barbet at Inguavuma, KwaZulu. Sclater (1901) recorded such an incident by stating that this mouse '. . . appears to be arboreal, as a rule making a nest or occupying that of a bird such as a barbet'.

It has been stated by others that populations of some tropical rodents are relatively stable. Trapping results from five areas in Uganda that were visited four times annually for three years, suggest that population level of

169

*Thamnomys dolichurus* was relatively stable over the period of study (Rahm 1967, *in* Golley, Petrusewicz & Ryszkowski 1975).

**Predators**
Unknown.

**Reproduction**
This species constructs its nest in hollow tree trunks lined with fine leaves and mosses which it may build itself. On the other hand, some piracy on certain species of birds may occur as in the case of the occupied barbet's nest (Sclater 1901) referred to above. Hanney (1975) also refers to the 'Climbing Rat' *Thamnomys dolichurus* which '. . . has been known to rear its young in the suspended nest of a golden weaver'.

Whether breeding occurs throughout the year or '. . . only during a restricted season' is unknown (Smithers 1975). Misonne (1974) states that about one litter is born every five to six weeks with two to five per litter, while Smithers (1975) reports females as having litters of up to five or six. Smithers (1975) reports females as having litters of up to five or six. Delany (1971) records the average litter size in Uganda to be 2,7. In the Transvaal Museum, specimen no. TM 24633 was a pregnant female on 19 September 1974 with three embryos (two in the left and one in the right uterine horn). It is stated that this species, like other arboreal mice, carries its young attached to its mammae. This observation, according to Sclater (1901), was made by Peters as early as 1852 during his visit and sojourn to Mozambique. He aptly referred to this species as *Mus arborarius*.

**Parasites**
According to De Meillon *et al.* (1961), 14 different flea species (representing five genera and three families) are known from *T. (G.) dolichurus*. These include: Pulicidae – *Xenopsylla cheopis*, *X. brasiliensis*, *X. hamula*, *X. zumpti*; Hystrichopsyllidae – *Ctenophthalmus acanthurus*, *C. ansorgei*, *C. eximius*, *Dinopsyllus ellobius*, *D. grypurus*, *D. lypusus*, *Listropsylla agrippinae*, *L. basilewskyi*; Ceratophyllidae – *Nosopsyllus incisus* (De Meillon, Davis & Hardy 1961). *Ctenophthalmus calceatus* (Hystrichopsyllidae) was also reported by Zumpt (1966). Blowflies (Calliphoridae) also occur and this family includes *Cordylobia ruandae*. Intranasal and intradermal chiggers also occur. From *T. (G.) dolichurus* four species of trombiculids (Trombidiformes) are known, including *Schoutedenichia berghei*, *S. cordiformis*, *S. paradoxa* and *S. pilosa* (Zumpt 1961a).

Finally, immature forms of two species of ticks have also been recorded (Theiler 1962) from the Woodland Mouse, *viz*. *Amblyomma cohaerens* and *Haemaphysalis leachii leachii*.

**Relations with man**
Not easy to determine, but the fact that they act as hosts to ticks such as *Haemaphysalis leachii* may contribute to the spread and transmission of *Babesia canis*, *Nuttallia felis*, *Coxiella burnetii*, *Rickettsia conorii* and *R. pijperii*.

**Prehistory**
*T. (G.)* cf. *dolichurus* has been identified from the Rodent Cave and the Limeworks Dumps at Makapansgat (De Graaff 1960a). It was also identified from Phase II breccia at the Kromdraai Faunal Site (Kromdraai A) which was originally collected at Kromdraai by Draper in 1895 (De Graaff 1961).

**Taxonomy**
Meester *et al.* (1964) have put is succinctly: 'Ellerman *et al.* (1953) recognise *dolichurus* and *ruddi* (a *Thallomys*) as species, but *cometes* only as subspecies. Roberts (1951) places southern African *Thamnomys* in the subgenus *Grammomys*, as does Davis *(in press)*. Ellerman *et al.* (1953) give *Grammomys* generic rank.' Meester *et al.* (1964) did not include *Grammomys* at all and list *Thamnomys dolichurus* and *T. cometes*. Misonne (1974) accepts the genus *Thamnomys* and his classification keys out to the subgenera *Thamnomys* and *Grammomys*.

To date the following subspecies of *dolichurus* have been described: *Thamnomys (Grammomys) dolichurus dolichurus*: eastern Cape to southern KwaZulu. *T. (G.) d. baliolus*: eastern Transvaal. *T. (G.) d. tongensis*: Manaba, northern KwaZulu. *T. (G.) d. vumbaensis*: Vumba and Inyanga, eastern Zimbabwe.

Ellerman *et al.* (1953) also include *cometes* and *silindensis* (with *vumbaensis* as a synonym) in *dolichurus*, while Roberts (1951) regards *vumbaensis* as being specifically distinct (Meester *et al.* 1964).

# *Thamnomys (Grammomys) cometes* Thomas & Wroughton, 1908

**Mozambique Woodland Mouse**

**Mosambiek Woudmuis**

The generic name is derived from the Greek *thamnos* = thicket or shrub, in which they commonly occur, while *mys* = mouse. The Greek word *kometes* = a dweller, indicates that this species is a dweller of thickets. The Greek word *gramme* = a mark or a line in writing. In this case, the postero-internal cusps has been deleted, distinguising *Grammomys* from *Thamnomys*.

This species was originally described as *Thamnomys cometes* by Thomas and Wroughton from Inhambane on the southern coast of Mozambique. Meester (1976) lists it in *The South African Red Data Book* as rare, uncommon and with a limited distribution in the RSA.

Due to the paucity of material in southern African collections, no values for the usual parameters could be determined. As was indicated above, *T. (G.) cometes* is not easily separated from *T. (G.) dolichurus*. *T. (G.) cometes* is somewhat larger as can be seen from the following values from specimens of the same region quoted by Misonne (1974):

|       | cometes      | dolichurus   |
|-------|--------------|--------------|
| HB    | 112–124 mm   | 95–129 mm    |
| T     | 142–194 mm   | 146–193 mm   |
| HF    | 23–25 mm     | 20,5–26 mm   |
| Skull | 31–33,2 mm   | 26,6–32,5 mm |

## Outline of synonymy

1908 *Thamnomys cometes* Thomas & Wroughton, *Proc. zool. Soc. Lond.*: 549. Inhambane, Mozambique.

1941 *Grammomys cometes* (Thomas & Wroughton). *In* Ellerman, *The families and genera of living rodents.*

1951 *Thamnomys (Grammomys) cometes* Thomas & Wroughton. *In* Roberts, *The mammals of South Africa.*

1953 *Grammomys dolichurus cometes* (Thomas & Wroughton). *In* Ellerman *et al.*, *Southern African mammals.*

1964 *Thamnomys cometes* Thomas & Wroughton. *In* Meester *et al.*, *An interim classification of Southern African mammals.*

1974 *Thamnomys (Grammomys) cometes* Thomas & Wroughton. Misonne, *in* Meester & Setzer (eds), *The mammals of Africa: an identification manual.*

## Identification

Externally, the genus is characterised by its conspicuously dark-coloured, long, slender and bristled tail, averaging 140% of the head-body length. The pelage (Meester *et al.* 1964) is silky to the touch, with a reddish-brown to clayish coloration dorsally and the overall colour is greyer than in the closely related *R. (G.) dolichurus*. The dorsal fur is short and the individual hairs are a dark slate colour basally extending along the shaft of the hair for about two thirds of its length. The remainder of the shaft shows a buff coloration. A certain proportion of black hairs are scattered through the pelage. The ventral surface is white and clearly demarcated from the flanks. The hands and feet are white. The claws on the latter are short and curved. Some individuals show a white subauricular patch but its occurrence is variable (plate 9).

Ellerman (1941) states that the feet as a rule are less obviously specialised for arboreal life than is found in either *Thamnomys* or *Thallomys,* while Meester *et al.* (1964) have described the hindfeet as slender.

Roberts (1951) describes the mammary formula as 1–2=6, i.e. with a single pair of pectoral and two pairs of inguinal mammae, while Meester *et al.* (1964) give an alternative formula of 0–2=4. Ellerman (1941) states that it can be either.

## Skull and dentition

On average, it appears that skulls of *T. (G.) cometes* are larger than *T. (G.) dolichurus*, with a condylobasal length exceeding 27 mm (Meester *et al.* 1964), while the entire skull (including the bullae and teeth) tends to be smaller than the related *Thallomys*. The bullae are more or less equal in length or longer than the molar row (Rosevear 1969), while the braincase may be referred to as being broad. Supraorbital ridges are either absent or poorly developed in contrast to the jugal element of the zygoma, which is realtively long.

In the five-rooted $M^1$ all the cusps are present with the exception of the vestigial postero-internal cusp (t7), while the t1 is about as large as t3. In the $M^2$ the t3 is small, the t7 virtually absent. The $M^3$ portrays a moderately developed t3 of the first lamina, followed by the usual t4, t5 and t6 in the second lamina, with a small portion representing the third lamina. The centre row of cusps of the upper molar tooth row of this genus tends to be greatly enlarged (Ellerman 1941) (fig. 76).

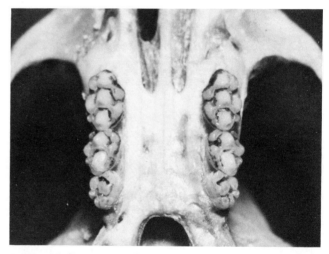

**Figure 76** Cheekteeth of the Mozambique Woodland Mouse *Thamnomys (Grammomys) cometes.* The toothrow is usually more than 5 mm in length (i.e. slightly longer than in the closely related *T. (G) dolichurus*).

Though a milk dentition is not normally seen in rodents, Rosevear (1969) mentions a specialisation of incisors in newly born *Thamnomys* (as is the case in *Aethomys* too). This was reported upon by Lawrence (1941) and Petter & Genest (1964). The upper and lower incisors are bifurcate at the tips and/or flared apart. Both curve laterally outwards to make a gap, facilitating a sound grip on the mother's nipple, and reducing the danger of the young being torn away and lost when carried through tangled vegetation. The forked tips disappear at about the 15th day and normal wear takes place at the tips until the normal chisel shape is assumed. As Rosevear correctly points out, the teeth themselves are not shed and replaced by a second set '. . . for these are only partial and temporary modications of form', the primary milk dentition as normally understood, being absent in rodents.

## Distribution
The distribution of the Mozambique Woodland Mouse is limited to northeastern southern Africa, from northern KwaZulu along the Mozambique coast into eastern Zimbabwe (i.e. in the Southern Savanna Woodlands biotic zone east of the Great Escarpment) (map 50).

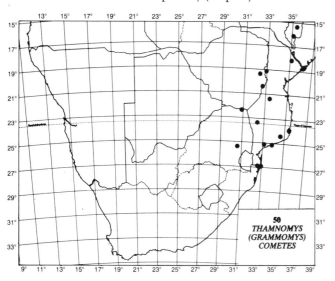

50
THAMNOMYS
(GRAMMOMYS)
COMETES

## Habitat
The Mozambique Woodland Mouse usually occurs sympatrically with *T. (G.) dolichurus*, but it seems to show a preference for dense forest. Compared to the related Blacktailed Tree Rat *(Thallomys paedulcus)*, it fancies evergreen forests in contrast to *Thallomys* which prefers the drier *Acacia* stands (Roberts 1951). Smithers & Lobão Tello (1976) record it in forest fringes and thickets.

## Diet
No data available.

## Habits
I know of no field study hitherto undertaken for this predominantly nocturnal species. Roberts (1951) may be quoted as follows: 'Members of this genus construct large nests of dry grass and leaves in matted tangles of bush, in the hollows of large trees or even in the forks of smaller trees; in these nests they live during daylight and rear their young.' Roberts also states that he has often come across family parties '. . . consisting of the adult parents, immature and juvenile young (of two litters) all in the same nests'. Ellerman (1941) says that this species, like several other arboreal murids, carry their young around attached to the nipples of the female.

## Predators, reproduction and relations with man
No information available at present.

## Parasites
I am aware of two species of fleas. On *T. (G.) cometes*, *Leptopsylla aethiopica* (Leptopsyllidae) and *Nosopsyllus incisus* (Ceratophyllidae) have been reported by De Meillon *et al.* (1961). The former is a rare species while the latter is known from Zaïre.

## Prehistory
This species has hitherto not been indentified from fossiliferous deposits in southern Africa.

## Taxonomy
Roberts (1951) and Misonne (1974) place the southern African material representing *Thamnomys* in the subgenus *Grammomys* in contrast to Ellerman *et al.* (1953) who recognise the generic rank of *Grammomys* while demoting *cometes* to a subspecies of *dolichurus*.

On the specific level, it is thought that *T. (G.) cometes cometes* occurs in northern KwaZulu and Mozambique, while *T. (G.) silindensis* is found near Mount Selinda in southeastern Zimbabwe. Ellerman *et al.* (1953) regard both *cometes* and *silindensis* as races of *T. (G.) dolichurus*, while Roberts goes to the other extreme and treats both *cometes* and *silindensis* as valid species.

GENUS *Dasymys* Peters, 1875

This animal was first described as *Mus incomtus* by Sundevall in 1846 and was originally collected at 'E Caffraria prope portum Natal' (which implies the present-day city of Durban). According to Sclater (1901), the type of the species was originally obtained by the well-known Swedish collector, Wahlberg. Subsequently in 1875, Peters described another specimen as *Dasymys gueinzii*, named after the collector Gueinzius, and this specimen was obtained from the interior of Port Natal. According to modern taxonomic procedure, by monotypy, *D. gueinzii* now equals *M. incomtus* as *D. incomtus*. Ellerman (1941) states that it appears to be allied to the *Arvicanthis* series of African rats, but Rosevear (1969) points out that Matthey (1958) has found no cytological evidence in favour of such an association.

They are medium-sized animals, with a well-devel-

172

oped pelage which is long and soft to the touch. Delany (1975) states that the fur is 'untidy' and this has undoubtedly contributed to the vernacular name of 'Shaggy Rat' which is sometimes applied to this species. In contrast to the body, the tail is sparsely haired. Seen head-on, the moderately sized eyes are nevertheless conspicuous and are set relatively close together.

Where they occur, their distribution seems to be intermittent and they can, therefore, not be described as common although they may be plentiful in localities where they occur (Rosevear 1969). Rosevear also points out that they are seemingly limited to a fairly specialised habitat (i.e. a swampy or wet environment where water is available and possibly covered with lush grass or reeds) and that it is a remarkably even genus throughout its range of distribution.

# *Dasymys incomtus* (Sundevall, 1847)

**Water Rat**

**Peters' Water Rat**

**Shaggy Rat**

**Waterrot**

The generic name is obviously derived from the Greek words *dasys* = hairy, shaggy, combined with *mys* = mouse. The specific name is the Latin word *incomtus* = unadorned, given with reference to the dull pelage.

The Water Rat is found over a good deal of Africa south of the Sahara (Rosevear 1969). It ranges from Sierra Leone in the west to Ethiopia in the east and southwards to the Cape Province. Despite its wide distribution, it is apparently limited to suitable habitat which, in this case, implies swampy environments, or as Rosevear (1969) puts it, '. . . at least rather wet ground, not necessarily near free water but covered with lush grass or reeds'. Davis (1962) states that it is a swamp rat which appears to be losing its hold in southern Africa and Meester (1976) has described it as rare but '. . . fairly widespread, but seldom caught, and probably occurring in low numbers' in the *South African Red Data Book*.

## Outline of synonymy

1847 *Mus incomtus* Sundevall, *Oefvers. K. Svenska Vet. Akad. Förh. Stockholm* 3 (5):120. 'E Caffraria prope portum Natal' = Durban.

1870 *Mus (Isomys) nudipes* Peters, *J. Sci. Math. Phys. Nat. Lisbon* 3:126. Huilla, Angola.

1875 *Dasymys gueinzii* Peters, *Monatsb. K. Preuss. Akad. Wiss. Berlin.* 12. (Type species *M. incomtus,* Sundevall.) Port Natal = Durban.

1897 *Dasymys incomtus fuscus* De Winton, *Proc. zool. Soc. Lond.* for 1896:804. Mazoe, Zimbabwe.

1899 *Dasymys incomtus* (Sundevall). Sclater, *Ann. S. Afr. Mus.* 1:218.

1936 *D. i capensis* Roberts, *Ann. Transv. Mus.* 18:254. La Plisante, Wolseley, CP.

1939 *Dasymys incomtus incomtus* (Sundevall). *In* Allen, *Bull. Mus. comp. Zool. Harv.* 83:382.

1951 *Dasymys incomtus incomtus* (Sundevall). *In* Roberts, *The mammals of South Africa.*

1953 *Dasymys incomtus* (Sundevall). *In* Ellerman *et al., Southern African mammals.*

1974 *Dasymys incomtus* (Sundevall). Misonne, *in* Meester & Setzer (eds), *The mammals of Africa: an identification manual.*

### TABLE 46
Measurements of male and female *Dasymys incomtus*

|  | Parameter | Value (mm) | N | Range (mm) | CV (%) |
|---|---|---|---|---|---|
| Males | HB | 159 | 9 | 139–190 | 8,3 |
|  | T | 144 | 9 | 133–157 | 4,2 |
|  | HF | 33 | 9 | 32–35 | 2,6 |
|  | E | 21 | 9 | 19–22 | 3,7 |
|  | Mass: | 124,6 g | 5 | 102–152 g | 12,7 |
| Females | HB | 153 | 8 | 139–175 | 5,9 |
|  | T | 128 | 8 | 112–150 | 7,5 |
|  | HF | 33 | 8 | 29–35 | 4,9 |
|  | E | 19 | 8 | 18–21 | 4,3 |
|  | Mass: | 89 g | 6 | 78–100 g | 7,1 |

The values obtained for weight measurements vary according to locality and size of sample. Wilson (1975) reported the weight of two males as 98 g and 92 g respectively. Delany (1975) obtained an average weight of 102,1 g (N = 7, observed range 64–120 g) in Uganda, while Smithers (1971) found the mean weight to be 128,5 g (N = 11 observed range 107,3–164,1 g) in Botswana. For females, Delany's figures are 81,8 (N = 10 observed range 59–105 g) for Uganda and those for Botswana by Smithers are 127,0 g (N = 6, observed range 102,3–161,2 g).

## Identification

*Dasymys incomtus* is stoutly built. The dorsal colour is dark grey to buffy-brown or black, with individual hairs dark slate at the base with black to brownish tips. The fur is rough though soft to the touch and according to Delany (1975), shaggy-looking. Under favourable conditions of illumination it shows iridescence (Rosevear 1969). Rosevear (1969) has made some interesting observations on the pelage of this species which I have paraphrased here. He notes that the composition of the pelage of *Dasymys incomtus* is markedly different from that of the majority of African murines. In the West African representative of the species the midback hairs may be 17–23 mm long, interspersed with longer weak bristles reaching some 27 or 28 mm. The texture of the pelage is affected by the very fine, almost straight underfur, playing a dominant role instead of a minor one as in, say *Thamnomys*. The relatively few hairs which are broadened into sub-bristles are indistinguishable (except under the microscope) from the underfur for about their basal two thirds, only then becoming gradually and slightly expanded, tapering again to a slender tip for the terminal 4 mm. The overall result is a long, fine coat devoid of the springiness or harshness produced in other species by a greater or smaller mixture of broadened convaco-convex gutter hairs. The composition of the belly fur is much the same as on the dorsal surface.

The flanks are paler in colour than the back, while the underside is dull grey (the individual hairs tipped white against a buffy grey base) and the coloration is not sharply demarcated from top to bottom. According to Roberts (1951) melanism can also be encountered. The buffy-brownish hands and feet are almost naked, covered with very short hair while the claws are whitish. The first and fifth digits of the pes are small but functional, while the first digit of the manus is vestigial. The ears are moderate, broad and rounded, well furred on the inside surfaces of the pinnae and these hairs project beyond the rims of the pinnae giving the ears a fringe-like appearance in the live animal. The tail is sparsely haired when compared to the rest of the body and well scaled (± 10 cm according to Roberts (1951)). It is slightly shorter than the overall head-body length (plate 9).

Juveniles are darker than adults, lacking the warm brown suffusion of the adults dorsally, while the pelage is even woollier (Smithers 1975).

The mammary formula conforms to 1 − 2 = 6 (one pectoral pair and two pairs of mammae inguinally).

## Skull and dentition

The skull is stoutly constructed with a very conspicuous interorbital constriction (fig. 77). Ellerman (1941) states it to be 12,6% of the occipitonasal length, a measurement usually less than that of any group of *Rattus*. From above one can see the outer edges of the molars. The supraorbital ridges are strong and well developed. The braincase is flat and broad. The anterior edge of the zygomatic plate curves concavely inwards in the middle, resulting in a pronounced hook at the top anterior corner. The infraorbital foramina are large. The muzzle is relatively short, while the palatine foramina are slitlike and reach the level of the $M^1$. The palate is narrow and conspicuously ridged. The bullae are small to moderate and the occipital region angular, and well ridged. The zygomatic arches are strong and well developed and the jugal element longer than normal (Ellerman 1941).

The dentition of *Dasymys incomtus* is interesting (fig. 78). The molars are large and heavily cusped with each cusp considerably raised when freshly erupted. The triple row of tubercles (of which the centre row is the largest, i.e. t2, t5 and t8), are obliterated in extreme age and the spaces between the original laminae may isolate as enamel islands. The $M^1$ is five-rooted with the t1 small, the t7 absent and the t9 vestigial. A similar arrangement is found in the $M^2$ except that t2 is absent and t3 small. In the $M^3$ there is a well-developed t1 representing the first lamina, followed by two tricuspidate laminae. In other words, the t7 has not been reduced in the $M^3$ and *Dasymys* is, therefore, an exception to the rule. The posterior lamina of the $M^3$ is not appreciably narrower than the middle (preceeding) lamina (differing in this respect from *Arvicanthis*) (Ellerman 1941). In *Dasymys incomtus* the $M^3$ is usually not smaller than the $M^2$ and with wear it may gradually become longer than the $M^2$. The combined length of the molars is nearly a fifth of the skull length and more than the greatest diameter of the bullae (Roberts 1951).

The orthodont upper incisors are broad, not grooved, strong and orange-coloured on the front outer surfaces. The lower incisors also show this coloration, but are more slenderly built and their roots tend to show on the outer surface of the mandible to a greater extent than is usual.

In the lower jaw there is no terminal heel to the $M_1$ and the $M_2$ but there is a large supplementary cusp in front of and between the first two cusps of the $M_1$. Occasionally, another large, additional cusp occurs alongside the antero-external cusp of $M_2$ (Rosevear 1969).

## Distribution

This Southern Savanna Woodlands and South West Cape biotic zone species occurs in the Transvaal, Natal and ranges south and westwards along the coast of the Cape Province. In South Africa, it shows a relict distribution pattern either at the subspecies level *(capensis)* or at local population level (Davis 1974) and there is a marked gap along the coast of the southern Indian ocean (map 51). Davis predicted that this gap is more apparent than real and Avery (1977) reported on specimens collected in the Cango Valley (3322 Ac) near Oudtshoorn, while Swanepoel *(pers. comm.)* has collected the species on the banks of the Lottering River (3323 Dc) in the Humansdorp Dis-

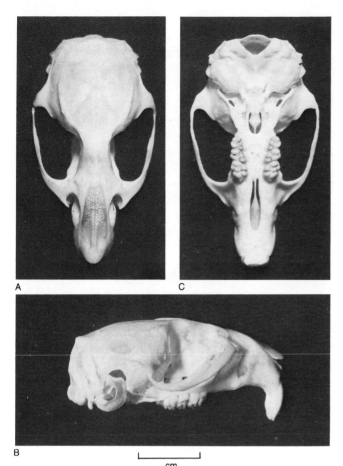

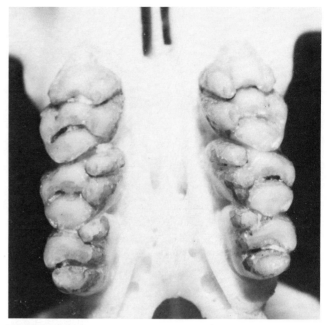

**Figure 78** Cheekteeth of the Water Rat *Dasymys incomtus.* Note the large and heavily cusped molars. In the five-rooted M$^1$ the t1 is small, the t7 absent and the t9 vestigeal. The M$^3$ is subequal to the M$^2$ in size. The combined length of the M$^{1-3}$ is usually more than the greatest diameter of the bulla.

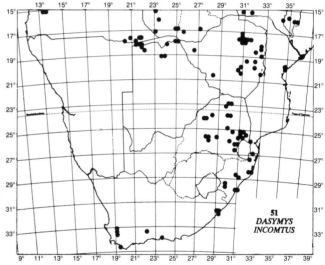

**Figure 77** Dorsal (A), lateral (B) and ventral (C) views of the skull of the Water Rat *Dasymys incomtus.* The supraorbital and occipital ridges are well developed with the interorbital constriction rather far forward. The muzzle is short. The interior palatine foramina slitlike, reaching level of M$^1$ while the palate is ridged and relatively narrow. Bullae small to moderate. The zygomatic arch is strong with a large jugal. The anterior edge of the zygomatic plate is concave. The braincase is not pronouncedly domed.

trict during June 1979. It is also encountered in Zimbabwe, the northern parts of Botswana and in the Caprivi Zipfel. North of the geographical limits of this work it ranges into Angola and the Congo and Liberia on the west and to Ethiopia and southern Sudan in the east.

### Habitat

This species frequents streams, rivers, reed beds and swamps and is partly aquatic. In the Wankie National Park in Zimbabwe, it lives in long grass close to water (Wilson 1975) and in Uganda it inhabits river valleys (Delany 1975). In West Africa it makes regular pathways through reeds which take the form of tunnels among the dead and rotting vegetation (Booth 1970). In Botswana it is closely associated with reedbeds or semi-aquatic vegetation on fringes of marshes '. . . or the back waters of rivers' (Smithers 1971). It shares the same habitat with *Pelomys fallax, Otomys irroratus, O. laminatus* and *O. angoniensis* (Davis 1974).

This would imply that it tends to be limited to lower lying country. This, however, is not necessarily the case. In Africa it has been collected on both the Central Ruwenzori Mountains and Mount Cameroon at altitudes of 2 700 m or more (Rosevear 1969). In the western Cape it also occurs on mountain ground which is often fairly waterlogged and where the vegetation generally holds a good amount of water derived from downpours or from mist.

Hanney (1975) reports that it is normally seen in or near water, but on the high plateau of Malaŵi it was

175

trapped among rocks and bracken up to 1 km from water. In the lowland area all '... catches were made within 20 m of a river or lake'.

## Diet

The diet is largely vegetarian but the Water Rat also consumes some insects (Smithers 1975) and other animal food (Hanney 1975). In Uganda they are often found in moist situations on cultivated lands (Delany 1975). Booth (1970) states that the food is thought to be young shoots and waterplants, while Smithers (1975) describes the food as fresh green shoots of reeds and aquatic grasses, the succulent stems of grass, as well as grass seeds. In Botswana they go for new shoots of *Phragmites* spp. (Smithers 1971). Captive animals have taken fruit, insects, cabbage and cereal (Hanney 1965).

## Habits

Although these rats are fairly common where they occur locally, much remains to be learnt about their habits. They are mainly nocturnal (Shortridge 1934), although Smithers (1975) reports them to be diurnal to a large extent. They are terrestrial and partly amphibious. They swim well and take to water readily (Booth 1970) and are consequently as at home on shallow inundated ground as they are on dry land (Smithers 1975). Shortridge (1934) states that they dive readily but do not remain submerged for long. Distinct runs (not so clear as those made by *Otomys*) are formed radiating from nesting sites to feeding areas where the remains of fresh reeds or grass stems on which they feed are found in little piles (Smithers 1975). Nests are made of grass in rank vegetation and holes are dug in banks well above the water level (Roberts 1951).

Mud tracks show the digits in the spoor widely separated (Shortridge 1934).

According to specimen labels and records in the Transvaal Museum, some specimens (e.g. no. 24706) were trapped in association with other genera, including *Otomys* and *Dendromus*, as well as the insectivore *Crocidura*.

## Predators

Hanney (1975) states that skulls regularly appeared in pellets of owls collected on the shores of a Malaŵi swamp. Hair identification has revealed them as prey items of the Long-crested Eagle *Lophaerus occipitalis* (Jarvis & Crichton 1977).

## Reproduction

Hanney (1975) describing his experiences with rodents in Malaŵi, states that this species seems to have no permanent nest or burrow '... but during the breeding season it uses a combination of both to form what must be one of the most elaborate structures used by any rodent'. He describes a grass nest as a '... 200 mm ball' consisting of

two compartments, one on the surface and the other in a shallow excavation immediately below. These chambers were so arranged that a predator could search the surface nest without realising the existence of a basement. 'When the complete nest was lifted from the shallow hole, it was seen that a tunnel extended through the soil in the direction of the swamp.' This was also observed in other nests and in all cases the tunnels were first over 1 m in length with the blind end filled with water. Fleeing occupants therefore escape by forcing their way through the mud into the water of the swamp.

Delany (1975) has recorded that *Dasymys* in Uganda constructs a straight, unbranched burrow about 2 m in length, which gradually descends to a maximum depth of 30 cm. Runs radiate from the burrow entrance in which the nest is located.

Smithers (1975) records that in Zimbabwe domed nests are built of cut grass and other vegetation matter at ground level or on slightly elevated dry ground with a short refuge tunnel being burrowed out in close proximity. The nests are very often infested with laelaptid mites (Zumpt 1961).

From the meagre data available, it seems that breeding takes place during the warm, wet summer months (August to March) in Botswana and Zimbabwe (Smithers 1971, 1975). The average number of pups is five, with an observed range of two to nine. A nest of seven juveniles at an average weight of 4 g each was taken in March. Females breed at quite a small size: two females from Botswana bred at 118,8 g and 89,4 g respectively. In the Kruger National Park a female with three suckling young was noted in April (Pienaar *et al.* 1980).

The post-natal development has been assessed by Dieterlen (1967) for pups from Zaïre. At age 30 days, a weight of 30 g was attained, while the 75 g mark was reached between the 60th and 80th days.

## Parasites

The parasites known to be associated with *D. incomtus* form an impressive array. The mites include ten genera and 28 different species grouped into four families. These include the following (Zumpt 1961a): Laelaptidae – *Laelaps giganteus*, *L. muricola*, *L. roubaudi*, *Haemolaelaps dasymys*, *H. labuschagnei* and *H. murinus*. Ereynetidae – *Speleognathopsis galliardi*. Trombiculidae – *Trombicula bruynoghei*, *T. jadini*, *T. panieri*, *T. quasigiroudi*, *T. rodhaini*, *Helenicula thomasi*, *Schoengastia o. oubanguiana*, *S. o. bicalar*, *S. r. radfordi*, *Euschoengastia laurenti*, *Schoutedenichia andrei*, *S. audyi*, *S. p. paradoxa*, *S. penetrans*, *S. pilosa*, *S. schoutedeni*, *Gahrliepia dureni*, *G. lawrencei*, *G. longiscutullata*, *G. philipi* and *G. traubi*. Listrophoridae – *Listrophoroides dasymys*.

The fleas include seven species, consisting of four genera and two families. These species are listed in De Meillon *et al.* (1961) as follows: Pygiopsyllidae – *Stivalius torvus*. Hystrichopsyllidae – *Ctenophthalmus ansorgei*, *C. cal-*

*ceatus, Dinopsyllus dirus, D. longifrons, D. lypusus* and *Listropsylla dolosa.*

Only one species of tick, i.e. *Haemaphysalis l. leachii* (adult and immature forms), has been recorded from *D. incomtus* by Theiler (1962).

Plathyhelminthes are also known from the Water Rat. These include *Inermicapsifer congolensis,* to whom *D. incomtus* is a definitive host and *I. madagascariensis* also known from the Congo, both cases reported by Collins (1972).

## Relations with man

It is apparently not a pest of cultivation, although in West Africa it could become so if swamp rice becomes a popular crop (Booth 1970). Shortridge (1934) states that it is not particularly attracted by cultivation. It carries intranasal and intradermal chiggers, but the effects of these infections on man is negligible.

## Prehistory

A new species of *Dasymys, D. bolti,* has been reported from Bolt's Farm near Sterkfontein (Broom, not described and thus not yet confirmed) while *D. incomtus* has been recorded from both Sterkfontein and Makapansgat (Lavocat 1957; De Graaff 1960a).

## Taxonomy

Roberts (1951) recognises two subspecies for *D. incomtus,* i.e. *D. i. incomtus* and *D. i. capensis* (the latter from the Wolseley area in the southwestern Cape), as well as another species *D. nudipes* (occurring in the Okavango region and the Caprivi Strip). The first two have been maintained by Ellerman *et al.* (1953), while the latter has been included in *incomtus* as a subspecies. Meester *et al.* (1964) followed Ellerman *et al.* (1953) in this respect. Misonne (1974) has interpreted *capensis* and *nudipes* as included in *incomtus.*

GENUS *Pelomys* Peters, 1852

SUBGENUS *Pelomys* Peters, 1852

According to Misonne (1974) the status of this genus is unsatisfactory. It has been subdivided (as suggested by Ellerman (1941)) into three subgenera, *viz. Komemys, Desmomys* and *Pelomys,* the former two being extralimital to this work (occurring in Uganda and Ethiopia respectively).

It is close to yet another extralimital genus, *Arvicanthis,* as well as to *Lemniscomys,* while there is a superficial resemblance to *Dasymys.*

The distinctions are based on the length of the D5 of the manus which has a claw in *Komemys* and the tail

length is longer than the head-body length.

In *Pelomys* the D5 has a claw lacking (as in *Desmomys*) but the absent claw is replaced by the presence of a nail and the tail is shorter than the length of the head-body. An additional distinction between *Pelomys* and *Desmomys* portrays deeply grooved upper incisors in the former and only faintly so in *Desmomys* (i.e. the grooves are almost obsolete).

The outstanding difference of this genus, however, is the presence of grooved upper incisors which immediately separates it from all other murine genera.

# *Pelomys (Pelomys) fallax* Peters, 1852

**Creek Rat**

**Groove-toothed Swamp Rat**

**Spruitrot**

**Groeftandrot**

The generic name is derived from the Greek words *pellos = pelos =* dusky, dark coloured and *mys =* mouse. It

apparently refers to the overall colour of the pelage. The specific name is coined from the Latin word *fallo =* to deceive and implies the possibility of confusing this species with either *Dasymys* or *Arvicanthis.*

This species resembles the extralimital *Arvicanthis,* but they can be told apart by the conspicuously grooved incisors of *Pelomys.* This genus is also related to *Rhabdomys* and *Lemniscomys,* which also occur in southern Africa. *Pelomys fallax* was originally described by Peters in 1852 on specimens collected in the Zambezi Valley in Mozambique.

## Outline of synonymy

1852 *Mus (Pelomys) fallax* Peters, *Reise nach Mossambique. Säugeth.*: 157. Caya district, Zambezi River, Mozambique.

1896 *Golunda fallax* De Winton, *Proc. Zool. Soc. Lond.*: 804.

1913 *Pelomys australis* Roberts, *Ann. Transv. Mus.* 4:90. Beira, Mozambique.

1929 *Pelomys fallax australis* Roberts, *Ann. Transv. Mus.* 13:118.

1929 *Pelomys fallax rhodesiae* Roberts, *Ann. Transv. Mus.* 13:118. Machile River, Zambia.

1939 *Pelomys (Pelomys) fallax fallax* (Peters). *In* Allen, *Bull. Mus. comp. Zool. Harv.* 83:408.

1946 *Pelomys fallax vumbae* Roberts, *Ann. Transv. Mus.* 20:320. Vumba, Zimbabwe.

1953 *Pelomys fallax* (Peters). *In* Ellerman et al., *Southern African mammals.*

1974 *Pelomys (Pelomys) fallax* (Peters). Misonne, *in* Meester & Setzer (eds), *The mammals of Africa: an identification manual.*

### TABLE 47
Measurements of male and female *Pelomys (Pelomys) fallax**

|  | Parameter | Value (mm) | N | Range (mm) | CV (%) |
|---|---|---|---|---|---|
| Males | TL | 292 | 20 | 220–365 | — |
|  | T | 145 | 19 | 114–183 | — |
|  | HF | 37 | 20 | 32–40 | — |
|  | E | 18 | 22 | 15–20 | — |
|  | Mass: | 141,5 g | 10 | 100,5–170,4 g | — |
| Females | TL | 287 | 30 | 239–330 | — |
|  | T | 146 | 29 | 131–175 | — |
|  | HF | 36 | 43 | 31–41 | — |
|  | E | 17 | 48 | 15–20 | — |
|  | Mass: | 117,0 g | 7 | 100,3–149,8 g | — |

*Smithers (1971)

Very few specimens of this species exist in the museum collections in South Africa. Data on measurements are not available, so I have quoted the measurements of *Pelomys fallax rhodesiae* which Smithers (1971) obtained from a good series collected in Botswana.

## Identification

The dorsal surface is a rusty brown, the individual hairs being a dark slaty grey at their bases while their tips are a conspicuous buff. The fur is somewhat rough to the touch but it cannot be termed spiny. It is much harsher than in the silky-haired *Dasymys*. There is a variable dark vertebral line. Smithers (1971) states that this dorsal line, running from the occipital region to the base of the tail, is very distinct in young specimens (8 or 9 g in weight) from Botswana. At a body weight of 20 g the long pelage has absorbed the distinctness of the dorsal stripe and in some cases it is hardly perceptible.

When seen in direct sunlight, the fur is glossy and shiny. Smithers (1971) states the pelage shows a distinct green or blue iridescence.

Towards the hind quarters the rusty colour becomes more perceptible, almost light ochre near the base of the tail. The tail is blackish along the top and dirty white below, fairly well haired with short bristles. It is somewhat longer than the head-body length in South African forms (Roberts 1951), but is always slightly shorter in Ugandan specimens (Delany 1975).

The ears which are of normal size show the same coloration as the dorsal surface and are not fringed as in *Dasymys*. The sides of the head, throat, flanks, upper arms and thighs are buffy, rather than rusty, merging into a dirty-white belly.

The feet are broad with well-developed claws, though not as well-developed as in arboreal murids. The clawed D5 of the pes is very short, subequal in lustre to the hallux. The D5 of the manus is also shortened and does not reach the base of the 4th digit, while the pollex is vestigial. The dorsal surfaces of the hands and feet are covered with short, dense hairs (plate 9).

The mammae are described by both Shortridge (1934) and Roberts (1951) as two pectoral pairs followed by two inguinal pairs.

## Skull and dentition

There is a marked arching of the skull from the nasal region to the occipital region. The skull is strongly built, the rostrum broad, supraorbital ridges present and the interorbital constriction slight. The skull gives an impression of angularity. The bullae are fairly large. The palate is narrow. The palatal foramina reach the level of the lamina of the first molars, but narrow down posteriorly. The zygoma is robust and broadened in the centre (fig. 79).

The teeth, especially the molars (usually exceeding 6 mm in total length), are heavy and broadened. The cusps are very prominent, especially the centre row and the inner row. In both the $M^1$ and $M^2$ the t3 is smaller than t1 and the t9 reduced. The $M^3$ has a well-developed t1 with a tricuspidate middle lamina behind it. This in turn is followed by a posterior but small lamina. The $M^3$ is scarcely smaller than the $M^2$ (fig. 80). The lower molars are laminate with the terminal heels of $M_1$ and $M_2$ reduced. The mandible is robust and, as in *Dasymys*, the roots of the ungrooved lower incisors are clearly visible on the outer surfaces. The upper incisors are grooved, an unusual condition among the murines.

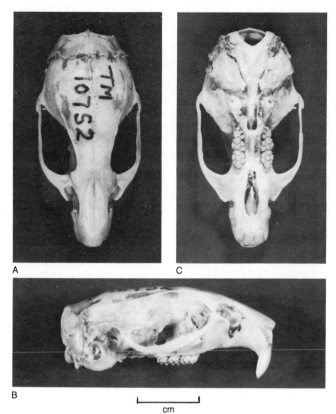

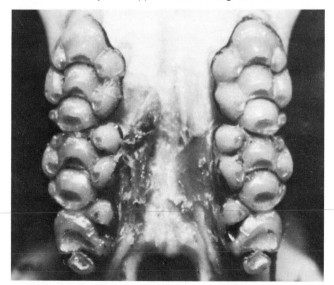

**Figure 79** Dorsal (A), lateral (B) and ventral (C) views of the skull of the Creek Rat *Pelomys (Pelomys) fallax*. Both the rostrum and interorbital constriction are wide while the skull is fairly heavily ridged. The palatine formina narrow posteriorly and reach the molars. The palate tends to be narrow and the bullae fairly large. The dorsal outline of the skull is pronouncedly arched while the strong zygomatic arch is often broadened in the middle. The anterior edge of the zygomatic plate tends towards concavity. The upper incisors are grooved.

**Figure 80** Cheekteeth of the Creek Rat *Pelomys (Pelomys) fallax*. The molars are heavy and broad with prominent cusps, especially the lingual and central rows. The $M^3$ is scarcely smaller than the $M^2$.

## Distribution

In the southern African context as defined south of the Cunene and Zambezi rivers, this species occurs in the Southern Savanna Woodlands to latitude 21°S. It is therefore found in the northern areas of Mozambique, northeastern area of Zimbabwe, as well as in northern Botswana along the Okavango River and in the Caprivi Zipfel (map 52). Further north it ranges into Zambia, the Congo, Malaŵi, Tanzania, southwestern Uganda and southern Kenya (Misonne 1974).

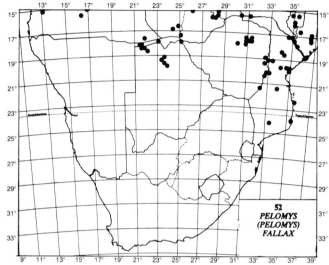

## Habitat

These rather common animals inhabit streams and rivers, vleis, swamps, reed beds, riverbanks – in fact all places where water is available and where it tends to be damp and where reeds, '. . . matted aquatic grasses, sedges, or other lush vegetation' occurs (Smithers 1971). They are also partial to grassveld occurring along rivers, near cultivated land and on the outskirts of forests. In Mozambique they occur in grassland or shrub on the fringes on swamps or the backwaters of rivers and streams (Smithers & Lobão Tello 1976).

## Diet

The Creek Rat is predominantly vegetarian, utilising the young shoots of reeds and other vegetation growing in swampy areas. Smithers (1971) remarks on the small piles of cut grass stems and scraps of thin reed which are found at intervals along their runs. Smithers (1975) also states that they eat grass and other seed.

## Habits

According to Shortridge (1934) *Pelomys (P.) fallax* is very similar to *Dasymys incomtus* in habits. He states that it is exclusively diurnal and that the two species are equally aquatic. Shortridge also reports that, according to the local inhabitants of the Okavango area, these animals live above the ground – no holes of any description were observed in the swamps. However, Smithers (1971) quotes Liversedge who states that at Sepopa there were many *Pelomys* holes in the dry ground on the riverbank. In Zimbabwe (Mashonaland), where agricultural lands adjoin wet, vlei areas, this species is frequently turned up in ploughing, indicating that they may dig their burrows at a shallower level than other murids (Smithers 1975).

They are good swimmers and can often be seen sunning themselves or feeding on the matted vegetation along the edges of rivers (Smithers 1971) confirming that they are predominantly diurnal, in contrast to Misonne (1974) who interprets them as being mainly nocturnal. Smithers also points out that they are difficult to trap.

## Predators
No information available.

## Reproduction
There appears to be no fixed breeding season. It constructs a nest of grass or leaves on the ground (Rahm 1967).

Indications are that parturition occurs in the warmer, wetter months of the year, i.e. August to April, in both Botswana and Zimbabwe (Smithers 1971, 1975). Very few records of litters are available, but such as there are, indicate a litter size varying from two to four young.

## Parasites
The following mites have been taken from *P. (P.) fallax:* Laelaptidae: *Laelaps giganteus.* Ereynetidae: *Speleognathopsis galliardi.* Trombiculidae: *Schoengastia oubanguiana bicalar, S. r. radfordi, S. rubi rodentis, Schoutedenichia p. paradoxa, S. p. gilleti, S. penetrans* and *Gahrliepia traubi.* Sarcoptidae: *Mysarcoptes paucipilis* (Zumpt 1961a).

The fleas (according to Zumpt 1966) include the following: Pulicidae: *Echidnophaga gallinacea.* Pygiopsyllidae: *Stivalius alienus, S. torvus.* Hystrichopsyllidae: *Ctenophthalmus atomus, C. ominosus, C. evidens, C. ansorgei, C. calceatus, C. verutus, Dinopsyllus dirus, D. longifrons, D. lypusus* and *D. pringlei.*

As fare as the ticks are concerned, adults of *Ixodes nairobiensis* have been collected from the Creek Rat (Theiler 1962).

## Relations with man
This species is not regarded as destructive and is often eaten by the Black tribes in Angola.

## Prehistory
Specimens of *Pelomys* cf. *fallax* have been recorded from both Sterkfontein and Makapansgat (Lavocat 1957; De Graaff 1960a). In fact, I found it common. The recent species appears to be on the same evolutionary level as its fossil counterparts.

## Taxonomy
Ellerman (1941) thought it to be unlikely that there is more than one valid species in the typical subgenus and this interpretation has been upheld to the present. Both Roberts (1951) and Ellerman *et al.* (1953) list the following subspecies: *P. (P.) fallax australis:* described from Mazambeti, Beira, Mozambique; *P. (P.) f. rhodesiae:* described from Machile River, a northern tributory of the Zambezi River; and *P. (P.) f. vumbae:* described from Vumba, southeastern Zimbabwe. *P. (P.) f. australis* may be a synonym of *P. (P.) f. rhodesiae,* an opinion expressed by Smithers & Lobão Tello (1976).

GENUS *Aethomys* Thomas, 1915
SUBGENUS *Aethomys* Thomas, 1915

The group of murids now included in *Aethomys* is rather difficult to classify. In 1941, Ellerman stated that it is one of the rather numerous African 'borderline' genera, overlapping on the one hand to some extent with *Rattus* and with *Arvicanthis* on the other.

As understood today, it is a large genus and Ellerman (1941) has listed 30 forms of six species. Over the years, many nomenclatural changes were proposed and implemented and for the sake of clarity the excellent açcount by Rosevear (1969) on the taxonomy of the genus is paraphrased below.

The earlier forms of the mice now assigned to *Aethomys* were originally ascribed to *Mus* Linn. and subsequently with the narrowing of that genus, to *Epimys* Trouessart in 1881. Gradually *Epimys* became as cumbersome as the previously all-embracing *Mus* and Thomas again tried to simplify the handling of *Epimys.* Consequently, in 1915 he proposed *Aethomys,* together with others, as a subgenus, raising it later to full generic status which was retained by Ellerman (1941) but was once more relegated to subgeneric status by Ellerman *et al.* (1953). Lundholm (1955a) and Meester *et al.* (1964) treat it as a full genus once again and Matthey (1958), from karyological data, also considered that it merited separate generic standing. Davis (1965) dealt with it in similar fashion and broadened it to cover *Stochomys* and *Micaelamys* as subgenera.

In his contribution to the Smithsonian Institution's *Preliminary identification manual for African mammals,* Davis (1968) followed Davis (1965) in placing *Stochomys* as a subgenus of *Aethomys*. By 1975, however, he concluded that there was every justification for removing *Stochomys* from *Aethomys*. '*Aethomys* is now divided into subgenus *Michaelamys* (with species *namaquensis* and *granti*) and the nominate subgenus *Aethomys*.' The latter contains a 'mixed bag' of forms, including *silindensis*, *nyikae* and *chrysophilus* which occur in southern Africa (Davis 1975b).

Members of this genus are large to medium-sized terrestrial rats, which may also inhabit crevices between rocks or in tree trunks. The light brown tips of the individual hairs allow the slate-grey colour of the bases to show through, giving the fur a pleasant mottled grey to ochre appearance which is gradually replaced by white-tipped hairs on the belly.

The fur is fairly long and well developed, not lying close to the body, and the animal is devoid of any special markings. The tail is brownish, long and conspicuous, scaly and covered with short, fine hairs through which the scales are clearly visible. The feet are not broadened.

The palatal foramina penetrate between the molars. The cusps of the first lamina of the $M^1$ are not marked by distortion, while the t7 on the $M^1$ is absent. There is no pronounced interorbital constriction, usually less than 17% of the greatest length of the skull. The skull is strongly built. The zygomatic plate has its front edge slightly concave with the antorbital foramen widely open above.

## Key to the southern African species of *Aethomys*
(After Davis (1975b))

This genus is distributed over much of the continent of Africa, south of the Sahara, and is especially common in central and eastern Africa. Although our interests are focused primarily on species occurring in southern Africa, the key by Davis (1975b) which I intend using to determine the southern African species, has the southern African forms so intertwined with other species from the rest of Africa, that it makes sense to use the key in its entirety.

1 Tail relatively shorter, less than 95% of HB length (70–95%); $M^1$ four-rooted ............................ 2
 Tail relatively longer, more than 95% of HB length (95–150%); $M^1$ either four-or ve-rooted..4

2 Incisors opisthodont (greatest length – condylobasal length = *ca* 2,5 mm); width of $M^1$ 2,2 mm or less; skull narrower, zygomatic width under 55% of greatest length........................................ 3
 Incisors orthodont (greatest length – condylobasal length = *ca* 1,5 mm); width of $M^1$ 2,3 mm and over; skull broader, zygomatic width over 55% of greatest length ..................... *Aethomys (Aethomys) kaiseri* (extralimital)

3 Five plantar pads; anterior palatal foramina extending to midline of inner alveoli of $M^{1-1}$ (Angola)...... *Aethomys (Aethomys) thomasi* (extralimital)
 Six plantar pads; anterior palatal foramina extending to, or just beyond, anterior margin of alveoli of $M^{1-1}$ (equatorial belt east and southeast of the Eastern Rift Valley) .................. *Aethomys (Aethomys) (hindei* group) (extralimital)

4 Size larger, greatest length of skull up to 43 mm ..... 5
 Size smaller, greatest length of skull not over 39 mm................................................. 7

5 Tail longer than HB length (108–120%) ............. 6
 Tail either slightly shorter or longer than HB length (95–110%) (equatorial belt west and north of Eastern Rift Valley)......... *Aethomys (Aethomys) hindei (medicatus* group) (extralimital)

6 Ear longer, about 25 mm (24–28); $M^1$ five-rooted (in type and a few others) or four-rooted (some in British Museum (Natural History); mammary formula: 0 – 2 = 4 ................... *Aethomys (Aethomys) bocagei* (extralimital)
 Ear shorter, 21 mm; $M^1$ five-rooted (in paratype): mammary formula unknown................ *Aethomys (Aethomys) silindensis* (Selinda Rat) Page 187

7 Size smallest, greatest length of skull under 35 mm; anterior median cusp (Sm) present and usually prominent.................................................. 8
 Size larger, greatest length of skull over 35 mm; anterior median cusp (Sm) absent, sometimes represented by an anterior cusplet (here called Smc)........................................................ 9

8 Tail longer than HB length (135%); tail bristles paler, less dense or dark towards tip; belly hairs white to roots or with greyish-white or grey tips, with grey base; incisors strongly opisthodont (greatest length – condylobasal length = *ca* 3,5 mm); Sm well developed as a rule (some exceptions in southern Cape) .... *Aethomys (Michaelamys) namaquensis* (Namaqua Rock Mouse) Page 188
 Tail slightly longer than HB length (115%); tail bristles darker, becoming rather denser towards tip; belly hairs grey-based, grey or greyish white at the tip; incisors moderately opisthodont (greatest length – condylobasal length = *ca* 2,5 mm); Sm always well developed ..... *Aethomys (Michaelamys) granti* (Grant's Veld Rat) Page 191

9 Tail longer than HB length (120%); skull narrower; zygomatic width less than 51% of greatest length; width of M$^1$ invariably 2,0 mm or less; incisors strongly opisthodont (greatest length – condylobasal length = *ca* 3,5 mm) ....................
*Aethomys (Aethomys)*
*chrysophilus*
(Red Veld Rat)
Page 182
Tail slightly shorter than HB length (95% – S. Malaŵi) or longer (115% – Zambia); width of M$^1$ 2,0 – 2,2 mm; incisors orthodont (greatest length – condylobasal length = 1,5 mm) ....................10
10 Hindfoot longer, *ca* 29 mm ... *Aethomys (Aethomys)*
*nyikae*
(Nyika Veld Rat)
Page 185
Hindfoot shorter, *ca* 26 mm ... *Aethomys (Aethomys)*
*(nyikae ?) dollmani* (extralimital)

# *Aethomys (Aethomys) chrysophilus* (De Winton 1897)

**Red Veld Rat**

**Rooi Veldrot**

The generic name is coined from the Greek *aithos* = sunburnt, the presumed root of the name Ethiopia for Africa in general, combined with the Greek *mys* = mouse. The species name is derived from a combination of the Greek *chrysos* = gold and *philos* = having affinity for. De Winton was evidently impressed with the warm 'golden' coloration shown by the dorsal surface.

The type specimen originally described as *Mus chrysophilus* is housed in the British Museum (Natural History), London, and was collected at Mazoe, Mashonaland, in eastern Zimbabwe. Allen (1939) states that the name *M(us) muscacardinus* Wagner, 1843 *(in Schreber's Säugethiere)* possibly supersedes *chrysophilus*, but this is presently undeterminable.

This species is larger than *A. (M.) namaquensis* and *A. (M.) granti*. Its distribution in southern Africa is not as extensive as that of *A. (M.) namaquensis* and hitherto the Red Veld Rat has not been recorded from the Cape Province south of the Orange River, the OFS or Lesotho (with few exceptions). Roberts (1951) has found that the eastern populations are larger and darker, and at high altitudes they are dark.

## Outline of synonymy

1897 *Mus chrysophilus* De Winton, *Proc. zool. Soc. Lond.* for 1896:801. Mazoe, Zimbabwe.

1908 *Mus chrysophilus ineptus* Thomas & Wroughton, *Proc. zool. Soc. Lond.:* 546. Tete, Mozambique.

1908 *M. c. acticola* Thomas & Wroughton, *Proc. zool. Soc. Lond.:* 547. Beira, Mozambique.

1909 *M. c. tzaneenensis* Jameson, *Ann. Mag. nat. Hist.* (8) 4:460. Tzaneen, Transvaal.

1913 *M. c. pretoriae* Roberts, *Ann. Transv. Mus.* 4:85. Pretoria, Transvaal.

1915 *Aethomys* Thomas, *Ann. Mag. nat. Hist.* (8) 16:477. Genotype *Epimys hindei*.

1926 *Aethomys chrysophilus magalakuini* Roberts, *Ann. Transv. Mus.* 11:254. Magalakwen River, Transvaal.

1926 *A. c. capricornis* Roberts, *Ann. Transv. Mus.* 11:254. Newgate, Soutpansberg, Transvaal.

1927 *A.c. imago* Thomas, *Proc. zool. Soc. Lond.:* 387. Stampriet, SWA/Namibia.

1931 *A. c. tongensis* Roberts, *Ann. Transv. Mus.* 14:235. Manguzi Forest, Natal.

1939 *A. chrysophilus chrysophilus* (De Winton). *In* Allen, *Bull. Mus. comp. Zool. Harv.* 83:368.

1946 *A. chrysophilus fouriei* Roberts, *Ann. Transv. Mus.* 20:319. Oshikanga, SWA/Namibia.

1946 *A. c. harei* Roberts, *Ann. Transv. Mus.* 20:320. Waterberg, SWA/Namibia.

1951 *A. c. chrysophilus* (De Winton). *In* Roberts, *The mammals of South Africa*.

1953 *Rattus (Aethomys) chrysophilus* De Winton. *In* Ellerman *et al., Southern African mammals*.

1975 *Aethomys chrysophilus* (De Winton). Davis, *in* Meester & Setzer (eds), *The mammals of Africa: an identification manual*.

## Identification

The Red Veld Rat is tawny-buff above with a sprinkling of black hairs showing through on the surface. The individual hairs are slate-coloured basally. In direct sunlight and when seen from a distance, the soft pelage has a golden to red hue, giving rise to both its scientific and vernacular names. The cheeks, throat, flanks and thighs are paler. The distal parts of both hands and feet are pure white, covered with fine, small hairs. The belly is a dirty white with the belly-hairs dark grey proximally (plate 10). Smithers (1971) points out that all-white-bellied individuals occur in a series of specimens from Botswana.

Roberts (1951) has also mentioned the occurrence of occasional erythrism in individuals.

The tail is subequal to or somewhat longer than the head-body length (120%, according to Davis (1975b)), almost naked but with a few short hairs increasing in number and length (very slightly) towards the tip of the tail. The tail is more coarsely scaled than that of the Namaqua Rock Rat *A. (M.) namaquensis*.

The fifth digit of the hindfoot is much shorter than the other digits.

The mammae consist of one pectoral and two inguinally situated pairs.

### TABLE 48
Measurements of male and female *Aethomys (Aethomys) chrysophilus*

|  | Parameter | Value (mm) | N | Range (mm) | CV (%) |
|---|---|---|---|---|---|
| Males | HB | 146 | 14 | 139–160 | 4,0 |
|  | T | 170 | 22 | 146–193 | 7,4 |
|  | HF | 29 | 24 | 27–33 | 5,4 |
|  | E | 21 | 22 | 20–27 | 8,1 |
|  | Mass: | 101 g | 6 | 87–112 g | 9,4 |
| Females | HB | 138 | 20 | 120–150 | 8,0 |
|  | T | 162 | 22 | 147–186 | 6,8 |
|  | HF | 28 | 23 | 26–32 | 5,0 |
|  | E | 20 | 24 | 17–24 | 9,0 |
|  | Mass: | 89,5 g | 7 | 82–95 g | 6,0 |

### Skull and dentition

The greatest length of the adult skull is more than 35 mm (Davis 1975b) and it is slightly arched from nasals to occipital when seen from the side.

The supraorbital ridges are strongly developed, while the interorbital constriction is not pronounced. The rostrum is heavy and deep. The zygomatic plate is cut back strongly and the anterior border is straight, or slightly concave in some specimens (fig. 81).

The anterior palatine foramina are very long, extending posteriorly to well between the first upper molars. The palate is broad, not ridged. The zygomatic arches are not particularly strongly built and the zygomatic width is less than 51% of the greatest skull length. The bullae are large, being 17–21% of the occipitonasal length. The greatest diameter of the bulla is subequal to or slightly more than the length of the upper tooth row.

The molars are broad, tending to be heavy and angular (fig. 82). The central cusps of the upper molars are enlarged. The t7 is absent. The $M^3$ is slightly reduced and often not very much smaller than the $M^2$. (In *A. (M.) namaquensis* the $M^3$ is constantly smaller than the $M^2$.) In the $M_1$ there is no anterior median cusp (cf. *A. (M.) namaquensis*) but this element is sometimes presented by a very small cusplet (Sm and Smc respectively).

As in *A. (M.) namaquensis*, the upper incisors of *A. (A.) chrysophilus* are also strongly opisthodont.

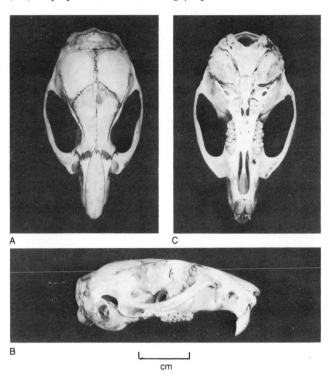

**Figure 81** Dorsal (A), lateral (B) and ventral (C) views of the skull of the Red Veld Rat, *Aethomys (Aethomys) chrysophilus*. Note the deep and heavy rostrum, the well-developed supraorbital and occipital ridges, the zygomatic plate which is cut back dorsally with an anterior border tending towards concavity, the broad palate and the relatively large bullae. The greatest diameter of the bulla is subequal to the length of the toothrow or longer. The incisors are strongly opisthodont. The interorbital constriction is gradual. The greatest length of the skull is over 35 mm.

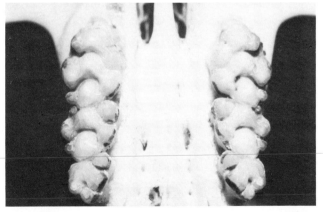

**Figure 82** Cheekteeth of the Red Veld Rat *Aethomys (Aethomys) chrysophilus*. The molars are heavy and angular with the four-rooted $M^1$ invariable less than 2 mm in width. The centre row of cusps is pronounced. The $M^3$ is hardly smaller than the $M^2$.

## Distribution

The distribution of this Southern Savanna Woodlands species can be described as the savanna bushveld and semi-desert scrub from southern Natal in the east, northwards through the Transvaal and Zimbabwe, ranging into eastern and northern Botswana and the northern parts of SWA/Namibia (map 53). It also occurs in Angola, Zambia, Malaŵi and further north into Kenya. In Mozambique it is widely distributed and common from the extreme south of the country to the northern parts of the Tete district (Smithers & Lobão Tello 1976).

In the west, these rats hardly occur south of latitude 23°S and west of longitude 21°E. Davis (1974) confirms Winterbottom (1973) who noted that latitude 23°S also marks the most southerly distribution of certain South West Arid birds. This observation also applies to *Praomys (Mastomys) natalensis* and *Lemniscomys griselda* apart from *A. (A.) chrysophilus*, but why this should be so is not yet clear.

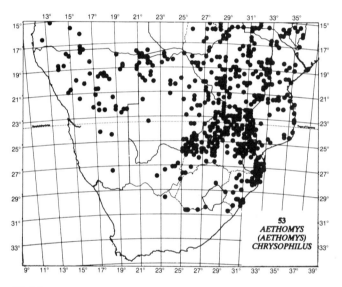

53
AETHOMYS
(AETHOMYS)
CHRYSOPHILUS

## Habitat

The Red Veld Rat is found in sheltering bush on the plains of the eastern interior of southern Africa, where it lives in burrows with inter-communicating runways under vegetation or rocks. It is also said that old termite mounds with their subterranean chambers are favourite points of convergence, especially if there is bush or shrub in the vicinity. In Botswana it occurs in grassland with open shrub associations, in open woodland, frequently occurring on the fringes of pans (Smithers 1971). Smithers also says that they may be taken in areas of sandy ground, or in sandy alluvium, or in areas where the ground is hard. The species in Zimbabwe occurs in dry *Acacia* scrub, as well as in the fringe vegetation of evergreen forest (Smithers 1975).

## Diet

The little information available is gleaned from Smithers (1971, 1975). As in so many cases, one has to rely heavily on the field work Smithers has done in both Botswana and Zimbabwe for information on the dietary requirements and preferences of many species of small mammals. In this case, they are seedeaters, feeding on grass seeds as well as other types of seed which they feed on *in situ*, but according to Smithers (1975) they will also gnaw on the kernels of nuts or drupes of wild fruits which have fallen to the ground. Roberts (*in* Shortridge 1934) states that they subsist mainly on grain but that they are omnivorous, for they will take any kind of bait in traps, of which they are less shy than most species. According to Jacobsen (1977) they also feed on the seeds of *Hibiscus engleri*.

## Habits

The Red Veld Rat is predominantly terrestrial and nocturnal. They are very active rats and Roberts (1951) maintains that they are more terrestrial in habits than *A. (M.) namaquensis*. Smithers (1975) says that while they are not adverse to burrowing, they do prefer to find refuge and nesting places in piles of debris or rocks, '. . . or in other types of substantial cover above ground'. Smithers (1971) has also observed that they seem to be particularly associated with thorn fences surrounding agricultural lands. Nestbuilding behaviour by *A. (A.) chrysophilus*, under experimental conditions, was compared with that of *Praomys (Mastomys) natalensis* and *Rhabdomys pumilio* by Stiemie & Nel (1973).

## Predators

No specific information is available. Kingdon (1974) states that they are '. . . presumably killed by small carnivores, by owls, hawks and snakes'.

## Reproduction

According to Smithers (1975) the Red Veld Rat breeds throughout the year in Zimbabwe, two to seven pups being born in a litter. In Botswana, gravid females have been taken in January, March, April, June, October and November, with an average number of fetuses 3,9 (N = 10, observed range 3–6) (Smithers 1971). The gestation period lasts 21–23 days and new-born young show nipple-clinging (Kingdon 1974; Choate 1972). Kern (1977) states that breeding appears to occur throughout the year in the Kruger National Park with a peak in summer.

## Parasites

The list of ecto- and endoparasites known from *A. (A.) chrysophilus* is formidable. The mites include the following: Laelaptidae – *Laelaps giganteus, L. muricola, L. simillimus, L. vansomereni, Haemolaelaps labuschagnei, H. taterae, Androlaelaps theseus, A. marshalli* and *A. zuluensis;* Trombiculidae – *Schoengastia radfordi radfordi;* Listrophoridae – *Listrophorus bothae* and *Gliricoptes lepidotus* (Zumpt 1961a). The brachycercid flies are also associated with this rodent and include the calliphorid *Cordylobia anthropophaga* (Zumpt 1966).

The fleas associated with *Aethomys (Aethomys) chrysophilus* are many. These comprise some five families, grouped as follows by Zumpt (1966): Pulicidae – *Echidnophaga gallinacea*, *E. larina*, *Ctenocephalides felis*, *Xenopsylla cheopis*, *X. frayi*, *X. geldenhuysi*, *X. hipponax*, *X. nubica*, *X. philoxera*, *S. phyllomae*, *X. bechuanae*, *X. brasiliensis*, *X. crinita*, *X. scopulifer*, *X. syngenis* and *X. erilli*. Pygiopsyllidae – *Stivalius torvus*; Hystrichopsyllidae – *Ctenophthalmus ominosus*, *C. acanthurus*, *C. ansorgei*, *C. calceatus*, *Dinopsyllus dirus*, *D. ellobius*, *D. grypurus*, *D. lypusus*, *D. pringlei*, *Listropsylla dorippae* and *L. prominens*; Leptopsyllidae – *Leptopsylla aethiopica thalia*; Chimaeropsyllidae – *Epirimia aganippes*, *Chiastopsylla rossi*, *C. godfreyi* and *C. octavii*.

The ticks include adult (A) or immature (I) forms of the following species (Theiler 1962): *Ornithodoros zumpti* (A,I), *Ixodes elongatus* (A), *I. pilosus* (I), *Aponomma latum* (I), *Haemaphysalis leachii leachii* (A,I), *H.l. muhsami* (A,I), *Rhipicephalus capensis* (I), *R. evertsi* (I), *R. pravus* (I) and *R. simus* (A,I).

The endoparasites, especially the cestodes, have been recorded by Collins (1972). These include *Catenotaenia compacta*, *C. lucida* and *Inermicapsifer madagascariensis*. Furthermore, the genera *Heligmonoides stellenboschius*, *Protospirura* and *Subulura* have also been collected from *A. (A.) chrysophilus* (Collins *pers. comm.*).

## Relations with man

On account of the well-developed ectoparasite fauna found on these rats, they must surely be reservoirs or vectors in the spread of disease and/or pathological conditions detrimental to man and his domestic stock. It has been described as the type host for the fleas *Xenopsylla phyllomae* and *Leptopsylla aethiopica thalia*, apart from being associated with fleas which are responsible for the distribution of plague, like *Chiastopsylla numae* form *rossi* (De Meillon, Davis & Hardy 1961). They sometimes enter houses, and can be very destructive, focussing attention on stored foodstuffs (Smithers 1975). Under laboratory conditions, they are good breeders although handling them is difficult (Keogh & Isaäcson 1978).

## Prehistory

Specimens referred to *Aethomys* cf. *chrysophilus* (*inter alia* a lower jaw with a full molar complement as well as two isolated upper first molars) have been recorded by De Graaff (1960a) from Middle Stone Age layers at the Cave of Hearths in the Makapansgat Valley.

## Taxonomy

Three subspecies, spread over the range of the African distribution of the species have been listed by Davis (1975b). There are *A. (A.) chrysophilus chrysophilus* (southern Africa, Angola, Zambia, southwestern Tanzania and Malaŵi), *A. (A.) c. singidae* (central and northern Tanzania) and *A. (A.) c. voi* (northeastern Tanzania and southeastern Kenya).

The following nine South Africans forms, described as subspecies for *chrysophilus*, are not allocated by Davis, but are probably synonyms: *acticola*, *capricornis*, *fouriei*, *harei*, *imago*, *ineptus*, *pretoriae*, *tongensis* and *tzaneenensis*.

# *Aethomys (Aethomys) nyikae* (Thomas, 1897)

**Nyika Veld Rat**

**Nyika Bosrot**

For the meaning of *Aethomys*, see *Aethomys (Aethomys) chrysophilus*. The species is named after the Nyika Plateau in northern Malaŵi.

I have not seen this species, nor have I handled study skins. Consequently, the following remarks have been taken from published information and additional background information was obtained from Dr D.H.S. Davis, by means of personal communication.

The known distribution of *A. (A.) nyikae* is described as northeastern Zambia, Malaŵi and eastern Zimbabwe. 'The status of *A. nyikae* has been clarified by Ansell & Ansell (1973), who note that it shares certain characteristics with *dollmani*, hitherto regarded as a subspecies of *A. chrysophilus*. Whether *dollmani* can be regarded as a subspecies of *nyikae* is still debatable, but the two taxa are much closer to each other than either is to *chrysophilus*' (Davis 1975b).

The inclusion of eastern Zimbabwe in the distribution of this species is based on a juvenile specimen trapped with similarly aged *A. chrysophilus* in the East Ingorima Reserve in Zimbabwe. These were sent to Davis by Dr R.H.N. Smithers for identification in 1965. Smithers reports *(pers. comm.)* that in spite of intensive subsequent collecting no additional specimens have been acquired. Davis (1975b) points out that since this is the first record of *A. nyikae* south of the Zambezi River, it '... must be accepted with reservation until a series is available'.

In 1897, Thomas described a new species of murid from the Nyika Plateau in northern Malaŵi as *Mus nyikae* (1897a). This was followed in 1907 by the description of *Mus walambae* by Wroughton, introducing this species as new to science from the vicinity of the Msofu River in the north of Zambia. In 1941, Ellerman discussed the genus *Aethomys* and as understood in that work, the genus was restricted to the species *thomasi*, *kaiseri*, *bocagei*, *chrysophilus*, *stannarius* and *walambae* and their immediate allies. It did not include the *namaquensis* group (as understood today) which was referred to the genus *Thallomys*. In Ellerman's listing *Mus walambae* appeared as *Aethomys walambae walambae* Wroughton, together with *A. w. amalae* Dollman (described from Kenya in 1914), *A. w. pedester* Thomas (described from Uganda in 1911) and *A. w. hintoni* Hatt (described from southern Zaïre in 1934)). Also listed as subspecies (name forms according to Ellerman 1941) were *Aethomys chrysophilus dollmani* Hatt (described from Katanga in southern Zaïre in 1934) and *A. c. nyikae* Thomas, previously known as *Mus nyikae*.

In 1953, Ellerman, Morrison-Scott & Hayman again relegated Thomas' subgenus *Aethomys* to subgeneric rank after Ellerman (1941) had previously retained it at full generic rank. It was again interpreted as a subgenus of *Rattus* and was interpreted to contain the following taxa: *Rattus (Aethomys) kaiseri, R. (A.) nyikae, R. (A.) chrysophilus, R. (A.) paedulcus* and *R. (A.) woosnami*.

Ellerman *et al.* (1953) state that *R. (A.) nyikae* '... is better known as *walambae;* but apparently *nyikae* belongs to what has hitherto been regarded as *walambae,* and it is the prior name'. They continue by listing three subspecies for *nyikae* as *Rattus nyikae nyikae* Thomas, 1897, *R. n. walambae* Wroughton, 1907 and *R. n. hintoni* Hatt, 1934.

The genus *Aethomys* was discussed by Davis (1975b). Firstly, he points out that Ansell & Ansell (1973) established for the first time that *nyikae* is a good species based on a series of specimens from the Mafringa Mountains in northeastern Zambia. Secondly, he reiterates that Ellerman *et al.* (1953) included *dollmani* as a 'race' of *chrysophilus.* Davis proposed the removal of *dollmani* from *chrysophilus* following Ansell & Ansell (1973), who suggest that it may be conspecific with *nyikae.* Davis consequently treated *dollmani* as a possible subspecies of *nyikae,* '... but a final decision as to its true status must be awaited'. Ansell & Ansell (1973) have pointed out that Hanney (1965) must have been dealing with a mixture of *chrysophilus* and *nyikae* under the name *chrysophilus* in his study of the murids of Malaŵi in which he understood *nyikae* to be a synonym. A re-examination of specimens obtained by G.C. Shortridge in southern Malaŵi by Davis and Coetzee (*pers. obs.*) confirms the conclusion that *nyikae* is separate from *chrysophilus.*

It must also be noted that *dollmani* and *chrysophilus* are allopatric, while *nyikae* and *chrysophilus* are sympatric species.

## Outline of synonymy
1897 *Mus nyikae* Thomas, *Proc. zool. Soc. Lond.*: 431. Nyika Plateau, Malaŵi.
1907 *Mus walambae* Wroughton, *Mem. & Proc. Manchester Lit. and Phil. Soc.* 51 (5):21. Msofu River, Zambia.
1941 *Aethomys chrysophilus nyikae* (Thomas). *In* Ellerman, *The families and genera of living rodents.*
1953 *Rattus (Aethomys) nyikae* (Thomas). *In* Ellerman *et al., Southern African mammals.*
1975 *Aethomys nyikae* (Thomas). Davis, *in* Meester & Setzer (eds), *The mammals of Africa: an identification manual.*

## Identification
I am not in a position to give data on dimensions of the head-body, tail, hindfoot and ear. The hindfoot has the fifth digit scarcely longer than the hallux and barely reaching the base of the fourth. The tail is relatively long, more than 95% of the head-body length (95–150%), but shorter than encountered in *A. (A.) chrysophilus* (where the corresponding value attains about 120%) (plate 10).

The hindfoot has been described as terrestrial with the fifth digit relatively short. The mammae consist of two inguinally situated pairs.

## Skull and dentition
According to Ellerman *et al.* (1953) the zygomatic width in 'South African' skulls is not below 19 mm, averaging 54% of the occipitonasal length. I take it that this would have to read 'southern African' skulls, for the species does not occur in South Africa. Davis (1975b) states that the skull is more than 35 mm long. The supraorbital ridges are well developed.

The upper incisors are orthodont. In the $M_1$, the anterior median cusp (Sm) is absent, sometimes represented by a small anterior cusplet (Smc).

## Distribution
As here understood *A. (A.) nyikae* occurs in eastern Zimbabwe, extending northwards to northern Malaŵi and westwards to northeastern Zambia (map 54).

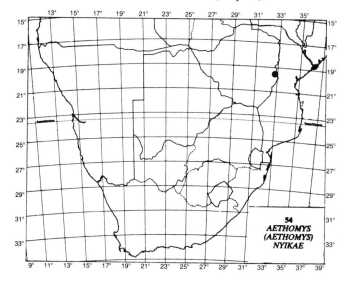

54
AETHOMYS
(AETHOMYS)
NYIKAE

## Habitat, diet, habits, predators, reproduction, parasites, relations with man, and prehistory
Little information is available. *A. (A.) nyikae* is believed to be communal or semi-communal, terrestrial and nocturnal and it frequents termite hills in open woodland.

## Taxonomy
This aspect has been dealt with in the introductory section to this species. It is sufficient to state that the status of

*dollmani* as a subspecies of *nyikae* is doubtful at present and therefore Davis (1975b) refers to it as *A. (A.) (nyikae?) dollmani* to accentuate this uncertainty.

# *Aethomys (Aethomys) silindensis* Roberts, 1938

**Selinda Veld Rat**

**Selinda Bosrot**

For the meaning of *Aethomys,* see *Aethomys (Aethomys) chrysophilus.* The species is named after Mount Selinda, where two specimens were taken in 1938 by A.G. White of the Transvaal Museum.

This is yet another species occurring in the southern African region, which is known from one or two specimens only. This species is founded on two specimens, the type and paratype respectively, both males (Davis 1975b). Despite intensive fieldwork near the Chirinda Forest on Mount Selinda and environs (where the type material was collected) no additional specimens have to date been collected and the status of the species can be properly resolved only with the aid of further material (Smithers 1975). The type was first described as *Aethomys silindensis* by Roberts in 1938 and though only known from the type locality it doubtless extends northwards along the escarpment in eastern Zimbabwe (Roberts 1951).

It is considerably larger than the other *Aethomys* species and it looks very much like *A. (A.) chrysophilus* externally – to such an extent that Smithers (1975) calls it 'indistinguishable'. Differences which exist, pertain to the skull and dentition. In this species the $M^1$ is five-rooted compared to the four-rooted condition of *chrysophilus.*

## Outline of synonymy

1938 *Aethomys silindensis* Roberts *Ann. Transv. Mus.* 19:245. Mount Selinda, Zimbabwe.

1951 *Aethomys silindensis* Roberts. *In* Roberts, *The Mammals of South Africa.*

1953 *Rattus chrysophilus silindensis* (Roberts). *In* Ellerman *et al., Southern African mammals.*

1964 *Aethomys silindensis* Roberts. *In* Meester *et al., An interim classification of southern African mammals.*

1975 *Aethomys silindensis* Roberts. Davis, *in* Meester & Setzer (eds), *The mammals of Africa: an identification manual.*

## Identification

Only the type and a paratype specimen exist and external measurements of the latter were not taken when it was collected. The type specimen measures as follows: HB 160 mm, T 194 mm, HF (s.u.) 32 mm, E 21 mm (after Roberts 1951). The dorsal colour is buffy-brown with a few black-tipped hairs over the upper parts of the head, neck and rump. The flanks are paler and not sharply marked off from the buffy-white ventral surfaces. The hands and feet are also off-white. The tail is brownish above, lighter below, with coarse rings and sparse bristling (plate 10). The tail is longer than the head-body length (108–120%) (Davis 1975b).

The mammary formula is unknown.

## Skull and dentition

The greatest length of the stoutly built skull exceeds 40 mm, while the zygomatic width exceeds 20 mm (Meester *et al.* 1964). The ridges on the skull (or 'beading' as Roberts (1951) calls it), are pronounced and extend posteriorly along the frontals and parietals to a point where it meets the downward occipital crest. The $M^1$ is five-rooted (in the paratype), while the molars are large with the anterior cusp of the $M^3$ (t1) conspicuous and isolated.

## Distribution

Known only from Mount Selinda in eastern Zimbabwe, but it may extend northwards along the escarpment (Roberts 1951) (map 55).

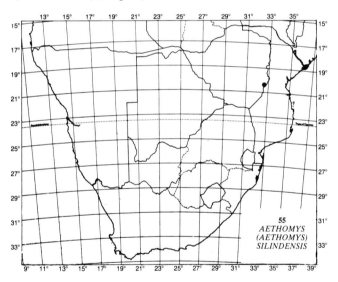

## Habitat, diet, habits, predators, reproduction, parasites, relations with man, and pre-history

No data available.

## Taxonomy

Roberts (1951) lists *A. silindensis* as a separate species. Ellerman *et al.* 1954) regard ths species as a subspecies of *A. (A.) chrysophilus,* but Davis points out that it should preferably be regarded as a separate species on account of its larger size and the five-rooted condition of the $M^1$.

*Micaelamys* Ellerman, 1941

# *Aethomys (Micaelamys) namaquensis* (A. Smith, 1834)

**Namaqua Rock Mouse**

**Namakwa Klipmuis**

For the meaning of *Aethomys*, see *Aethomys (Aethomys) chrysophilus*. The species was originally collected in Namaqualand as the specific name indicates.

The type specimen collected by Smith (1834) in Namaqualand and originally named *Gerbillus namaquensis* is housed in the British Museum (Natural History), London. This species outnumbers all other small rodents in rocky situations in SWA/Namibia (Shortridge 1934).

Its presence is often denoted by the crevices in and between rocks being stuffed with long pieces of grass stems and sticks, behind which it constructs a nest of grass shreds (Roberts *in* Shortridge 1934).

Smithers (1971) correctly writes that the species is in need of revision. It shows a very extensive geographical distribution, although conspicuously absent from the Great and Upper Karoo. Consequently, this species is represented in South African study collections in considerable numbers.

This species was originally placed under *Praomys* by Thomas (1915b) and consequently referred to *Aethomys* by Thomas during the same year. Ellerman (1941) regarded the *namaquensis* group as belonging to the genus *Thallomys,* stating that *namaquensis* closely connects *Thallomys, Aethomys* and *Praomys* with *Rattus* and with each other. He considers that all true African rats are closely allied, '. . . their relationships frequently being hidden under a bewildering number of generic names which are for the most part little more than well-marked specific groups of *Rattus*'.

## Outline of synonymy

1834 *Gerbillus namaquensis* A. Smith, *S. Afr. Quart. Journ.* 2:160. Little Namaqualand, CP.

1836 *Mus lehocla* A. Smith, *App. Rept. Exped. Explor. S. Afr.*: 43. Near Kuruman, CP.

1852 *Mus arborarius* Peters, *Reise nach Mossambique. Säugeth.*: 152. Tete, Mozambique.

1896 *Mus namaquensis* De Winton, *Proc. zool. Soc. Lond.*: 798.

1897 *Mus auricomis* De Winton, *Proc. zool. Soc. Lond.* for 1896:802. Mazoe, Zimbabwe.

1906 *Mus auricomis centralis* Schwann, *Proc. zool. Soc. Lond.*: 107. Deelfontein, CP.

1908 *Mus namaquensis lehocla* Thomas & Wroughton, *Proc. zool. Soc. Lond.*: 548.

1908 *Mus namaquensis centralis* Thomas & Wroughton, *Proc. zool. Soc. Lond.*: 548.

1908 *Mus avarillus* Thomas & Wroughton, *Proc. zool. Soc. Lond.*: 547. Tete, Mozambique.

1909 *Mus albiventer* Jentink, *Zool. Jahrb. Syst.* 28:246. Mosselbay, CP.

1909 *Mus namaquensis monticularis* Jameson, *Ann. Mag.nat. Hist.* (8)4:461. Johannesburg, Transvaal.

1915 *M. n. grahami* Roberts, *Ann. Transv. Mus.* 5:118. Grahamstown, CP.

1917 *Epimys (Aethomys) namaquensis monticularis* Roberts, *Ann. Transv. Mus.* 6:112.

1926 *Aethomys namaquensis calarius* Thomas, *Ann. Mag. nat. Hist.* (9)17:184. Lehutitung, Botswana.

1926 *A. n. siccatus* Thomas, *Proc. zool. Soc. Lond.*: 304. Ruacana Falls, Cunene River, SWA/Namibia.

1926 *Praomys namaquensis capensis* Roberts, *Ann. Transv. Mus.* 11:254. Lormarin's, Paarl, CP.

1926 *P. n. klaverensis* Roberts, *Ann. Transv. Mus.* 11:254. Klaver, CP.

1926 *P. n. drakensbergi* Roberts, *Ann. Transv. Mus.* 11:254. Utrecht, Natal.

1926 *P. n. lehochloides* Roberts, *Ann. Transv. Mus.* 11:255. Magalakwen River, Transvaal.

1938 *Aethomys namaquensis waterbergensis* Roberts, *Ann. Transv. Mus.* 19:239. Waterberg, SWA/Namibia.

1946 *Aethomys namaquensis namibensis* Roberts, *Ann. Transv. Mus.* 20:320. Karub, SWA/Namibia.

1951 *Aethomys namaquensis namaquensis* (A. Smith). *In* Roberts, *The mammals of South Africa.*

1953 *Rattus (Praomys) namaquensis* (A. Smith). *In* Ellerman *et al., Southern African mammals.*

1975 *Aethomys namaquensis* (A. Smith). Davis, *in* Meester & Setzer (eds), *The mammals of Africa: an identification manual.*

### TABLE 49

**Measurements of male and female *Aethomys (Micaelamys) namaquensis***

| | Parameter | Value (mm) | N | Range (mm) | CV (%) |
|---|---|---|---|---|---|
| Males | HB | 121 | 28 | 110–137 | 5,2 |
| | T | 151 | 31 | 127–168 | 6,8 |
| | HF | 25 | 31 | 21–28 | 5,1 |
| | E | 17 | 29 | 15–21 | 8,5 |
| | Mass: | 48,0 g | 30 | 38–75 g | 18,5 |
| Females | HB | 118 | 29 | 105–130 | 6,7 |
| | T | 157 | 29 | 132–168 | 7,1 |
| | HF | 23 | 31 | 23–27 | 4,6 |
| | E | 17 | 28 | 15–20 | 7,2 |
| | Mass: | 44,1 g | 28 | 33–57 g | 14,5 |

The only other comparative weights to be found in the southern African literature are given by Wilson (1975) on material from the Wankie National Park in Zimbabwe. He found males to weigh 67 g (N=6, observed range 52–81 g) and females 70 g (N=7, observed range 52–85 g).

## Identification

This species is smaller than the Selinda Rat *Aethomys (A.) silindensis* which is also larger than the intermediate-sized Red Veld Rat *A. (A.) chrysophilus*. Both these species have also been discussed individually in previous sections. Compared to its close relative, the Red Veld Rat, it is a much more lightly built species with a smaller body, head, a thinner and much longer tail '. . . which is well over half the overall length, the scales finer and more closely spaced' (Smithers 1975). However, the characteristic feature that strongly distinguishes it from *A. (A.) chrysophilus*, is the pure white underparts. In some cases there are traces of grey-based hairs on the upper chest and flanks.

The dorsal colour is reddish-brown, while the ventral surface is grey-white or pure white. The belly hairs are white to the roots or may be grey to the base with greyish-white or grey tips. The fur is fairly long and soft. The chin, lower neck, throat and limbs are also white (plate 10).

The ears are medium-sized and broad with a thin covering of hair on the outer surface. The tail is longer than the head-body length with fine bristles covering the distinct scales, which are arranged in close succession.

Shortridge (1934) mentions a variation in tail length among different populations in SWA/Namibia.

The feet are well developed, not particularly large, while the fifth digit is relatively long.

The mammae consist of one pectoral pair and two inguinal pairs.

## Skull and dentition

The greatest length of the skull is less than 35 mm and the interorbital constriction is not pronounced. The supraorbital ridges (like other 'ridges' on the skull) are slight. The bullae are small and flattened and their greatest diameter is less than the length of the upper molar tooth row. The anterior edge of the zygomatic plate is usually vertical but occasionally tends towards concavity. The antorbital foramen is wide (fig. 83).

The molars are well cusped and relatively heavy (fig. 84). In the $M^1$ the t2 is very prominent but smaller than the t2 shown by Grant's Rat *A. (M.) granti*. The $M^3$ is consistently smaller than the $M^2$ (this is not always the case in the Red Veld Rat *A. (A.) chrysophilus*).

The $M_1$ has a well-developed anterior median cusp (Sm).

The upper incisors are strongly opisthodont (fig. 83).

## Distribution

This species occurs throughout southern Africa, except the central Karoo (map 56).

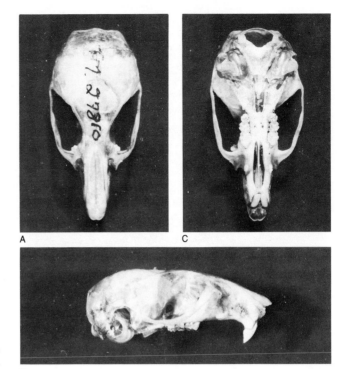

**Figure 83** Dorsal (A), lateral (B) and ventral (C) views of the skull of the Namaqua Rock Mouse *Aethomys (Micaelamys) namaquensis*. The supraorbital and occipital ridges are slight. The greatest diameter of the bulla is less than the molar row length. The anterior edge of the zygomatic plate tends to be concave. The incisors are slightly opisthodont.

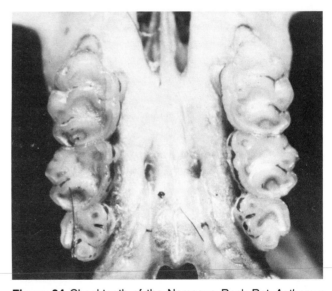

**Figure 84** Cheekteeth of the Namaqua Rock Rat *Aethomys (Micaelamys) namaquensis*. The molars are well cusped and heavily built. The t2 in the four-rooted $M^1$ is prominent. The $M^3$ is consistently smaller than the $M^2$.

It extends north-east across the Zambezi River into Zambia and southern Malaŵi, while in the northwest it occurs to the north of the Cunene River well into central

189

Angola. According to Smithers (1971) it has not yet been taken in the Caprivi Strip; it has been collected on the fringes of the Okavango Delta, but not within the delta itself. Mozambique records occur from the extreme south of the Maputo and Inhambane districts northwards along the eastern border of Mozambique to the Tete district. There are no records from the northern coastal or central areas (Smithers & Lobão Tello 1976).

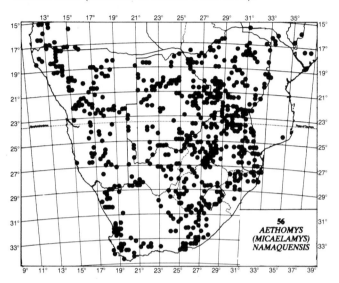

## Habitat

The wide geographical distribution of this species indicates a wide habitat tolerance. It lives in cracks and crevices of rocky koppies and outcrops, or in piles of stones in the veld, lowlying ridges and stony country and is often plentiful in old ruins. In 1926 Thomas (1926b) described a new subspecies, *siccatus*, from the Ruacana Falls on the Cunene River. Its range includes the Kaokoveld and the northern part of SWA/Namibia. This subspecies tends to inhabit more level country, spreading into many parts of the open sandveld.

A second subspecies, *calarius*, also described by Thomas (1926a) from Lehutitung in Botswana, has also adapted to open sandveld. Shortridge (1934) states that large domed nests are constructed of small sticks and grass and '... even the nests of weaver-birds being occasionally dragged from low branches and brought into use'. In Botswana *A. (M.) namaquensis* occurs in areas of rocky koppies (in the east), in open shrub with scattered trees (in the southwest), in riverside woodland, on fringes of pans where there are trees, bushes or calcareous outcrops (Smithers 1971). In Zimbabwe they also prefer rocky places. When rocks are not available, they will use holes in trees, particularly *Acacia* and *Sclerocarya* (Smith 1975).

## Diet

The species is granivorous (Smithers 1971). Roberts (1951) thought that it may be omnivorous as it can be trapped with almost any kind of bait. In SWA/Namibia,

Shortridge (1934) mentions that they feed largely on the seed of camelthorn and other leguminous shrubs, and that they are not attracted to cultivation. In the Nylsvley Nature Reserve in the central Transvaal they feed on the seeds of *Diodea natalensis* and *Pavonia transvaalensis* (Jacoben 1977).

## Habits

Namaqua rock rats are communal, partly arboreal but mainly terrestrial. They are predominantly nocturnal, leaving their abodes at dusk. The nests are built of grass stalks and other grasslike material stuffed into the rock crevices, by which the presence of the species is immediately given away. In other cases it piles straw up round the base of trees in which '... there are crevices, or fills the crevices quite high up the stems' (Roberts 1951). These crevices may be occupied by a number of family parties. The entrances to these tangled structures are holes just large enough for them to slip through.

The subspecies *calarius* builds large communal nests over burrows in the sand. These nests may measure 120 cm in diameter at the base and may attain a height of some 60–90 cm. Shortridge (1934) says that these nests must inevitably be destroyed by veld-fires in which case the inmates can retreat into the subterranean burrows. Stray individuals of the Multimammate Mouse *(Praomys (M.) natalensis)* occasionally find themselves within these nests '... associating on apparently friendly terms ...' with the usual inhabitants.

Additional interesting data have been published by Wilson (1975) on trapping results in the Wankie National Park in Zimbabwe; seven specimens were taken from under a single rock over a period of four days. Each individual was trapped between 22h00 and 06h00, with no sign of activity before 22h00.

The Namaqua Rock Rat seems to be more communal than the Red Veld Rat *A. (A.) chrysophilus*, ten or even more utilising the same hollow (Smithers 1975) and they are possibly more active as well (Shortridge 1934).

## Predators

Virtually nothing is known. Shortridge (1934) has mentioned probable predation by snakes and mongooses.

## Reproduction

Shortridge (1934) suggests that they breed throughout the year. In contrast, Smithers (1971) states that gravid females were taken in October and from January to May, without any signs of breeding during June to September (winter) in Botswana. Two to five pups per litter have been recorded (August to September) in Great Namaqualand and a gravid female in the Kaokoveld in March (Shortridge 1934). Smithers has recorded litters of two to seven pups. Wilson (1975) reports gravid females taken in October and February in Wankie National Park, Zimbabwe. In the subspecies *calarius*, the young cling to the

Plate 9

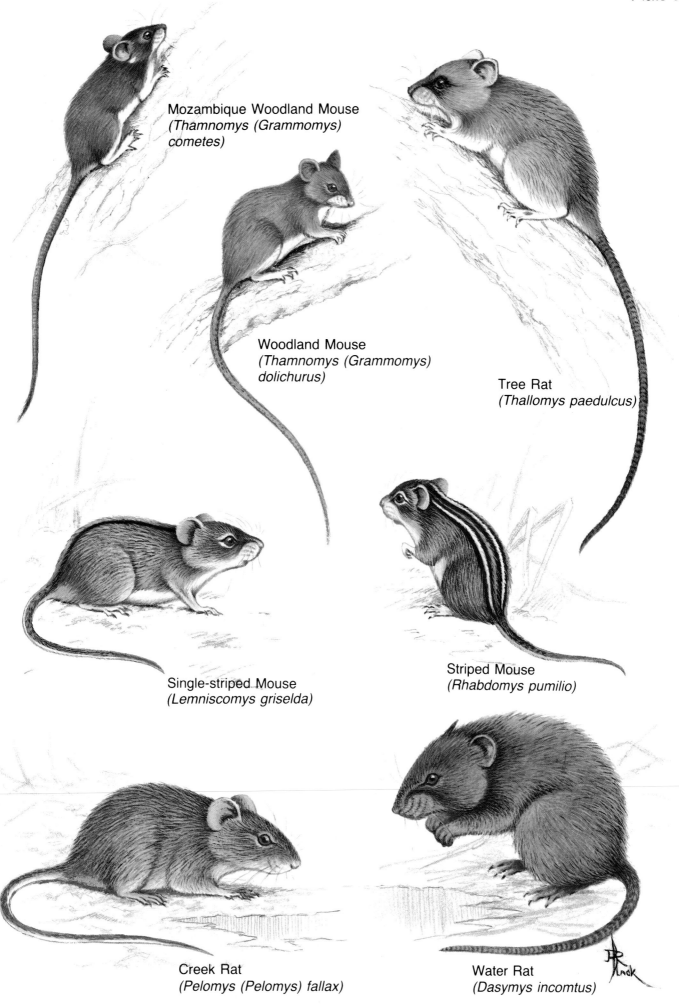

Mozambique Woodland Mouse
*(Thamnomys (Grammomys) cometes)*

Woodland Mouse
*(Thamnomys (Grammomys) dolichurus)*

Tree Rat
*(Thallomys paedulcus)*

Single-striped Mouse
*(Lemniscomys griselda)*

Striped Mouse
*(Rhabdomys pumilio)*

Creek Rat
*(Pelomys (Pelomys) fallax)*

Water Rat
*(Dasymys incomtus)*

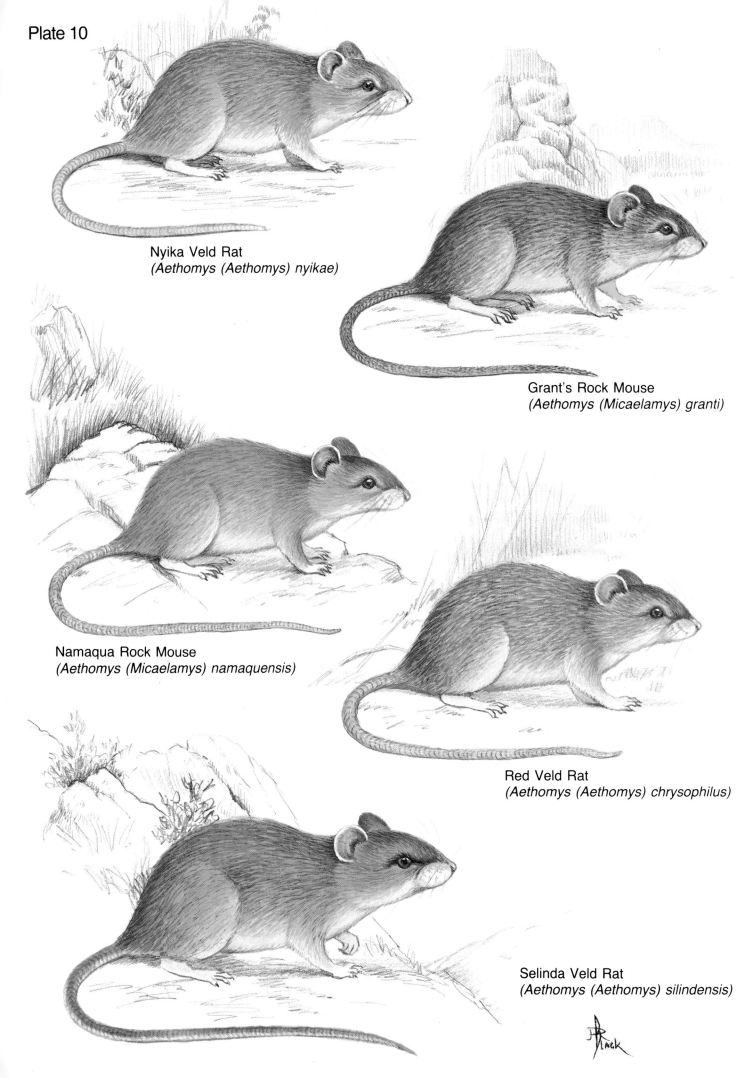

Plate 10

Nyika Veld Rat
(*Aethomys (Aethomys) nyikae*)

Grant's Rock Mouse
(*Aethomys (Micaelamys) granti*)

Namaqua Rock Mouse
(*Aethomys (Micaelamys) namaquensis*)

Red Veld Rat
(*Aethomys (Aethomys) chrysophilus*)

Selinda Veld Rat
(*Aethomys (Aethomys) silindensis*)

mammae while the female is on the move (Shortridge 1934; Choate 1972). Specimen cards in the Transvaal Museum show data on two embryos in a female in May and four and five embryos in two females in October from two localities in the northwestern Transvaal (Huwi 2327 Db and Thabazimbi 2427 Da) respectively. Meester (1958) has also reported on a litter of *A. (M.) namaquensis* born in captivity.

## Parasites

The list of ectoparasites collected from this species is impressive. The mites are represented by two families, the Laelaptidae and the Trombiculidae. The laelaptids include *Laelaps giganteus, L. muricola, L. simillimus, L. vansomereni, Haemolaelaps labuschagnei, Androlaelaps marshalli* and *A. zuluensis*. The trombiculids are known by *Euschoengastia aethomyia, E. longispina, Gahrliepia nana, Acomatacarus lawrencei* and *A. thallomyia* (Zumpt 1961a).

The fleas of *A. (M.) namaquensis* include 34 species. These include: (i) the Pulicidae – *Echidnophaga gallinacea, Ctenocephalides connatus, C. felis, Xenopsylla cheopis, X. hipponax, X. philoxera, X. piriei, A. trifaria, X. versuta, X. bechuanae, X. brasiliensis, X. scopulifer* and *X. zumpti,* (ii) the Hystrichopsyllidae – *Ctenophthalmus calceatus, Dinopsyllus abaris, D. ellobius, D. lypusus, Listropsylla agrippinae, L. aricinae, L. cerrita, L. dorippae* and *L. prominens,* (iii) the Leptopsyllidae – *Leptopsylla segnis* and (iv) the Chimaeropsyllidae – *Epirimia aganippes, Demeillonia granti, Chiastopsylla nama, C. rossi, C. carus, C. coraxis, C. gariepensis, C. godfreyi, C. mulleri, C. octavii* and *Praopsylla powelli* (Zumpt 1966).

Ticks (both soft and hard) have been recorded by Theiler (1962) and include adults (A) and/or immatures (I) of the following species: *Ornithodoros zumpti* (A,I), *Ixodes* sp. (A,), *I. alluaudi* (A,I), *Haemaphysalis l. leachii* (A,I), *H. l. muhsami* (A,I), *Hyalomma rufipes* (I), *Rhipicephalus capensis* (I), *R. oculatus* (I), *R. pravus, R. sanguineus, R. simus, R. theileri,* and *R. tricuspis*.

The following endoparasites (i.e. cestodes) have been recorded by Collins (1972): *Catenotaenia lucida, Inermicapsifer madagascariensis* and *Raillietina (R.) trapezoides*. Unpublished Onderstepoort records also include *Cysticercus parva* and *Capillaria hepatica* (Boomker *pers. comm.)*

## Relations with man

Judging from the list of ectoparasites, this species could be a reservoir of diseases spread by fleas and ticks. The species has not been successfully bred in the animal house of the South African Institute for Medical Research (Keogh & Isaäcson 1978), although it seems to adapt to captive conditions.

## Prehistory

An isolated left M$_1$ and three isolated M$^1$'s were identified as *Aethomys* cf. *namaquensis* by De Graaff (1961) from Sterkfontein and the Makapansgat Limeworks Dumps

respectively. It is strange that this species, now so widespread, should be so poorly represented in these breccias.

## Taxonomy

Roberts (1951) lists 14 subspecies, while Ellerman *et al.* (1953) list 15 subspecies. The present subdivision is not final and more subspecies might prove to be valid than *A. (M.) namaquensis namaquensis* and *A. (M.) n. arborarius* recognised by Meester *et al.* (1964). The former (i.e. for populations with grey base to the belly hairs) occurs in the southern Transvaal, OFS, Natal, Cape Province and the southern parts of SWA/Namibia while the latter (i.e. white-coloured belly hairs) occurs in the north of SWA/ Namibia, Botswana, northern Transvaal, Zimbabwe and Mozambique. Davis (1975b) has included them all under *A. (M.) namaquensis* and it therefore contains *arborarius, calarius, capensis, centralis, drakensbergi, grahami, klaverensis, lehocla, 'lehochloides, monticularis, namibensis, siccatus* and *waterbergensis* as synonyms.

# Aethomys (Micaelamys) granti (Wroughton, 1908)

**Grant's Rock Mouse**

**Grant se Klipmuis**

For the meaning of *Aethomys*, see *Aethomys (Aethomys) chrysophilus*. The subgeneric name was proposed by Ellerman (1941) named after a character in the opera 'Carmen'. The species name refers to Mr C.H.B. Grant, who collected the type specimen (an adult female) on 2 February 1902.

The status of this species is uncertain and the information available about its biology is meagre. The type specimen was originally described by Wroughton in 1908 from Deelfontein in the Cape Colony as *Mus granti*.

Meester *et al.* (1964) have succinctly described the rather stormy taxonomic passage this species has had.

Allen (1939) placed it in the genus *Myomys* (=*Praomys*), while Ellerman (1941) has created the subgeneric rank *Micaelamys* for it within the genus *Rattus*. Roberts (1951) considers the animal to be a synonym of *Mastomys natalensis natalensis* (now *Praomys (Mastomys) n. natalensis),* while Ellerman *et al.* (1953) list it as being of uncertain status without attempting to assign it to any particular group. Meester *et al.* (1964) list it as a separate species under *Aethomys*, while Davis (1975b) has again accepted the validity of *Micaelamys* by referring to it as *Aethomys (Micaelamys) granti*.

Part of the uncertainty as to how this species is to be interpreted taxonomically, stems from the fact that three rather closely related species of murids occurring together at Deelfontein, were represented in the British Museum collections. These include 'Rattus' natalensis (=Praomys (Mastomys) natalensis), Rattus namaquensis (=Aethomys (Micaelamys) namaquensis) and 'Rattus' granti (=A. (M.) granti). As indicated above, Roberts (1951) referred granti to the synonomy of Mastomys natalensis on account of the fact that its dental characteristics do not appear to differ very much from Mastomys and that it was assumed to have ten mammae only. This was refuted by Ellerman et al. (1953), who pointed out that Roberts had not seen the original series on which the name granti was based and that one specimen of this original series was labelled as having ten mammae. In Mastomys there are usually more than six pairs of mammae in contrast to two to three pairs in namaquensis. Furthermore, the dentition of granti is closer to that of namaquensis than to natalensis and it also agrees with namaquensis in having a black, well-haired tail in contrast to natalensis, where the tail is not black and hairy. It does approximate natalensis with its shorter tail, rather than namaquensis, which has a long tail. The suggestion then put forward by Ellerman et al. (1953) is that '. . . the original series might possibly have been hybrids between R. natalensis and R. namaquensis, both of which occur in the same locality, as it seems so precisely intermediate between them in a number of ways . . .'

## Outline of synonymy

1908 Mus granti Wroughton, Ann. Mag. nat. Hist. 1:257. Deelfontein, CP.

1915 Aethomys Thomas, Ann. Mag. nat. Hist. (8) 16:477. (Genotype Epimys hindei Thomas.)

1939 Myomys granti (Wroughton). In Allen, Bull. Mus. comp. Zool. Harv. 83:406.

1941 Rattus (Micaelamys) granti (Wroughton). In Ellerman, The families and genera of living rodents.

1951 Mastomys natalensis natalensis (part). In Roberts, The mammals of South Africa.

1953 Mus granti Wroughton. In Ellerman et al., Southern African mammals. (Of uncertain status.)

1964 Aethomys granti (Wroughton). Meester et al., An interim classification of southern African mammals.

1975 Aethomys (Micaelamys) granti (Wroughton). Davis, in Meester & Setzer (eds), The mammals of Africa: an identification manual.

This species is poorly represented in South African study collections, whether through misidentification or paucity of material, is not clear. I came across a small series of this species in the collection of the Port Elizabeth Museum and am satisfied that this sample (four females, one male) collected at Gouna Farm, some 86 km (54 miles) east of Calvinia may have been correctly identified as Aethomys (M.) granti.

TABLE 50
Measurements of male and female Aethomys (Micaelamys) granti

| | Parameter | Value (mm) | N | Range (mm) | CV (%) |
|---|---|---|---|---|---|
| Male | HB | 126 | 1 | — | — |
| | T | — | | — | — |
| | HF | 24 | 1 | — | — |
| | E | 18 | 1 | — | — |
| | Mass: No data available | | | | |
| Females | HB | 96 | 3 | 88–105 | — |
| | T | 101 | 3 | 91–108 | — |
| | HF | 23 | 3 | 22–24 | — |
| | E | 16,5 | 3 | 16–17 | — |
| | Mass: No data available | | | | |

## Identification

The dorsal colour is grey-brown, the belly-hairs grey based, with their tips grey or greyish-white. The tail is relatively long, slightly longer than the head-body length (115%) and reasonably well haired with the dark-coloured bristles becoming denser towards the tip (plate 10).

The mammae consist of three pectoral pairs and two inguinal pairs according to Ellerman (1941), while Roberts (1951) described it as to consist of one pectoral and two inguinal pairs.

## Skull and dentition

The greatest length of the skull does not exceed 35 mm. The supraorbital ridges are weak and the size of the bullae is approximately 17% of the occipitonasal length. The zygomatic plate is cut back above, somewhat reminiscent of the condition in Rattus.

The molars are broad, heavy, complex in cusp structure and angular, unlike any other 'Rattus' (fig. 85).

Ellerman (1941) provided a good description of the molar structure. The four-rooted $M^1$ has the t1, t2 and t3 well developed lying more or less in a straight line with the t2 very prominent. The same applies to the second lamina of the $M^1$ which is broad and which has t4 and t6 jutting outwards lingually and buccally respectively. The t8 on lamina three is broadened with the t9 of medium size.

In the $M^2$ the t1 is large and the second lamina repeats the structure as in the $M^1$, only more exaggerated. The t9 of the third lamina also tends to project outwards. The entire $M^2$ is somewhat broader than long.

The $M^3$ is nearly as large as the $M^2$ with its centre lamina bent backwards on each side (t4 and t6), overlapping the last lamina which is two-cusped.

It is not surprising that this pattern resists wear and

persists well into old age. The molar tooth row is longer than the greatest diameter of the bulla.

The lower molars develop strong subsidiary cusps, while the terminal heel of both $M_1$ and $M_2$ tend to be suppressed. An anterior medium cusp (Sm) is present in the $M_1$ and is usually prominent.

The upper incisors are moderately opisthodont and fairly strongly constructed.

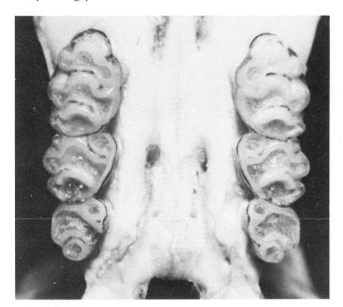

**Figure 85** Cheekteeth of Grant's Rock Mouse *Aethomys (Micaelamys) granti*. The molars are broad, heavy and angular, with a very large t1, t2 and t3 on the four-rooted $M^1$. The t9 is of medium size.

### Distribution
Grant's Rock Mouse occurs in the South West Arid zone and is restricted to the Upper and Great Karoo (map 57). *A. (M.) granti* is described from the same type locality (Deelfontein) as *A. (M.) namaquensis centralis,* the two species having been found living together in the same koppie (Davis 1974). This provides a unique opportunity for a critical study of ecological niches of two species coexisting in the same habitat.

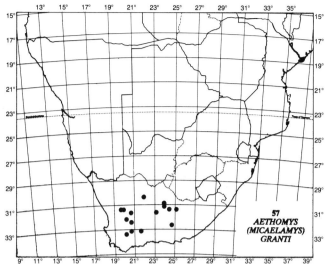

57 AETHOMYS (MICAELAMYS) GRANTI

### Habitat, diet, habits, predators, reproduction, parasites and relations with man
No data available.

### Prehistory
The species has not hitherto been identified in any of the fossil-bearing localities anywhere in southern Africa. The teeth are so characteristic and robust (facilitating preservation over a long span of time) that it seems unlikely to have been misidentified or that its remains could not have been fossilised.

### Taxonomy
This has been dealt with in some detail earlier, when the present status of the species was discussed. At present the species is regarded as monotypic and is likely to be interpreted accordingly for the forseeable future.

GENUS *Thallomys* Thomas, 1920

The interesting genus *Thallomys* was separated from the genus *Rattus* by Thomas in 1920 because of a greater complexity and angularity of the molars than is usual in *Rattus*, as well as certain external specialisations towards arboreal life which prompted Thomas to elevate the group to generic rank. Whether these arguments distinguish the genus from all forms of *Rattus,* is doubtful. Ellerman (1941) has pointed out that the genus has been retained mostly for convenience. Smithers (1975) says that it ranks amongst the more attractive rats with its pure white ventral surface and pale yellowish to greyish- yellow dorsal pelage. In addition, there are two other outstanding phenotypic features: a dark patch or stripe running between the nose and the eye which often continues as a dark line towards the base of the ear; and the conspicuous dark-coloured tail, giving rise to its vernacular name of Black-tailed Tree Rat or 'Swartstertrot' in Afrikaans. Apart from these characteristics, the genus is also characterised by the length of its well-ringed long tail, while the short hands and feet are provided with strongly developed curved, sharp claws, characteristic of arboreal rats and mice.

193

# Thallomys paedulcus (Sundevall, 1846)

**Tree Rat**

**Boomrot**

The generic name is derived from the Greek *thallos* = a branch or a young shoot and *mys* = mouse. It evidently refers to the preferred habitat of the animal among branches of trees. The specific name is probably derived from the Latin *pedis* = a foot, referring to the relatively broad hindfoot.

The type specimen, described as *Mus paedulcus* by Sundevall in 1846, is housed in the Stockholm Museum, Sweden. The specimen was collected by Wahlberg in South Africa and the locality was given as 'In Caffraria interiore, prope tropicum'.

After Thomas (1920) had separated this genus from *Rattus*, Ellerman (1941) retained it as a separate genus '... mostly for convenience'. The type species he took to be *Mus nigricauda* described by Thomas in 1882 on a specimen collected at the Hou(n)top River in Great Namaqualand, SWA/Namibia. He furthermore divided the genus into two groups: the *nigricauda* group and the *namaquensis* group. The former contained the species *nigricauda, damarensis, moggi* and *paedulcus* as species (together with subspecies equalling some 17 named forms), while the latter group contained the *namaquensis* species (a total of 13 named forms). The *namaquensis* group was referred to *Aethomys* by Thomas after originally placing it in *Praomys*, and *namaquensis* is today accepted as a full species of *Aethomys*. However, Ellerman (1941) retained it under *Thallomys*, with a note by Roberts in Shortridge (1934) that if the *namaquensis* group cannot be referred to *Praomys* neither can it be referred to *Aethomys*, because of the heavier molar structure as well as features of the hindfoot and tail and that the matter should be looked into again.

In 1951, however, Roberts interpreted the *namaquensis* group as a full species of *Aethomys* (together with *silindensis* and *chrysophilus*), while *Thallomys* contained *nigricauda, shortridgei, damarensis* and *moggi* as species. The species described as *Mus paedulcus* was mentioned in the synonymy of *Praomys (Mastomys) coucha breyeri* Roberts 1926 as a synonym. Roberts (1951) states the original description of the locality 'In Caffraria interiore, propre tropicum', was interpreted by Sclater (1901) to imply Pondoland in Transkei and Lydenburg. The original description by Sundevall was not available and the description given in some detail by Sclater (1901) is puzzling and pointed to it being a *Mastomys*. Allen (1939) places it as an *Aethomys* but Roberts knew of no *Aethomys* which could fit Sclater's description.

Roberts concluded that if it were to be a *Mastomys*, the name would displace *P. (M.) natalensis breyeri*. In 1953

Ellerman *et al.* point out that a syntype skull of *Rattus paedulcus* Sundevall exists in the British Museum (Natural History) under that name, but hitherto not positively identified. The syntype skin bears no measurements, but the skull has the unusually large bullae characteristic of *Thallomys* while the shape of the hindfoot with its relatively long fifth toe '... enables us to suggest that *paedulcus* is in reality the prior name for the group to which *nigricauda, damarensis* and *moggi*, all here considered conspecific, belong'. Consequently, these animals were interpreted as *Rattus (Aethomys) paedulcus* in the systematic section which followed in Ellerman *et al.* (1953).

## Outline of synonymy

1846 *Mus paedulcus* Sundevall, *Oefvers. K. Svenska Vet. Akad. Förh. Stockholm* (4):120 (1847). 'Interior of Kaffirland.'

1882 *Mus nigricauda* Thomas, *Proc. zool. Soc. Lond.*: 266. Hou(n)top River, Great Namaqualand, CP.

1897 *Mus damarensis* De Winton, *Ann. Mag. nat. Hist.* 19:349. Damaraland, CP (exact locality unknown).

1908 *Thamnomys ruddi* Thomas & Wroughton, *Proc. zool. soc. Lond.*: 549. Tete, Mozambique (= *Grammomys ruddi*).

1911 *Epimys nigricauda kalaharicus* Dollman, *Ann. Mag. nat. Hist.* (8) 8:544. Molopo River, northern CP.

1913 *Mus moggi* Roberts, *Ann. Transv. Mus.* 4:85. Soutpan, Pretoria District, Transvaal.

1915 *Mus moggi acaciae* Roberts, *Ann. Transv. Mus.* 5:120. Woodbush, Transvaal.

1920 *Thallomys nigricauda* Thomas, *Ann. Mag. nat. Hist.* (9) 5:141. (Genotype *Mus nigricauda* Thomas.)

1923 *T.(hallomys)longicauda* Thomas, *Proc. zool. Soc. Lond.*: 493. *Lapsus calami* for *T. nigricauda.*

1923 *Thallomys shortridgei* Thomas & Hinton, *Proc. zool. Soc. Lond.*: 492. Louisvale, CP.

1926 *T. leuconoë* Thomas, *Proc. zool. Soc. Lond.*: 303. Osohama, Etosha Pan, SWA/Namibia.

1926 *T. herero* Thomas, *Proc. zool. Soc. Lond.*: 303. Ondongwa, SWA/Namibia.

1931 *T. moggi lebomboensis* Roberts, *Ann. Transv. Mus.* 14:234. Ubombo District, Natal.

1933 *Thallomys leuconoë molopoensis* Roberts, *Ann. Transv. Mus.* 15:269. Between Setlagoli & Molopo River at Pitsani, Botswana.

1933 *T. stevensoni* Roberts, *Ann. Transv. Mus.* 15:269. Bembesi, Zimbabwe.

1933 *T. leuconoë bradfieldi* Roberts, *Ann. Transv. Mus.* 15:268. Quickborn Farm, Okahandja, SWA/Namibia.

1951 *Thallomys nigricauda nigricauda* (Thomas). *In* Roberts, *The mammals of South Africa.*

1953 *Rattus paedulcus paedulcus* (Sundevall). *In* Ellerman *et al., Southern African mammals.*

1953 *Rattus paedulcus robertsi* nom. nov. *In* Ellerman *et al., Southern African mammals.*

1955 *T. zambeziana* Lundholm, *Ann. Transv. Mus.* 22:279.

1955 *T. davisi* Lundholm, *Ann. Transv. Mus.* 22:279.

1964 *Thallomys paedulcus* (Sundevall). *In* Meester *et al.*, *An interim classification of southern African mammals.*

1974 *Thallomys paedulcus* (Sundevall). Misonne, *in* Meester & Setzer (eds), *The mammals of Africa: an identification manual.*

### TABLE 51
Measurements of male and female *Thallomys paedulcus*

|  | Parameter | Value (mm) | N | Range (mm) | CV (%) |
|---|---|---|---|---|---|
| **Males** | HB | 138 | 15 | 124–153 | 5,7 |
|  | T | 170 | 14 | 142–210 | 10,0 |
|  | HF | 26 | 15 | 24–28 | 4,0 |
|  | E | 24 | 12 | 21–27 | 6,1 |
|  | Mass: | 128,5 g | 1 | — | 6,1 |
| **Females** | HB | 143 | 33 | 125–157 | 6,5 |
|  | T | 167 | 32 | 145–193 | 8,0 |
|  | HF | 25 | 32 | 23–28 | 5,6 |
|  | E | 23 | 30 | 20–25 | 8,6 |
|  | Mass: | 107,75 g | 4 | 97–125 g | 9,8 |

Smithers (1971) records a weight of 81 g for males (N = 5, observed range 46–100 g) and 80 g (N = 5, observed range 63–91 g) for females in specimens from Botswana.

### Identification

The upper parts are light grey to yellowish-grey. The individual hairs are tipped yellowish-brown and contrast strongly with the dark slaty colour of the basal portions of the hairs. The fur is soft and well dressed, giving the animal a neat appearance. The black mark on the face, extending from the nose to the eye (surrounding it) and spreading in a more diffuse manner in the direction of the ear, is very conspicuous in this species. The chin, throat and entire belly portion are pure white, gradually merging into the darker coloration of the upper parts along the flanks. The dorsal surfaces of the hands and feet are pure white, thickly clad with dense, short hair. The hindfoot is broad, the toes short with the fifth digit well developed and extending beyond the base of the fourth toes. Both hands and feet have strong, short claws. The tail is considerably longer than the head-body length (as a rule somewhat less than 140% but exceptions do occur) and relatively well covered with blackish hairs making the tail conspicuous. The tail is closely ringed with scales. The ears are large and wide open and in live animals stand away from the head (plate 9).

The colour in this species can vary geographically, as was documented by Smithers (1971).

Thomas (1920) has given the mammary formula as 0–2=4 (i.e. two pairs of inguinally situated mammae only) whereas Shortridge (1934) states that the 'normal number is 6 (1–6=6)' (i.e. an additional pectoral pair being present), a figure also given by Roberts (1951).

### Skull and dentition

The skull is *Rattus*-like in general appearance. Where the frontals meet the parietals on the dorsal surface of the braincase, the area tends to be flattened, thereby reducing the slight bulge that the parietals usually make on the top of the braincase. The interorbital constriction is fairly well developed. The supraorbital ridges are moderately marked. The anterior palatine foramina are large (longer than half the length of the diastema) and penetrate between the first two upper molars. The palate is broad and not ridged. The anterior margin of the zygomatic plate is vertical and the dorsal portion is cut back, exposing a large antorbital foramen. The zygomatic arches are not robust. The bullae are unusually large, usually more than 6 mm in greatest diameter (Meester *et al.* 1964). The external auditory meatus is large. The paroccipital process is small but sturdily built (fig. 86).

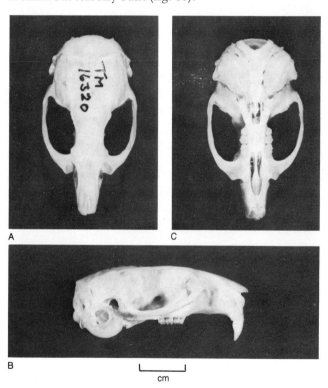

**Figure 86** Dorsal (A), lateral (B) and ventral (C) views of the skull of the Tree Rat *Thallomys paedulcus*. The interorbital constriction is pronounced while the supraorbital and occipital ridges are moderate. The anterior palatine foramina penetrate between the molars and are longer than half the length of the diastema. The anterior border of the zygomatic plate is cut back dorsally. The skull shows a flattened profile with large bullae, usually more than 6 mm in diameter. The incisors are orthodont.

The M$^1$ is a five-rooted tooth with the t7 absent, giving it eight cusps in total, with the centre row of cusps prominent. In the M$^2$ both the t2 and t7 are absent, making it a seven-cusped tooth. The M$^3$ is smaller than the M$^2$ but not greatly reduced. The molars make a cusped, angular impression and the teeth are surprisingly small for an animal the size of *Thallomys* which is on average larger than the Namaqua Rock Rat *(Aethomys (Micaelamys) namaquensis)* (fig. 87). The upper incisors are smooth, orthodont.

The lower molars have prominent and sharply defined cusps. The terminal heels of the M$_1$ and M$_2$ are usually obliterated. A row of additional small cusps or small ridges often occur on the labial side of the lower molars as extra elements.

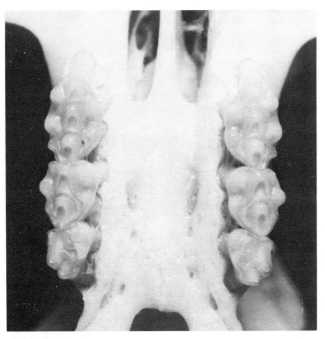

**Figure 87** Cheekteeth of the Tree Rat *Thallomys paedulcus*. The molars are small, cusped and angular. The centre row of cusps of the five-rooted M$^1$ is pronounced.

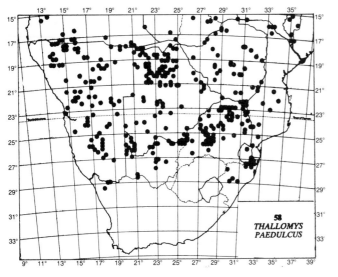

## Distribution

The species occurs in the South West Arid and Southern Savanna Woodlands biotic zones, but is absent in the Southern Savanna Grassland zone. It is, therefore, found throughout SWA/Namibia, Botswana, Zimbabwe, the extreme north and extreme south of Mozambique, northern Cape Province, western, northern and eastern Transvaal, Swaziland and Natal. It is conspicuously absent from the Cape Province south of the Orange River, the OFS and Lesotho (map 58).

## Habitat

In Botswana it is widely distributed throughout the country from the very dry areas (mean annual rainfall less than 200 mm) to higher rainfall areas with a precipitation of more than 750 mm annually (Smithers 1971). Generally, however, it prefers drier environments, inhabiting *Acacia* trees and living in crevices in the trunks, under loose strips of bark or in holes in the ground between the roots of the tree. In the Kalahari Gemsbok National Park where I have trapped this species, it occurs in both *Acacia erioloba* and *A. reficiens*. Nel (1978) found that *Thallomys* in the southern Kalahari has no small mammal competitors in the top layers of *Acacia* trees and that it only marginally overlaps with *Rhabdomys* closer to the ground, but at a different time of day. According to Shortridge (1934) it inhabits *Acacia*, wild fig and more seldom Mopane trees in the Kaokoveld in SWA/Namibia, while Van Rooyen (1955) reports it as frequenting 'cabbage trees' (presumably *Cussonia* sp.).

It has also been recorded occasionally nesting in large *Aloe* sp. (e.g. specimen no. 6122 in the Transvaal Museum), sometimes staining its pelage with the pollen and nectar of the flowers (information from specimen card for specimen no. 7220).

Woosnam *(in* Shortridge 1934) mentions that it seems to be more numerous near water.

Smithers (1971) has pointed out that there is a distinct difference in the length of the hair and general weight of the pelage between those specimens in Botswana collected from warmer and wetter, as opposed to colder and drier months of the year. In the former (November to April) the pelage is lighter and shorter, while in the latter (June to October) the pelage is longer and woollier with a dense undercoat. He also records them to be present in *Acacia* woodland, '. . . or other types of woodland providing *Acacia* sp. are present'.

## Diet

The Tree Rat feeds largely on *Acacia* seeds, and other vegetable matter like young *Acacia* leaflets, but may infrequently also take insects or meat. Other items on the menu may include '. . . "wacht-een-bietjie" berries, probably "gemsbok" roots and the gum of thorn trees' (Bradfield *in* Shortridge 1934). Smithers (1971) gives more specific information by stating that they take the outer green coating of *Acacia tortilis* pods, or the coating itself when dry. They are also partial to young leaves of the Witgat *Boscia albitrunca*, as well as to the green outer coating of the Buffalo Thorn *Ziziphus mucronata*.

According to Bradfield (*in* Shortridge 1934) they do not go to water to drink. This is certainly true for certain populations occurring in the Kalahari Gemsbok National Park where I know of resident populations which are at least 5 km from the nearest available water.

## Habits
They build large, untidy 'nests' in the upper strata of *Acacia* trees. These are constructed of large twigs, grass and other foliage intertwined to form a very conspicuous structure near the outer branches of *Acacia* trees, especially during the winter months when these trees are devoid of their normal leaf load. They are shy animals and are not easily flushed from their shelters (cracks, crevices or hollows, in tree trunks) by day. They are predominantly nocturnal and emerge at dusk to run about the branches with great agility in search of food. Certain horizontal branches in particular trees are used as regular highways en route to lower lying branches where they forage.

The climbing ability of the arboreal *Thallomys* was studied in the laboratory by Earl & Nel (1976). In contrast to the scansorial Multimammate Mouse *Praomys natalensis* and the terrestrial Pouched Mouse *Saccostomus campestris*, *Thallomys* was morphologically and psychologically better adapted to arboreal life. On the test apparatus, branches were marked by means of perineal dragging against the branches. This marking behaviour would be of importance to *Thallomys* under natural circumstances, facilitating direction and movement along branches during darkness by means of scent.

These arboreal rats live in small communities, usually consisting of a pair of adult animals with their latest offspring and other progeny in various stages of immaturity. The entrances to the cracks or crevices where they hide during the day are devoid of accumulated debris with the exception of the telltale evidence of faeces scattered around the bases of tree trunks.

Shortridge (1934) states that they are inquisitive and will peek out of their hollows if the tree is tapped. In order to collect them, they may be shot or smoked out of their dwellings. I have tried the latter procedure, but find it to be ineffective. Smithers (1971) speaks of a similar experience: they are reluctant to leave the shelters and will perish in a fire rather than leave them. I have been fairly successful in trapping them in Sherman collapsible traps affixed to horizontal branches by means of masking tape baited with a mixture of peanut butter and oats.

The conspicuous 'nests' can, from a distance, be mistaken for the nests of Buffalo Weavers *Bubalornis albirostris*, although these structures hardly attain the size and extent of the nests built by the birds. The amount of material could possibly fill a normal-sized wheelbarrow, and Shortridge (1934) points out that these structures are not nests in the true sense of the word and that they are not occupied by day. I have observed this as well, and they may serve as playgrounds or shelters while roaming at night. Some populations (e.g. in Ovambo) do not construct these large, conspicuous masses. Smithers (1971) has not come across them in Botswana and he even questions whether *Thallomys* build these structures themselves. The actual nests consist of sticks and grass which are collected and compacted in hollow tree trunks, some 3 to 4 m above the ground (Van Rooyen 1955). Pienaar *et al.* (1980) have recorded their nests built in the Umbrella Thorn *Acacia tortillis*, the Tambotie *Spirostachys africana* and in a Nyala Tree *Xanthocercis zambesiaca*.

Smithers (1971) also points out that they will use holes in the tree trunks of the Witgat *Boscia albitrunca*, the Leadwood *Combretum imberbe* and the Mopane *Colophospermum mopane* for this purpose.

Near Upington they inhabit the stunted *Acacia* belt alongside the Orange River, which is often partly submerged when the river is in flood, forcing them to stay among the upper branches of the trees for days on end (Shortridge 1934).

Nel & Rautenbach (1977) have found that *Thallomys paedulcus* have a higher mean body temperature than other rodents occurring in the Kalahari Gemsbok National Park. Why this should be so, is not clear.

## Predators
Little information is available, apart from the fact that they are taken by the Giant Eagle Owl *Bubo lacteus* (Pitman & Adamson 1978).

## Reproduction
In SWA/Namibia gravid females have been taken at Gobabis between the months of September and December, while another was taken near Kovares in April (Shortridge 1934). This tallies with observations recorded by Smithers (1971) that gravid females were collected in October, February, March and April and that they appear to drop their young from October through May.

According to Wahlberg, who collected the type specimen, the female carries her young attached to her nipples when running or climbing. A similar observation has been described on specimen card no. 3885 in the Transvaal Museum.

## Parasites
The calliphorid fly *Cordylobia anthropophaga* is known to be associated with the Tree Rat (Zumpt 1966). Other ectoparasites include fleas represented by three families, *viz.* the Pulicidae, Hystrichopsyllidae, and Chimaeropsyllidae. The pulicids are represented by the following species: *Echidnophaga gallinacea, Parapulex echinatus, Xenopsylla cheopis, X. philoxera, X. versuta, X. brasiliensis, X.*

graingeri, X. robertsi and X. zumpti. The hystrichopsyllids include Dinopsyllus ellobius, D. lypusus, Listropsylla aricinae, L. dorippae and L. prominens. The chimaeropsyllids are represented by a single species only, Chiastopsylla rossi (Zumpt 1966).

Mites are represented by the laelaptid Androlaelaps marshalli and the listrophorid Listrophoroides mastomys (Zumpt 1961a). Ticks recorded by Theiler (1962) include immatures of Haemaphysalis leachii leachii and H. l. muhsami. The cestode Inermicapsifer madagascariensis was identified by Collins (1972) in specimens of Thallomys paedulcus.

## Relations with man

The 'stick tight' flea Echidnophaga gallinacea has a worldwide distribution and is frequently found on domestic chickens. The Tree Rat acts as a reservoir for this ectoparasite capable of transmitting plague. It is also one of the principal hosts of Xenopsylla brasiliensis, an important plague vector (De Meillon, Davis & Hardy 1961). Keogh & Isaäcson (1978) have assessed this animal as a laboratory animal. It handles and breeds poorly in the laboratory although it becomes adapted to these man-made conditions.

## Prehistory

A maxillary fragment, a mandibular fragment and an isolated M$^1$ of Thallomys cf. paedulcus was identified from fossiliferous breccia collected at the Cave of Hearths in the Makapansgat Valley (De Graaff 1960a). It was also reported among the mammalian microfauna collected at Kromdraai by Draper in 1895 (De Graaff 1961).

A fossil species, Thallomys debruyni was reported from the Taung deposit (Broom 1948a) which had yielded the juvenile australopithecine Australopithecus africanus some 23 years earlier (Dart 1925).

## Taxonomy

Meester et al. (1964) uphold Thallomys paedulcus paedulcus (southeastern Botswana, southern Mozambique, Swaziland and KwaZulu); T. p. nigricauda (SWA/Namibia, southwestern Kalahari); T. p. shortridgei (vicinity of Upington) and T. p. damarensis (along the Zambezi River, Zimbabwe, Ngamiland, the Caprivi and northern SWA/Namibia) as acceptable subspecies. T. p. paedulcus is taken to include moggi, acaciae, lebomboensis and perhaps molopoensis. T. p. nigricauda includes kalaharicus, leuconoë, bradfieldi (= robertsi) and davisi. T. p. damarensis includes ruddi (= Thamnomys (Grammomys) ruddi Thomas & Wroughton 1908), herero, stevensoni and zambesiana. Van Rooyen (1955) made a taxonomic study of the genus in southern Africa and regarded shortridgei of the Orange River as a well-marked distinct form (Davis 1965). Misonne (1974) includes all the southern African forms named above, as well as loringi, nitela, quissamae, rhodesiae, scotti and somalensis from elsewhere in Africa under Thallomys paedulcus.

GENUS *Lemniscomys* Trouessart, 1881

This genus is found over most of Africa and it shows different patterns of coloration throughout its range.

# *Lemniscomys griselda* (Thomas, 1904)

**Single-striped Mouse**

**Single-striped Grass Rat**

**Eenstreepmuis**

The name is coined from the Greek lemniscos = a band or a ribbon and mys = mouse, in reference to the conspicuous dorsal stripe. The reason for griselda is unknown, but Rosevear (1969) says it can only be surmised that it was inspired by the grizzled coloration of the pelage.

This genus consists of medium-sized, coarse-haired, terrestrial and predominantly diurnal mice characterised by their entirely striped dorsal surface (whence the vernacular name 'zebra mouse'). The stripes may be interrupted to form spots or there may be a single black line along the vertebral column from the base of the neck to the root of the tail. In southern Africa, the single-striped form is found, the spotted and multistriped forms occurring further north in Africa. According to Kingdon (1974) the genus Lemniscomys is a complex of different evolutionary levels. *Lemniscomys griselda* seems to be closely related to *Pelomys minor* (extralimital to this work) and is also closely related to *Rhabdomys* (Misonne 1974). There are strong indications that *L. griselda* is a species in retreat, e.g. in the central Kalahari (Davis 1962); it is rare enough through most of East Africa, although sporadically common in *Brachystegia* woodland communities. Otherwise, it is abundant throughout the rest of southern Africa.

The species was first described as *Mus dorsalis* by A. Smith in 1845, based on a specimen or specimens collected to the north of '. . . the Great Orange River'. The name *dorsalis* evidently referred to the conspicuous stripe on the mid-dorsal line. The type no longer exists and Roberts (1917) assumed that it was collected somewhere in the western Transvaal. In 1881 the taxon *Lemniscomys* was erected by Trouessart as a subgenus of *Mus* and by designation he used the type specimen of *Mus barbarus* as described by Linnaeus in 1766 on specimens from Morocco. Subsequent to 1881 and even in the second edition of Trouessart's *Catalogue of Mammals*, the mice that we presently associate with *Lemniscomys* were commonly referred to as *Arvicanthis* Lesson, until Thomas (1916a) revised the name. In 1904, Thomas described a murid specimen collected at Muene Costri, Juila country, northern Angola and named it *Arvicanthis dorsalis griselda*, the type of which is in the British Museum. The name *griselda* replaced *dorsalis* as the specific name since the latter was preoccupied by *Mus dorsalis* Fischer, 1814.

During 1916, Thomas erected *Lemniscomys griselda spinalis* from South Africa, assumed to be the western Transvaal, which in turn has *dorsalis* Smith, 1845 as a preoccupied synonym. The correct naming of species in this genus was taken up by Thomas (1927).

## Outline of synonymy

1845 *Mus dorsalis* A. Smith, *Illustr. Zool. S. Afr. Mammals*, pl.46, fig. 2. 'North of the Orange river'; Western Transvaal according to Roberts 1917. Not of Fischer 1814.

1881 *Lemniscomys barbarus* Trouessart, *Bull. Soc. Scient. Angers* 10(2):124. (Genotype *Mus barbarus* Linnaeus.)

1896 *Arvicanthis dorsalis* De Winton, *Proc. zool. Soc. Lond.*: 803.

1904 *Arvicanthis dorsalis griselda* Thomas, *Ann. Mag. nat. Hist.* (7)13:414. Muene Coshi, Angola.

1908 *Arvicanthis dorsalis calidior* Thomas & Wroughton, *Proc. zool. Soc. Lond.*: 545. Tambarara, Gorongoza Mountains, Mozambique.

1916 *Lemniscomys griselda spinalis* Thomas, *Ann. Mag. nat. Hist.* (8)18:69. South Africa, probably western Transvaal. Replaces *dorsalis* A. Smith, preoccupied.

1927 *Lemniscomys griselda sabulata* Thomas, *Proc. zool. Soc. Lond.*: 385. Sandfontein, Gobabis district, SWA/Namibia.

1931 *L. griselda zuluensis* Roberts, *Ann. Transv. Mus.* 14:235. Manaba, Natal.

1932 *Lemniscomys griselda fitzsimonsi* Roberts, *Ann. Transv. Mus.* 15:11. Kaotwe Pan, Botswana.

1946 *Lemniscomys griselda sabiensis* Roberts, *Ann. Transv. Mus.* 20:231. Gravellote, Transvaal.

1951 *Lemniscomys griselda spinalis* Thomas. *In* Roberts, *The mammals of South Africa.*

1953 *Lemniscomys griselda* (Thomas). *In* Ellerman *et al.*, *Southern African mammals.*

1974 *Lemniscomys griselda* (Thomas). Misonne, *in* Meester & Setzer (eds), *The mammals of Africa: an identification manual.*

### TABLE 52
Measurements of male and female *Lemniscomys griselda*

| | Parameter | Value (mm) | N | Range (mm) | CV (%) |
|---|---|---|---|---|---|
| Males | HB | 130 | 29 | 113–150 | 6,6 |
| | T | 137 | 28 | 120–154 | 6,6 |
| | HF | 28 | 28 | 22–31 | 8,3 |
| | E | 17 | 26 | 15–19 | 6,5 |
| | Mass: | 64 g | 26 | 48–80 g | 14,0 |
| Females | HB | 132 | 30 | 122–143 | 4,5 |
| | T | 130 | 29 | 114–147 | 6,4 |
| | HF | 28 | 31 | 25–30 | 5,6 |
| | E | 17 | 28 | 14–20 | 6,6 |
| | Mass: | 60 g | 30 | 44–80 g | 17,0 |

Comparative measurements and weights have been taken from Wilson (1975). These read as follows (mm):

| | Males | | Females | |
|---|---|---|---|---|
| Overall length | 227 (5) | 215–247 | 221 (4) | 219–250 |
| T | 120 (5) | 110–134 | 117 (4) | 108–133 |
| HF c/u | 28 (5) | 27–30 | 29 (4) | 27–31 |
| E | 15 (5) | 14–17 | 15 (4) | 14–17 |
| Weight | 59 g (5) | 51–73 | 60 g (4) | 50–76 |

Four males recorded by Smithers (1971) weighed 72, 70, 54 and 46 g respectively, while a single female weighed 56 g.

## Identification

These medium-sized mice attain an adult length of 27 mm and are covered with a rather thin, coarse fur which is not spiny. The dorsal surface is pale buff or reddish-orange, profusely pencilled with liver-brown, resulting in a 'pepper and salt' effect, conspicuously traversed by the single, liver-brown dorsal stripe extending from the occiput to the root of the tail (plate 9). The flanks, sides of the head and the upper surfaces of the limbs are the same colour as the dorsal surface. The chin, throat and belly are rusty white, while the muzzle can be rusty ochre yellow. The ears are covered with short, reddish-yellow hair. There is a pronounced colour cline from west to east over southern Africa; paler coloured populations occur in the west, darker forms in the east. The tail length is more than 94% of the head and body length, scaly and thickly clad in short hairs. The clawless D5 of the hand is reduced and does not reach the base of the D4 so that the manus only has three functional digits, the D1 or pollex being reduced as well. In the hindfoot the D5 is subequal to the hallux with the three centre digits elongated, giving the entire foot a rather narrow appearance.

The skin differs from that of *Rhabdomys* in being white on its inner surface instead of being slaty-black, as is found in *Rhabdomys* (Shortridge 1934).

The mammary formula reads 2–2=8, implying the presence of two pectorally and two inguinally placed pairs of mammae.

## Skull and dentition

Seen in profile, the skull is somewhat arched with the anterior edge of the zygomatic plate occasionally slightly convex and the position of the plate itself about the height of the muzzle. It is a large skull, rather reminiscent of the *Arvicanthis* type. The supraorbital and supraoccipital ridges are well developed. The anterior palatal foramina are well behind the henselion and narrow backwards to a point between the first two molars (fig. 88).

In the upper tooth row, the centre cusps of the relatively broad molars are large. The $M^1$ is eight-cusped (the t7 being absent, the t9 reduced). The $M^2$ is seven-cusped (with the absence of t2 and t7, and the t9 also small). The $M^3$ is relatively large. The details of the $M^3$ (which is similar to the $M^2$ but on a reduced scale) as noted and utilised by Thomas (1904) when he separated *Lemniscomys* and *Rhabdomys* from *Arvicanthis* do not, according to Ellerman (1941) appear to be of generic value. The combined length of the upper tooth row more or less equals the diameter of the bulla (fig. 89).

When the skull is removed from the skin, it is seen to be covered by a flimsy black membrane. It has been said that this serves as a protective mechanism against the action of the sun, in view of the fact that this species is diurnal. This reasoning is difficult to accept since Petter, Chippaux & Monmignaut (1964) found that *Lemniscomys* usually avoids the midday sun. It spends the greater part of its time in the shadow of dense grass (Rosevear 1969). Hill (1942) reported the same condition in *Rhabdomys* collected in Uganda, as well as in the diurnal genera *Pelomys* and the extralimital *Arvicanthis*.

## Distribution

The distribution of this species occurs throughout the Savanna Woodlands and northwards to East Africa, while it is also encountered in the northern South West Arid region. As a species it seems to be retreating in the central Kalahari. In southern Africa it funnels down as far south as Natal, the bushveld of the northern and western Transvaal and westwards to Damaraland in SWA/Namibia. It also occurs in Swaziland, Mozambique, Zimbabwe and central and northern Botswana where it is uncommon. It is absent from the Cape Province, the OFS and Lesotho (map 59).

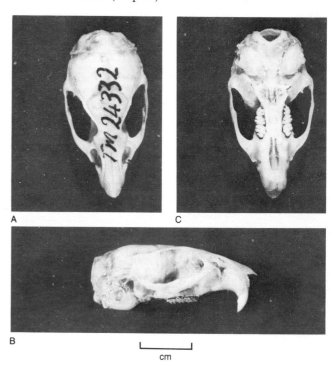

**Figure 88** Dorsal (A), lateral (B) and ventral (C) views of the skull of the Single-striped Mouse *Lemniscomys griselda*. The supraorbital and occipital ridges are pronounced. The slightly convex anterior border of the zygomatic plate is about half that of the height of the muzzle. It is a large skull, somewhat arched in profile.

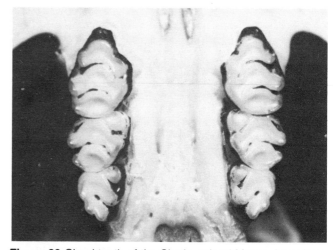

**Figure 89** Cheekteeth of the Single-striped Mouse *Lemniscomys griselda*. The centre cusps of the relatively broad molars are large. In the five-rooted $M^1$, the t7 is absent and the t9 rounded.

200

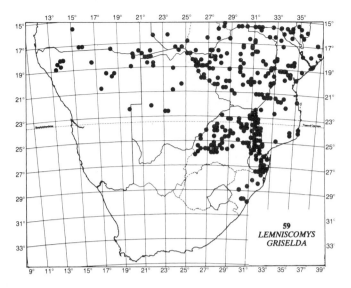

**59
LEMNISCOMYS
GRISELDA**

## Habitat

The Single-striped Mouse inhabits grass or low bushy vegetation in dry *Acacia* veld in the tropics. It apparently needs good grass cover, not necessarily associated with damp ground (Smithers 1971) and in Botswana it has been taken in a wide variety of habitats near swamps to *Terminalia-Acacia* scrub on red sand. In Mozambique it is widely distributed and common from Maputo in the south to the Tete district in the north, occurring in grassland and in any type of savanna with a good grass cover (Smithers & Lobão Tello 1976). Likewise, in Zimbabwe it has a wide distribution (Smithers 1975), being primarily associated with dry grassland and it appears to be more common in the higher rainfall areas from about the midlands eastwards, than in the drier west. Similarly, Wilson (1975) has reported a wide habitat tolerence for specimens collected in the Wankie National Park in Zimbabwe. Shortridge (1934) mentions that it is often outnumbered by other rodents in SWA/Namibia and in the northeast of that territory it ranges all over the Grootfontein district, entirely replacing *Rhabdomys* there.

## Diet

As far as is known, it subsists largely on grass and seeds. Rosevear (1969) also includes 'vegetable matter' in the dietary list. Roberts (1951) writes that it may occasionally be trapped with cheese or meat as bait. In the Kruger National Park, I have trapped it near Pretoriuskop with a mixture of peanut butter and oats, where incidentally it also seems to replace *Rhabdomys*.

## Habits

*Lemniscomys griselda* is a terrestrial, diurnal, crepuscular species, living in holes which it may burrow and excavate itself and from which runways radiate towards the feeding grounds. The home range of the male exceeds that of the female significantly (Swanepoel 1972). Its runways are well demarcated, like those of *Rhabdomys*. It also re-

sembles *Rhabdomys* in habits, but seems to be more restricted to long grass or damp vleis from which it seldom wanders for any great distance. Its choice of habitat was well described by Shortridge (1934) who stated it to be somewhat between that occupied by *Pelomys* on the one hand and *Rhabdomys* on the other. It is a nervous animal and may consequently die of stress on being handled. It sheds the skin of its tail easily if held (Jacobsen 1977). It is reluctant to bite and can be manipulated freely in spite of well-developed incisors. In my experience in the field, the animal occasionally tends to sham death or appears immobile when touched.

The species normally dwells on the surface, breeding in grass nests. How it survives the frequent and common grass-fires so typical of large tracts of Africa is not clear, but it may take refuge in the burrows of other animals under these circumstances (Rosevear 1969).

Shortridge (1934) says that *Lemniscomys* has an unusually long tail for an essentially terrestrial mouse. He believes it to be a more active species than *Rhabdomys*. It readily enters traps in the runways.

## Predators

According to Kingdon (1974) raptors of the genera *Melierax, Falco* and *Buteo* are important predators on this species, while they are also taken by the Long-crested Eagle *Lophaetus occipitalis* (Jarvis & Crichton 1977) and the White-faced Owl *Otus leucotis* (Worden & Hall 1978) in Zimbabwe.

## Reproduction

Very little is known about the breeding habits of *Lemniscomys*. It constructs nests lined with fine, shredded grass in which five to seven young are born (Jacobsen 1977), in March. A gravid female with three fetuses has been taken in October in Botswana (Smithers 1971) while in Zimbabwe pregnant females have been collected from October to March, each with three to five fetuses (Smithers 1975). On the other hand, two gravid females were recorded in Wankie and the Kruger National Park during May, carrying four young in both cases (Wilson 1975; Pienaar *et al.* 1980). Sixty-six percent of males had testes descended till April and after November with no reproductive activity in winter while females showed the same reproductive pattern in the Kruger National Park (Kern 1977). Swanepoel (1972) gives the female breeding season as September to February and the male season as September to May based on field work in a similar lowveld situation near Pongola, northern KwaZulu.

## Parasites

The mites include the following species: Laelaptidae – *Laelaps giganteus, L. muricola, L. parvulus, L. tillae, Haemolaelaps glasgowi, H. zulu* and *Androlaelaps marshalli*. Listrophoridae – *Listrophoroides lemniscomys* (Zumpt 1961a).

201

The following species of flea are known to be associated with this species: Pulicidae – *Echidnophaga gallinacea, Xenopsylla hipponax,* and *X. brasiliensis*. Hystrichopsyllidae – *Ctenophthalmus calceatus, Dinopsyllus ellobius* and *D. lypusus*. Chimaeropsyllidae – *Chiastopsylla rossi* (Zumpt 1966).

The ticks include immatures of the following species: *Haemaphysalis leachii leachii* and *Rhipicephalus simus* (Theiler 1962).

According to Collins (1972) the helminths include the following: *Inermicapsifer congolensis* and *I. madagascariensis.*

### Relations with man

Because these animals are readily trapped and abundant, they form a common article of diet over large areas of Africa (Rosevear 1969). They are sometimes attracted to cultivation (Shortridge 1934) and often frequent undisturbed grassland round agricultural land. There are no indications, however, that they tend to become agricultural pests. They do not have population explosions like some other species; nor do they enter buildings as a rule (Smithers 1975).

### Prehistory

Fossil specimens of this genus have tentatively been recorded from Makapansgat by Lavocat (1957).

### Taxonomy

Basically this genus can be divided into three groups:

(a) The *griselda* group occurring in central and southern Africa, having a single liver-coloured mid-dorsal stripe.
(b) The *barbarus* group occurring in North Africa, having a mid-dorsal stripe in addition to many parallel stripes along the back and towards the flanks.
(c) The *striatus* group occurring in West Africa, having the body stripes broken up into rows of spots.

Ellerman (1941) lists ten subspecies of *griselda* of which seven occur in southern Africa and two subspecies of *macculus* (a species occurring in Central Africa).

Roberts (1951) describes the southern African subspecies as *spinalis* (upper Limpopo Valley), *fitzsimonsi* (central Kalahari), *sabulata* (SWA/Namibia), *zuluensis* (northeastern KwaZulu), *sabiensis* (eastern Transvaal) and *calidior* (Mozambique). Ellerman, *et al.* (1953) also list these forms in addition to *L. g. griselda* from Angola. Misonne (1974) includes all seven subspecies as conspecific under *L. griselda*. Wilson (1975) states that there is a wide variation of colour within restricted areas and it is questionable whether the validity of some of the subspecies will stand the test of time. In contrast, Smithers (1971) recognises *L. g. fitzsimonsi* from Botswana as valid until the relationship between *L. g. spinalis*, found in Zimbabwe in the east and *L. g. sabulata*, found at Gobabis in SWA/Namibia in the west, is cleared up by an in-depth taxonomic study.

# GENUS *Rhabdomys* Thomas, 1916

This genus is very common in southern African and it may even be referred to as ubiquitous. It is closely related to the Grass Mouse, *Arvicanthis* (extralimital, East and Central Africa).

# *Rhabdomys pumilio* (Sparrman, 1784)

**Striped Mouse**

**Streepmuis**

The name is derived from the Greek *rhabdos* = a rod, obviously referring to the pattern of the dorsal pelage which reminds one of seven parallel-lying rods, and *mys* = mouse. *Pumilio* is the Latin for pygmy, also *pumilo*, while *pumilus* refers to diminutive or dwarfish. It may refer to the small or diminutive fifth digit of the hand.

*Rhabdomys pumilio* can be easily recognised for it is prettily marked along its back. A narrow composite band of seven alternating stripes, four black and three whitish, runs from the occipital region of the head to the base of the tail.

This species was first described by Sparrman in 1784 from a locality known as the Slangen River in the present Humansdorp district of the Cape Province. It was named *Mus pumilio*. With the passing of the years it became the custom to refer to this species as *Arvicanthis* and in 1916 it was eventually given separate generic rank as *Rhabdomys* by Thomas, who designated the species *pumilio* as the type specimen for his new genus.

The species occurs discontinuously throughout southern Africa and in many localities it is the dominant mouse. Beyond the borders of southern Africa as defined in this book, it occurs in Angola and the southern part of Zaïre on the west of the African continent, while it also ranges northwards through Malaŵi into Tanzania, Kenya and Uganda.

The Striped Mouse has probably received more attention than many of the other African rodents because of its economic importance and overall abundance. Nevertheless, the first comprehensive investigation of its ecology was compiled only during 1974 by Brooks who did a study on a natural *Rhabdomys* population on the Maria van Riebeeck Nature Reserve near Pretoria. This work is an important contribution to our knowledge of *Rhabdomys* ecology. This species was also studied in captivity and in the field by Choate (1971, 1972).

## Outline of synonymy

1784 *Mus pumilio* Sparrman, *Oefvers. K. Svenska Vet. Akad. Förh. Stockholm*: 236. 'Sitzicamma Forest- on Slangen River.' Humansdorp, CP.

1827 *M. pumilio* var. *major* Brants, *Het Geslacht der Muizen*: 105. Cape of Good Hope. Not of Pallas 1779.

1827 *M. donavani* Lesson, *Manual de Mammologie*: 268. Cape of Good Hope.

1829 *M. lineatus* F. Cuvier, *in* Geoffroy & Cuvier. *Hist. Nat. des Mamm.* pl. 161.

1842 *M. vittatus* Wagner, *Arch. f. Naturgesch.* 8(1):11. Cape of Good Hope.

1845 *M. septemvittatus* Schinz, *Syst. Verzeichn. d. Säugeth.* 2:155. Cape of Good Hope.

1892 *Isomys pumilio bechuanae* Thomas, *Proc. zool. Soc. Lond.*: 551. Rooibank, SWA/Namibia.

1897 *Arvicanthis pumilio dilectus* De Winton, *Proc. zool. Soc. Lond.*: 803. Mazoe, Zimbabwe.

1899 *A. pumilio* W. Sclater, *Ann. S. Afr. Mus.* 1:219.

1904 *A. pumilio cinereus* Thomas & Schwann, *Proc. zool. Soc. Lond.*: 2:5. Klipfontein, Little Namaqualand, CP.

1905 *A. p. intermedius* Wroughton, *Ann. Mag. nat. Hist.* (7)16:635. Deelfontein, CP.

1905 *A. p. chakae* Wroughton, *Ann. Mag. nat. Hist.* (7) 16:636. Sibudeni, Natal.

1905 *A. p. meridionalis* Wroughton, *Ann. Mag. nat. Hist.* (7) 16:632. Tokai, Cape Town, CP.

1905 *A. p. moshesh* Wroughton, *Ann. Mag. nat. Hist.* (7) 16:638. Maseru, Lesotho.

1905 *A. p. griquae* Wroughton, *Ann. Mag. nat. Hist.* (7) 16:632, Kuruman, CP.

1910 *A. p. deserti* Dollman, *Ann. Mag. nat. Hist.* (8) 6:399. Lehutitung, Botswana.

1916 *Rhabdomys pumilio* Thomas, *Ann. Mag. nat. Hist.* (8) 18:69. (Genotype *Mus pumilio* Sparrman.)

1926 *R. p. namibensis* Roberts, *Ann. Transv. Mus.* 11:255. Swakopmund, SWA/Namibia.

1946 *R. p. cradockensis* Roberts, *Ann. Transv. Mus.* 20:323. Cradock, CP.

1946 *R. p. algoae* Roberts, *Ann. Transv. Mus.* 20:323. Centlivres, near Port Elizabeth, CP.

1946 *R. p. orangiae* Roberts, *Ann. Transv. Mus.* 20:321. Goodhouse, CP. (Probable synonym of *cinereus*.)

1946 *R. p. namaquensis* Roberts, *Ann. Transv. Mus.* 20:322. Berseba, Fish River, SWA/Namibia. (Probable synonym of *griquae*.)

1946 *R. p. prieska* Roberts, *Ann. Transv. Mus.* 20:322. Prieska, CP. (Probable synonym of *orangiae*.)

1946 *R. p. fouriei* Roberts, *Ann. Transv. Mus.* 20:322. Ondonga, SWA/Namibia.

1946 *R. p. vaalensis* Roberts, *Ann. Transv. Mus.* 20:322. Bloemhof, Transvaal. (Probable synonym of *griquae*.)

1946 *R. p. griquoides* Roberts, *Ann. Transv. Mus.* 20:322. Fourteen Streams, northern Cape. (Probable synonym of *griquae*.)

1946 *R. p. bethuliensis* Roberts, *Ann. Transv. Mus.* 20:322. Bethulie, OFS. (Probable synonym of *vittatus*.)

1951 *R. p. pumilio* (Sparrman). *In* Roberts, *Mammals of South Africa*.

1953 *R. pumilio* (Sparrman). *In* Ellerman *et al.*, *Southern African mammals*.

1974 *R. pumilio* (Sparrman). Misonne, *in* Meester & Setzer (eds), *The mammals of Africa: an identification manual*.

## TABLE 53
### Measurements of male and female *Rhabdomys pumilio*

| | Parameter | Value (mm) | N | Range (mm) | CV (%) |
|---|---|---|---|---|---|
| Males | HB | 109 | 45 | 99–124 | 5,0 |
| | T | 86 | 43 | 71–101 | 8,1 |
| | HF | 21 | 45 | 19–24 | 10,0 |
| | E | 13 | 42 | 11–15 | 8,5 |
| | Mass: | 43,1 g | 22 | 41–53 g | 8,3 |
| Females | HB | 107 | 33 | 91–120 | 8,0 |
| | T | 86 | 33 | 78–94 | 4,7 |
| | HF | 21 | 33 | 20–24 | 4,8 |
| | E | 13 | 33 | 11–15 | 9,5 |
| | Mass: | 41,4 g | 13 | 36–51 g | 9,6 |

## Identification

This species is of medium size. Its fur is slightly harsh in texture. The ground coloration is yellowish grey-brown to speckled bufffy overlain on the dorsal surface by three light and four dark bands. The result is a prettily marked and conspicuous little animal. The front of the face and the cheeks are somewhat darker than the flanks. The ears have an orange to yellowish tinge to them and are of normal size. The feet are well haired, while the D5 in both the hand and the foot barely reaches the base of the adjoining D4. Although their digits are small, they are still functional, which is not the case in the closely related *Lemniscomys*. The lower parts of the body, i.e. the chin, throat and belly, are whitish with the bases of the individual hairs dark grey (plate 9).

The tail is approximately subequal to the length of the head and body, scaly, clad in short hairs, dark on the dorsal surface and lighter below. Davis (1962) remarked that the western, semi-desert forms tend to be longtailed, while the eastern forms tend to have shorter tails. This matter was further pursued by Coetzee (1970), who discussed the relative tail lengths in relation to climate (especially annual rainfall and temperature).

When preparing these animals as study skins, it is noticeable that the skin is comparatively thick and shows a slaty-black coloration on its inner surface.

The mammae consist of two pectoral and two inguinal pairs.

## Skull and dentition

The structure of the skull is reminiscent of that of a small *Arvicanthis*. Supraorbital ridges are usually present. The incisors are relatively large.

The rostrum tends to be shortened slightly and the anterior edge of the zygomatic plate is more or less vertical. The anterior part of the zygomatic arch is fairly flat, i.e. does not tilt upwards (fig. 90).

The incisors are not grooved. The molar teeth show a trend towards the development of powerful, coneshaped cusps, but less so than in *Lemniscomys*. The dentition is also of a lighter build than that of *Aethomys*. The centre row of cusps is large. In both the $M^1$ and the $M^2$, the t9 are reduced in size. The $M^3$ is somewhat smaller than the $M^2$, but not greatly reduced (fig. 91). In the M1 and M2 the terminal heels are reduced.

## Distribution

Davis (1962) discussed the distribution of southern African murids in terms of four major biotic zones of Moreau (1952) called the South West Cape, South West Arid, Southern Savanna Woodland and Southern Savanna Grassland zones. Davis described *Rhabdomys* as primarily a savanna form, classifying it as near endemic to the Southern Savanna zone, but ranging into the South West Cape and South West Arid zones, where he accords it marginal status, although pointing out that in the latter, species recorded as marginal often had substantial ranges within it. Furthermore, Brooks (1974) also states that *Rhabdomys* is absent from the montane and subtropical evergreen forests which are distributed in the Savanna and South West Cape zones.

This species is generally distributed throughout southern Africa with the exception of certain more tropical

woodland savannas as is found over large tracts of country in the lowveld of the Transvaal, Mozambique, the Zambezi Valley, Zimbabwe and eastern Botswana. It is common in the rest of southern Africa (map 60). In Zimbabwe it occurs at Bulawayo and ranges eastwards on to the highveld throughout the eastern districts (Smithers 1975). It occurs only on the western borders of Mozambique in the Vila Pery districts, east of Gorongoza (Smithers & Lobão Tello 1976). In Botswana it is associated with grassland or *Acacia* shrub where grass is present. It frequents thorn fences erected around cultivated ground (Smithers 1971). In the Kalahari Gemsbok National Park, it has a wide distribution also favouring areas with a good grass cover and it often lives near fallen tree trunks.

To the north of the Cunene and Zambezi rivers, its distribution towards East Africa is discontinuous and patchy, which may suggest that this species is either on the decline, or is very specialised. There is little evidence to support the latter view.

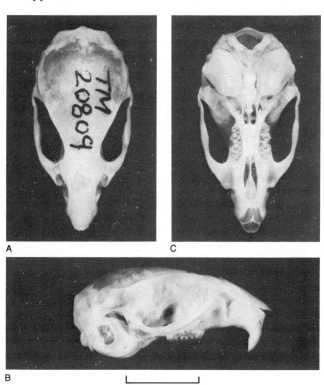

**Figure 90** Dorsal (A), lateral (B) and ventral (C) views of the skull of the Striped Mouse *Rhabdomys pumilio*. The rostrum tends to be shortened to a degree. The zygomatic arch is fairly narrow and slender. The anterior edge of the zygomatic plate is vertical.

## Habitat

According to Brooks (1974), *Rhabdomys* occupies a wide avariety of habitat types and may thus be considered a broad-niche species. It prefers grassland, although its habitat includes bushy and semi-dry vlei country as well as dry riverbeds, high grassveld areas, the edges of forests and the bases of hills. Its absence from the central Karoo area may be explained by the lack of grass which has been replaced by the typical karroid vegetation. It is apparently displaced by *Lemniscomys* in the northeastern section of SWA/Namibia, as well as in the Caprivi Zipfel.

Haim & Fourie *(pers. comm.)* have assessed the effect of acclimation to cold and long scotophase on heat production in *Rhabdomys pumilio* and their physiological and biochemical studies stress the importance of photoperiodicity in thermoregulation.

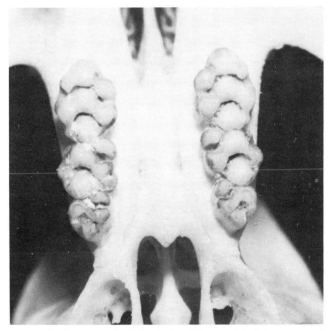

**Figure 91** Cheekteeth of the Striped Mouse *Rhabdomys pumilio*. The cusps on the teeth are not as strongly developed as in *Lemniscomys*. The t9 is reduced in the five-rooted M¹ and the M².

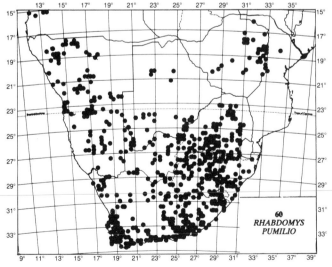

60
*RHABDOMYS PUMILIO*

### Diet

The Striped Grass Mouse has been called an opportunistic omnivore (Perrin *in press* b). It feeds predominantly on grass seeds, but it will also take any form of grain (Smithers 1975). In Botswana it readily eats the seeds and pods of *Acacia* spp., as well as the dried outer covering of the berries produced by *Ziziphus mucronata* (Smithers 1971). Roberts (*in* Shortridge 1934) states that a species of *Amaranthus* is also a favoured food of *Rhabdomys*. It also takes snails, worms, eggs and the nestlings of birds when the occasion arises (FitzSimons *in* Shortridge 1934). Other foods incude roots, fibres and bark. The eating of the latter causes considerable damage to forest plantations (Davis 1942; Mackellar 1952) especially during winter, suggesting a nutrient-deficiency response (Brooks 1974). Cannibalism may occur, for Brooks found large pieces of tissue with *Rhabdomys* hairs attached in the contents of one stomach. Captive animals often eat dead cage companions and Brooks concluded that this wild individual had probably raided a nearby snap-trap on the experimental grid. It does not seem to be very dependent on surface water, for I have trapped specimens in the Kalahari Gemsbok National Park which were at least 7 km from the nearest available water. It may obtain its moisture from dew and from plants with a high water content.

*Rhabdomys pumilio* may thus be described as an unspecialised omnivore exploiting transient but nutritions foods, which may account for its seasonal breeding and its fluctuations in density (Perrin *in press* b). It may thus be termed an *r*-strategist (Perrin *in press* a).

### Habits

*Rhabdomys* is not as shy as *Lemniscomys* and it is therefore seen more often and is better known. Greater awareness of the existence of this species is probably also enhanced by the conspicuous pattern on its dorsal surface.

It is predominantly diurnal, being very active from 05h00–08h30 and 14h30–17h30, as observed in Botswana (Smithers 1971). Coetzee (1970) stated that they also exhibit crepuscular activity. It moves around to some extent on warm, moonlit nights (Shortridge 1934). They are not communal animals. When trapping these animals in the vicinity of Dankbaar in the Kalahari Gemsbok National Park, I have trapped them in association with *Tatera* and *Zelotomys*. They burrow into the ground at a slanting angle (Shortridge 1934), the entrances usually partly hidden under vegetation, to a depth of 50 cm below the surface. These burrows may have an excavated chamber lined with soft grass in which they nest and according to Smithers (1975) the young are born here. These burrows apparently rarely have more than one entrance (Shortridge 1934). Radiating from the burrow entrances are well-defined runways through the grass cover. On other occasions, they will also use holes in termite mounds (Smithers 1971) for refuge, or they will hide in small, spherical nests reminiscent of bird nests, made in clumps of grass. Brooks (1974) described two such nests which were rounded and constructed of

205

lengths of chopped grass 5–8 cm long and lined with finer shredded grass. Each had a horizontal diameter of about 12 cm and a height of about 9 cm and each showed one rather diffuse exit. One of the nests was situated on the ground in grass standing one metre high, while the other was about 10 cm above groundlevel in a grass tussock of *Hyparrhenia dregeana*. In Natal, Maclean *(in litt.)* found that it bears its young in a small spherical grass nest on the ground among grass.

They never venture far from cover and marked individuals have revealed that members of this species can range over an area of some 20 ha (Kingdon 1974), which is an impressive surface area for such a small mammal. When handled they may utter a sharp, metallic sounding chirp. Shortridge (1934) states that they show regional differences in behaviour. In certain parts of Namaqualand they have been recorded as climbing low thorn trees and occasionally sleeping in the nests of birds. This behaviour is unknown elsewhere. Otherwise, *Rhabdomys* seems to show much the same habits as *Lemniscomys*. They are predominantly veld mice and seldom enter houses (Smithers 1975).

## Predators

According to FitzSimons (1920), the Puffadder *Bitis arietans,* the Rinkhals *Hemachatus haemachatus* and the Common Molesnake *Pseudaspis cana* all feed on *Rhabdomys.* To this list, Burton (1963) has added the Python *Python sebae.* Under experimental conditions, Brooks (1974) also found that the Spotted Grass Snake *Psammophylax rhombeatus* readily takes young *Rhabdomys,* up to three weeks old. These rodents are often taken by the Barn Owl *Tyto alba* (Vernon 1972; Dean 1977), the Marsh Owl *Asio capensis* and the Spotted Eagle Owl *Bubo africanus* (Dean 1977). Other avian predators include the Secretary Bird *Sagittarius serpentarius* (FitzSimons 1920), the Long-crested Eagle *Lophaetus occipitalis* (Jarvis & Crichton 1977), the Black-shouldered Kite *Elanus caeruleus* (Bell 1978) and the White-faced Owl *Otus leucotis* (Worden & Hall 1978).

This mouse is part of the diet of the Cheetah *Acinonyx jubatus* (Burton 1963), the Caracal *Felis caracal* (Bothma 1965), the Black-backed Jackal *Canis mesomelas* (Grafton 1965), as well as the Wild Cat *Felis lybica* and unspecified mongooses (FitzSimons 1920).

## Reproduction

According to Smithers (1975) they breed throughout the year with 3–9 pups a litter. The figure of 4–12 young per litter is given by Kingdon (1974), who also reports that they are capable of breeding at an age of three months. Brooks (1974) reported a mean litter size of 5,9 with a range of 2–9 pups. In SWA/Namibia, Shortridge (1934) refers to females collected in September and March to April with six fetuses and 6–7 fetuses respectively. In Botswana, Smithers (1971) recorded pregnant females in January, February, June and July. Very young specimens were taken during January, February, May and June. The average litter was five (N = 11), with the observed range of 3–9. The implantation is irregular. Smithers also refers to the Medical Ecology Centre in Johannesburg, who has breeding records for all months of the year except during August and September. Records available in the Transvaal Museum show pregnant females in November taken at Bethal in the eastern Transvaal. The *Rhabdomys* population studied by Brooks (1974) near Pretoria has shown a breeding season extending from September to April, followed by a four-month anoestrus. This pattern may not apply throughout the range of *Rhabdomys*, however, as reproduction has been shown to be affected by a number of environmental factors (Delany 1972), which may be different even though geographical separation is slight. Nel *(pers. comm.)* noticed that most large males in the Kalahari Gemsbok National Park were non-scrotal and most females non-perforate in May and July.

The average gestation period was 25,4 days and the minimum 23 days, as determined by Brooks (1974). Females undergo a post-partum oestrus (Choate 1971).

The young can walk effectively at ten days and are weaned by the 16th day. The incisors erupt relatively late on the sixth day which is an adaptation to the practice of transporting young by maternal nipples – i.e. they hang onto the mother when she moves around (Meester & Hallett 1970). For additional information and a good account of the post-natal development of *Rhabdomys*, see Brooks (1974), who has discussed behavioural development in detail.

## Parasites

Field tests have shown *Rhabdomys pumilio* to be susceptible to the bluetongue virus (Du Toit & Goosen 1949), while laboratory tests have indicated susceptibility to melioidosis (Findlayson 1944), *Yersinia pestis* (Mitchell 1927) and *Listeria monocytogenes* (Pirie 1927). It has also been shown that *Rickettsia conorii* (Wolstenholme & Harwin 1951) and *Coxiella burnetii* (Wolstenholme 1952) are endemic infections in these mice.

Mites and chiggers are also frequently associated with these striped mice. Hitherto they are known by three families and seven genera and 16 species which include the following: Laelaptidae – *Laelaps giganteus, L. lamborni, L. muricola, L. parvulus, L. peregrinus, L. transvaalensis, Haemolaelaps glasgowi, H. labuschagnei, H. murinus, H. rhabdomys, Androlaelaps marshalli* and *Macronyssus roseinnesi;* Trombiculidae – *Euschoengastia alticola* and *E. rhabdomyia;* Listrophoridae – *Listrophoroides mastomys* and *Trichoecius hollidayi* (Zumpt 1961a).

The fleas are represented by five families, 12 genera and 41 species. The Pulicidae include *Echidnophaga gallinacea, Pulex irritans, Ctenocephalides connatus, C. craterus, Xenopsylla eridos, X. mulleri, X. philoxera, X. piriei, X. versuta, X. brasiliensis* and *X. hirsuta.* The Hystrichopsyllidae include *Ctenophthalmus calceatus, C. eumeces, C. stenurus, C.*

Plate 11

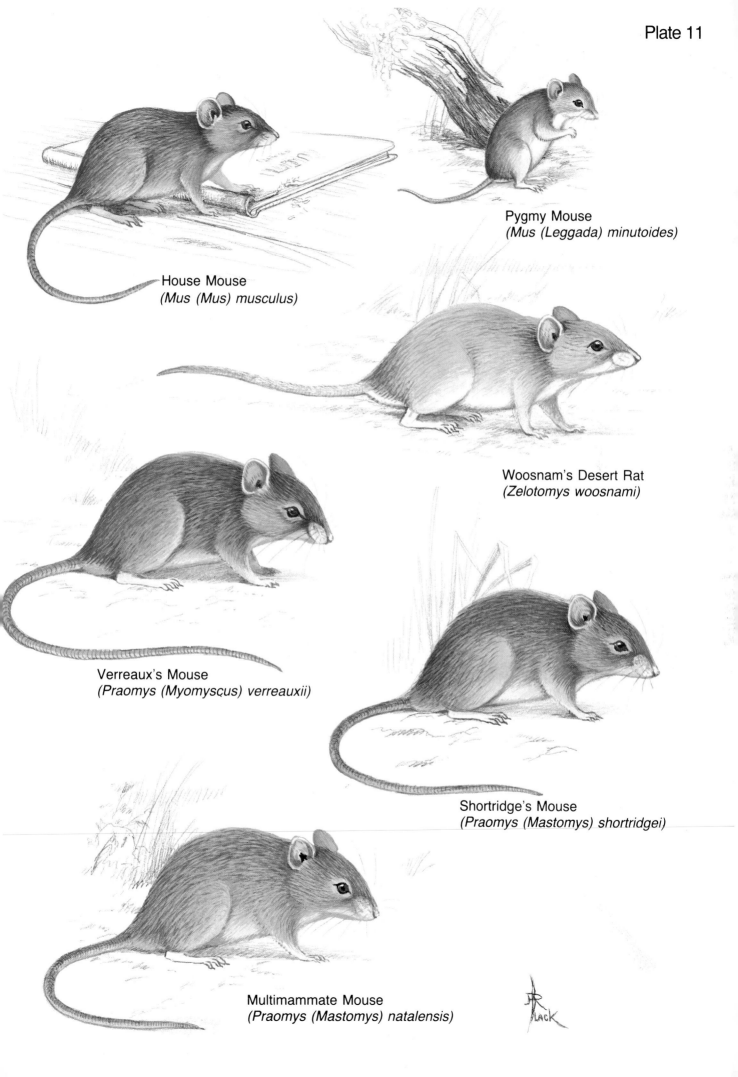

House Mouse
*(Mus (Mus) musculus)*

Pygmy Mouse
*(Mus (Leggada) minutoides)*

Woosnam's Desert Rat
*(Zelotomys woosnami)*

Verreaux's Mouse
*(Praomys (Myomyscus) verreauxii)*

Shortridge's Mouse
*(Praomys (Mastomys) shortridgei)*

Multimammate Mouse
*(Praomys (Mastomys) natalensis)*

Plate 12

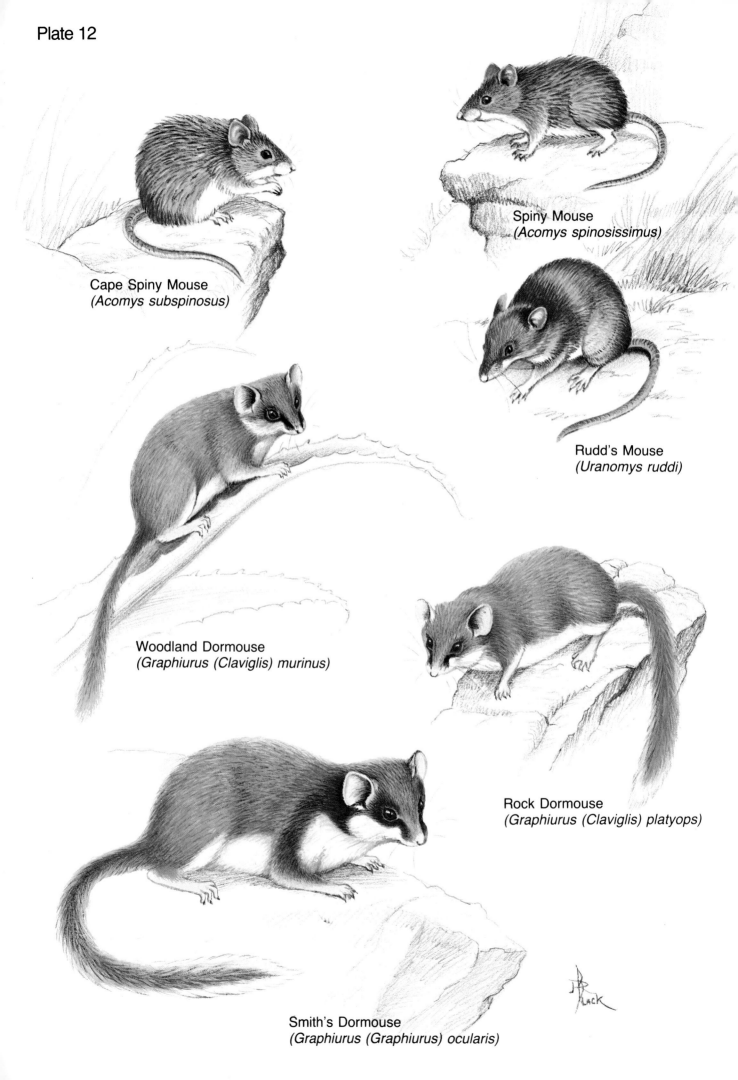

Cape Spiny Mouse
*(Acomys subspinosus)*

Spiny Mouse
*(Acomys spinosissimus)*

Rudd's Mouse
*(Uranomys ruddi)*

Woodland Dormouse
*(Graphiurus (Claviglis) murinus)*

Rock Dormouse
*(Graphiurus (Claviglis) platyops)*

Smith's Dormouse
*(Graphiurus (Graphiurus) ocularis)*

*verutus, C. cophurus, C. singularis, Dinopsyllus abaris, D. echinus, D. ellobius, D. longifrons, D. lypusus, D. tenax, Listropsylla agrippinae, L. basilewskyi, L. cerrita, L. chelura, L. dorippae* and *L. prominens*. The Leptosyllidae and Ceratopsyllidae are represented by *Leptopsylla segnis* and *Nosopsyllus incisus* respectively. The Chimaeropsyllidae include *Hypsophthalmus temporis, Epirimia aganippes, Chiastopsylla numae, C. rossi, C. capensis, C. carus, C. coraxis, C. godfreyi, C. mulleri* and *C. pitchfordi* (Zumpt 1966).

One species of louse, *Polyplax arvicanthis*, is known from *Rhabdomys pumilio* (Ledger *pers. comm.,* in Brooks 1974).

The ticks include adult (A) and immature (I) specimens of the following species: *Ornithodorus zumpti* (A,I), *Ixodes alluaudi* (A,I), *Amblyomma hebraeum* (I), *Haemaphysalis leachii leachii* (A,I), *H. l. muhsami* (A,I), *Hyalomma rufipes* (I), *Rhipicephalus appendiculatus* (I), *R. arnoldi* (I), *R. capensis* (A,I), *R. simus* (I) and *R. tricuspis* (I) (Theiler 1962).

The cestodes include *Hymenolopis fraterna* (Boomker *pers. comm.*), *H. nana* (Brooks 1974), *H. microstoma* (Collins 1972), *Catenotaenia lobata, Inermicapsifer congolensis, I. guineensis, I. madagascariensis, Raillietina (R.) trapezoides* (Collins 1972) and *Skrjabinotaenia baeri* (Brooks 1974). The nematodes include *Trichuris* sp. (Brooks 1974). *Heligmospiroides spira, Longistriata (L.) capensis, Protospirura bonnei, Subulura* sp., *Capillaria hepatica* and unidentified trematodes (Boomker *pers. comm.*).

## Relations with man

It is attracted to cultivation and they often outnumber other rodents of the veld in the vicinity of cultivation. They are known to cause considerable damage to young trees (especially conifers in commercial plantations in South Africa (Davis 1942; Hechter-Schulz 1951)). Their burrows are often made in mealie lands where they feed on the fallen seeds after harvesting. However, these animals are predominantly terrestrial and do not climb the mealie stalks to get to the cobs while still attached to the plants (Smithers 1975). They are often found near the granaries and huts of Blacks, as well as in urban gardens. As far as is known, they do not build up to large numbers and there is no evidence of population explosions. Consequently, they do not assume the proportions of a serious pest.

Although it has been reported as a transmitter of plague (De Meillon *et al.* 1961), it is not an important vector and they are of greater significance to man through damage to crops. Under laboratory conditions this rodent does not breed freely and it is active and awkward to handle (Keogh & Isaäcson 1978).

## Prehistory

In southern Africa, this species is known as a fossil from the different Pleistocene localities. It is listed in the faunal list of the Kromdraai material collected by Draper in 1895 (De Graaff 1961), while it has also been reported from Sterkfontein as well as the Makapansgat sites (De Graaff 1960a). It is also known from Pleistocene deposits at Olduvai in East Africa.

## Taxonomy

This genus is closely related to the *Lemniscomys-Pelomys-Arvicanthis* group on the one hand and the extralimital *Hybomys* on the other (Misonne 1969).

In 1946 Roberts described nine races of *Rhabdomys pumilio* from different localities in South Africa. This list included *orangiae* (from Goodhouse, lower Orange River), *namaquensis* (Berseba, Namaqualand), *prieskae* (Prieska), *fouriei* (Ondonga, Ovambo), *vaalensis* (Bloemhof), *griquoides* (Fourteen Streams), *bethuliensis* (Bethulie), *cradockensis* (Cradock) and *algoae* (Centlivres, near Port Elizabeth). In 1951 Roberts listed the following additional forms: *pumilio* (Humansdorp), *vittatus* (Cape of Good Hope) (= *meridionalis*), *moshesh* (Lesotho), *chakae* (Kwa-Zulu), *intermedius* (Deelfontein), *cinereus* (Little Namaqualand), *griquae* (Kuruman), *bechuanae* (Rooibank near Walvis Bay), *deserti* (Lehutitung), *namibensis* (Swakopmund) and *dilectus* (Mazoe, Mashonaland in Zimbabwe). The latter was basically the list given by Ellerman in 1941, who also listed *nyassae, angolae* and *diminutus* from Malaŵi, Angola and Kenya respectively. Many of these described forms are of doubtful validity. There may be a valid eastern subspecies (in areas with an annual rainfall of 400 mm) and a western subspecies (with an annual rainfall below 400 mm), but until the entire species has been revised, very little else can be added at this stage. Misonne (1974) has grouped the whole lot as conspecifics under *Rhabdomys pumilio*.

GENUS *Praomys* Thomas, 1915

This genus, and certainly the subgenus *Mastomys,* is undoubtedly the commonest and most widespread indigenous rodents of tropical and temperate Africa extending from the southern and eastern edges of the Sahara right down the continent to Plettenberg Bay on the south coast of South Africa.

The name *Praomys* was originally proposed as a subgenus of *Epimys* Trouessart (= *Rattus* Fischer, 1803) by Thomas (1915b), using *Epimys tullbergi* Thomas as the type species. It was raised to full generic rank by Thomas in 1926 (1926a). Thomas' genera were widely accepted until Ellerman (1941) revised them. In his excellent overview of the history of taxonomy of the Murinae, Rosevear (1969) reiterates that Ellerman again placed the taxonomic emphasis almost exclusively on cusp pattern of the upper molars and he brought almost all African genera together once more '. . . in company with a multitude of rats from Europe, Asia and Australia, as subgenera of *Rattus* Fischer'. Thus constituted, the genus *Rattus* contained about half the known murines totalling slightly more than 550 described forms.

Since the publication of Ellerman's *magna opera* in 1940 and 1941, there has been a tendency to swing away from the options offered in Ellerman's monograph. New taxonomic techniques are being developed and cytology, ethology and serology are being taken into account alongside features of the dentition, in order to determine the correct status of a taxon.

Consequently the taxonomy of these genera or subgenera is clouded by a diversity of opinions. To paraphrase Rosevear (1969), Ellerman, judging from molar pattern, interpreted *Praomys* as identical to *Rattus*: Matthey (1958),

interpreting chromosomes, is equally positive that they are different. Both Setzer (1956) and Davis (1965) have questioned Ellerman's *Rattus* – Setzer gives both *Mastomys* and *Praomys* full generic rank (with *Myomys* and *Hylomyscus* as subgenera of *Praomys*), while Davis treats *Hylomyscus*, *Mastomys* and *Myomys* (as *Myomyscus* Shortridge) as subgenera of *Praomys*.

The genus *Praomys*, as here understood, ranges over tropical Africa from Sierra Leone and Guinea to Ghana in the west, to southern and eastern Zaïre, Uganda, Kenya and Tanzania in the east and southwards to Zambia and Malaŵi. It is therefore extralimital to this work. Following Misonne (1974) it is here separated from *Rattus*, but is extended to include the subenera *Mastomys*, *Hylomyscus* and *Myomyscus*, a position already adopted by Davis (1962, 1965). Misonne (*op. cit.*) has already pointed out that all these smaller rats with rather longer incisive foramina '. . . certainly from a natural group', different from *Rattus*. Under the subgenus *Mastomys* are included short-tailed species with broader molars; under the subgenus *Hylomyscus* (extralimital) longtailed species with narrow molars; between these two extremes lie the subgenera *Praomys* (extralimital) and *Myomyscus*. The former is composed of forest species with a somewhat stronger set of molars, while the latter includes rock and grass species which in general have weaker molars than those of the *Praomys* group. They are, however, rather close together and are keyed together for practical reasons.

---

**Key to the subgenera of *Praomys***
(Modified after Misonne (1974))

1 Tail usually shorter than HB length, or slightly longer; molars broad . . . . . . . . . . . . . . . . . . . . . *Mastomys*
(Multimammate Mouse)
Page 209
(Shortridge's Mouse)
Page 215
Tail longer than HB length, usually over 105%; molars narrower . . . . . . . . . . . . . . . . . . . . . . . . . . . . . . . . . 2
2 Molars very narrow . . . . . . . . . . . . . . . . *Hylomyscus**
Molars narrow . . . . . . . . . . . . . . . . . . . . . . . *Praomys**
*Myomyscus*
(Verreaux's Mouse)
Page 216

* extralimital

**Key to the southern African *Praomys* subgenus *Mastomys***
(Modified after Meester *et al.* (1964))

1 Mammae 12 or more, not in pairs or divided into pectoral or inguinal pairs; pterygoids narrow; anterior palatal foramen reaches beyond the inner root of M$^1$; in Okavango region of northern SWA/Namibia, colour lighter than *shortridgei* in same area . . . . . . . .
*P. (M). natalensis*
(Multimammate Mouse)
Page 209

Mammae ten; pterygoids wider; anterior palatal foramen shorter; colour darker than *natalensis* occurring in same area in the Okavango region . . . . . . . . . . . . . . . . . . . .
*P. (M.) shortridgei*
(Shortridge's Mouse)
Page 215

# SUBGENUS *Mastomys* Thomas 1915

## *Praomys (Mastomys) natalensis* (A. Smith, 1834)

**Multimammate Mouse**

**Moedermuis**

The generic name is derived from the Greek *praos* = soft and *mys* = mouse, evidently referring to the soft pelage which has also earned them the vernacular name of soft furred rats. The subgeneric name is a continuation of the Greek *mastos* = breast and *mys* = mouse in reference to the large number of mammae in females. The specific name refers to the province of Natal where this species was initially collected by Smith in 1834.

The type species for the subgenus as designated by Thomas (1915b) is *Mus coucha* A. Smith, 1836 (= *Mus marikquensis* A. Smith, a race of *Mus natalensis* A. Smith, 1834). The original *Mus natalensis* (also illustrated in 1847) was collected in the neighbourhood of Port Natal (= Durban) and the type specimen is housed in the British Museum (Natural History), London.

Because of its frequent and close association with man, the Multimammate Mouse is fairly well known. It is commonly captured and the large number of mammae allows positive identification by layman and specialist alike. Ten to 12 mammae on both sides of the belly, from breast to groin, gives the female a high suckling capacity and gravid females carrying up to 24 fetuses are known (Smithers 1975). Multimammate mice are smaller than ordinary rats and are therefore not easily confused with *Rattus*. Although they occur predominantly in the veld, they have become commensal with man and are the common 'House Mouse' of many areas in southern Africa.

### Outline of synonymy

1834 *Mus natalensis* A. Smith, *S. Afr. Quart. Journ.* 2:156. 'About Port Natal' (= Durban).

1834 *Mus caffer* A. Smith, *S. Afr. Quart. Journ.* 2:157. 'Cafferland'.

1836 *Mus coucha* A. Smith, *App. Rept. Exped. Explor. S. Afr.*: 43. Type locality taken by Roberts as Kuruman, CP.

1836 *Mus marikquensis* A. Smith, *App. Rept. Exped. Explor. S. Afr.*: 43. Type locality taken by Roberts as Marico River, western Transvaal.

1842 *M. silaceus* Wagner, *Arch. f. Naturgesch.* 8(1):11, Cape of Good Hope.

1843 *M. muscardinus* Wagner. *In* Schreber's *Säugeth.* Suppl. 3:430. Kaffraria, CP.

1852 *M. microdon* Peters, *Reise nach Mossambique. Säugeth.*: 149. Tete, Mozambique.

1890 *M. fuscus* Bocage, *Journ. Sci. Math. Phys. Natl. Lisbon* 2:14. Angola.

1905 *M. coucha zuluensis* Thomas & Schwann, *Proc. zool. Soc. Lond.* 1:268. Umfolozi, Natal. (Probably the same as *M. c. microdon* (see Thomas & Wroughton, *Proc. zool. Soc. Lond.* 1908:546) in which case *natalensis* supersedes.)

1909 *M. illovoensis* Jentink, *Zool. Jahrb. Jena. Syst.* 28:248. Illovo, Natal.

1913 *Mus socialis* Roberts, *Ann. Transv. Mus.* 4:88. Transvaal. Not of Pallas, 1773.

1914 *Mus limpopoensis* Roberts, *Ann. Transv. Mus.* 4:183. Sand River, Transvaal.

1915 *Mastomys* Thomas, *Ann. Mag. nat. Hist.* (8) 16:477. As a subgenus of *Epimys (= Rattus)*. (Genotype *Mus coucha* A. Smith 1836.)

1915 *Mus breyeri* Roberts, *Ann. Transv. Mus.* 5:120. Potgietersrust, Transvaal.

1926 *Mastomys coucha komatiensis* Roberts, *Ann. Transv. Mus.* 11:258. Etosha Pan, SWA/Namibia.

1926 *M. coucha bradfieldi* Roberts, *Ann. Transv. Mus.* 11:257. Okahandja, SWA/Namibia.

1934 *M. coucha sicialis* (sic) (Roberts). *In* Shortridge, *The mammals of South West Africa.*

1939 *Mastomys coucha natalensis* (A. Smith). *In* Allen, *Bull. Mus. comp. Zool. Harv.* 83:400.

1951 *Mastomys natalensis natalensis* (A. Smith). *In* Roberts, *The mammals of South Africa.*

1953 *Rattus natalensis* (A. Smith). *In* Ellerman *et al.*, *Southern African mammals.*

1974 *Praomys (Mastomys) natalensis* (A. Smith). Misonne, *in* Meester & Setzer (eds), *The mammals of Africa: an identification manual.*

### TABLE 54
**Measurements of male and female *Praomys (Mastomys) natalensis***

| | Parameter | Value (mm) | N | Range (mm) | CV (%) |
|---|---|---|---|---|---|
| | HB | 117 | 61 | 100–145 | 8,9 |
| | T | 117 | 56 | 107–156 | 8,3 |
| Males | HF | 23 | 66 | 20–26 | 6,5 |
| | E | 18 | 63 | 16–22 | 7,7 |
| | Mass: | 62 g | 5 | 56–70 g | 9,7 |
| | HB | 115 | 46 | 102–133 | 8,8 |
| | T | 113 | 47 | 101–143 | 9,1 |
| Females | HF | 22 | 44 | 20–25 | 6,6 |
| | E | 18 | 42 | 16–20 | 6,3 |
| | Mass: | 54 g | 10 | 49–62 g | 8,0 |

In the Wankie National Park, Zimbabwe, Wilson (1975) reported an average weight of 63 g and 60 g for males and females respectively (N = 25 in both cases, observed range 50–77 g in males and 51–88 g in females). Smithers (1971) recorded similar figures for populations in Botswana: in the case of males, the mean value equalled 64,5 g (N = 14, observed range 51,3–77,8 g) and 56,8 g for females (N = 16, observed range 383,–74,5 g).

## Identification

The fur of the species is moderately long and comparatively soft to the touch. The dorsal colour (in adults) is buffy, suffused with black hairs. The dorsal fur is variable in colour even within a population from any one locality. Botswana specimens have been described as indifferently coloured, generally dark grey dorsally or with a distinct brownish tinge (Smithers 1975). Smithers (1971) has found co-inhabitants varying from dark grey to predominantly brown. Rosevear (1969) points out the existence of a number of long, fine and dense underfur, as well as the presence of a number of dark-tipped gutter hairs as well as a few rather longer guard hairs, both of which taper to very fine bases and project somewhat beyond the general contour of the coat. The underparts vary from white to darkish grey, the individual hairs being grey at their bases, often with whitish tips. There is a distinct difference in colour between juvenile and older animals. The former are dull, smoky grey, becoming a paler drab as they age, to become more rusty-coloured and grizzled when they mature (plate 11).

The tail is relatively short, rarely as long as or slightly longer than the head-body length. It is sparsely covered with short, rigid hairs, brownish above and a dull white below. The rings on the tail are fairly close to each other. The length of the tail can also vary considerably in animals from the same area.

The hands and feet are fairly narrow. In the hindfoot, the hallux falls just short of the base of the second digit. The D5 reaches to more than the base of the D4.

The ears are of moderate size, ovate and thinly covered with sparse, short hair.

Usually there are at least eight pairs of mammae and according to Smithers (1975), the rows of mammae are clearly visible in adult females as each teat is usually ringed with lighter coloured hairs. The mammary formula, therefore, reads 8–12=16–24. These teats do not always occur in pairs.

## Skull and dentition

Skulls show a great structural variation according to age. The supraorbital ridges are weakly developed and the interorbital constriction is variable. According to Rosevear (1969), two features should be looked at to determine a *Mastomys* skull with certainty:

(i) The palatine never extends further forward than the

first lamina of the M$^2$ (more often it lies posterior to this point).

(ii) The anterior palatine foramina always end well between the molar rows, usually to the median (lingual) root of the M$^1$.

Apart from these features, the palate is never more than twice the width of the M$^1$, while the internal nares are generally narrowed, as are the pterygoids. The bullae are variable in size, but never large (fig. 92).

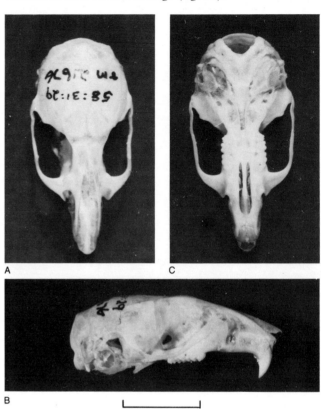

**Figure 92** Dorsal (A), lateral (B) and ventral (C) views of the skull of the Multimammate Mouse *Praomys (Mastomys) natalensis*. The ridges on the skull are poorly developed. The anterior palatine foramina always extend well between the molars, usually to the median root of the M$^1$. The suture between the maxilla and the palatine never lies further forward than the first lamina of the M$^2$. The choanae are narrowed (narrower than the width of the M$^1$) and the bullae never large. The pterygoids are narrow. The edge of the zygomatic plate is vertical. The incisors are orthodont.

Rosevear (1969) remarks that the skull in the female is less crested and not as angular as in the male.

*P. (M.) natalensis* has a more slender and smaller skull than its close ally *P. (Myomyscus) verreauxii*.

The M$^1$ is three-rooted with the t7 lacking. The t1 is often shifted well to the rear and it looks as if it has a closer association with the t5 of lamina two than with its adjoining t2. The t4 often vanishes with wear. The basal portion of the t2 is often projected considerably forward. The M$^2$ is smaller than the M$^1$ with the t2 (as well as the t7) absent and the t3 small to vestigial. The M³ in turn is also smaller than the M² and may be moderately or strongly reduced. Nevertheless, in young animals it can still be demonstrated to possess cusps t1, 4, 5, 6, 8 and 9, arranged in three laminae. On the whole, the molars are well cusped (fig. 93).

There is nothing special about the lower dentition and this also applies to the orthodont upper incisors.

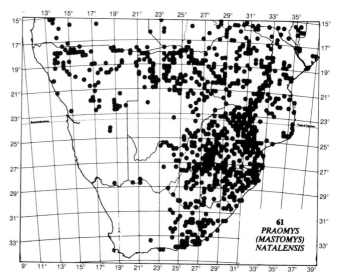

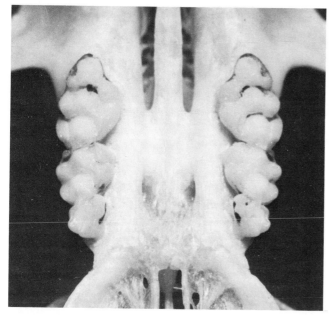

**Figure 93** Cheekteeth of the Multimammate Mouse *Praomys (Mastomys) natalensis*. The basal portion of the t2 of the three-rooted M¹ often projects forward considerably. The M² is smaller than the M¹ but markedly bigger than the M³. The molars are well cusped.

## Distribution

The Multimammate Mouse occurs in the eastern Cape Province, including King William's Town and Pondoland in Transkei, northeastwards into Natal, KwaZulu and to Swaziland and the Transvaal along the escarpment, as well as in the northern Cape Province, Lesotho and the OFS (map 61). It occurs throughout SWA/Namibia (except in the Namib Desert and in the more arid parts of the north), into the Caprivi Zipfel, Botswana, Zimbabwe and Mozambique in the east where it is widely distributed '... and the commonest and most widespread species occurring' (Smithers & Lobão Tello 1976). As far as biotic zones are concerned, this species frequents both the Southern Savanna Woodlands and Grasslands as well as the moister (250 mm and over) parts of the South West Arid (Davis 1974).

## Habitat

The Multimammate Mouse has a wide habitat tolerance and occurs from sea level to high lying ground. However, it is absent from really dry or arid areas. In Mozambique, Smithers & Lobão Tello (1976) reports them oc-

curring in rainfall areas varying between 400 mm to 1 800 mm annually. They are fond of grassland where there is some cover of low scrub. In Botswana they are often encountered in dry water courses, on the fringes of swamps in Ngamiland, in *Acacia* and mopane scrub and woodland, as well as in *Terminalia* scrub (Smithers 1971). They are well-known commensals with man and tend to be abundant where population concentrations are high. Consequently, they are often found in scrub fences erected around cultivated lands. Smithers states that they also frequent the fringes of pans, especially where there are calcareous outcrops nearby. They are also partial to sandy ground, overgrown with scrub and grass, where they may excavate their own burrows. Alternatively, they will use any sort of cover which happens to be available, e.g. spaces under fallen logs, crevices between rocks, cavities inside piles of stones or debris or even holes in termite mounds. In Zimbabwe they also show a wide habitat range. They may occur under conditions ranging from dry grassland with *Terminalia* scrub on Kalahari sand to areas close to water on basalt soils with mopane woodland (Wilson 1975).

The ecological role played by *Praomys (Mastomys) natalensis* in the variety of habitats in which it occurs, has been discussed by Meester, Lloyd & Rowe-Rowe (1979). It lives under diverse climatic and geographic circumstances, is common in cultivated areas and around human habitation and it coexists from place to place with a variety of other small mammal species. Meester *et al.* (1979) therefore interpret the Multimammate Mouse as a species with generalised ecological requirements. Field work has shown that where it occurs, *P. (M.) natalensis* (with its wide habitat tolerance and adaptability) is likely to be 'one of the first species to become established in an area which is recovering from some form of habitat destruction'. These processes involve destruction by fire as well as habitats modified by man. While these conditions are suboptimal for other small mammals, (e.g. the establishment of pioneer secondary succession) the greater plant

211

species diversity, often with a high proportion of weed and exotics, provides a varied choice of food and shelter for the Multimammate Mouse. As succession proceeds, *P. (M). natalensis* seems to be replaced by specialist species of small mammals, because the latter are more closely adapted to more specialised and homogeneous plant communities, which is characteristic of later stages of plant succession. In grassland, for instance, where succession proceeds relatively fast, *P. (M) natalensis* may be replaced within months by grassland 'specialists' such as the Striped Mouse *Rhabdomys pumilio* and at a somewhat later stage, by vlei rats, *Otomys* spp. As such, Meester *et al.* (1979) point out that *P. (M.) natalensis* can be regarded as a pioneer species '... and its. presence as indicating an area in an early stage of secondary succession after habitat damage of some sort'.

## Diet

Smithers (1971) states that the Multimammate Mouse eats grass and other seeds, including *Acacia* seeds. They also take the dry pods as well as the pulpy exterior of wild fruits. When populations reach high levels they will eat dry grass stems and tree bark and may become cannibalistic (Smithers 1971). They appear to be more common where water is available, but are apparently not dependent on it.

Their commensalism with man is not surprising because they show a marked preference for grain (Roberts 1951). Shortridge (1934) has also stated that they are mainly granivorous, but that they also occasionally feed on locusts and other insects. This observation is borne out by work done by Delany (1964), who analysed the contents of some 25 stomachs containing remnants of ants, orthopterans and beetles. Rosevear (1969) states that the species prefers hard foods to soft. Near the Main Camp area in the Wankie National Park, they have been observed to feed on seeds which had been passed out undigested in elephant droppings (Wilson 1975) while Wilson also reports them as feeding on the seeds of *Acacia* spp., *Colophospermum mopane* and the fruits of *Ziziphus mucronata, Diospyros mespiliformis* and *Terminalia sericea.*

## Habits

Multimammate mice are communal, terrestrial and nocturnal. They freely enter huts and other human abodes, where they breed under floors, in the thatch of roofs or in holes in the walls. In fact, they will shelter in any kind of hole in the ground or under the foundation of houses. Swanepoel (1972) has found that the home ranges of males significantly exceed those of females.

Although in Africa they are very closely associated with man, they do appear to give way to invading house rats *Rattus rattus* or brown house rats *Rattus norvegicus* whenever these foreigners appear on the scene. Rosevear (1969) makes the interesting observation that although

they are house-loving, they seem to avoid towns, possibly to avoid meeting unwelcome competition.

They will make their own burrows, but Shortridge (1934) believes that they use the existing tunnels of gerbils and other subterranean rodents extensively. Such a burrow consists of a network of galleries without a central chamber. Rosevear (1969) states that these galleries may be shared with other murines. What he calls a 'breeding gallery' is usually lined with soft material made up of long sections of leafblades, stems of grasses and other (types of soft) debris.

In captivity they are clean animals and they can often be observed grooming themselves. They urinate and defaecate in set toilet areas under laboratory conditions and are peaceful and tolerant of other rodents. Yet they are wary animals and are easily alarmed. The alarm or defence note is a harsh screech. They twitter when content in company of others. They are good swimmers and fairly competent climbers. In captivity they become fairly tame, breed freely and are less quarrelsome than most other African rodents. For a discussion on the behaviour of the Multimammate Mouse, see Veenstra (1958).

Under natural conditions, Shortridge (1934) believes them not to be gregarious. He also pointed out, that in his experience in SWA/Namibia, only a particularly small percentage of fully adult animals are trapped, the bulk of the specimens being in various stages of immaturity. For this reason, he warns, measurement tables of *Mastomys* could be misleading, even when several large series are obtained.

## Predators

The predators include various species of owls, including the Giant Eagle Owl *Bubo lacteus,* the White-faced Owl *Otus leucotis* (Worden & Hall 1978), the Barn Owl *Tyo alba* (Vernon 1972; Davis 1973; Dean 1977), the Marsh Owl *Asio capensis,* the Spotted Eagle Owl *Bubo africanus* (Dean 1977) and the Grass Owl *Tyto capensis* (Vernon 1972, Davis 1973; Dean 1977), the Long-crested Eagle *Lophaetus occipitalis* (Jarvis & Grichton 1977), smaller carnivores and snakes. They are also taken by domestic cats and Booth (1970) states that their numbers could be kept down in domestic dwellings in West Africa, if only the West Africans would keep cats for this purpose.

## Reproduction

According to Smithers (1975) the average number of pups in a litter varies from ten to 16, depending on the conditions. They tend to breed throughout the year, although less so during the colder months of the year, from May to July. In Botswana the percentage of gravid females form a decided peak in March and there is a marked drop in gravid females from March to September. In Zimbabwe, Wilson (1975) notes that pregnant females with full-term fetuses were taken as follows: August (12), September (7), October (18), February (21), May (180),

June (71), and July (19), thus confirming that breeding activities occur throughout the year. In SWA/Namibia, two females (each with nine fetuses) were collected in the Gobabis district in December. The breeding season of *P. (M.) natalensis* in the Transvaal highveld was studied by Coetzee (1965), while Meester (1960) studied the early post-natal development. An earlier laboratory study on mortality, fecundity and intrinsic rate of natural increase was done by Oliff (1953).

The first litters are usually small and the young can commence breeding at an age of approximately 3–3½ months. After a gestation period of 23,12 ± 0,66 days (Johnston & Oliff 1954) the young are born helpless and hairless, covered by a pink and translucent skin. These altricial young are blind and the ear pinnae are positioned close to the contour of the head. Their eyes open on the 15th to 16th day. The young stay in the nests until capable of fending for themselves. The number of young reared at a time can be anything from 12 to 20 (Roberts 1951). Under normal circumstances they construct rounded nests in subterranean nesting chambers, but given half the chance in human habitation they will nest anywhere, their chosen sites ranging from drawers and bookshelves to stacks of grain bags and piles of rubble.

Indications are that they can start breeding when they are approximately 100 days old. Giving birth to pups can occur regularly at intervals of 25–27 days, implying that their numbers can increase markedly to unprecedented levels in a fairly short time. When this occurs, this phenomenon is described as a population explosion.

It is interesting to take cognisance of how such an explosion develops and one such occurrence was documented by Smithers (1971). He relates that 1964 and 1965 marked the last two years of a serious drought which affected not only Botswana, but virtually the entire southern section of Africa. This four-year drought was terminated by heavy rains which fell rather patchily throughout Botswana in October and November 1965. Many areas, particularly the Lake Ngami area, benefited as it had not done for four seasons. A return to more favourable conditions was reflected in a lush growth of grasses and annuals forming a thigh-deep cover over large tracts of country. By July 1966 the populations of multimammate mice in the Lake Ngami area were building up and in September 1966 some populations of *P. (M.) natalensis* at Kumana on the Botletle River, 160 km east of Lake Ngami, had already reached extraordinarily high levels. 'Soon after dark they swarmed everywhere even entering the tents, making it necessary to securely pack all soft goods and provisions and to place them out of reach.' Smithers also notes that this explosion of *P. (M.) natalensis* was paralleled to a lesser degree by the Bushveld Gerbil *Tatera leucogaster* and the Pygmy Mouse *Leggada* sp.

This build-up of the Multimammate Mouse continued in the northwestern and western Okavango between April and August, 1967. Their numbers were so vast that they masked the position as far as other species were concerned, although *Leggada* sp. was very plentiful. Towards the end of this period many of the *P. (M.) natalensis* specimens that were collected were diseased – showing swollen lumps on the feet, on the reproductive tract and scrotal sacs, with uteri often grossly deformed. During July 1967 the populations remained at very high levels and it was '... impossible to set a 100m trap-line from 5.30 pm onwards for, on 15 or 20 traps being set, those at the start of the line were already being set off'. At this time cannibalism was rife.

At the peak of these explosions, the animals literally ate themselves out of house and home. They became cannibalistic and diseased, and other natural causes reduced the population to far below the normal level.

A similar population explosion was documented by Wilson (1975) at Robin Camp in the Wankie National Park in Zimbabwe. He reports that every trap set was sprung within one hour (after 18h00). Mice were found in every conceivable cranny and many showed signs of disease (swollen and deformed eyes).

**Parasites**

Laboratory tests have shown *Praomys (Mastomys) natalensis* to be susceptible to neurotropic virus (African horse sickness) (Alexander 1936), melioidosis (Finlayson 1944), *Listeria monocytogenes* (Pirie 1927), *Borrelia duttoni* (Zumpt 1959) and *Trypanosoma brucei* (Curson 1928). Periodic outbreaks of sylvatic plague *(Yersinia pestis)* also occur in natural populations (Mason & Amies 1948).

The mites and chiggers associated with the Multimammate Mouse have been listed by Zumpt (1961a). These include the following: Laelaptidae – *Laelaps giganteus, L. lamborni, L. muricola, L. tillae, L. transvaalensis, L. vansomereni, Haemolaelaps glasgowi, H. labuschagnei, H. murinus, H. taterae, Androlaelaps marshalli* and *A. theseus;* Ereynetidae – *Speleognathopsis bakeri;* Psorergatidae – *Psorergates oettlei;* Trombiculidae – *Trombicula mastomyia mastomyia, T. sicei, T. youhensis, Leptotrombidium legaci, Schoengastia radfordi radfordi, Ascoschoengastia benuensis, Schoutedenichia benuensis, S. brachiospissi, S. cordiformis, S. panai panai, S. p. luberoensis, S. pilosa, S. pirloti, Gahrliepia hypoderma* and *G. traubi;* Listrophoridae – *Listrophoroides africanus, L. mastomys* and *Myocoptes musculinus;* Sarcoptidae – *Notoedres alepis.*

According to Zumpt (1966), the calliphorid fly *Cordylobia anthropophaga* is also associated with the Multimammate Mouse, as are approximately 50 species of flea. These are represented by the following families: Pulicidae – *Echidnophaga gallinacea, Pulex irritans, Ctenocephalides connatus, C. felis, Procaviopsylla creusae, Xenopsylla bantorum, X. cheopis, X. frayi, X. hipponax, X. nubica, X. philoxera, X. phyllomae, X. piriei, X. versuta, X. bechuanae, X. brasiliensis, X. morgandaviesi, X. scopulifer, X. syngenis* and *X. cryptonella;* Pygiopsyllidae – *Stivalius alienus* and *S.*

torvus; Xiphiopsyllidae – *Xiphiopsylla levis;* Hystrichopsyllidae – *Ctenophthalmus acanthurus, C. evidens, C. ansorgei, C. calceatus, C. eumeces, C. gilliesi, C. eximius, C. phyris, Dinopsyllus apistus, D. dirus, D. ellobius, D. grypurus, D. longifrons, D. lypusus, D. pringlei, D. wansoni, Listropsylla agrippinae, L. chelura, L. dolosa, L. dorippae, L. fouriei* and *L. prominens;* Leptopsyllidae – *Leptopsylla aethopica* and *L. segnis;* Ceratophyllidae – *Nosopsyllus fasciatus* and *N. incisus;* and finally the Chimaeropsyllidae – *Hypsophthalmus campestris, Epirimia aganippes, Chiastopsylla rossi* and *C. godfreyi.*

The ticks associated with *Praomys (Mastomys) natalensis* have been recorded by Theiler (1962). These include adults (A) or immatures (I) of the following species: *Ornithodoros zumpti* (A,I), *Ixodes* sp. (A), *I. auriculaelongae* (A), *I. elongatus* (A), *I. nairobiensis* (A), *Haemaphysalis leachii leachii* (A,I), *H. l. muhsami* (A,I), *Rhipicephalus appendiculatus* (I), *R. oculatus* (I), *R. pravus* (I), *R. sanguineus* (I), *R. simus* (I) and *R. tricuspis* (I).

The known endoparasitic worms of the Multimammate Mouse are as follows: *Catenotaenia lobata* (= *C. capensis*), *Hymenolepis diminuta, H. microstoma, H. uncispinosa, Inermicapsifer congolensis, I. madagascariensis, Raillietina (R.) baeri* (Collins 1972), *Trichuris mastomysi* (Verster 1960), *T. natalensis, Syphacia* sp., *Protospirura numidica, Mastophorus numidica* and *Capillaria hepatica* (Boomker *pers. comm.*). Laboratory tests have also shown susceptibility to *Schistosoma mansoni* and *S. bovis* (Lurie & De Meillon 1956).

## Relations with man

The animals are excellent gnawers and when they enter human dwellings, they can be very destructive. In West Africa, they do many thousands of pounds' worth of damage annually, to crops, stored foods and household goods (Booth 1970).

As a commensal with man, it is the common 'House Mouse' of Mozambique, occurring in stores, outhouses and huts built by the local inhabitants where they nest in the thatch or under the floors (Smithers & Lobão Tello 1976). In other African countries it destroys cotton, groundnuts and other crops such as rice fields in Liberia (Rosevear 1969). In Botswana it is common in and on the fringes of cultivated crops such as peanuts, maize and sorghum where they can be a pest, especially when their numbers are high (Smithers 1971).

The population explosions referred to earlier, can be very destructive when they occur in agricultural areas. Smithers (1975) has pointed out that in endemic plague areas their numbers imply a corresponding increase in the number of fleas. As the explosion peak passes, many mice die off and a multitude of fleas, possibly infected with the plague bacillus, abandon their dead hosts to seek a meal of blood on other mammals, including man.

*Mastomys* has become of increasing importance in medical research. They have been used in plague research studies since 1940, in bilharzia research due to their susceptibility to *Schistosoma mansoni* (Randeria 1978) and the species is most susceptible to osteo-arthritis (Sokoloff, Snell & Stewart 1967) while spontaneous lesions of glomerulonephritis, comparable to that in man, has been described (Snell & Stewart 1967). A large variety of spontaneous tumours are often found in older laboratory specimens (stomach, lung, liver, ovary and thymus) which make them well suited for cancer research. For a more detailed review of laboratory uses of *Mastomys*, see Randeria (1978).

Apart from being well known as a plague reservoir in southern Africa, it also appears to be a reservoir of an apparently newly emerging disease in forested West Africa. Lassa fever is a virus infection and in Sierra Leone an attack rate of 2,2 per thousand people was measured with a case-fatality of 38% among hospitalised patients (Arata 1975). The ecology and systematics of the Multimammate Mouse and its relationship to the epidemiology of Lassa fever are presently being studied (Robbins 1978).

## Prehistory

*Praomys (M.) natalensis* has been identified in Pleistocene deposits in Zimbabwe (Zeally 1916) as well as in South African mammalian breccias from Taung, Sterkfontein and Makapansgat (De Graaff 1960a). Remains of this animal were also identified by De Graaff (1961) from the Kromdraai Faunal Site (= Kromdraai A) where breccia was collected by Draper in 1895.

Avery *(in litt.)* has not found *Praomys (Mastomys) natalensis* in any of the upper Pleistocene to Holocene localities which she investigated along the southern Cape Coast and it must, therefore, have arrived subsequently, '... possibly in the wake of European farmers who settled there around 200 BP' (Avery 1977) in view of the fact that *P. (M.) natalensis* is a semi-domestic species and present distribution is possibly dependent on having followed early human population movements (Davis 1962).

## Taxonomy

This is a very difficult taxon to arrange satisfactorily into species and subspecies. Allen (1939) uses the name *coucha* for this species, but Roberts (1944) pointed out that the prior name should be *natalensis*. Roberts (1951) interprets *Mastomys* as a full genus, while Ellerman *et al.* (1953) regard *Mastomys* as a subgenus of *Rattus*. Davis (1962, 1965) construes *Mastomys* as a subgenus of *Praomys* (as does Misonne 1974).

According to Meester *et al.* (1964), several subspecies have been described but are doubtfully valid or not recognised. Some 31 different forms supposedly occuring in Africa are listed in Misonne (1974), while Roberts (1951) recorded seven different subspecies occurring in southern Africa alone.

Formal taxonomy often breaks down when dealing with biologically distinct yet morphologically identical species. Recently, Taylor & Gordon (1978) have demonstrated the presence of generic species in the taxon *P. (M.) natalensis* in Zimbabwe. Two chromosome types 2n=32 (species A) and 2n=36 (species B) occur and electrophoresis of haemaglobin revealed specific haemaglobin phenotypes correlated with chromosome number. There is evidence of positive assortive mating and the existence of good biological species is indicated.

# *Praomys (Mastomys) shortridgei* (St Leger, 1933)

**Shortridge's Mouse**

**Shortridge se Vaalveldmuis**

The derivation of *Praomys* and *Mastomys* has been given under the Multimammate Mouse *Praomys (Mastomys) natalensis*. This species was named after Capt. G.C. Shortridge, a former Director of the Kaffrarian Museum in King William's Town, and who has contributed significantly to our knowledge of southern African mammals.

This species is very similar to the Multimammate Mouse, *Praomys (Mastomys) natalensis*. The type specimen is housed in the British Museum (Natural History) and was taken at the Okavango-Omatako Junction in the Grootfontein district of SWA/Namibia.

## Outline of synonomy

1915 *Myomys* Thomas. *Ann. Mag. nat. Hist.* (8) 16:477. As a subgenus of *Epimys* (= *Rattus*), type *Epimys colonus* (A. Smith) = *Mus colonus* Brants.

1933 *Myomys shortridgei* St Leger, *Proc. zool. Soc. Lond.*: 411. Okavango-Omatako Junction, Grootfontein district, SWA/Namibia.

1939 *Myomys colonus shortridgei* St Leger. *In* Allen, *Bull. Mus. comp. Zool. Harv.* 83:405.

1951 *Myomys shortridgei* St Leger. *In* Roberts, *The mammals of South Africa*.

1953 *Rattus angolensis legerae* nom. nov. *In* Ellerman *et al., Southern African mammals.*

1974 *Praomys (Mastomys) shortridgei* (St Leger). Misonne, *in* Meester & Setzer (eds), *The mammals of Africa: an identification manual.*

## Identification

This species is darker than *P. (M.) natalensis* which occurs in the same area (Meester *et al.* 1964) and Smithers (1971) has reported that some specimens are almost black. The general colour is smoky-grey, darker on the mid-dorsal surface. The individual hairs on the back are iron-grey with short, buffy tips. The buffy tips become longer and more noticeable laterally and on the flanks. The belly is white with the individual hairs slaty-grey for about half their length from the base. The tail is brownish, finely scaled and haired. The hands and feet are white (plate 11).

The mammary formula reads 3–2=10 compared to the six or more pairs of mammae in *P. (m.) natalensis*. Shortridge (1934) states that there is a slight gap between pectoral and inguinal mammae while Smithers (1971), on the other hand, writes that the mammae are not separated into pectoral and inguinal pairs.

TABLE 55
Measurements of male and female *Praomys (Mastomys) shortridgei*

|  | Parameter | Value (mm) | N | Range (mm) | CV (%) |
|---|---|---|---|---|---|
| Males | TL | 113 | 16 | 106–127 | 3,4 |
| | T | 104 | 15 | 96–114 | 4,2 |
| | HF c/u | 27 | 15 | 25–28 | 2,6 |
| | E | 18 | 14 | 17–19 | 3,0 |
| | Mass: | 45 g | 15 | 35–67 g | 15,0 |
| Females | TL | 116 | 10 | 106–144 | 5,4 |
| | T | 106 | 10 | 99–111 | 2,9 |
| | HF c/u | 26 | 9 | 24–27 | 3,4 |
| | E | 19 | 19 | 17–20 | 4,5 |
| | Mass: | 40 g | 10 | 36–74 g | 26,0 |

## Skull and dentition

The pterygoidal elements of the skull are wider in *Praomys (Mastomys) shortridgei* than in *P. (M.) natalensis*, while the anterior palatinal foramina are shorter in the former, i.e. not reaching beyond the inner root of the $M^1$ (fig. 94).

## Distribution

Shortridge's mice are known from the Okavango region of northern SWA/Namibia and they also occur in the northwestern corner of Botswana (map 62). They are restricted to the Southern Savanna Woodland biotic zone (Rautenbach 1978a).

## Habitat

In Botswana they prefer rocky areas on raised banks of the Okavango River at Shakawe. They are also found in the dry underbrush on the fringes of vleis on the SWA/Namibian border (Smithers 1971). They are plentiful among reedbeds and in swamp grass in SWA/Namibia (Shortridge 1934).

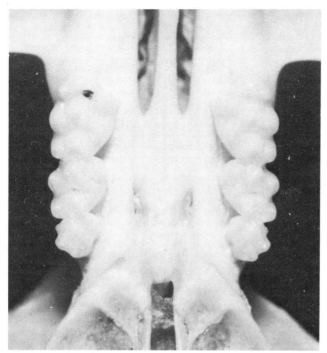

**Figure 94** Cheekteeth of Shortridge's Mouse *Praomys (Mastomys) shortridgei.*

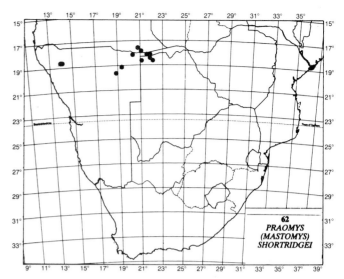

### Diet

Very little is known, but they are graminivorous, while a tendency towards omnivory has been mentioned by Smithers (1971).

### Habits

The species is terrestrial and nocturnal and commensalism with man is suggested. They certainly occur around houses with *P. (M.) natalensis* where, according to Smithers (1971), the latter is the commoner of the two species. Shortridge's Mouse is more essentially a swamp animal, but resembles the Multimammate Mouse in other respects. Shortridge (1934) states that it does not seem to live in burrows and that it often associates with *Otomys, Dasymys* and other rodents in swamp-like habitats.

### Predators, reproduction, parasites, relations with man, and prehistory

No information available.

### Taxonomy

Roberts (1951) and earlier authors (e.g. St Leger 1933) place this species in the genus *Myomys* which Ellerman *et al.* (1953) regard as a synonym of *Rattus* with *shortridgei* as a synonym of *Rattus angolensis* (extralimital to this work). Consequently, *shortridgei* becomes a *nomen nudem* being a homonym of *Rattus* (i.e. *Thallomys paedulcus shortridgei*). Ellerman *et al.* (1953) therefore proposed the name *legerae* to replace it. However, as Meester *et al.* (1964) point out, as *shortridgei* is here placed in *Praomys,* while *paedulcus shortridgei* is again designated by the generic name *Thallomys,* the name is no longer preoccupied and hence valid, with *legerae* as a synonym. Furthermore, it has also been separated specifically from *R. angolensis* by Davis (1965).

---

SUBGENUS *Myomyscus* Shortridge, 1942

# *Praomys (Myomyscus) verreauxii* (A. Smith, 1834)

**Verreaux's Mouse**

**Verreaux se Vaalveldmuis**

The meaning of *Praomys* has been given under *Praomys (Mastomys) natalensis. Myomyscus* is derived from the Greek *myodes* = like a mouse. The Latin suffix *-cus* is added to noun stems denoting possession. The generic name therefore points to the mouselike disposition of this rodent. The species is named after Pierre Jules Verreaux, naturalist and taxidermist who lived at the Cape between the years 1818 and 1838. He acted as temporary curator of the South African Museum in Cape Town during the absence of Sir Andrew Smith, who was on an expedition to the interior between 1834 and 1836.

The genus *Myomyscus* was proposed by Shortridge in 1942 but has been accepted as a subgenus of *Praomys* in more recent times with *Mus verreauxii* as the type species.

The previous name of *Myomys* is abandoned because its original type species *Mus colonus* cannot be defined precisely and the type specimen appears to be lost (Misonne 1974). In the *South African Red Data Book* its status is given as rare, but in no apparent danger of extinction (Meester 1976).

According to Roberts (1951) the species comes near to *Zelotomys woosnami*, although it is somewhat smaller, has a slightly longer tail and is more thickly haired.

## Outline of synonymy

1827 *Mus colonus* Brants, *Het Geslacht der Muizen:* 124. 'Eastern Parts of the Cape Colony.' (Algoa Bay, according to Allen, 1939.)

1834 *Mus verroxii* A. Smith, *S. Afr. Quart. Journ.* 2:156. Vicinity of Cape Town.

1901 *Mus verreauxi* Sclater, *The mammals of South Africa* (emendation).

1939 *Myomys verroxii* (A. Smith). *In* Allen, *Bull. Mus. comp. Zool. Harv.* 83:406.

1942 *Myomys (Myomyscus) verroxii* Shortridge, *Ann. S. Afr. Mus.* 36:93.

1951 *Myomys colonus verreauxi* (A. Smith). *In* Roberts, *The mammals of South Africa.*

1953 *Rattus (Praomys) verreauxi* (A. Smith). *In* Ellerman *et al., Southern African mammals.*

1964 *Praomys verrauxi* (A. Smith). *In* Meester *et al., An interim classification of southern African mammals.*

1974 *Praomys (Myomyscus) verreauxi* (A. Smith). Misonne, *in* Meester & Setzer (eds), *The mammals of Africa: an identification manual.*

### TABLE 56
Measurements of male and female *Praomys (Myomyscus) verreauxii*

| | Parameter | Value (mm) | N | Range (mm) | CV (%) |
|---|---|---|---|---|---|
| Males | HB | 106 | 15 | 90–118 | 8,7 |
| | T | 143 | 15 | 124–156 | 7,1 |
| | HF | 24 | 15 | 20–28 | 8,1 |
| | E | 18 | 11 | 17–20 | 6,5 |
| | Mass: | 44 g | 4 | 41–54 g | 14,8 |
| Females | HB | 110 | 10 | 102–133 | 9,2 |
| | T | 143 | 16 | 132–157 | 5,0 |
| | HF | 24 | 16 | 22–26 | 4,9 |
| | E | 10 | 14 | 16–21 | 9,2 |
| | Mass: | 38 g | 5 | 36–42 g | 6,2 |

## Identification

The general dorsal colour is buffy-greyish, rather dark in appearance on account of the presence of dark-tipped hairs. The flanks are lighter, shading into grey-white un-

derparts from chin to anus. Individual hairs of both the dorsal and ventral surfaces are slaty-grey at their bases. The hands and feet are white. The D5 of the hindfoot is fairly long. The whiskers and ears are long and obvious. The tail is long, averaging 120% of the head-body length. It is darker above than below. Young individuals are said to be greyer than adults (Roberts 1951) (plate 11).

The mammary formula is given as 3–2=10 (i.e. three pairs of pectoral and two pairs of inguinal mammae).

## Skull and dentition

The skull is slightly broader than that of *Praomys (Mastomys) natalensis*. As a rule, there are no supraorbital ridges. The zygomatic plate is slightly cut back above. The anterior palatine foramina reach only to the anterior edge of the inner root of the $M^1$. The internal nares are narrow and this also applies to the width between the pterygoids. The pterygoids themselves are slender and narrow. The bullae are approximately 15–17% of the occipitonasal length (fig. 95).

The dentition is not robust and the t3 of the $M^1$ is clearly defined. An obvious feature is the strongly reduced size of the $M^3$ (fig. 96).

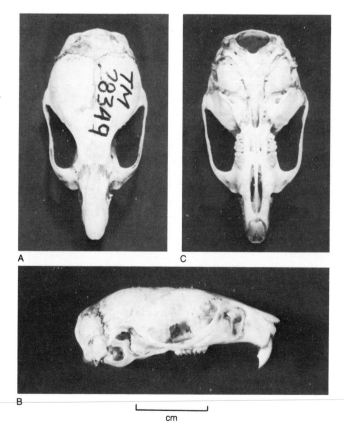

**Figure 95** Dorsal (A), lateral (B) and ventral (C) views of the skull of Verreaux's Mouse *Praomys (Myomyscus) verreauxii.* Supraorbital ridges absent as a rule; the anterior palatine foramina reach only to the anterior edge of the inner root of the $M^1$. The edge of the zygomatic plate is cut back dorsally. The choanae are narrower in width than the width of the $M^1$.

217

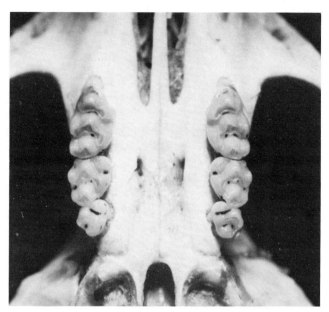

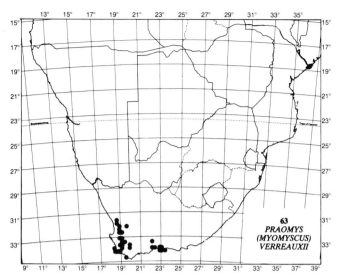

**Figure 96** Cheekteeth of Verreaux's Mouse *Praomys (Myomyscus) verreauxii*. The t1 of the three-rooted M¹ is large. The M³ is strongly reduced in size.

## Distribution

Verreaux's Mouse occurs in the southwestern Cape only, ranging from the Olifants River and Klaver in the northwest to Knysna in the east. It appears to be restricted to the coastal areas of the South West Cape biotic zone (map 63).

## Habitat

In the vicinity of Knysna, it appears to frequent river banks near the sea. It can also be found in damp, meadow grass. It occurs in forests in the Knysna area, where it favours fallen trees as well as grass-filled vleis of the open veld (Grant *in* Shortridge 1934).

## Diet, habits, predators, reproduction, relations with man, and prehistory

No information available.

## Parasites

The endoparasitic worms found in Verreaux's Mouse are *Heligmonoides stellenboschius* (Ortlepp 1962a) and *Trichiuris* sp. (Collins *pers. comm.*) The fleas as recorded by Zumpt (1966) include two hystrichopsyllids *(Dinopsyllus abaris* and *D. ellobius)* and the chimaeropsyllid *Epirimia aganippes*. Theiler (1962) has listed immature specimens of two tick species, including *Ixodes* sp. and *Haemaphysalis leachii muhsami*.

## Taxonomy

Roberts (1951) and earlier authors place this species in the genus *Myomys* which Ellerman *et al.* (1953) regard as a synonym of their subgenus *Rattus (Praomys)*.

GENUS *Rattus* Fischer, 1803

The main taxonomic stumbling block in the subfamily Murinae is undoubtedly the sorting out of subdivisions of the genus *Rattus* which was first defined by Fischer in 1803. For many years the great majority of murines which were obviously either 'rats' or 'mice' were referred to the genus *Mus*. This policy continued until 1881, when Trouessart proposed the erection of the subgenus *Epimys* to cover the more typical 'rats' as opposed to 'mice'. In this move, however, he was unaware of Fischer's separation of *Rattus* from *Mus*. When Trouessart's scheme was implemented, only 27 forms were left in the typical *Mus*, while 140 forms were placed in *Epimys* which became an unwieldy taxon to handle conveniently.

In 1910, G.S. Miller raised the subgenus *Epimys* to generic rank, but in 1916 Hollister pointed out the overlooked priority of *Rattus*. Consequently, all *Epimys* have been returned to *Rattus* and *Epimys* no longer has any standing taxonomically.

In 1915, Thomas thought to simplify the position (1915b) by separating a few African *Rattus*-like groups as subgenera which he finally raised to generic rank in 1926. In this he relied on the rather unsatisfactory criterion of the mammary formulae. The genera proposed by Thomas were widely accepted until the beginning of the 1940s. In his now classic work *The families and genera of living rodents,* Ellerman (1941) placed the taxonomic em-

218

phasis almost completely on the cusp pattern of the upper molars, thereby bringing almost all genera together once again in company with a multitude of rats from Europe, Asia and Australia as subgenera of *Rattus* Fischer. Thus reconstituted, *Rattus* includes 554 forms, about half the known murines.

Rosevear (1969) points out that tooth pattern is of profound systematic importance, but questions whether it is the sole criterion for generic separation. He concedes that *Rattus* might have some clearly marked characteristics, associating the 554 forms with some certainty. But, in Ellerman's own words, it was an exceedingly difficult genus to define. The question is not so much whether the distinctions drawn by Thomas and others exist with reasonable constancy '... as the degree of importance that should be attached to them'. Rosevear questions whether these distinctions amount to anything more than specific differences, and probably not of such moment as to support genera or even subgenera as Thomas suggested in 1915 and 1926. Normally, a species defines itself by its inability to interbreed with other species but '... what constitutes a genus is purely a matter of opinion' (Rosevear 1969). The important outcome of the matter is that modern thought is swinging away from Ellerman's wide-reaching apprehension, especially when data based on cytogenetics, serology, ethology and ecology and other newer methods reveal evidence that cannot be ignored and that must be taken into account alongside information based on dental morphology.

### Key to the species of *Rattus*
(Modified after Roberts (1951))

1 Size smaller, HB length 150–200 mm, but tail longer in proportion, 185–245 mm; skull greatest length 38–44 mm, width 18,5–21,5 mm; braincase wider; antero-external cusp of M$^1$ not reduced........
*Rattus rattus*
(House Rat)
Page 219

Size larger; HB length 210 mm and more, but tail shorter, less than 210 mm; skull greatest length 45 mm and more, width 23–25 mm; braincase narrower; antero-external cusp of M$^1$ reduced ...........
*Rattus norvegicus*
(Brown House Rat)
Page 225

# *Rattus rattus* (Linnaeus, 1758)

**House Rat**

**Black Rat**

**Huisrot**

The word *rattus* is the mediaeval Latin for the 'Black Rat', which was an animal unknown to the Romans themselves residing in Italy.

The House Rat *Rattus rattus,* has become established in most parts of the world, including southern Africa. Like the Brown House Rat *Rattus norvegicus,* it has become cosmopolitan and has been carried all over the world by ships from European harbours. This could convey the impression that the Black House Rat is indigenous to Europe, which in fact, it is not. It has been established in Europe since the 13th century.

The literature on this species is vast and there are well over 2 000 works on the behaviour of *R. rattus* alone (Kingdon 1974). Originally, it appears to have been a native of India or Burma whence it must have been carried to the eastern Mediterranean region in the course of commerce, probably by camel caravan (Rosevear 1969). As was indicated above, it was unknown in northern and western Europe until the 13th century AD where it started spreading across the continent by shipping as well as other ways, which also included the Crusades. These were a series of wars undertaken with papal authorisation by the Christians of western Europe from 1095 to about 1450, for the purpose of recovering the Holy Sepulchre at Jerusalem from the Muslims.

The origin of this species therefore seems to be in the tropics or warmer climates of the earth (see section on 'Distribution'). The species is rarely completely black and the vernacular name is misleading. The commonest colour of the typical form is a slaty-grey, while it can also occur in different shades of brown. Rosevear (1969) laconically points out that the Brown House Rat as a vernacular is also not acceptable because of the development in Europe of a black Brown House Rat.

The Black House Rat occurs in three colour varieties. Specimens belonging to the Black House Rat *Rattus rattus rattus* usually have blackish to dark slaty-grey or a deep sepia-brown backs, with grey bellies. The type specimen, no longer in existence, was described from Upsala, Sweden, by Linneaus in 1758. The Alexandrine Rat, *R. r. alexandrinus* (also known as the Grey Rat), is coloured grey-brown or light sepia-brown dorsally and also shows a grey belly. The type specimen, at one time thought to have been in the Natural History Museum in Paris, was described from the Egyptian city of Alexandria by Geoffroy in 1803. The third variety, the Frugivorous Rat (also known as the White-bellied Rat) *R. r. frugivorus* has a rather thicker and heavier belly fur and a whiter belly than either *rattus* or *alexandrinus* and the type (probably not in existence) was described from Sicily by Rafinesque in 1814. *Rattus r. frugivorus* appears to be the most ancestral form of the three varieties and prefers vegetation around homesteads. When crosses are made in the laboratory, Feldman (1926 *in* Kingdon 1974) found that *R. r. rattus* was genetically dominant over *R. r. frugivorus,* which was in turn dominant over *R. r. alexandrinus.*

The skull of the *Rattus rattus* group shows moderately cusped molars in the young, while the molars are not excessively heavy either. Ellerman (1941) also points out that the bullae are relatively large (17–20% of the occipitonasal length) with the $M^3$ comparatively little reduced.

The tail is usually longer than the head and body, the mammae varying between 3–3=12 or 2–3=10. The animals are of moderate size, head-body length 144 mm or more, but usually under 200 mm.

This group is principally Indo-Malayan in its geographical distribution, although a few races have now become more or less cosmopolitan.

## Outline of synonymy

*Rattus rattus rattus*
1758 *Mus rattus* Linnaeus, *Syst. Nat.* ed. 10, 1:61. Upsala, Sweden.
*Synonyms*
1803 *Rattus rattus* Fischer, *National Mus. Naturg. Paris.* 2:128. (Genotype *Mus norvegicus* Erxleben, 1777.)
1833 *Mus tectorum* var. *fuliginosus* Bonaparte, *Iconogr. Fanna Ital.* 1: fasc. 3. Italy.
1867 *Rattus domesticus* Fitzinger, *Sitzb. k. Akad. Wiss. Wien.* 56(1):64. Germany.
1881 *Epimys rattus* Trouessart, *Bull. Soc. Etudes Sci. Angers.* 10:117. *(Mus rattus* Linn.)
1902 *R. alexandrinorattus* Fatio, *Rev. Suisse. Zool.* 10:402. Switzerland.
1905 *R. rattus ater* Millais, *Zoologist* (4) 9:205. London.

*R. r. alexandrinus*
1829 *Mus alexandrinus* E. Geoffroy & Audouin, *Deser. de l'Egypte.* 2:733. Alexandria, Egypt.
*Synonyms*
1837 *R. asiaticus* Gray, *Ann. Mag. nat. Hist.* 1: 585, India.
1841 *R. sylvestris* Pictet, *Mem. Soc. Phys. Hist. Nat. Genève:* 153. Switzerland.
1852 *R. tettensis* Peters, *Reise nach Mossambique. Säugeth.:* 156. Tete, Mozambique (doubtless an introduced animal).

1871 *R. novaezelandiae* Buller, *Trans. New. Zeal. Inst.* 3:1. New Zealand.
1877 *R. samharensis* Heuglin, *Reise n. Ost. Afr.* 2: 67. Eritrea.
1882 *R. intermedius* Ninni, *Atti. Inst. Venet.* 5 (8): 571. Venice, Italy.
1883 *R. tamarensis* Higgins & Petterd, *Proc. Roy. Soc. Tasmania.* 185. Tasmania.

*R. r. frugivorus*
1814 *Rattus frugivorus* Ranesque, *Prec. des Découv. et. Travaux Somiologiques:* 13. Sicily.
*Synonyms*
1825 *R. tectorum* Savi, *Nuovo Giorn. de Lett. Pisa.* 10:74. Italy.
1827 *R. siculae* Lesson, *Man. de Mamm.:* 274.

*General*
1835 *R. latipes* Bennett, *Proc. zool. Soc. Lond.:* 89. Asia Minor.
1838 *R. insularis* Waterhouse, *Zool. Voy. Beagle:* 35. Asia Minor.
1839 *R. galapagoensis* Waterhouse, *Zool. Voy. Beagle:* 65. Galapagos.
1842 *R. subcaerulus* Lesson, *Nouv. Tabl. Regn. Anim.:* 138. France.
1863 *R. arboricola* Gould, *Mamm. Austr.:* 35. Australia.
1881 *R. tompsoni* Ramsay, *Proc. Linn. Soc. N.S.W.* 6: 763. New South Wales.
1897 *R. doriae* Trouessart, *Cat. Mamm.* 1:472. New Guinea.
1932 *Rattus (Epimys) rattus rattus* (Linn.). *In* Allen, *Bull. Mus. comp. Zool. Harv.* 83:410.
1951 *Rattus rattus rattus* (Linn.). *In* Roberts, *The mammals of South Africa.*

### TABLE 57
Measurements of male and female *Rattus rattus*

| | Parameter | Value (mm) | N | Range (mm) | CV (%) |
|---|---|---|---|---|---|
| Males | HB | 183 | 8 | 170–205 | 5,8 |
| | T | 196 | 7 | 183–217 | 6,8 |
| | HF | 33 | 14 | 28–37 | 8,5 |
| | E | 23 | 12 | 20–27 | 9,9 |
| | Mass: | 175 g | 1 | — | — |
| Females | HB | 181 | 12 | 164–201 | 6,0 |
| | T | 205 | 12 | 185–226 | 7,1 |
| | HF | 35 | 13 | 30–39 | 6,2 |
| | E | 22 | 13 | 20–25 | 5,8 |
| | Mass: | 159 g | 6 | 153–167 g | 3,2 |

## Identification

The House Rat is rather thickset (head-body length,

150–200 mm) and its face is sharper than that of the Brown House Rat *Rattus norvegicus*. *Rattus rattus* is dark grey all over with a considerable admixture of long, black hairs on its dorsal surface. This overall coloration may have a pronounced brown suffusion throughout. The fur on the back is long and slightly harsh to the touch. The hands and feet are a little paler than the back. The feet are heavy. In the hand the rudimentary thumb has a flat nail. The D2 is slightly shorter than D4, while the D3 is the longest. The D5 reaches about halfway the length of the D4. In the pes, digits 2, 3 and 4 are subequal in length, the D1 barely reaching the base of D2. It is shorter than the D5 which reaches to the end of the first joint of the D4.

The tail is proportionally longer (185–245 mm) than the length of the head and body, covered with short bristles and not increasing in length towards the tip (cf. *R. norvegicus*). The tail rings are also finer in *R. rattus* than in *R. norvegicus*. The ears are larger than in *R. norvegicus*, thin, virtually naked, and they can cover the eyes when pressed forward (plate 8).

The mammary formula conforms to 3–2=10, i.e. three pairs of pectoral and two pairs of inguinal mammae.

As was indicated in the Introduction to this work, four kinds of hair are distinguishable under magnification and the description given by Rosevear (1969) – which incidentally also applies to the Brown House Rat – is worthwhile paraphrasing at this point. In both the Black House Rat and the Brown House Rat, there is a very fine and short (9–10 mm) undulate underfur which forms the major portion of the coat. Interspersed are a number of longer hairs (12–13 mm), fine at the base, but expanded and fusiform in the distal half of the shaft generally showing a yellowish coloration distally. In addition there are a number of much broader, channelled bristles which are distinguishable with the naked eye because of their obviously greater breadth and their whitish colour for most of their length (15–17 mm), but showing a sepia coloration towards the tip which tapers to a long, drawn-out, fine blackish point. Finally, there are a few extra long (28–30 mm) subcircular sectional guard hairs, which project beyond the coat and are blackish-brown throughout their lengths.

The ventral surface has hair of a similar configuration to that of the dorsal surface, but guard hairs are absent.

## Skull and dentition

The skull of the House Rat *Rattus rattus* resembles that of the Brown House Rat *R. norvegicus*, but it tends to be somewhat shorter in the nasal region. The overall profile is gently curved with the rostrum fairly deep, the braincase rounded and the short posterior portion of the braincase dips at a more marked angle than in other murids. The post-temporal ridges are strongly developed. The anterior edge of the zygomatic plate tends to be slightly concave near the ventral border. The interorbital constriction is moderately narrowed, while the supraorbital

ridges are pronounced and they extend posteriorly over the lateral braincase. When the skull is viewed from the ventral aspect, the rostrum is slightly deflected, yet narrow. The palatinal foramina are short (slightly longer than the length of the tooth row) barely reaching the root of the $M^1$. The palate is fairly broad (about twice as wide as the width of the $M^1$) and it terminates well posterior to the $M^3$. The bullae are well developed (measuring some 17–20% of the occipitonasal length) (fig. 97).

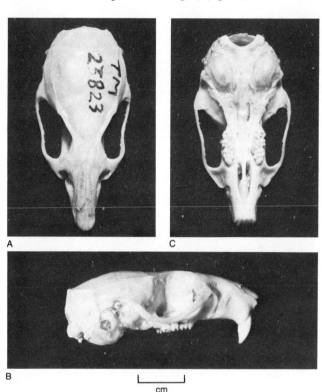

**Figure 97** Dorsal (A), lateral (B) and ventral (C) views of the skull of the House Rat *Rattus rattus*. Supraorbital and occipital ridges well developed; anterior palatine foramina slightly larger than the length of the toothrow; the rostrum is fairly deep and short; palate broad, terminating well beyond the level of the $M^3$; anterior edge of the zygomatic plate is slightly concave; the posterior part of the braincase dips with a marked angle.

The dentition is fairly small and the individual teeth relatively narrow. In the upper molars the central cusps (t2, 5, 8) are prominent. In the $M^1$ (which is five-rooted) the t1 is not markedly distorted backwards in relation to t2 and t3, the t7 being absent, while the t9 is small. In the $M^2$ the t3 is small or vestigial (occasionally missing) while the t8 is usually closely associated with the t9. In the $M^3$ the t1 is prominent, while the t4, 5 and 6 in the second lamina are concurrent. The t8 and t9 of the third lamina form a single cusp. The $M^3$ is not markedly reduced compared to $M^2$ (fig. 98).

In the lower jaw, a terminal heel is developed in the $M_1$ and $M_2$. There is no development of an additional anteroexternal cusp on the $M_1$, but small additional cusps may occur on the labial side between the second and third lamina of the $M_1$, as well as adjacent to the first and second lamina of the $M_2$.

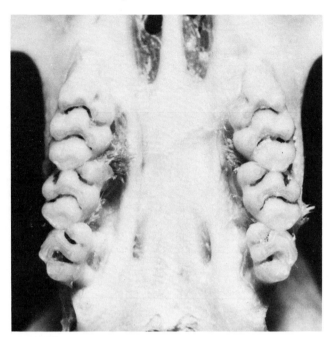

**Figure 98** Cheekteeth of the House Rat *Rattus rattus*. The molars are relatively narrow and small for the size of the skull. The t3 of the $M^1$ is not reduced.

## Distribution

The distribution of the House Rat is cosmopolitan. It has become commensal with man and has been dispersed by means of rail and shipping transport to practically all points where man has been established. Its original home is taken to be the subcontinent of India and by ± 50 BC it had spread to Libya (i.e. the eastern Mediterranean) through trade between the Roman Empire and India. It gradually strengthened its foothold there and by 540 AD it was a well-known animal in the eastern parts of the Roman Empire. During that year the first recorded great plague epidemic erupted in that part of the world (the great plague of Justinian). Hanney (1975) has followed the subsequent extension of the geographical distribution of this rat and states that by 547 AD it had reached France, possibly by ship from Egypt to Marseilles. Between the 6th and 14th centuries, there are no authentic records of plague in Europe and its introduction to the British Isles occurred in 1187 while the plague first reached England in 1348 '... travelling from the port of Weymouth via Oxford to London' (Hanney 1975), and it is believed to have killed about 20% of the population of London. By the 14th century, *R. rattus* was certainly a familiar pest in Britain and a major pest in virtually all European countries. Another well-known epidemic struck London in 1665 and reduced the capital's population to half its normal size. When it reached the New World is unknown, but it is certain to have occurred

during the Elizabethan period (1558–1603). Zinsser (1942) records that Bermuda was invaded by the House Rat in 1615.

It is likely that rats were introduced to many Pacific islands through the voyages of Cook who in 1785 recorded that his ships were heavily infested. In southern Africa, it was also introduced by marine traffic, although the date of entry is uncertain. In South Africa, it tends to spread inland from the coast while its range is still expanding, but Davis (1974) thinks it unlikely that it will establish itself in areas with less than 500 mm annual rainfall, except in towns. Consequently it is absent from most of the Karoo and northern Cape. Its absence from Botswana (Smithers 1971) is explained accordingly. In South Africa it has spread along with settlement by Whites, by shipping and rail and road transport to many inland cities, towns and villages as well as the larger farm homesteads (Roberts 1951) (map 64).

The commonest form in Zimbabwe is the grey variety of the House Rat *R. r. alexandrinus* and the bulk of this introduction has probably taken place through coastal ports in Mozambique (Smithers 1975). The House Rat *R. r. rattus* is less common in Zimbabwe and has been found along the railwayline or in centres served by rail and motor services. *Rattus r. rattus* entered Uganda as an immigrant species around 1910 (Delany 1975).

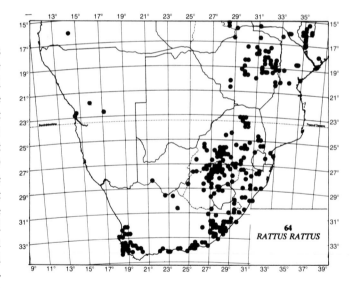

## Habitat

Being predominantly a climbing species, as well as being exceptionally unafraid of man, the House Rat is more or less ubiquitous in cities, towns, villages and farming centres except in the alien, northwestern portion of South Africa. It is commensal with man in dwellings, stores and other buildings. It is a common and ready invader of ships and is therefore often referred to as the Ship Rat. Rosevear (1969) states that this calls for a good head for heights while running along the ropes from ship to shore.

222

## Diet

The House Rat is omnivorous. It will eat all types of human food and as Kingdon (1974) puts it '... and much else besides including soap, hides, paper and beeswax'. It seems to be dependent on water or at least on food with a rather high moisture content. Vegetable food is probably preferred, but Booth (1970) states that when it is hungry, it will eat anything from lizards and cockroaches to candles and soap. It sometimes hoards food and often carries food items back to the nest where it is eaten. In Uganda, Delany (1975) records its liking for all kinds of grain, groundnuts, cotton seeds, meat, potatoes and the like. Hanney (1975) states that it needs plenty of drinking water and that it is unable to survive in arid regions. He also makes the interesting point that new territories may have been opened up for the House Rat by the development of irrigation projects in drier areas of the world.

## Habits

The House Rat is a communal and predominantly terrestrial species, although it climbs with exceptional facility. It will surmount obstacles where there appear to be few toe-holds and it will run along slippery pipes or girders with confidence (Smithers 1975). It is chiefly nocturnal and not as aggressive as the Brown House Rat *R. norvegicus*. It is not as suspicious and wary of contact with man as the Brown House Rat. It is an able and willing climber, and in its original state, may have been an arboreal species (Rosevear 1969). It therefore tends to inhabit the upper levels of floors, roofs and ceilings of buildings. Within these man-made structures, it uses habitual paths (even in open rooms), and these pathways are often betrayed at corners and crannies by greasy smears. It favours shelters under floors or in lofts where food and other potentially edible goods are stored. In order to get to food or whatever it is that attracts it, it will gnaw its way through any substance.

House rats are cunning animals and, according to Roberts (1951), soon learn the dangers of traps and poisons from the experiences of their companions. It is normal in poisoning campaigns to establish feeding stations for a few days with plain bait and then to introduce the poison on the same substratum (Smithers 1975). Kingdon (1974) states that domestic cats are probably the best means of keeping their numbers in check and they may even stop reinvasions after the rats have been eliminated. However, cats have been shown to be unable actually to eliminate rats in the first place. In Uganda, the House Rat competes with *Praomys (Mastomys) natalensis,* the Multimammate Mouse, which is also a commensal of man and which it has largely displaced from dwellings in towns (Delany 1975).

Among the multitude of forms which have been placed within the genus *Rattus* in the past (rightly or wrongly), the House Rat is one of the most active representatives of its family and is approached in agility among the Murinae

of SWA/Namibia only by the Tree Rat *Thallomys paedulcus* (Shortridge 1934). Shortridge (1934) also mentions that it leads a feral life in the Perie *(sic)* and Pondoland forests in Transkei.

## Predators

According to Kingdon (1974), wild predators of the House Rat are numerous in East Africa and include many species of carnivores, birds of prey and snakes. At Barberspan in the Transvaal, Dean (1977) has reported the House Rat as a prey item for the Barn Owl *Tyto alba* and the Spotted Eagle Owl *Bubo africanus.*

## Reproduction

House rats are prolific breeders, which is undoubtedly the outcome of the commensal life they lead with man, where a good and nourishing food supply is almost always available. Nutritional levels influence the number of young and the number of litters produced annually. According to Kingdon (1974), oestrus is very short and recurs every ten days. The gestation period lasts between 21 and 30 days. Post-partum oestrus is common, allowing continuous breeding. The number of young averages six to seven but a litter may occasionally consist of as many as 17 pups. They are born blind and naked with ears sealed down at birth, but they grow rapidly. The eyes open on the 14th day and weaning is usually completed after a month. Sexual maturity is attained at three to four months by which time they weigh from 70–140 g. The turnover of population is very rapid and few wild rats live much longer than one year. Their rate of proliferation is astounding and within one year a single pair of rats may be responsible for more than 100 offspring if no steps are taken to check their reproductive output. Nests are constructed of paper, cloth or any soft material found indoors. In the wild, they use dried leaves, moss and feathers (Rosevear 1969). The young are usually born in these nests and a number of instances have been recorded of 'king rats' resulting from the young getting their tails so inextricably tangled together with the nest materials, that they form a compact bunch from which they seem unable to free themselves. Consequently these individuals cannot forage for themselves and other rats continue to feed them for the rest of their lives (Roberts 1951).

In Zimbabwe, they breed throughout the year, the females having litters of up to seven or eight pups (Smithers 1975). This continual breeding has also been observed in Uganda by Delany (1975) who states that at least 37% of all females are pregnant at any one time. The average number of fetuses from a sample of 54 pregnant females, was six. In Kenya breeding is apparently at a peak in the rainy season and least in the dry season.

In West Africa, they also breed throughout the year. Rosevear (1969) reports the birth of five to six litters annually with some ten pups as an average in each litter.

In captivity they can attain an age approximating four years.

**Parasites**

Research by the South African Institute for Medical Research (SAIMR) has shown *Spirillum minus* to be endemic in populations of the House Rat (SAIMR 1948). Sporadic incidences of *Salmonella braenderup* have been recorded (Gear, Roux & Bevan 1942) while periodic outbreaks of plague *Yersinia pestis* are described in annual reports of the SAIMR since 1935. Laboratory tests have also shown the House Rat to be susceptible to *Listeria monocytogenes* (Pirie 1927) and *Clostridium botulinum* (types C and D) (Theiler & Robinson 1927; Robinson 1929). Endemism of *Rickettsia typhi*, *R. conorii* and *Coxiella burnetii* in populations of the House Rat in South Africa has been documented by Harrington & Young (1944), Gear *et al.* (1951–1957) and Wolstenholme (1960).

The mites and chiggers include the following species, as listed by Zumpt (1961a): Laelaptidae – *Laelaps giganteus*, *L. muricola*, *L. nuttalli*, *Haemolaelaps casalis*, *H. labuschagnei*, *H. rhodesiensis*, *Macronyssus bacoti*, *Hirstionyssus latiscutatus* and *Dermanyssus muris*; Myobiidae – *Radfordia ensifera*; Trombiculidae – *Trombicula mastomyia giroudi*, *T. sicei*, *Leptotrombidium legaci*, *Schoengastia lavoipierrei*, *S. l. lucassi*, *Schoutedenichia congolensis*, *S. cordiformis*, *S. nana nana*, *S. paradoxa paradoxa*, *Gahrliepia nasicola*, *G. wansoni*, *Listrophoroides expansus* and *L. womersleyi*; Sarcoptidae – *Notoedres alepis* and *N. muris*.

The calliphorid fly *Cordylobia anthropophaga* is usually associated with the House Rat (Zumpt 1966). The fleas are the following: Pulicidae – *Echidnophaga gallinacea*, *Pulex irritans*, *Ctenocephalides canis*, *C. crataepus*, *C. felis*, *Xenopsylla cheopis*, *X. nubica*, *X. philoxera*, *X. brasiliensis*, *X. crinita*, *X. georychi*, *X. robertsi* and *X. lobengulai*; Pygiopsyllidae – *Stivalius torvus*; Xiphiopsyllidae – *Xiphiopsylla hyparetes*; Hystrichopsyllidae – *Ctenophthalmus evidens*, *C. calceatus*, *C. eumeces*, *C. debrauweri*, *Dinopsyllus apistus*, *D. ellobius*, *D. longifrons*, *D. lypusus*, *Listropsylla agrippinae* and *L. dolosa*; Leptopsyllidae – *Leptopsylla aethiopica* and *L. segnis*; Ceratophyllidae – *Nosopsyllus fasciatus* and *N. londiniensis*; Chimaeropsyllidae – *Epirimia aganippes*, *Chiastopsylla rossi*, *C. godfreyi* and *C. octavii* (Zumpt 1966).

The ticks associated with *Rattus rattus* have been recorded by Theiler (1962). She listed adult (A) and immature (I) specimens of the following species: *Ornithodoros erraticus* (A), *Ixodes* sp. (A), *I. rasus* (A,I), *Amblyomma variegatum* (I), *Haemaphysalis leachii leachii* (A,I), *Hyalomma dromedarri* (I), *Rhipicephalus appendiculatus* (A), *R. sanguineus* (I), and *R. simus* (I).

The endoparasitic worms found in *Rattus rattus* have been documented by Collins (1972). These include *Hymenolepis diminuta*, *Inermicapsifer congolensis*, *I. guineensis* (=*I. zanzibarensis*), *I. madagascariensis* and *Raillietina (R.) baeri*. *Protospirura bonnei* and *Protospirura* sp. have been reported

associated with the House Rat by Boomker *(pers. comm.)*. *Cysticercus fasciolaris* occurs sporadically in the House Rat while *Hepaticola hepatica*, *Capillaria hepatica* and *Longistriata (L.) capensis* are mentioned by Boomker *(in litt.)*.

**Relations with man**

There is no doubt that the House Rat can be described as a worldwide domestic pest. It is semi-parasitic on man, living in houses and aided by humans. Roberts (1951) has aptly summarised the position by pointing out the enormous damage which this species causes to goods and food and the immeasurable harm it does by acting as a reservoir for bubonic plague, which is transmitted to man by fleas in all countries of the world. Consequently, officials are often engaged in trying to control the spread and increase of these rats. Special legislation for their destruction and control has been promulgated nearly everywhere in the world. They are captured by their thousands by health authorities alone, without special extermination campaigns.

It is an extremely destructive species, not only in the amount of foodstuffs and grains which it consumes, but in the amount it spoils through spillage from damaged containers, or which it fouls by defaecation or urination (Smithers 1975). Economically, nothing is safe from attack (Rosevear 1969). It will gnaw through electric cable (even when coated with lead) and the mechanical damage they do to buildings by gnawing through woodwork, or even by excavating in concrete floors or brick walls, is considerable (Smithers 1975). Smithers (1975) also points out that this group of rats *(R. r. rattus)* travelling in ships, were the main culprits in introducing plague to southern Africa. The flea *Xenopsylla cheopis* transmits the disease to the Multimammate Mouse and thus to the gerbils which (in southern Africa) act as a reservoir of the bacillus *Yersinia pestis* which causes plague in man. The chain of transmission in southern Africa has been observed to be *Tatera* sp. → *Praomys (Mastomys) natalensis* → *Rattus rattus* → *Homo* (Kingdon 1974).

Many species of rodents act as reservoirs for organisms responsible for serious diseases afflicting mankind (Chanlett 1973). Of all the ills to which man is susceptible, plague is one of the most terrifying, causing a mortality rate of up to 95% among those affected (Hanney 1975). The 'Black Death' is a bacillus discovered in 1894 and the roles of the flea and the rat had not been fully elucidated until 1914. If a flea feeds on a plague-infected rat '. . . its stomach becomes blocked by a blood clot containing a pure culture of bacillus, and it dies within about three days. If, in the meantime, the infected flea attempts to feed on another animal, the bacteria are regurgitated and infect the new host' (Hanney 1975). Most epidemics of bubonic plague, caused by *Yersinia pestis*, involve the House Rat as well as the Brown House Rat, whose fleas are capable of penetrating human skin with their biting mouthparts. If rats succumb to the disease, the fleas leave their bodies and will attack humans if the latter are in the vicinity.

Kingdon (1974) states that the relatively recent arrival of *R. rattus* (as well as *R. norvegicus*) has had considerable impact in Uganda where 60 000 people died of plague between 1917 and 1942. Booth (1970) maintains that it should be slaughtered on sight.

## Prehistory
This species has not been identified from any of the fossiliferous localities in southern Africa, because it was introduced to this part of the world only within the last 100 years or so.

## Taxonomy
Colour in the genus *Rattus* is very variable and is not at all a reliable indicator. As Rosevear (1969) points out, the House Rat is only rarely completely black, usually a slaty-grey. It also occurs in a number of other colour variations, mostly different shades of brown. In these guises, one may confuse it with the Brown House Rat *R. norvegicus,* also known as the Brown Rat, a vernacular name which should be rejected because of its ambiguity.

However, there are three colour varieties of *R. rattus* which occur in southern Africa. Originally regarded as subspecies, they are now interpreted as colour phases which may appear in the same litter. These may be separated by the following key:

> **Key to the subspecies of *Rattus rattus***
> (After Rosevear (1969))
>
> 1 Belly fur pure white or bright lemon-yellow...........
> *frugivorus*
>   Belly fur dark grey or dingy-brown ................... 2
> 2 Back blackish, dark slaty-grey or deep sepia-brown...
> *rattus*
>   Back grey-brown or light sepia-brown..*alexandrinus*

# *Rattus norvegicus* (Berkenhout, 1769)

**Brown House Rat**

**Bruin Huisrot**

For the meaning of *Rattus,* see *Rattus rattus.* The specific name refers to Norway from where specimens were presumably introduced to the British Isles and subsequently described by Berkenhout in 1769.

The Brown House Rat *Rattus norvegicus* is very different in character and to some extent in appearance to the House Rat *R. rattus.* The former attains a much larger size and can always be distinguished by its stouter body as well as by the shortness of its ears, length of tail (shorter than length of head-body) and its coarser fur. Like the House Rat, it is also cosmopolitan because of the activities of man. Its original home appears to have been western China whence it spread to Europe early in the 18th century. It therefore originated in a colder climate than the more tropical *R. rattus.* According to Sclater (1901), it reached England around 1728. Consequently it started its wanderings much later than *R. rattus* and was unknown to the Western world until it was originally described as *Mus decumanus* by Pallas in 1778 on specimens from China. This specific name is an adjective of the Latin which implies the largest of its kind. However, it appeared that Pallas' description was forestalled by Berkenhout some nine years earlier, when the latter described *R. norvegicus* in 1769 in a publication entitled *Outlines of the natural history of Great Britain and Ireland.* Berkenhout apparently described the species from a specimen collected in the British Isles which may have been introduced earlier from Norway. The type specimen no longer exists.

In contrast to the House Rat, the Brown House Rat is more cautious and wary of direct contact with man. Yet, when cornered, it can become aggressive and develop into a dangerous fighter (Rosevear 1969). It has a preference for filth and this habit makes it even more repulsive to man than its fellow species *R. rattus.* On the other hand, *R. norvegicus* is the genotype of the well-known albino laboratory rat, used in thousands for experimental purposes in laboratories throughout the world. Consequently, there is a vast literature available on this species.

The *Rattus norvegicus* group differs from the *R. rattus* group in respect of the relatively shorter tail, which is normally shorter than the length of the head and body. It also portrays a narrower braincase. Furthermore, the antero-external cusp of the $M^1$ is more reduced in *R. norvegicus.* The mammae are usually present as $3-3=12$.

Ellerman (1941) states that the two forms are closely related, notwithstanding the differences described above. In fact, Tate (1936), who has published what amounts to a revision of the Indo-Malayan and New Guinea forms of *Rattus,* has referred the *norvegicus* rats to the *rattus* group in contrast to Miller (1923), who referred the two species to distinct subgenera.

*Rattus norvegicus* is more or less cosmopolitan nowadays but seems to be primarily Palaearctic.

## Outline of synonymy
1769 *Mus norvegicus* Berkenhout, *Outl. Nat. Hist. Gt. Britain & Ireland* 1:5. Great Britain.
    *Synonyms*
    1777 *Mus norvegicus* Erxleben, *Syst. Regn. Anim.* 1:381. Norway.
    1778 *Mus decumanus* Pallas, *Nov. Spec. Quad. Glir. Ord.:* 91. Western China.

1779 *M. surmolottus* Severinus, *Tentatem Zool. Hungaricae:* 73.

1837 *R. hibernicus* Thompson, *Proc. zool. Soc. Lond.:* 52. Ireland.

1839 *R. maurus* Waterhouse, *Zool. Voy. Beagle:* 31. America.

1842 *P. leucosternum* Rüppell, *Mus. Senckenb.* 3:108. Eritrea.

1848 *R. maniculatus* Wagner, *Arch. f. Naturg.* 14:186. Egypt.

1934 *Mus decaryi* Grandidier, *Bull. Mus. Nation. d'Hist. Nat. Paris.* (2) 6:478. Madagascar.

1951 *Rattus norvegicus norvegicus* (Berkenhout). *In* Roberts, *The mammals of South Africa.*

### TABLE 58
Measurements of male and female *Rattus norvegicus**

|  | Parameter | Value (mm) | N | Range (mm) | CV (%) |
|---|---|---|---|---|---|
| Males | HB | 239 | — | 222–255 | — |
|  | T | 196 | — | 175–218 | — |
|  | HF | 43 | — | 41–45 | — |
|  | E | 20 | — | 20–21 | — |
|  | Mass: No data available |  |  |  |  |
| Female | HB | 237 | 1 | — | — |
|  | T | 211 | 1 | — | — |
|  | HF | 37 | 1 | — | — |
|  | E | 20 | 1 | — | — |
|  | Mass: No data available |  |  |  |  |

*Roberts (1951)

### Identification

The species is a reddish-brown to a greyish-brown on the back, often darker medially. The fur is woolly, harsher to the touch than in *R. rattus*, while the guard hairs are not as long and frequent as in *R. rattus*. In addition the channelled bristles are less numerous and not as broad as in *R. rattus* (Rosevear 1969). The flanks are a more greyish-brown which turns into a whitish ventral pelage. The individual hairs on the belly are slate-coloured at their bases. The head-body size, in adult specimens, is usually in excess of 210 mm. The limbs and feet are stout and strong, while the dorsal surfaces of these appendages are of a lighter hue.

The thumb is rudimentary, with a flat nail. The D5 reaches halfway along the D4. Digits 4 and 2 are more or less subequal in length but shorter than D3. In the broad and strong hindfoot, digits 2, 3 and 4 are more or less subequal in length. The hallux reaches to the base of the D2 while the D5 reaches to the end of the first joint of the D4. The tail is shorter than the length of the head and body, almost naked with about 25 rings to the inch

(Sclater 1901). The ears are small and when pressed forward, do not reach the edge of the eye (plate 8).

The mammary formula is $2-3=10$ or $3-3=12$.

### Skull and dentition

The skull of *R. norvegicus* is larger than that of *R. rattus*. The supraorbital ridges are well developed and extend posteriorly as temporal ridges. Vertical post-temporal ridges in the occipital region are strongly developed. Seen from above, the braincase is decidedly narrower and more oblong than in *R. rattus*. The ventral aspect of the skull shows no specific differences, apart from the fact that the zygomatic arches are more strongly developed. The bullae are large. The anterior edge of the zygomatic plate is more or less perpendicular, while the dorsal portion projects slightly forward. The antorbital foramen is narrow in its lower reaches. In profile, the rostrum (which is not as deep as in *R. rattus*) dips at a sharper angle below the frontals.

The $M^1$ is five-rooted. The overall cusp pattern of *R. norvegicus* is very much like that found in *R. rattus*, but differs in the following respects:

(i) In the $M^1$ the t3 tends to be more reduced.

(ii) The t2 of the $M^1$ can carry an additional supplementary ridge on its anterior face.

(iii) The t1 of the $M^1$ is displaced backwards, lying against the t5 and is often connected with the t2 by a small ridge (fig. 99).

The upper incisors are orthodont. As in many species of mammals, the number of incremental layers in tooth cement and dentine has been found to be a useful criterion of age. Klevezal & Kleinenberg (1967) have reported on this phenomenon in *R. norvegicus*.

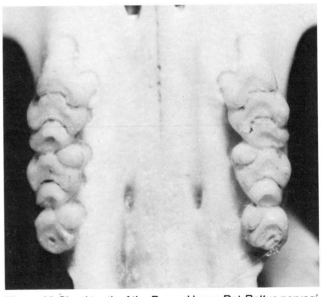

**Figure 99** Cheekteeth of the Brown House Rat *Rattus norvegicus*. The t3 of the $M^1$ is reduced and the front edge of the t2 often carries an additional supplementary ridge. The t1 is displaced backwards, lying against the t5 of the second lamina.

## Distribution

In southern Africa it is confined to ports and larger coastal towns and their immediate vicinity (map 65). Davis (1974) states that it is known to inhabit the shoreline south of Durban (e.g. Isipingo Beach) and also a similar habitat near Hout Bay in the Cape Peninsula.

Like the House Rat, the Brown House Rat is now also cosmopolitan and is found all over the world near sea ports. It originated in the more northerly temperate regions of Asia, possibly Siberia or China and spread westwards, largely aided by its fierce and powerful nature (Rosevear 1969). Apparently it was introduced to western Europe and, more specifically, to Copenhagen, during a visit of the Russian fleet to Denmark in 1716. From there it was reported in England (1728), Paris (1750), Norway (1762) and Spain (1880). It was first reported in the USA in 1775 (Hanney 1975).

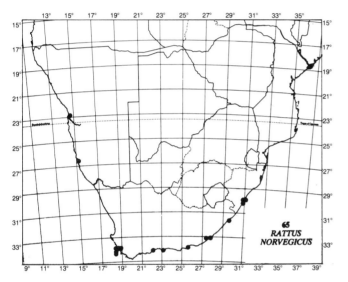

65
RATTUS
NORVEGICUS

## Habitat

It has been suggested that the tropics (and possibly temperate areas as well) are unsuitable for this species and that it acclimatises less readily than *R. rattus* to hot countries. In addition to its aversion to warmer climates, it is very partial to water and can often be seen swimming or playing in water either in the sea or in rivers at the coast. Pied mutants of the albino laboratory strain are occasionally located in Johannesburg, living 'wild' on domestic premises but '. . . are obviously escapees from pet shops or pet-owners' (Davis 1974). Roberts (*in* Shortridge 1934) states that it is essentially a sewer rat and more ferocious than any other species.

## Diet

Experience points to the fact that this species needs a copious supply of drinking water and is therefore unable to survive in arid regions. Like the House Rat, the diet of the Brown House Rat is catholic. It will gnaw and nibble at virtually anything. It readily consumes fruit, grain, meat and any kind of human food. Sclater (1901) reports

it to be carnivorous and that it will devour trapped or otherwise disabled compatriots.

## Habits

The Brown House Rat is a hardy species and well adapted to colonise colder countries. According to Hanney (1975) it has even established itself at the whaling station of South Georgia in the Antarctic. It is chiefly nocturnal and is rather suspicious of strange objects. Consequently, in comparison to *R. rattus,* it is rather more difficult to trap. It is a more powerful, stronger and fiercer rat than the House Rat and in many parts of the world it has driven out or supplanted the once prevalent House Rat, but it is a less agile climber. In the original state it was more terrestrial and even more subterranean than *R. rattus*. It is consequently found on ground floors, cellars and in colder and damper environments of drains and sewers (Rosevear 1969). Rosevear (1969) also states that it often elects to live away from towns by small streams, making its home in holes dug in the banks. *R. norvegicus* therefore tends to dig and swim, while *R. rattus* is more inclined to climb.

Roberts (1951) states that *R. norvegicus* is known to attack and kill domestic cats which have been put into buildings to destroy them and that even dogs have been known to be killed by a number of these rats combining to attack the dog.

## Predators

Brown House Rats are probably taken by smaller carnivores, owls and snakes.

## Reproduction

This species is also a very prolific breeder and a female can produce several litters annually. The number of young in a litter varies from six to 12. Breeding can commence at the age of three months. The gestation period is from 20–26 days and the pups are born blind, pink and naked. The eyes open between days 14 and 17.

## Parasites

Laboratory tests have shown the Brown House Rat to be susceptible to African horse sickness (Nieschulz 1932), Rift Valley fever (Kaschula 1961), *Yersinia pestis* (Powell 1925), *Listeria monocytogenes* (Pirie 1927), *Clostridium botulinum* (types C and D) (Theiler & Robinson 1927; Robinson 1930b), *Rickettsia prowazekii* (Findlayson & Grobler 1940), *R. typhi* (Pijper & Dau 1933; Gear, De Meillon & Davis 1944) and *R. conorii* (Mason & Alexander 1938).

Laboratory tests have also brought to light that *Rattus norvegicus* can be infested by trypanosomes. The presence of *Trypansoma brucei* (Curson 1928), *T. congolense* (Robinson 1930a) and *T. equiperdum* (Haig & Lund 1948) was demonstrated.

Schistosomes may also be present in the Brown House Rat. Le Roux (1929b) has shown them to be susceptible to *Schistosoma mattheei* under laboratory conditions.

Fleas include the following families and species: Pulicidae – *Echidnophaga gallinacea*, *Pulex irritans*, *Ctenocephalides felis*, *Xenopsylla aequisetosa* and *X. cheopis;* Hystrichopsyllidae – *Listropsylla agrippinae;* Leptopsyllidae – *Leptopsylla segnis;* Ceratopsyllidae – *Nosopsyllus fasciatus* (Zumpt 1966).

The mites that have been taken from *Rattus norvegicus* include Laelaptidae – *Laelaps muricola* and *L. lamborni;* Myobiidae – *Radfordia davisi;* Listrophoridae – *Listrophoroides expansus* and *L. womersleyi* (Zumpt 1961a).

### Relations with man

Like the House Rat, the Brown House Rat has become a worldwide domestic pest (Rosevear 1969) and it is difficult to control. Once it is familiar with traps and poisoned bait it becomes extremely wary of them (Kingdon 1974). It is the most important pest in present-day Europe and North America (Hanney 1975). The era of the rat seemed to have ended when the Wisconsin Alumnus Research Foundation gave its initials to a new anti-coagulant poison (Warfarin) which, when ingested, reduces the clotting properties of blood and leads to fatal haemorrhage. However, since 1958, Warfarin-resistant strains have been reported (Hanney 1975) and the struggle between man and rat has started anew.

Like *R. rattus*, *R. norvegicus* is also involved in most epidemics of bubonic plague (Roberts 1951; Hanney 1975), although according to Kingdon (1974) *R. norvegicus* is not implicated in plague transmission. It is, however, an agent for human leptospiral jaundice and trichinosis, while it also contaminates foodstuffs with *Salmonella*, which in turn is responsible for food poisoning. In contrast to *R. rattus*, it does more harm to small livestock (Roberts 1951).

### Prehistory

As is to be expected, this species is unknown in either the fossil or subfossil state from any of the southern African fossiliferous localities.

### Taxonomy

Ellerman (1941) states that *norvegicus* and *rattus* agree in many essential characteristics. Tate (1936) refers the *norvegicus* race to the *rattus* group while Miller (1923) refers the two species to distinct subgenera. Ellerman has subdivided the genus *Rattus* into 38 groups of which *rattus* is a separate group, as is *norvegicus*. Within the *norvegicus* group, four subspecies are listed apart from the nominate race of *norvegicus* (the latter containing ten synonyms), while *Rattus humiliatus* (with four subspecies), *R. tyrannus*, *R. burrus*, *R. burrulus* and *R. burrescens* are also included as species in the *norvegicus* group (Ellerman 1941).

GENUS     *Mus*   Linnaeus, 1758

SUBGENUS   *Mus*   Linnaeus, 1758

As understood here, the genus *Mus* Linnaeus contains two subgenera. This includes the nominate subgenus *Mus* and the subgenus *Leggada*, in this respect following the classification as interpreted by Rosevear (1969). *Mus* is a large genus and Ellerman (1941) lists 130 different forms which have been described from all over the world and divided into two subgenera *(Mus and Leggadilla)*. *Mus* contained 117 species of which 98 were listed as polytypic. *Leggadilla* on the other hand, consisted of 13 species of which two were polytypic. Ellerman (1941), in following Miller (1910), emphasises that there are no characteristics which distinguish 'Leggada' (tropical species) from *Mus (musculus* group). In his opinion, there is not the slightest need to regard *Leggada* as anything but a synonym of *Mus*. He consequently splits the 130-odd species into six groups with *musculus*, *booduga*, *bufo* (including *triton*), *minutoides* and *tenellus* as groups of the subgenus *Mus*, while the subgenus *Leggadilla* contains the *platythrix* group.

Apart from the typical *Mus*, the genus *Leggada* was erected by Gray in 1837 for an Indian species, mainly because of taller molars with rather convex crowns, a characteristic now regarded as having no diagnostic value (Rosevear 1969).

The status of *Leggada* was criticised by Miller (1910), who expressed the view that it was impractical to maintain two generically distinct groups. Thomas (1919), on revising the murines of India, believed that *Mus* and *Leggada* were distinct natural groups. He thought that the distinction lay in the somewhat longer rostrum in *Mus* and by the fact that in *Mus* the front edge of the zygomatic plate lay somewhat forward of the midpoint of the anterior palatal foramina, whereas in *Leggada* it lay at this point or posterior to it. Thomas recognised that not every individual skull would answer to these tests. As was indicated above, Ellerman was strongly in favour of sinking *Leggada*. As a result of chromosome studies on the African *minutoides*, Matthey (1958) has shown that *Mus* and *Leggada* are clearly separable '... though he leaves open the question of whether the distinction should be of a generic or only subgeneric order' (Rosevear 1969). Petter (1963) correctly sounds a warning: he points out that the situation is not that simple, as a large number of forms are involved, both in Asia and in Africa, and it is not certain that Matthey's results are equally applicable to all. He also maintains that it is virtually impossible to draw up a diagnosis which will separate all reputed *Mus* from all reputed *Leggada* and from this point of view it is pointless to maintain them as separate genera. Rosevear (1969) maintains that from a global point of view, the argument proposed by Petter would hold, but in the West African material it is quite easy to distinguish between the local *Mus* and the local *Leggada,* given a fairly powerful lens.

Examination of Barn Owl pellets collected in 1964 at Essexvale and Sinoia Cave (Zimbabwe) and subsequently also at Makapansgat near Potgietersrust in the northern Transvaal, points to the co-occurrence of two species of Pygmy Mouse. Apart from *Mus (Leggada) minutoides,* the second species is provisionally reported as a form of *Mus sorella,* known from central Africa (Pocock 1974). More work should be carried out (especially trapping) to clarify the situation and if the provisional identification turns out to be correct, the range of *M. sorella* is thus extended into southern Africa by several thousand kilometres.

Given the present confused situation of the taxonomy of *Mus (Leggada)*, it is heartening to know that a Pan-African revision of this subgenus was commenced at the end of 1976 by Vermeiren (1979). The program started in West Africa and material from southern Africa is being studied. The research is edging in from both ends of sub-saharan Africa and areas in between are continually filled in. This will lead to an overall interpretation of this taxon in Africa.

---

**Key to the subgenera of *Mus***
(Modified after Rosevear (1969))

Fur of long, soft, fine hairs with few slightly shorter bristles; zygomatic plate with anterior margin straight and vertical, passing more or less abruptly into the rostrum with little forward curve; masseteric knob at lower anterior corner of plate; $M^1$ with anterior root emerging from bone at a slope so that it is visible from above, the base of the anterior lamina not much expanded and without a transverse notch; $M^3$ small, but generally signs of two laminae with a fairly well-developed antero-internal cusp showing; $M_1$ consisting of five large cusps and one very small one as a subsidiary of the large centrally placed anterior one ....... *Mus*
(House Mouse)
Page 229

Fur of a mixture of very fine and relatively very broad bristly hairs, the latter dominating the forms, stiffened and longitudinally channelled ('gutter hairs'); zygomatic plate with anterior margin usually boldly convexly curved and passing into the rostrum below with a wide forward sweep; masseteric knob situated near the middle of the plate; $M^1$ with anterior root emerging from the bone more at right angles and thus largely or completely hidden from above by the forward bulge of the lamina, the front face of which carries a small transverse cusp (sometimes not detectable in old teeth); $M^3$ simple with little sign of lamination; $M_1$ clearly with six cusps, four large posterior ones and two smaller anterior ones .......................... *Leggada*
(Pygmy Mouse)
Page 233

---

# *Mus (Mus) musculus* Linnaeus, 1758

**House Mouse**

**Huismuis**

The generic name *Mus* is Latin, derived from the Greek *mys* = mouse. *Musculus* is Latin, the diminutive of *mus* and evidently refers to the small size of this species.

The House Mouse is another foreign species (like *Rattus rattus* and *Rattus norvegicus*) which has found a foothold in southern Africa. Like the House Rat and Brown House Rat, it is now cosmopolitan. It is very likely that the House Mouse *Mus musculus* may have been the first rodent to become intimately associated with man. It is believed to have evolved from a wild subspecies, *Mus musculus wagneri*, which is still widely distributed in the dry steppes of central Asia between the Caspian Sea and the Himalayas (Hanney 1975). The species evidently took advantage of man's new agricultural techniques and the resultant more stable way of life. It consequently spread to the Middle East and the shores of the Mediterranean and thence northwards into virtually the whole of Europe. It is not certain when it made its appearance in western Europe, but it has been in the British Isles since Neolithic times at least.

It is conceivable that since Britain, France and the Netherlands founded their overseas possessions in the 17th century, the northern race of the House Mouse, *Mus musculus domesticus,* was inadvertently transported on ships en route to far off destinations, thereby establishing itself elsewhere, especially in North America, while the southern race, *M. m. brevirostris,* went to Latin America in Spanish and Portuguese vessels. It is possible that *Mus musculus* was introduced to the southern tip of Africa by either Portuguese or Dutch maritime activities. It has also colonised the subantarctic Marion Island, apparently inadvertently introduced, with resultant thriving populations (Berry, Peters & Van Aarde 1978).

The type specimen of *Mus musculus* is no longer in existence.

## Outline of synonymy

1758 *Mus musculus* Linnaeus. *Syst. Nat.* ed. 10, 1:62. Sweden.

1826 *Mus orientalis* Cretzschmar, *Rüppell's Atlas z.d. Reise im nördliche Afrika:* 76. Ethiopia.

1843 *M(us) modestus* Wagner, *Arch. f. Naturg.* 8(1):14; Schreber's *Säugeth.* Suppl. 3:432. Cape of Good Hope.

1845 *Mus vignaudi* Prevost. *In Lefebvre's Voy. en Abysinnia.* Abyssinia.

1877 *Mus pallescens* Heuglin, *Reise in nord-ost Afrika.*

1939 *Mus musculus musculus* Linnaeus. *In Allen, Bull. Mus. comp. Zool. Harv.* 83:402.

1951 *Mus musculus musculus* Linneaus. *In Roberts, The mammals of South Africa.*

1975 *Mus musculus* Linnaeus. Petter, *in Meester & Setzer (eds), The mammals of Africa: an identification manual.*

### TABLE 59
Measurements of male and female *Mus (Mus) musculus*

| | Parameter | Value (mm) | N | Range (mm) | CV (%) |
|---|---|---|---|---|---|
| Males | HB | 85 | 15 | 75–103 | 8,7 |
| | T | 84 | 16 | 74–95 | 8,2 |
| | HF | 17 | 22 | 14–19 | 8,4 |
| | E | 14 | 18 | 12–17 | 10,4 |
| | Mass: | 20 g | 5 | 16–24 g | 14,2 |
| Females | HB | 87 | 10 | 77–107 | 9,6 |
| | T | 83 | 9 | 78–90 | 3,8 |
| | HF | 17 | 10 | 14–20 | 10,9 |
| | E | 13 | 8 | 11–15 | 9,7 |
| | Mass: | 18,7 g | 4 | 16–25 g | — |

## Identification

Members of this species are small mice with a head-body length of less than 90 mm (Delany 1975). The fur is short, crisp, soft, sleek and lying close together. The general colour dorsally is a dark to medium brownish-grey with a slight yellowish tinge. Individual hairs are slate-grey basally. This coloration becomes paler along the flanks and shades into lightish grey on the ventral surface, which may also be a warm, buffy brown. The eyes are prominent and beady, while the ears are medium to large in size, roundish, nearly naked, and, when pressed forwards, reach the eyes. The fine, almost naked tail is slightly shorter than the head and body length. It is thinly covered with hair, lighter below than above. The hands and feet are small, usually coloured grey with a pink suffusion on their dorsal surfaces. The D1 and D5 of both the hand and foot are much shorter than the other digits, the pollex being very little more than a stump on the manus (plate 11).

Both albino and melanistic forms occur. The white strain has become an important laboratory animal.

The mammary formula reads $3-2=10$.

## Skull and dentition

The skull of *Mus musculus* is slender and, seen in profile, rather flat-crowned. The zygomatic plate is very slightly $\lceil$-shaped along its anterior margin and there is a pronounced masseteric knob near its lower end. The anterior palatal foramina reach well into the area between the two first upper molars, usually as far as the middle root of the $M^1$. The supraorbital ridges are faint or absent. The muzzle is short and the size of the infraorbital foramen not enlarged. The bullae are not enlarged (fig. 100).

The dentition of *Mus musculus* is characteristic. The notched posterior surface at the distal end of the upper incisor (cut by the action of the lower incisors) is unmistakable and only mice belonging to the genus *Mus* show these incisors. In the three-rooted and eight-cusped $M^1$, the t1 is very much tilted towards the rear and has shifted backwards so far as to be in a straight line with t5 and t6 of the second lamina. It therefore occupies the position of the t4, which has similarly shifted to the back of the tooth. In the $M^2$ the t1 is prominent, but the t3 is very small while the t2 has been lost. The $M^3$ tends to become very small (fig. 101).

In the lower jaw, the two front laminae of the $M_1$ display an interesting variety in the arrangement of the cusps, a fact which has been used diagnostically by Petter (1963). The $M_1$ consists of five large cusps and one very small one as a subsidiary of the large, centrally placed anterior cusp.

## Distribution

The House Mouse is found in habitations of man throughout the world. Hanney (1975) describes how this came about by stating that when ancient civilisations

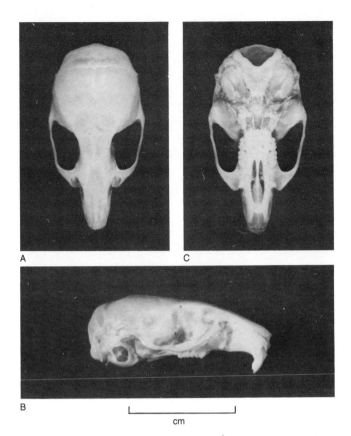

A        C

B

cm

**Figure 100** Dorsal (A), lateral (B) and ventral (C) views of the skull of the House Mouse *Mus (Mus) musculus*. Note the slight supraorbital ridges, the short muzzle and the anterior palatine foramina reaching well between the upper molars. The bullae are not enlarged. A pronounced masseter knob is situated at the lower end of the straight-edged zygomatic plate. The skull is flat crowned. The orthodont incisors are notched on their inner surfaces.

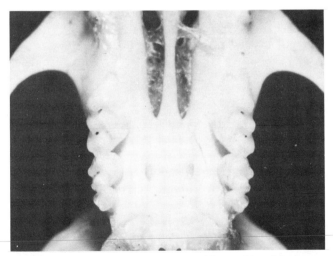

**Figure 101** Cheekteeth of the House Mouse *Mus (Mus) musculus*. The anterior root of the M$^1$ is situated far forward. The crown of the three-rooted M$^1$ is longer than the length of the M$^2$ and M$^3$ combined. The t1 is much tilted towards the rear, virtually in line with the t5 and t6 of the second lamina, occupying the usual position of the t4 which has almost moved back. The M$^3$ is small but traces of two laminae persist.

began trading with each other, many small mammals must have been carried unwittingly among baggage of caravans crossing the desert, in much the same way as they are transported in holds of ships and aircraft today. Hanney points out that it has been less successful in the colonisation of Africa, possibly because of the absence of permanent sources of grain in some areas and competition with *Praomys (Mastomys) natalensis*, a well-established species domestic to Africa.

In South Africa, it is distributed throughout, except in the Kalahari and environs (map 66). As is the case in *Rattus rattus,* the recorded distribution is sketchy, being based on museum specimens, flea-host records and literature (Davis 1974). In Zimbabwe it is fairly widespread, always in association with man, probably having made its way from the different ports on the East Coast of Africa (Smithers 1975). In Botswana feral populations have established themselves (Smithers 1971). Both Kingdon (1974) and Delany (1975) report that this species was recorded for the first time in Kampala (Uganda) in 1965 and, according to Kingdon, is on the increase there.

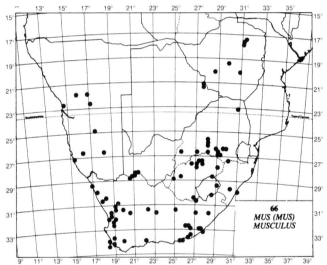

66
MUS (MUS) MUSCULUS

## Habitat

Hanney (1975) states that where climatic conditions are mild, the House Mouse may live out of doors, but elsewhere it is generally an indoor animal. It is consequently found near human habitations and cultivated fields. It makes nests from rags, paper, cloth, upholstery and the like (Delany 1975), and in thatch roofs of African huts (Smithers 1975). It appears to be better adapted to living in semi-arid situations than *Rattus rattus,* so it may have originated from a desert dwelling species. According to Shortridge (1934), it does not seem to take readily to water. Rosevear (1969) says that *M. musculus* is possibly not very able to compete with the existing West African fauna.

## Diet

House mice are omnivorous, though they subsist chiefly

231

on vegetable matter, including seeds, tubers, groundnuts or maize. Hanney (1975) states that they can survive for long periods without drinking, provided they have access to grain and can make some kind of a nest in which a high relative humidity can be maintained. Hanney also reports that they can live for months on end on dry seeds, while they can withstand considerable dehydration. Smithers (1971) states that they are graminivorous, but that they will take a wide range of household foods.

## Habits

These mice are nocturnal, terrestrial, agile and alert and occur alone, in pairs or in small family parties. They will live in amity in the same house as the Brown House Rat *Rattus norvegicus*, though the latter will, if possible, endeavour to get rid of the House Rat *Rattus rattus* if they happen to reside under the same roof (Sclater 1901). They are fair climbers and can scale a nearly vertical surface with ease. They are rather secretive (Rosevear 1969) but very adaptable. Hanney (1975) writes that they can live in cold stores kept at temperatures of −10°C (=14°F), feeding entirely on frozen meat. They even reproduce under these circumstances, although fertility is reduced and there is a high mortality rate among the young.

Their voices are high-pitched squeaks (Delany 1975) and they have a pronounced musky smell. Roberts (1951) says that these mice are remarkable for 'singing', uttering a faint but audible twittering song when in the safety of their retreats.

Detection of their abodes is not always easy, for they use very small holes.

In contrast to populations in Australia, house mice in southern Africa do not show the same tendency to become feral. Furthermore, in contrast to the House Rat, they are less dependant on buildings and may be found some 1 500 m away from buildings, living in the bush (Powell *in* Shortridge 1934), in shallow, self-excavated burrows. Occasionally, they may inhabit burrows of other rodents (e.g. gerbils), sometimes adding to the existing excavation themselves and thereby increasing the complexity of the subterranean structure.

House mice leap better than any of the other small Murinae (Sclater *in* Shortridge 1934). Their hearing is acute, but their eyesight is not remarkable (Rosevear 1969). Rosevear also states that they are sensitive to vibrations and have a highly developed sense of smell, sometimes their chief means of finding food. They can swim effectively. In West Africa, the nests are usually occupied by a single family, but when the population density becomes high, it may be held communally by two or three families.

## Predators

Rosevear (1969) states that in the tropics, the House Mouse is taken by snakes and driver ants. When emerging to feed in the wild, they are subject to a multitude of dangers posed by a wide variety of predators, whether mammal, night bird, reptile or insect. They are taken by the Barn Owl *Tyto alba* and the Spotted Eagle Owl *Bubo africanus* (Dean 1977).

## Reproduction

Small nests are constructed of any soft debris and any available material such as paper, rags and grass below floors, within hollow walls above ground level, or in any available sheltered corner (Smithers 1975).

Like so many murine species *Mus (Mus) musculus* is a prolific breeder. Females can start breeding at 40 days of age. Gestation lasts 19 days and the litter sizes vary between five and seven pups (Hanney 1975). Three to five litters may be had annually (Sclater 1901). An oestrus cycle lasts 4–6 days (Delany 1975) with the females fecund at 7,5 g and the males at 10 g. The young are weaned at 18 days of age. This high reproductive output has undoubtedly contributed to its success as a useful laboratory animal in research (Rosevear 1969). The young are born throughout the year (Smithers 1975). Smithers also states the average litter size to be between two and four young, which is considerably lower than the figure given by Hanney (1975). Rosevear (1969) mentions that litters may vary from 4–7 young with 5–10 litters annually, depending on the environment. A male and female seemingly stay together during the breeding process though there may be some degree of polygamy.

## Parasites

Research has disclosed that house mice are either hosts for or at least susceptible to an enormous diversity of other species with biological significance either to themselves or to *Homo sapiens*.

Among the viruses, it has been shown that *M. musculus* is susceptible to rabies under laboratory conditions (Neitz & Schulz 1949). The same applies to African horse sickness (Nieschulz 1932), Rift Valley fever, bluetongue, Wesselsbron virus disease and Middelburg virus disease (Neitz 1965).

Susceptibility to infections of protophytes under laboratory conditions has also been demonstrated. These include *Pseudomonas pseudomallei*, the causitive organism of melioidosis (Finlayson 1944), *Salmonella amersfoort*, the organism responsible for outbreaks of salmonellosis in domestic chickens (Henning 1937) and *Pasteurella multocida* var. *ictero-hepatides* which cause bovine bacterial hepatitis and bovine bacterial icterus, both occurring widely distributed in South Africa and SWA/Namibia (Neitz 1965). To this list *Yersinia pestis*, the plague organism, is also to be added, as well as *Sphaerophorus necrophorus*, a common infection of cattle, sheep and pigs, especially prevalent if these animals are kept under unhygienic conditions, although it may also manifest itself under ideal sanitary conditions (Neitz 1965). These mice are also susceptible to *Corynebacterium pyogenes* causing, *inter alia*, acute or chronic bovine mastitis. Laboratory tests have also shown susceptibility to *Listeria monocytogenes* (listeriosis), *Erysipelothrix insidiosa* (swine erysipelas), *Bacillus anthracis* (anthrax), *Clostridium septicum* (malignant oedema in cattle and sheep), *C. welchii* Type B (lamb dysentery, 'bloedpens'), *C. welchii* Type D (enterotoxaemia, pulpy kidney disease), *C. botulinum* Type C and Type D (lamsiekte, botulism) and *C. tetani* (lockjaw) (Neitz 1965). This species of mouse is also susceptible to ixodidborne borreliosis caused by the spirochaetid *Borrelia duttoni* as well as to *Leptospira canicola*, *Rickettsia prowazekii* (endemic typhus, louse typhus) and *R. conorii* (tick-bite fever).

When the chlamydiacid *Miyagawanella psittaci* is administered intraperitonally and *M. ornithosis* is administered intracerebrally, this species shows susceptibility to psittacosis and ornithosis respectively under laboratory conditions (Neitz 1965).

Among the thallophytes, chiefly the Fungi Imperfecti, it has been demonstrated that *Mus musculus* reacts to laboratory induced infections of *Histoplasma capsulatum*, the fungus often affecting man after exposure in caves. Field tests have also shown them to show dermatomycosis (ringworm) caused by the moniliacid fungus *Trichophyton mentagrophytes* (Neitz 1965).

Laboratory tests have also indicated a degree of susceptibility to the following species of protozoans: *Trypanosoma brucei* (the organism causing nagana), *T. congolense*, *T. equiperdum* (dourine) and the sarcosporidid *Toxoplasma gondii* (Neitz 1965).

The plathyhelminth worms which may affect this mouse adversely under laboratory conditions include *Schistosoma haematobium*, *S. mansoni* and *S. bovis* (Neitz 1965). Other flat worms include *Hymenolepis diminuta*, *H. microstoma*, *H. nana* and *Inermicapsifer madagascariensis* as reported by Collins (1972). The cestode *Oxysticercus fasciolaris* is also associated with this species (Boomker *pers. comm.*). Other helminths include *Aspiculuris tetraptera* and *Protospirura muris* (Verster & Brooker 1970).

The arthropods associated with *Mus musculus* in Africa south of the Sahara includes the following mites, as recorded by Zumpt (1961a): *Macronyssus bacoti* and *Allodermanyssus sanguineus* (Laelaptidae) and *Myobia murismusculi*, *Trombicula mastomyia giroudi* and *Myocoptes musculinus* of the Myobiidae, Trombiculidae and Listrophoridae respectively.

The fleas recorded from *Mus musculus* collected in Africa south of the Sahara include the following families, genera and species: Pulicidae – *Echidnophaga gallinacea*, *Pulex irritans*, *Ctenocephalides canis*, *Xenopsylla cheopis*, *X. piriei*, *X. brasiliensis*; Hystrichopsyllidae – *Dinopsyllus ellobius*, *Listropsylla agrippinae*; Leptopsyllidae – *Leptopsylla segnis*; Ceratophyllidae – *Nosopsyllus fasciatus*, *N. londiniensis*; Chimaeropsyllidae – *Chiastopsylla rossi* (Zumpt 1966).

The ticks taken from *M. musculus* have been recorded by Theiler (1962) as immature specimens of *Rhipicephalus sanguineus* and *R. simus*.

## Relations with man

The House Mouse does considerable damage to foodstuffs and domestic cats are probably the best means of keeping its numbers down (Roberts 1951). It is an extensive pest in Europe as well as in several other parts of the world, where it causes damage probably running into millions of rands annually. It is a serious pest in houses and food stores, and populations may attain plague proportions in certain parts of the USA and Australia. Although it is of minor economic importance compared to the house and brown house rats, a single individual can cause extensive damage in a short time just by gnawing books (Hanney 1975). On the credit side, mention must be made of the large captive albino populations which are specially bred in laboratories for experimental research.

## Prehistory

*Mus (Mus) musculus* was introduced into Africa, and is thus not found as fossils in the Pleistocene deposits in southern Africa.

Misonne (1969) has suggested that *Mus* immigrated from Eurasia independently of the original parent stock of the other African Muridae, probably at a later date. Misonne believes that there are three distinct groups within the genus *Mus* and that representatives of each group are found in both Africa and Eurasia, implying that three species of an isolated genus have managed to make their way into Africa independently. According to Kingdon (1974), this is a controversial conclusion.

## Taxonomy

The genus *Mus* contains small rodents which are related to the larger rats, now assigned to the genus *Rattus* Fischer, '. . . and until in 1881 Trouessart created *Epimys* (now replaced by *Rattus*) to accommodate these latter (i.e. the larger rats), it was customary to refer them all to *Mus*' (Rosevear 1969). However, the genera *Mus* and *Rattus* are clearly separable on dental characteristics and there is no justification for lumping them together in a single genus.

SUBGENUS *Leggada* Gray, 1837

# *Mus (Leggada) minutoides* A. Smith, 1834

**Pygmy Mouse**
**Dwergmuis**

According to Rosevear (1969), the generic name is said to be a derivation from *legyade*, a vernacular name of an

Indian species of mouse. The species name is derived from the Latin *minutus* = minute, obviously referring to the small size of this species.

This small and slender mouse, apart from being one of the smallest murid species, is also to be reckoned among the smallest mammals in the world. The type of this species was collected in Cape Town by Andrew Smith in 1834 and described as *Mus minutoides*. It is smaller than *Mus (Mus) musculus* and it has a shorter tail. The head and body length and width of the braincase are seldom more than 70 mm and 9 mm respectively and the adult mass is about 7,5 g (Willan & Meester 1978). It is closely related to the exotic *Mus (Mus) musculus* but it lives independently from man.

## Outline of synonymy

1834 *Mus minutoides* A. Smith, *S. Afr. Quart. Journ.* 2:157. Vicinity of Cape Town, CP.

1837 *Leggada* Gray, *Charlesworth's Mag. Nat. Hist.* 1:586. (Genotype *Leggada booduga* Gray.)

1852 *Mus minimus* Peters, *Monatsb. K. Preuss. Akad. Wiss. Berlin:* 274. Tete, Mozambique. (*nec Mus minimus* White, 1789, which is a *Micromys*.)

1939 *Leggada minutoides minutoides* (A. Smith). *In* Allen, *Bull. Mus. comp. Zool. Harv.* 83:388.

1951 *L. m. minutoides* (A. Smith). *In* Roberts, *The mammals of South Africa.*

1975 *Mus minutoides* A. Smith. Petter, *in* Meester & Setzer (eds), *The mammals of Africa: an identification manual.*

TABLE 60
Measurements of male and female *Mus (Leggada)*
*minutoides*

| | Parameter | Value (mm) | N | Range (mm) | CV (%) |
|---|---|---|---|---|---|
| Males | HB | 62 | 25 | 53–72 | 7,8 |
| | T | 46 | 24 | 38–52 | 8,5 |
| | HF | 13 | 24 | 11–17 | 9,9 |
| | E | 10 | 24 | 9–13 | 10,6 |
| | Mass: | 7,8 g | 21 | 6–12 g | 17,3 |
| Females | HB | 62 | 31 | 53–71 | 5,7 |
| | T | 45 | 31 | 37–54 | 8,5 |
| | HF | 13 | 32 | 11–16 | 9,2 |
| | E | 10 | 31 | 8–13 | 10,4 |
| | Mass: | 8,0 g | 29 | 6–11 g | 17,2 |

Smithers (1975) gave the live weight of this species as 5–6 g. In Uganda, Delany (1975) found that males weighed 7,6 g (N=31, observed range 2,5–12 g) and females weighed 7,3 g (N=24, obs. range 2,5–12,0 g).

## Identification

The dorsal colour is a brownish-buff, darkening somewhat along the midline through an admixture of black hairs. The fur is coarser than that of *Mus (M.) musculus* and individual hairs have pale slate-coloured bases. This coarseness of the fur is partially caused by the presence of relatively broad, bristly hairs which are stiff and channelled longitudinally and referred to as 'gutter hairs'. The different hairs can best be seen with the aid of a low power microscope. The flanks tend to be orange-buff in colour while the ventral surface is pure white; the individual hairs are white throughout. The white belly is sharply demarcated from the flanks. The ears are of a moderate size, rounded and nearly naked, a little darker along their margins. The short tail is less than half the total length of the animal (i.e. head-body length plus tail length). The hands and feet are scantily covered with white hairs, while the digits have well-developed claws (plate 11).

## Skull and dentition

The skull is small and lightly constructed. A number of features have been described as diagnostic, but these do not hold good for all skulls. For instance, the muzzle may be slightly longer in *minutoides* than in *musculus,* though this is not constant as was originally supposed. The width between the outer edge of the molar tooth row to the outer surface of the zygomatic arch on the opposite side is shorter than the length of the nasals in *minutoides,* whereas the length of the nasals is shorter in the case of *musculus.* Another example refers to the anterior edge of the zygomatic plate and the anterior palatal foramina. In *minutoides* the edge of the zygomatic plate lies midway along the anterior palatal foramen, or posterior to midway. Other features include the anterior edge of the zygomatic plate which is strongly convexly curved and it passes into the rostrum below with a pronounced forward sweep. In contrast to *musculus,* the masseteric knob lies closer to the middle of the zygomatic plate (fig. 102).

The main distinguishing characteristics of *M. (L.) minutoides* is to be found in its dentition. The notched inner surface of the upper incisors is very characteristic.

The anterior root of the three-rooted $M^1$ emerges from the maxilla at a greater angle in relation to the rest of the tooth (compared to the virtually in-line development in *M. (M.) musculus*). Consequently, this root is covered to a far greater extent by the overlying forward bulge of the first lamina which, incidentally, may occasionally show the development of a small cusplike protuberance. The arrangement of the rest of the cusps is very much like that of *M. (M.) musculus.* The $M^3$ is usually reduced to a small ringshaped peg without traces of any laminae (fig. 103). In the lower jaw the $M_1$ has six clear cusps – four large posterior ones and two smaller anterior ones.

The dentition in the subgenus *Leggada* is highly particular, the teeth being the most extreme and morphologically specialised in the whole genus.

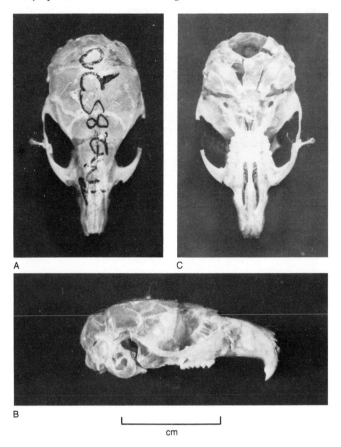

A    C

B

cm

**Figure 102** Dorsal (A), lateral (B) and ventral (C) views of the skull of the Pygmy Mouse *Mus (Leggada) minutoides*. The supraorbital ridges are weak and in contrast to *Mus (Mus) musculus* the muzzle tends to be somewhat longer and the masseter knob lies closer to the middle of the zygomatic plate which has a convex edge.

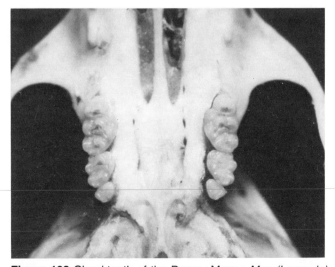

**Figure 103** Cheekteeth of the Pygmy Mouse *Mus (Leggada) minutoides*. The anterior root of the three-rooted M¹ is largely hidden by a forward bulge of the first lamina. The M³ is peglike without traces of laminae.

## Distribution

*Mus (Leggada) minutoides* occurs in the Southern Savanna Woodlands and Grasslands, the South West Cape and the South West Arid (Davis 1974). It is therefore widespread, if sparsely distributed, throughout southern Africa (map 67).

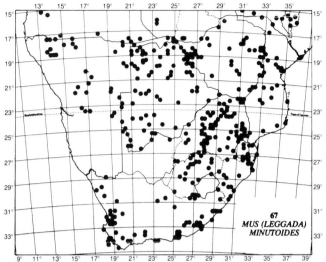

67
MUS (LEGGADA)
MINUTOIDES

## Habitat

These mice occur in virtually all types of vegetation. It does seem, however, as if the central Karoo is not particularly favoured. In Botswana they occur in a wide variety of habitats from arid shrub savanna of the extreme southwestern parts of the Kalahari with an annual rainfall of 200 mm, to rich riverine forests along the Chobe River at Kasane which receives 700 mm of rain annually (Smithers 1971). Shortridge (1934) states that they can be trapped in long grass along water holes near the Cunene, '. . . inhabiting fairly damp country where there is high grass, bush or other cover'. They frequent the edges of vleis, riverbanks and the stone walls of fields and kraals (Grant *in* Shortridge 1934). In the Wankie National Park, Zimbabwe, they occur in all habitat types from wet areas around Deteema Dam to very dry country at Mitswiri (Wilson 1975). This also includes *Baikiaea* woodland on Kalahari sand and mopane woodland on clay soils. Smithers & Lobão Tello (1976) also record that in Mozambique they tend to use holes in termite mounds and other refuges in areas of hard ground, while also living under debris, fallen tree trunks and any similar type of cover. In Uganda their habitat includes dry sandy ground, savanna and forest and they may be collected at altitudes of 2 450 m above sea level (Delany 1975). They will also enter houses and may occasionally adopt a domestic way of life in West Africa (Rosevear 1969).

## Diet

Pygmy mice are omnivorous and will eat food of both vegetable and animal origin. Arthropod remains have been found in the stomach contents in specimens col-

235

lected in Uganda and Malawi (Delany 1975). Shortridge (1934) reports them to take grasshoppers, termites and other insects readily. Captive specimens collected in grassland and disturbed farmland near Pietermaritzburg, Natal, readily ate *Tenebrio* larvae, mouse cubes, millet, fresh seedheads of *Panicum,* apples and tomatoes (Willan & Meester 1978). A tendency towards graminivory (Smithers 1971) is substantiated by Wilson (1975) who records pieces of the inflorescence of *Chloris virgata* with the seeds still attached, in a nest of a specimen taken in Wankie, while other captured specimens were found to have eaten seeds of *Cynodon dactylon, Setaria anceps* and *Andropogon gayanus.*

## Habits

These nocturnal, terrestrial little mice stay under any sort of cover (sheets of corrugated iron, stone, jute sacks, tarpaulins, the ground sheeting of tents or even the bedding of survey parties sleeping on the ground during the day (Smithers 1971)). In sandy areas they can burrow down to a depth of 45 cm to form a nest chamber after the fashion of *Malacothrix,* by constructing a separate outlet by which to escape if the pressure is on. The nests are made of strips of grass. Roberts (*in* Shortridge 1934) reports that they will sometimes enter burrows of other animals, where they will excavate a side chamber and fill it with shredded grass. Consequently, they can be trapped in the runways of other rodents or, according to Kingdon (1974), in close association with shrews. Smithers (1975) states that they do not burrow (although they may do so in sandy areas as related above) and that they prefer to construct their bell-shaped nests (made of soft grass or other fibres) under stones or any other substantial cover.

It has been noted that in areas where population explosions of the Multimammate Mouse *Praomys (Mastomys) natalensis* occur, the numbers of *M. (L.) minutoides* also rise, and fall when the numbers of *Praomys* decline. Vesey-Fitzgerald (1966) has suggested that the huge numbers of *Praomys* at such times provide a 'predator-shield' for *Mus* and other species that may also increase at this time, just when one might have thought that competition for resources was most keen. However, the quantity of food eaten by *minutoides* is a tiny fraction of that consumed by larger species and it might even benefit directly from the more wasteful feeders (Kingdon 1974).

Pygmy mice occasionally enter houses where they are often caught by domestic rats. They are active little animals but are reputed not to have the leaping or jumping powers of *M. (M.) musculus.* Smithers (1975) states that they are difficult to trap because of their small size. They can, however, be caught in pit traps and in tunnel traps similar to those described by Meester (1970). Pit traps can be 30 cm in diameter and 45 cm deep, while the tunnel traps (3x4x20 cm) are small and can be triggered by mass of 5 g (Willan & Meester 1978).

## Predators

Very little information is available. Snakes are possibly important predators and Kingdon (1974) states that pygmy mice are frequently recorded in the crops of birds of prey in East Africa. They are taken by the Barn Owl *Tyto alba* (Vernon 1972; Dean 1977), the Grass Owl *T. capensis* (Davis 1973) and the Marsh Owl *Asio capensis* (Dean 1977).

## Reproduction

In the wild the young are born in nests made of soft grass. In captivity the breeding biology and post-natal development have been discussed by Willan & Meester (1978). Twenty-seven litters were observed. The gestation period was 19 days or less with a mean litter size of 4,0 (1–7). At birth the pups weighed approximately 0,8 g and they were born naked and helpless. After 14 days all sensory abilities had developed, as well as the motor abilities. Weaning occur on day 17 while sexual activity commenced on day 32, the first successful mating being recorded at approximately 42 days. The first litter arrived on the 62nd day. The first appearance of adult behaviour patterns is delayed in *M. (L.) minutoides* relative to *M. (M.) musculus,* while other patterns (e.g. weaning) occur earlier in *minutoides.* Similar information is provided by Ansell (1963) who describes the naked and blind young as developing their skin pigments on day 3, the detachment of the ears on day 4 and the development of a thin coat by day 8. In Zimbabwe, Wilson (1975) reports that gravid females occur in February–October (with the exception of April and June in his material) and indications are that the species breeds throughout the year. Smithers (1971, 1975) also states that females may breed throughout the year, but that the young are mainly born during the summer months. The average number of fetuses was 4,9 (based on a sample of 17), with an observed range of 2–8. Implantation in either left or right uterine horn was irregular. In the Kruger National Park, a female with six newly born young was collected in February (Pienaar 1964).

## Parasites

According to Neitz (1965), *Mus (Leggada) minutoides* is susceptible to infections of plague *Yersinia pestis* under laboratory conditions. Likewise, they also react to infections of *Listeria monocytogenes,* the cause of listeriosis.

The helminths associated with this little mouse in southern Africa are poorly known. Collins (1972) lists *Hymenolepis microstoma* as an endoparasite.

The arthropods hitherto recorded associated with *M. (L.) minutoides* include a number of fleas. The Pulicidae are represented by *Xenopsylla cheopis, X. piriei* and *X. brasiliensis.* The Hystrichopsyllidae occur as *Ctenophthalmus acanthurus, C. calceatus, C. gilliesi, C. phyris, Dinopsyllus ellobius, D. grypurus, D. lypusus* and *Listropsylla prominens.* Finally, the Chimaeropsyllidae are represented by *Chiastopsylla rossi* (Zumpt 1966).

Immature specimens of the following ticks have been taken from *M. (L.) minutoides* (Theiler 1962): *Haemaphysalis leachii muhsami*, *Rhipicephalus appendiculatus* and *R. simus*.

## Relations with man

Pygmy mice are attracted to cultivation and are often caught in gardens around African villages. They will enter houses, but according to Smithers (1975), this is not their normal behaviour.

Like the Multimammate Mouse, they undergo population explosions from time to time, reaching astronomical numbers. Owing to their size and the fact that they do not normally carry fleas associated with plague, they are not considered a pest or health hazard (Smithers 1975).

## Prehistory

Specimens of *M. (L.) minutoides* have been recorded from the Taung and Makapansgat breccias by De Graaff (1960a). The fossil forms do not differ from extant specimens and there is an interesting upsurge in their numbers at the Cave of Hearths, which, indicentally, is the youngest fossiliferous locality analysed by De Graaff. A new form, *Leggada major,* was named by Broom from Bolt's Farm, but it has not yet been described. In addition, specimens of *Mus* cf. *Leggada* were also found among the rodent remains of the Draper collection of mammalian microfauna collected at Kromdraai in 1895 (De Graaff 1961).

## Taxonomy

Ellerman (1941), though not recognising *Leggada* as a valid genus (nor for that matter as a valid subgenus), has divided the '*Leggada*-like' species of African into three groups:

(i) The *triton-bufo* group (the *triton* examples are grey-backed, white-bellied, with an elongated $M^1$; the *bufo* examples are dark, brown-backed and buffy-bellied, while the $M^1$ is not so long. The *bufo* specimens are probably the most primitive and unspecialised members of the species).

(ii) The *Minutoides* group (the backs are brown, the bellies white, tails smooth and long, the rostrum short and the $M^1$ very long).

(iii) The *tenellus* group (the backs are brown, the bellies white, the tail short and hairy, the rostrum long and the incisors prominent. This probably includes the most specialised members of the species).

Petter (1963) broadly confirms this grouping and also supplies a key. However, he thinks a reduction of the 40-odd species possible and distinguishes some 15 species in Africa (Petter 1975). Rosevear (1969) states that the situation is without question highly complex and that active speciation seems to be in progress. Matthey (1966) has analysed the karyotypes of 213 specimens from various localities in Africa. His results '. . . indicate sympatric speciation resting on reproductive isolation inside a population through chromosomic mutations bearing on the sex-chromosomes'. He concludes that *M. (L.) minutoides* is a polymorphic 'super species' with chromosome complements ranging from 2n=18 to 2n=36 (Davis 1974). Some of the minor external differences that have been noted for similar forms of mice are often reinforced by chromosomal differences that ensure reproductive isolation (Kingdon 1974).

*Leggada* has been used as a generic name by Roberts (1951), but Ellerman *et al.* (1953) regard it as a synonym of *Mus*, an interpretation followed by Meester *et al.* (1964). Davis (1965) recognises *Leggada* as a subgenus. Both Roberts and Ellerman *et al.* list eight southern African subspecies which include *minutus* (Cape Town), *umbrata* (Wakkerstroom), *marica* (Beira), *sybilla* (Angola), *indutus* (= *deserti* as synonym) (Molopo), *valschensis* (Bothaville), *orangiae* (Viljoensdrift) and *pretoriae* (Pretoria). The status of the taxon is, however, uncertain. On chromosome counts two species seem to be represented. At Kirkwood (3325 AD), Uitenhage district, *minutoides* (2n=18) occurs with *indutus* (2n=36). An ecocytogenetic study of the pygmy mice of the Kalahari would be of great importance for *indutus* and *deserti* may be distinct and not synonyms as currently accepted (Davis 1974). Smithers (1971) also documented the possibility of the coexistence of two species in Botswana.

GENUS *Acomys* I. Geoffroy, 1838

These small rodents occur chiefly in Africa, although some species extend into the Sinai Peninsula and certain Mediterranean islands. They are fairly easily recognised by their dorsal fur being spiny or showing a bristly rigidity. They are typical inhabitants of dry, semi-arid areas.

The genus *Acomys* was first described in 1838 by Geof-

froy and because of its spiny pelage, was always accepted as a separate genus. The taxonomic argument over the years concerned the validity and acceptability of diverse species and subspecies which were described by many authors. As an example, Ellerman (1941) cited 38 forms, 26 as distinct species.

In order to establish some sort of baseline for this discussion of the genus, the interpretation suggested by Setzer (1975) is adhered to, while his key to the genus has also been consulted.

**Key to the southern African species of *Acomys***
(After Sclater (1901) and Setzer (1975))

1 Length of nasals 9,5 mm or less; tail shorter than HB length; greyish brown above ...... *A. subspinosus* (Cape Spiny Mouse) Page 238

Length of nasals 9,8 mm or more; tail longer than HB length; rufous brown above...... *A. spinosissimus* (Spiny Mouse) Page 240

# *Acomys subspinosus* (Waterhouse, 1838)

**Cape Spiny Mouse**

**Kaapse Stekelmuis**

The name is derived from the Greek *akn* = a sharp point and *mys* = mouse, referring to the sharp and bristle-like pelage of the animal; and the Latin *spinatus* = spined. The prefix *sub* indicates under or below, i.e. the bristles of the pelage are not quite spiny.

This species was first described by Waterhouse in 1838 as *Mus subspinosus,* based on a specimen presumably collected on Table Mountain. This mouse has also been listed in the *South African Red Data Book* (Meester 1976) as rare and limited '...in distribution and nowhere very abundant'.

## Outline of synonymy

1838 *Mus subspinosus* Waterhouse, *Proc. zool. Soc. Lond.* for 1837:104. Cape of Good Hope.

1838 *Acomys* I. Geoffroy, *Ann. des Sci. Nat. Zool.* (2) 10:126. (Genotype *Mus cahirinus* Desmarest.)

1939 *Acomys subspinosus* (Waterhouse). *In* Allen, *Bull. Mus. comp. Zool. Harv.* 83:366.

1951 *A subspinosus* (Waterhouse). *In* Roberts, *The mammals of South Africa.*

1953 *A. subspinosus* (Waterhouse). *In* Ellerman *et al., Southern African mammals.*

1975 *A. subspinosus* (Waterhouse). Setzer, *in* Meester & Setzer (eds), *The mammals of Africa: an identification manual.*

## Identification

The colour on the back is greyish-brown, becoming lighter and more rust-coloured on the flanks. It is a small rodent whose fur is decidedly harsh to the touch. The entire pelage consists of broad, toughened gutter hairs which reach their maximum development dorsally from the lumbar region to the root of the tail. The other hairs

### TABLE 61
Measurements of male and female *Acomys subspinosus*

|  | Parameter | Value (mm) | N | Range (mm) | CV (%) |
|---|---|---|---|---|---|
| Males | HB | 88 | 16 | 81–97 | 4,7 |
|  | T | 84 | 16 | 75–98 | 7,1 |
|  | HF | 17 | 17 | 13–19 | 9,1 |
|  | E | 13 | 17 | 11–16 | 11,2 |
|  | Mass: | 20 g | 9 | 17–23 g | 9,4 |
| Females | HB | 93 | 17 | 82–98 | 4,7 |
|  | T | 87 | 17 | 81–96 | 6,0 |
|  | HF | 17 | 18 | 14–19 | 7,8 |
|  | E | 14 | 18 | 12–15 | 7,4 |
|  | Mass: | 22 g | 17 | 20–25 g | 8,0 |

are softer to the touch but still bristle-like. According to Rosevear (1969) each individual bristle is concave on its dorsal surface and convex along its ventral surface. Although these bristles are sharp to the touch, they are not drawn out into long, fine unchannelled subcylindrical tips as gutter hairs elsewhere in the murid family. The dirty-white ventral surface (the change from the flanks to the belly is abrupt) consists of similar flat bristles which are shorter and narrower, and are consequently softer to the touch than the dorsal surface. The individual bristles are white throughout, as are the throat, cheeks and upper lips. Interspersed with these bristles on both dorsal and ventral surfaces, is a fine underfur as well as a few widely scattered guide hairs. The tail is rather stout, roughly as long as the length of the head and body and covered with short dark bristles above and white ones below (plate 12).

The white hands and feet are covered with fine hair. There are four digits on the hand, the thumb being reduced to a small stump with a nail rather than a claw. The foot has five digits, and the hallux is reduced though not to the same extent as the pollux.

The mammary formula is 1−2=6 (Sclater 1901).

## Skull and dentition

An obvious feature of the skull of this species is the relatively broad, rounded braincase. The anterior palatinal foramina are longer than half the length of the diastema and taper to a point well between the two first upper molars. A characteristic feature of this species is its very small teeth. The upper tooth row is less than 3,5 mm in length. The foremost space between the pterygoids (the mesopterygoidal fossa) is roofed in by bones, which are extensions of the palatine elements. The hamular components of the pterygoids fuse with the bullae posteriorly, which in turn are rather small. Consequently, the aperture of the nares interni is a small triangular hole. The foramen magnum is fairly large. The nasals are 9,5 mm or less in length. The anterior edge of the zygomatic plate is upright and slightly cut back above. The antorbital foramen is moderately open. The zygoma widens anteriorly. The jugal is long (fig. 104).

The very small molars show a *Mus*-like cusp configuration where the t1 of the first lamina of the $M^1$ lies in line with the t5 and t6 of the second lamina and the concomitant backward displacement of t4. In both the $M^1$ and $M^2$ the t7 is missing. The $M^3$ is small (fig. 105). The upper molars are plain and there is a tendency towards opisthodonty in the incisors. In the lower jaw the $M_1$ and $M_2$ have reduced terminal heels. The coronoid process of the mandible is poorly developed if compared to other murids.

## Distribution

In southern Africa, this species is confined mainly to the South West Cape and Forest biotic zones. It is known from Knysna in the east to Citrusdal in the west, mainly along the coast (map 68). It probably occurs '...throughout southern and eastern Africa but so far recorded only from South Africa, Kenya, Somalia, Uganda and Sudan' (Setzer 1975). It therefore occurs patchily to East Africa.

## Habitat

Very little is known about this aspect of the Cape Spiny Mouse. Its distribution resembles that of *Praomys (Myomyscus) verreauxii*. Kingdon (1974) states that in East Africa, it is a dry savanna species in grassy situations.

## Diet, habits, reproduction, predators, relations with man, and prehistory

No data available to date.

## Parasites

Zumpt (1966) in his work on the arthropod parasites of vertebrates in Africa south of the Sahara, lists a number of fleas which are known to occur on *Acomys*. However, because of the confused state of the taxonomy of this genus, as pointed out by Zumpt, identification of host species is often rendered doubtful in this genus. The following list therefore refers to *Acomys* sp.: Pulicidae – Cte-

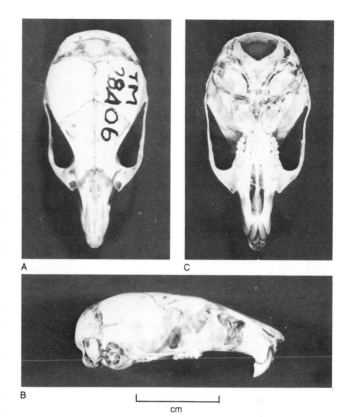

**Figure 104** Dorsal (A), lateral (B) and ventral (C) views of the skull of the Cape Spiny Mouse *Acomys subspinosus*. The supraorbital and occipital ridges of the skull are poorly developed. The foremost space between the pterygoids is roofed over by bone, being a posterior extension of the palatines. The bullae are small. The zygoma tends to broaden anteriorly and it has a large jugal element. The edge of the zygomatic plate is perpendicular. The incisors tend towards opisthodonty.

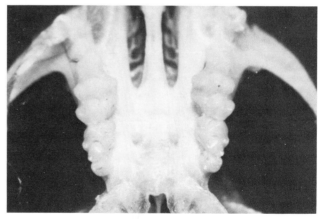

**Figure 105** Cheekteeth of the Cape Spiny Mouse *Acomys subspinosus*. The upper toothrow is less than 3,5 mm in length. In the three-rooted $M^1$ the t1 is displaced towards the rear, being in line with the t5 and t6 of the second lamina. In the $M^2$ the t2 is absent while the t3 is relatively small. The $M^3$ is small and bilaminate.

*nocephalides felis, Parapulex chephrenis, P. echinatus, Synosternus cleopatrae, S. somalicus, Xenopsylla cheopis, X. brasiliensis, X. morgandaviesi, X. debilis* and *X. difficilis;*

Hystrichopsyllidae – *Ctenophthalmus calceatus, Dinopsyllus abaris* and *D. lypusus*. The same remarks apply to Theiler (1962) who listed immature specimens of the tick *Hyalomma excavatum*, taken from an *Acomys* specimen.

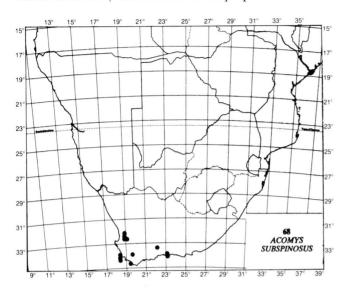

**68**
**ACOMYS SUBSPINOSUS**

## Taxonomy

Ellerman (1941) interprets *A. subspinosus* as being absolutely distinct from all other species in the abnormally narrowed teeth and strongly reduced M³. He therefore refers to a *subspinosus* group and a *cahirinus* group. The latter contains the rest of the genus with recognisable sections, which Ellerman called the *russatus* section, the *wilsoni* section and the *cahirinus* section.

Roberts (1951) interpreted *transvaalensis* (i.e. *A. spinosissimus*), a form described by him in 1926 as a separate species, as well as *A. selousi (A. spinosissimus)* described by De Winton in 1896. He also recognised *subspinosus* as a separate species, as did Ellerman *et al.* (1953) and Meester *et al.* (1964). Setzer (1975) also accepted it as a separate species and listed some ten subspecies. The following forms occur in northern, northwestern, eastern and southern Kenya respectively: *ablutus, enid, wilsoni* and *nubilus*. In eastern Sudan, *argillaceus* is found while *bovonei, louisae* and *umbratus* are encountered in southern, northern and northeastern Somalia respectively. The nominate race *subspinosus* is described from southern Africa.

It will be noted that *wilsoni*, placed under the *cahirinus* group as the *wilsoni* section of Ellerman's 1941 scheme, is here demoted to subspecies rank under *subspinosus*. In order to complete the picture, both *A. transvaalensis* and *A. selousi* have been demoted to subspecies rank under *A. spinosissimus (q.v.)*.

# *Acomys spinosissimus* Peters, 1852

**Spiny Mouse**
**Stekelmuis**

For the meaning of *Acomys*, see *Acomys subspinosus*. The specific name is the Latin word *spinosissimus* = thorniest, referring to the coarser spiny nature of the pelage (in contrast to *Acomys subspinosus*).

This species was first collected and described by Peters in 1852 from 'Tette and Buio, Portugese E. Africa' (Ellerman 1941) and described as *Mus spinosissimus*. Ellerman proposed that the genus *Acomys* in Africa could be classified into two species groups, the *cahirinus* group and the *subspinosus* group. The former was divided into a *russatus* section, a *wilsoni* section and a typical section (i.e. *cahirinus* section). Ten species were, however, not examined by Ellerman and he was unable to allocate them to any group or section. This list included *A. spinosissimus* Peters as well as *Acomys transvaalensis* Roberts.

In 1951, Roberts described the genus *Acomys* in southern Africa as consisting of *A. subspinosus* (Waterhouse), *A. transvaalensis* Roberts and *A. selousi* De Winton. (Incidentally, *A. selousi* was listed as a separate species under the *cahirinus* section of Ellerman (1941).) Roberts (1951) did not list *spinosissimus* about whose affinities he was uncertain and referred material from Zimbabwe and Mozambique to *A. selousi* (see De Winton 1896). Ellerman *et al.* (1953) state that a large number of recognised species in this genus could well be considered as subspecies of the earliest named *A. cahirinus* '...which was based on a commensal form from Egypt'. They concluded, however, that *A. russatus* (Egypt and Arabia), *A. wilsoni* (East Africa) and *A. subspinosus* (Cape Province) may seem valid as species and the last was accordingly accepted as such along with *A. cahirinus. A spinosissimus* was relegated to *A. c. spinosissimus, A. selousi* to *A. c. selousi* and *A. transvaalensis* to *A. c. transvaalensis*. In 1964, Meester *et al.* reinstated the species rank of *spinosissimus* and proposed that *A. s. spinosissimus* included the material from Mozambique and Zimbabwe, while *A. s. transvaalensis* was to be retained as a subspecies for material from the northern and eastern Transvaal, as well as that from southeastern Botswana. However, Setzer (1975), on accepting *A. spinosissimus* as a species, proposed to interpret *selousi* as a subspecies (to include material from western Zimbabwe and eastern Botswana) while *A. s. spinosissimus* was retained for material from western Mozambique and eastern Zimbabwe.

## Outline of synonymy

1852 *Acomys spinosissimus* Peters, *Reise nach Mossambique. Säugeth.*: 160. Tete and Buio, Mozambique.

240

1926 *A transvaalensis* Roberts, *Ann. Transv. Mus.* 11:252. Soutpansberg, Transvaal.

1939 *A. spinosissimus* Peters. *In* Allen, *Bull. Mus. comp. Zool. Harv.* 83:366.

1953 *A. cahirinus spinosissimus* (Peters). *In* Ellerman *et al.*, *Southern African mammals.*

1975 *A. spinosissimus* Peters. Setzer, *in* Meester & Setzer (eds), *The mammals of Africa: an identification manual.*

### TABLE 62
Measurements of male and female *Acomys spinosissimus*

|         | Parameter | Value (mm) | N  | Range (mm) | CV (%) |
|---------|-----------|------------|----|------------|--------|
| Males   | HB        | 91         | 23 | 80–100     | 6,3    |
|         | T         | 79         | 22 | 70–90      | 7,4    |
|         | HF        | 16         | 24 | 13–17      | 5,9    |
|         | E         | 14         | 20 | 12–16      | 6,3    |
|         | Mass:     | 23,4 g     | 36 | 20–30 g    | 11,7   |
| Females | HB        | 88         | 23 | 75–102     | 8,9    |
|         | T         | 75         | 23 | 62–86      | 7,5    |
|         | HF        | 16         | 21 | 13–19      | 8,0    |
|         | E         | 14         | 22 | 12–17      | 9,6    |
|         | Mass:     | 22,2 g     | 16 | 19–28 g    | 12,2   |

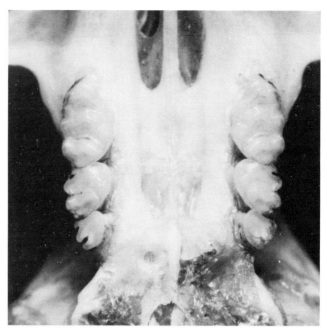

**Figure 106** Cheekteeth of the Spiny Mouse *Acomys spinosissimus*. The upper toothrow is more than 3,7 mm in length.

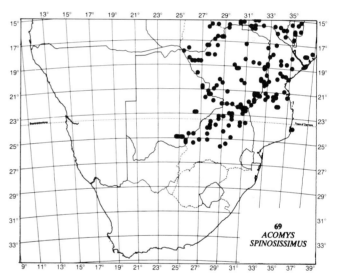

69
ACOMYS
SPINOSISSIMUS

## Identification
The colour of the spiny coat is normally smoky-grey which gradually fades to a reddish brown as the moult approaches. During the moult, the reddish brown spines are replaced by smoky-grey spines once again (Smithers 1975). Some specimens can be taken halfway through this stage, the body being half one colour, half the other. The lower surface of the body is pure white (plate 12).

## Skull and dentition
The skull is very much like that of *A. subspinosus*. In *A. spinosissimus,* however, the nasals are longer than in *A. subspinosus* (9,8 mm and more compared to 9,5 mm or less) while the upper tooth row usually attains a length of 3,7 mm or more in contrast to less than 3,5 mm in *subspinosus* (fig. 106).

## Distribution
The Spiny Mouse occurs in the Southern Savanna Woodlands as well as in the northeastern margin of the South West Arid (Davis 1974). It is, therefore, encountered in the Transvaal, southeastern Botswana, Zimbabwe and Mozambique (map 69).

## Habitat
*Acomys spinosissimus* is basically a dry woodland species in rocky sites. It inhabits koppies and other rocky places (Davis 1974) with *A. namaquensis* and elephant shrews as associates. It is also found in alluvium along rivers and on sandy ground with no rocky conditions (Smithers & Lobão Tello 1976). In Mozambique, it has been taken in miombo and mopane savanna and in *Androstachys* thickets. Smithers (1975) states that quite large numbers may occupy the shelter of a rocky ledge '. . . up to nine or ten being taken from a single cranny'.

## Diet
The Spiny Mouse feeds predominantly on grass and other seeds (Smithers 1975). Vesey-Fitzgerald (1966) found *A. s. selousi* eating beetles, ants, termites, spiders, millipedes and small snails. In the Kruger National Park it has been observed feeding on the fruits of the Jakkalsbessie *Diospyros mespiliformis* (Pienaar *et al.* 1980).

241

## Habits

Spiny mice are gregarious, ground dwelling and predominantly (not exclusively) nocturnal. They occur as solitary individuals, in pairs or in small parties (Smithers 1971). Smithers states that their presence can be determined by their scats which accumulate near the entrances to their shelters. The scats are cylindrical, their ends broken off at right angles and are normally slightly reddish in colour.

## Predators

Among 1 353 rodent skulls which Hanney (1975) extracted from Barn Owl *Tyto alba* pellets in Malaŵi, only one belonged to a Spiny Mouse (in all probability *A. s. selousi*). Since the species is common in the vicinity of the roost, the spines must act as a deterrent to the owls, there being little doubt that their passage down (or up) an oesophagus would be as painful for the swallower as for the swallowed! Worden & Hall (1978) lists the White-faced Owl *Otus leucotis* as a predator as well. Likewise, domestic cats seem to find these mice difficult to eat (Kingdon 1974).

## Reproduction

Full-term fetuses have been taken in February from a female in the Kruger National Park (Pienaar *et al.* 1980). Walker (1975) states that most births occur between February and September and that the young are born with their eyes open. Males are capable of breeding at the age of seven weeks.

Smithers (1975) states that grass and leaves are dragged into crannies and crevices between rocks where rough nests are built, in which the young are born. The litters number two to five. The young are usually dropped in summer from about November to April. In Botswana gravid females have been taken in December, January, March and April. The average number of fetuses equals three (N=12) with an observed range of two to five (Smithers 1971).

Dieterlen (1961) has described the breeding of *Acomys*

*cahirinus dimidiatus* in captivity. This genus has an exceptionally long gestation period and the young are born in a relatively advanced state of development, compared to other murines. Gestation lasts 36–40 days, averaging about 38. In other murines it is 21–28 days. Immediately after parturition the female comes into oestrus again and breeding may be virtually continuous, month after month for 12 litters or more without interruption. Under favourable conditions in the wild, natural populations may do the same. Dieterlen bred 179 animals from a stock of seven. This allowed him to watch many births of *A. cahirinus* and simultaneously he observed an unusual social behaviour amongst the females. Multiparous females frequently assist during parturition of other females by biting the umbilical cord and cleaning the newborn while the mother delivers the rest of the litter. There are no data on the breeding of South African *Acomys* species.

## Parasites

The remarks made appropriate to the preceding species *A. subspinosus*, are also applicable to *A. spinosissimus*.

## Relations with man

No data available.

## Prehistory

Specimens of *Acomys* (in all probability *spinosissimus*) have been tentatively identified from the Makapansgat breccia by De Graaff (1960a). It was also found among the mammalian microfauna collected at the Kromdraai (A) Faunal Site by Draper in 1895 (De Graaff 1961).

## Taxonomy

This has been dealt with under the introductory remarks to this species. The interpretation of the genus *Acomys* (and its species and subspecies) as proposed by Setzer (1975) has been adhered to and *A. spinosissimus* is accepted as a species with *A. s. selousi* and *A. s. spinosissimus* as subspecies.

Genus *Uranomys* Dollman, 1909

It is not a well-known genus and study material is non-existent in South Africa. As in *Acomys*, the mesopterygoid fossa is roofed in by bone. The external appearance is similar to that of the Harsh-furred Mouse *Lophuromys* (extralimital, East and West Africa), but the hands and feet of the latter are brownish in contrast to white in *Uranomys* (Walker 1975).

242

# *Uranomys ruddi* Dollman, 1909

**Rudd's mouse**

**Rudd se Muis**

The name is derived from the Greek word *ouranos* = palate and *mys* = mouse and evidently refers to the backward extension of the palatine bones in the roof of the mouth. Mr C. D. Rudd, a patron of Zoology responsible for extensive collections on the continent of Africa, is honoured by the specific name.

This '...rather strange rodent' as it was referred to by Kingdon (1974) was first described by Dollman in 1909 on a specimen collected in the southern foothills of Mount Elgon in Kenya. The type specimen is housed in the British Museum (Natural History), London.

The overall distribution of this species is south of the Sahara and extensive but scattered. Its stronghold seems to be in West Africa where it is known from the Gold Coast, Guinea, Togo and Nigeria. However, it also ranges through East Africa, including the northern parts of Zaïre, Kenya and Uganda, while it is also known to occur in Malaŵi and Mozambique. According to Smithers (1975) '...it has only in recent years been shown to occur in Mocambique and was recorded for the first time in 1969 in Southern Rhodesia'. According to Smithers it ranks as the rarest species in Zimbabwe and indeed, in spite of its wide range, there are very few specimens in collections anywhere in the world. This statement was also confirmed by Rosevear in 1969. I have not personally seen any specimens and the account which follows has been compiled from literature.

## Outline of synonymy

1909 *Uranomys ruddi* Dollman. *Ann. Mag. nat. Hist.* (8) 4: 552. Kenya.

1953 *Uranomys ruddi* Dollman. *In* Ellerman *et al.*, *Southern African mammals*.

1974 *Uranomys ruddi* Dollman. Misonne, *in* Meester & Setzer (eds), *The mammals of Africa: an identification manual*.

## Identification

They are moderate-sized mice. Smithers (1975) states that the adults attain an overall length of 15 cm, the tail making up about one third of this length.

The upper parts vary in colour from dark brown, sometimes tinged with light brown, to dark grey. The pelage is shiny. The individual hairs on the back have slaty-grey bases, orange-buff subterminal rings and dark brown tips. The hair is fairly long, about 17 mm on the back (Delany 1975), of a springy nature, each hair like a long, fine bristle. These hairs consist of broad, slightly concavely channelled gutter hairs. There is virtually no underfur while a few guard hairs are visible when the

## TABLE 63
### Measurements of male and female *Uranomys ruddi**

| | Parameter | Value (mm) | N | Range (mm) | CV (%) |
|---|---|---|---|---|---|
| Males | HB | — | 2 | 100, 102 | — |
| | T | — | 1 | 71 | — |
| | HF | — | 2 | 16, 17 | — |
| | E | — | 2 | 15, 16 | — |
| | Mass: | 41 g, 55 g | 2 | — | — |
| Females | HB | — | 3 | 95, 104 ,106 | — |
| | T | — | 3 | 50, 66, 71 | — |
| | HF | — | 3 | 16, 17, 18 | — |
| | E | — | 2 | 13, 15 | — |
| | Mass: | 53 g | 1 | — | — |

*After Delany (1975)

animal is held against the light (Rosevear 1969). The top of the head is rather darker than the body while the side of the face is a pale buff. The muzzle is long. The ears are of moderate size, dark and thinly clad in short hairs. The flanks are a pale buff. The underparts are much lighter than the back, the tips of individual hairs sometimes tipped white (Smithers 1975). The upper surfaces of the limbs are pale buff. The hindfoot is short, broad and strong and digits 2, 3 and 4 are subequal in length. Digits 1 and 5 reach only to the bases of 2 and 4 respectively. The thumb is rudimentary and has developed a nail rather than a claw. The D5 reaches to about halfway the basal joint of D1.

The tail is scaly and covered with short bristles (plate 12). Rosevear (1969) states that in study skins the tails are often broken.

The mammary formula, according to Rosevear (1969), reads 3−3=12.

## Skull and dentition

The skull is rather flat and the narrow rostrum curves down only slightly (Rosevear 1969). The interorbital constriction is apparent and the supraorbital ridges peter out about halfway along the length of the parietals. The braincase is broad. The anterior palatal foramina are very long and broad and often extend to the back of the $M^1$ or nearly to the $M^2$. The palate is broad and these cranial elements have developed bone-like flanges in a posterior direction whereby the roof of the palate is extended over the mesopterygoid fossa (Ellerman 1941). In this respect it approximates the condition found in *Acomys* and many authorities therefore suggest a close phylogenetic relationship between *Acomys* and *Uranomys*. The bullae are of moderate size. The condylobasal length is at least as long as, but usually longer than the occipitonasal length (Ellerman *et al.* 1953).

The zygomatic plate is upright with a convex anterior margin (Rosevear 1969) which is cut back above. The jugal element is relatively long.

The upper incisors are slightly pro-odont. The molars are close to those of *Acomys*. In the $M^1$ the t1 is displaced backwards, curving round the t5 of the second lamina. In turn, the t4 also curves backwards, but it is not as narrow as the t1. The $M^2$ has a small t1 and a very small t3. The $M^3$ is considerably reduced, but shows signs of two laminae (Rosevear 1969).

The lower incisors are long and their roots form a fairly strong ridge on the outer surface of the mandible (Ellerman 1941). Kingdon (1974) has drawn attention to the unique condition of the teeth in young animals. The anterior cusps of the $M_1$ are 'worn' ahead of any possibility of friction against the upper molars. This phenomenon has been called 'l'abrasion préalable' by Heim de Balsac (1967, *in* Kingdon 1974) and it may involve a curtailment of cusp development during intra-alveolar genesis or else '. . . the cusp tips are resorbed as soon as they are formed' (Kingdon *op. cit.*). The effect of the removal of the cusps is to allow the tubercles behind to grind directly against those of the upper teeth as they come into contact. Furthermore, these lower molars in young animals are also exceptional in having well-developed pulp cavities in which the pulp becomes progressively replaced by dentine '. . .so that an old worn tooth is almost solid ivory' (Kingdon *op cit.*).

## Distribution

According to Misonne (1974), this rare savanna species occurs in Mozambique, Malaŵi, northwards to East Africa and westwards to West Africa. It has also been taken in Zimbabwe (Smithers 1975) (map 70).

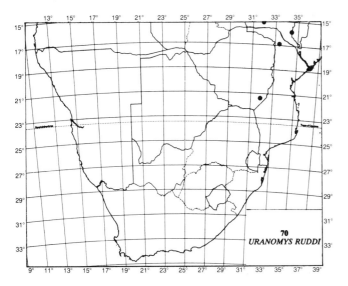

**70**
**URANOMYS RUDDI**

## Habitat

According to Kingdon (1974), all locality records for *Uranomys ruddi* are from grass savanna and secondary savanna to moist savanna and woodlands. In Mozambique it occurs in grassland on alluvial soils (Smithers & Lobão Tello 1976). In West Africa it is found in savannas accompanied by *Borassus* palms and it also readily colonises oil palm plantations (Bellier 1965, *in* Delany 1975).

## Diet

Rudd's mouse is reported to be mainly insectivorous (Kingdon 1974). Unfortunately no information is available on its diet from either Zimbabwe (Smithers 1975) or Mozambique (Smithers & Lobão Tello 1976).

## Habits

There is an apparent lack of information in the habits and behaviour of *Uranomys*. It is reported to be nocturnal (Kingdon 1974). It is usually caught singly or in pairs but has been captured by the hundreds '. . .in *Borassus* palm savanna' in the Ivory Coast (Kingdon 1974). Its burrows may contain an enlarged nesting chamber 10–15 cm in diameter, made approximately 15 cm below the surface of the soil. Each burrow apparently has two exits, surmounted by piles of ejected soil. A third blind tunnel goes down to a depth of 30 to 40 cm. The nest is lined with finely cut grass. Each burrow is inhabited by one or two animals only. There appears to be no storage of food underground (Bellier 1968).

## Predators

Virtually nothing is known, except that they are taken by the Barn Owl *Tyto alba* (Heim de Balsac & Lamotte 1958, *in* Kingdon 1974)

## Reproduction

In the Ivory Coast, Rudd's Mouse seems to breed throughout the year. The litters are bigger from September to December (4,0–5,7 pups), compared to the months of February to June when litter size fluctuates between 2,6 and 3,7 pups (Delany 1975).

## Parasites, relations with man and prehistory

No data available.

## Taxonomy

Over the years eight different forms of *Uranomys* have been named, all of specific rank, but they are now all regarded as races of a single species. Misonne (1974) included the following known forms under *U. ruddi*: *acomyoides* (from the Gold Coast), *foxi* (northern Nigeria), *oweni* (Gambia), *ugandae* (Uganda) and *tenebrosus*, *woodi* and *shortridgei* (from Malaŵi).

# Gliroidea Simpson, 1945

# Gliridae Thomas, 1897

The second superfamily of the suborder Myomorpha is the Gliroidea, which incorporates the dormice and allied forms. The third superfamily which is usually included in the myomorphs, the Dipoidoidea (which includes the jerboas), are extralimital.

The systematic position of the Gliroidea is by no means clear and has been one of the major problems in the classification of the order Rodentia. According to Ellerman (1941) they could be regarded as primitive and aberrant members of the superfamily Muroidea. However, the superfamily Gliroidea was erected by Simpson in 1945 and is more or less equivalent to the interpretation of the family Muscardinidae proposed by Palmer in 1899, and as was used by Ellerman (1941). The Gliroidea are now divided into three families, the Gliridae, the Platacanthomyidae and the Seleviniidae, of which only the African Gliridae are of concern to us here, while the latter two families are purely Asian. There has been some resistance to accepting the taxon Gliridae as proposed by Thomas in 1896, because it was preoccupied by Gliridae Ogilby, 1837. Consequently, the name Muscardinidae was applied more often. Simpson (1945) has argued for the return of the long established use of Gliridae Thomas, and since 1945 it has become the accepted family name of this group of rodents (Rosevear 1969).

The Gliridae are small, bushy-tailed, squirrel-like, predominantly arboreal rodents with short, rounded heads, large eyes and ears, consisting of seven genera and 23 species confined to the Old World. They occur throughout much of Africa south of the Sahara, Britain, Europe, southern Scandinavia, Asia minor and the Indo-Malayan region extending eastwards to Japan. In Europe they have been found as fossils in Oligocene deposits. They are not uncommon, but not as abundant as the muroids. The pelage is always soft, dense and short and there is no dorsal pattern. The tail is about as long as the length of the head and body. The feet are broad and strong. The hand shows four subequal digits with the pollex missing. The foot has five digits, with the D1 as the shortest, reaching the base of the D2, while the rest of the digits are of subequal length. The digits are clawed and adapted for climbing.

The skull shows the frontal elements much contracted and a wide elliptical infraorbital foramen. This foramen, though transmitting muscle, is comparatively unspecialised (Ellerman 1941). The zygomatic plate lies below the foramen. The palatine foramina are small while an interparietal element is present dorsally. The four cheekteeth ($Pm^4$, $M^{1-3}$) are brachydont with shallow, basinshaped occlusal surfaces, sometimes showing faint transverse ridging. The cutting edges of the incisors wear down in such a way that they form a backward directed angle. The lower jaw is elongated with a longer coronoid process.

They hibernate in colder regions, but not in warmer climates. They are nocturnal animals (compare the squirrels which are primarily diurnal), omnivorous to carnivorous and terrestrial to arboreal.

The family Gliridae has been divided into two subfamilies, viz. the Glirinae and the Graphiurinae. The former are Palaearctic in their distribution (Europe to the Far East), while the latter form the African dormice. It is now believed that they all belong to a single genus Graphiurus. This genus has, at times, also been split into four distinct genera, some of which have subsequently been retained as subgenera. As an example, the work by Thomas & Hinton (1925) can be referred to. At that time, three genera were accepted: Graphiurus (with a very small or minute and simple $Pm^4$) which would include the South African species ocularis, described by Smith. In the other two genera, Gliriscus and Claviglis (where the $Pm^4$ is not much smaller than the $M^{1-3}$) the other species were put, i.e. the South African platyops in Gliriscus and all the other species under Claviglis. However, Ellerman et al. (1953) saw no valid distinction between Gliriscus and Claviglis and consequently Graphiurus and Claviglis were retained. In 1964 Meester et al. pointed out that the genus Graphiurus is in need of a thorough revision, a task that is presently being undertaken by Channing (pers. comm.).

Roberts (1951) and earlier authors recognised three species: Graphiurus (for ocularis), Gliriscus (for platyops) and claviglis (for murinus). Ellerman et al. (1953) retain Claviglis as a subgenus for platyops and murinus while Meester et al. (1964) have placed all species under Graphiurus. Misonne (1974) follows Ellerman et al. (1953), an interpretation adhered to in this book as well.

GENUS *Graphiurus* Smuts, 1832

This dormouse was first described by A. Smith in 1829 as a squirrel, *Sciurus ocularis,* no doubt on account of its bushy tail. In 1832 Smuts placed it in the genus *Graphiurus*. The type specimen, collected in the vicinity of Plettenberg Bay, is housed in the British Museum (Natural History), London.

SUBGENUS *Graphiurus* Smuts, 1832

# *Graphiurus (Graphiurus) ocularis* (A, Smith, 1829)

**Smith's Dormouse**

**Spectacled Dormouse**

**Gemsbokmuis**

The generic name is derived from the Greek words *grapho* = to write, draw, combined with *oura* = tail. It refers to the cylindrical tail ending in long hairs, recalling an artist's paint brush. The specific name *ocularis* is the Latin referring to the eye. The facial markings in this species tend to accentuate the eyes.

This species is scarce and it shows a patchy distribution in southern Africa. It is a very attractive little animal. Its status is listed as rare, widespread but uncommon in the *South African Red Data Book* (Meester 1976). There is no reason to assume that it is threatened as a species.

**Outline of synonymy**

1829 *Sciurus ocularis* A. Smith, *Zool. Journ.* 4:439. Near Plettenberg Bay, CP.

1832 *Graphiurus* Smuts, *Enum. Mamm. Cap.*: 32. (Genotype *Graphiurus capensis* Smuts = *Sciurus ocularis* A. Smith.)

1832 *Graphiurus capensis* F. Cuvier & Geoffroy, *Hist. Nat. Mamm.*: 6.

1834 *Graphiurus typicus* A. Smith, *S. Afr. Quart. Journ.* 2:145. Renaming of *ocularis*.

1838 *Graphiurus elegans* Ogilby, *Proc. zool. Soc. Lond.*: 5. Damaraland (? Cape of Good Hope).

1898 *Graphiurus ocularis* De Winton, *Ann. Mag. nat. Hist.* (7) 2:3.

1939 *Graphiurus ocularis ocularis* (A. Smith). *In* Allen, *Bull. Mus. comp. Zool. Harv.* 83:312.

1951 *Graphiurus ocularis* (A. Smith). *In* Roberts, *The mammals of South Africa.*

1953 *G. ocularis* (A. Smith). *In* Ellerman *et al., Southern African mammals.*

1964 *G. ocularis* (A. Smith). *In* Meester *et al., An interim classification of southern African mammals.*

1974 *G. ocularis* (A. Smith). Misonne, *in* Meester & Setzer (eds), *The mammals of Africa: an identification manual.*

## TABLE 64
### Measurements of male and female *Graphiurus (Graphiurus) ocularis*

|         | Parameter | Value (mm) | N | Range (mm) | CV (%) |
|---------|-----------|------------|---|------------|--------|
| Males   | HB        | 133        | 7 | 120–148    | 6,6    |
|         | T         | 112        | 6 | 103–125    | 7,0    |
|         | HF        | 25         | 7 | 23–26      | 3,6    |
|         | E         | 20         | 7 | 15–25      | 17,6   |
|         | Mass:     | 83 g       | 2 | 81–85 g    | —      |
| Females | HB        | 124        | 2 | 121–128    | —      |
|         | T         | 103        | 2 | 100–106    | —      |
|         | HF        | 20         | 2 | 20         | —      |
|         | E         | 18         | 2 | 18         | —      |
|         | Mass: No data available | | | | |

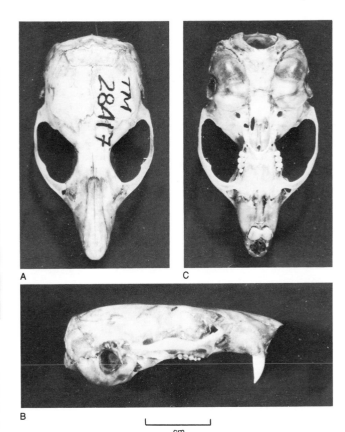

**Figure 107** Dorsal (A), lateral (B) and ventral (C) views of the skull of Smith's Dormouse *Graphiurus (Graphiurus) ocularis*. The interorbital constriction is moderate. The skull length varies between 32,8—37,1 mm (Meester *et al.* 1964) and is flattened.

## Identification

This is a large species with a head–body length of up to 150 mm. The colour pattern is specialised. The general colour is ashy grey. The fur is soft and thick, dark slaty at the base, dull white mixed with black at the apex. There is a patch of white on the snout, chin and cheeks, as well as in front of the shoulder and on top of the head at the base of the large, rounded ear pinna. The surface underneath, from the chest downwards and the sides of the body between the limbs, is dull white. The tail shows a striking black line down its midline on its ventral surface, with white hairs on each side of this. The tail is very bushy, more or less the same colour as the back. There is a conspicuous black stripe running from the eye to the ear (plate 12).

The mammary formula reads 2–2=8.

## Skull and dentition

In the skull, the frontal elements appear to be contracted by the moderately developed interorbital constriction. The infraorbital foramen has an elliptical shape. In southern African forms, the skull length is 32,8–37,1 mm (Meester *et al.* 1964) (fig. 107).

The molars are rooted and the single premolar (= Pm⁴) is very small. The crowns of all the cheekteeth are hollowed out, the rims formed by ridges of enamel. Feint infoldings or crossridgings can sometimes be detected (fig. 108).

## Distribution

This species occurs in the southern Cape, ranging northwards through the western Karoo to Little Namaqualand in the west, to Queenstown and northwards to the OFS and western Transvaal in the east (map 71).

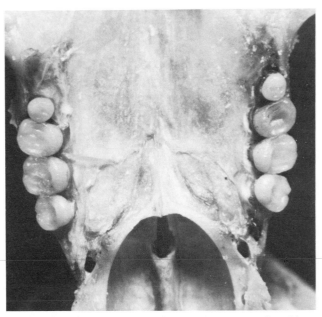

**Figure 108** Cheekteeth of Smith's Dormouse *Graphiurus (Graphiurus) ocularis*. The Pm⁴ is very small. All the molars are rooted, the crowns hollowed out and their rims formed by ridges of enamel. Infoldings can occasionally be seen, though weakly.

247

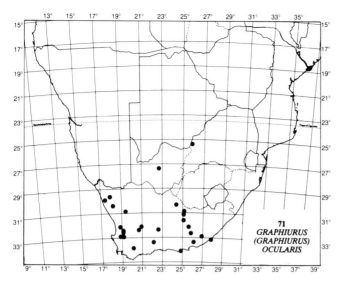

**71**
**GRAPHIURUS**
**(GRAPHIURUS)**
**OCULARIS**

## Habitat

It frequents crevices between rocks on and near koppies, while trees growing in close proximity to such outcrops are also occupied.

## Diet

Little is known about this animal's food but judging from the nature of its teeth, it '...may eat anything edible' (Roberts 1951). When kept in captivity it devours '...meat with avidity' (Sclater 1901).

## Habits

It is said to utter a loud, shrill call. Smith's dormice are savage little animals and will attack other small rodents which are placed with them in the same cage (Shortridge 1934). They bite readily when handled.

## Predators, reproduction and prehistory

No data available.

## Parasites

Very little is known about the associated parasites. A literature survey indicates that the following fleas have been associated with this dormouse: the hystricopsyllid *Listropsylla agrippinae*, the leptopsyllid *Leptopsylla segnis* and the chimaeropsyllids *Epirimia aganippes* and *Chiastopsylla octavii* (Zumpt 1966).

The ticks, as recorded by Theiler (1962), include immature individuals of *Rhipicephalus simus*.

## Relations with man

Jarvis (*in* Meester 1976) says that they are locally (the southern Cape Province) a pest in mountain club huts and that they adjust well to captivity.

## Taxonomy

This is a monotypic species and *Graphiurus (G.) ocularis* has not been divided subspecifically.

SUBGENUS *Claviglis* Jentink, 1888

# *Graphiurus (Claviglis) platyops* Thomas, 1897

## Rock Dormouse

## Klipwaaierstertmuis

For the meaning of *Graphiurus*, see *Graphiurus (Graphiurus) ocularis*. The name *claviglis* is coined from the Latin words *clava* = a club and *glis* = dormouse. It refers to the tail of the animal which looks clubshaped under certain conditions. The specific name is coined from the Greek *platos* = flat, wide and *opis* = appearance, evidently referring to the flattened skull of the species.

The type specimen is housed in the British Museum (Natural History), London and was collected by J. ffolliott Darling at Enkeldoring in Zimbabwe. It was subsequently described by Thomas (1897b). It is very much smaller than *G. (G.) ocularis,* but somewhat larger than

*G. (C.) murinus,* the species to be discussed in the next section.

## Outline of synonymy

1897 *Graphiurus platyops* Thomas, *Ann. Mag. nat. Hist.* (6) 19:388. Mashonaland, Zimbabwe.

1925 *Gliriscus.* Thomas & Hinton, *Proc. zool. Soc. Lond.:* 232. (Genotype *Graphiurus platyops* Thomas.)

1939 *Gliriscus platyops* (Thomas). *In* Allen, *Bull. Mus. comp. Zool. Harv.* 83:313.

1940 *Graphiurus platyops* Thomas. *In* Ellerman, *The families and genera of living rodents.*

1951 *Gliriscus platyops* (Thomas). *In* Roberts, *The mammals of South Africa.*

1953 *Graphiurus platyops* Thomas. *In* Ellerman *et al., Southern African mammals.*

1964 *Graphiurus platyops* Thomas. *In* Meester *et al., An interim classification of southern African mammals.*

1974 *Graphiurus platyops* Thomas. Misonne, *in* Meester & Setzer (eds), *The mammals of Africa: an identification manual.*

## TABLE 65
Measurements of male and female *Graphiurus (Claviglis) platyops*

|  | Parameter | Value (mm) | N | Range (mm) | CV (%) |
|---|---|---|---|---|---|
| Males | HB | 115 | 11 | 105–130 | 5,5 |
|  | T | 70 | 11 | 60–80 | 9,6 |
|  | HF | 20 | 11 | 18–23 | 7,2 |
|  | E | 16 | 11 | 14–18 | 7,0 |
|  | Mass: | 43,6 g | 3 | 39–52 g | — |
| Females | HB | 105 | 16 | 91–117 | 7,1 |
|  | T | 75 | 16 | 60–98 | 13,2 |
|  | HF | 20 | 19 | 17–22 | 6,4 |
|  | E | 15 | 17 | 13—18 | 7,7 |
|  | Mass: | 48,6 g | 3 | 40–65 g | — |

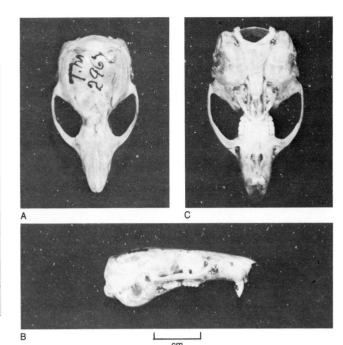

**Figure 109** Dorsal (A), lateral (B) and ventral (C) views of the skull of the Rock Dormouse *Graphiurus (Claviglis) platyops*. The flattened skull varies between 29–32 mm in length (Misonne 1974). It is also relatively broad, with a pronounced interorbital constriction.

## Identification

The colour pattern is not as bold as it is in *G. (G.) ocularis*. The overall colour is drab and it resembles *G. (C.) murinus* rather closely. It is, however, slightly larger and its tail is bushier with hairs 25–28 mm long, brown and broadly tipped with white. A whitish tail tip is usually prominent. In contrast to the smaller *murinus*, it has a hindfoot length of 18–23 mm (14–19 mm in *murinus*) (Misonne 1974). There may be a dark-coloured ring around the eye (plate 12).

## Skull and dentition

The skull is flattened, usually 29–32 mm in length (Misonne 1974). It is also rather broad and this feature distinguishes this species from the other dormice (Thomas 1897b). The muzzle appears broad and heavy, while the nasals are long. The interorbital constriction is narrow. The braincase is flattened, giving the skull a peculiar shape. The palatine foramina are situated rather further forward from the premolar (more than the length of the upper toothrow), than in *G. (C.) murinus*. The palate terminates at the level of the $M^3$ (fig. 109).

The molars are somewhat smaller than those found in the other species (fig. 110).

## Distribution

This species shows a discontinuous distribution and is found in Little Namaqualand, northwards to the Kaokoveld in SWA/Namibia and eastwards to eastern Zimbabwe (Misonne 1974) (map 72). Extralimitally, it occurs in Zambia, southern Malaŵi, eastern Angola and Zaïre.

## Habitat

The Rock Dormouse is often taken near or on rocky outcrops to which it seems to be closely confined, at least

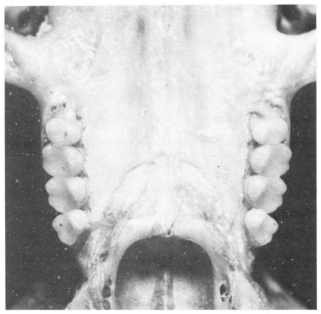

**Figure 110** Cheekteeth of the Rock Dormouse *Graphiurus (Claviglis) platyops*. The $Pm^4$ is not as reduced as in *G. (G.) ocularis*. The molars are somewhat smaller than those found in *G. (G.) ocularis* and *G. (C.) murinus*.

in Botswana (Smithers 1971). Other populations occurring in the northern and northeastern Transvaal are often found living in association with dassies (Hyracoidea) (Roberts 1951) while Smithers & Lobão Tello (1976) have reported the same association with *Procavia capensis* and

*Heterohyrax brucei* in Mozambique. Although it can thus be described as a rock-frequenting dormouse, Smithers & Lobão Tello have also taken it in the dry *Androstachys* scrub thickets of dry riverbeds in the Save Valley in Mozambique, frequenting trees.

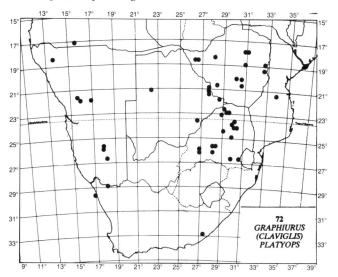

## Diet

Not much is known; Smithers (1971) reports that they feed on seeds. Traces of green vegetable matter and the chitinous remains of insects have also been found in stomach contents.

## Habits

The Rock Dormouse appears to be solitary, partly arboreal and partly terrestrial and fully nocturnal.

## Reproduction

Ansell (1960) states that breeding occurs during the months of November, December and February in Zambia. A female with three young was caught in a hollow stem of a large tree at Wonderboom near Pretoria (Roberts 1951).

## Parasites

Zumpt (1966) has recorded the occurrence of the chimaeropsyllid flea *Chiastopsylla nama* with *G. (C.) platyops*. It also occurs on other rock-living rodents, such as the Namaqua Rock Rat *Aethomys (Micaelamys) namaquensis* and the Pygmy Rock Mouse *Petromyscus collinus*.

## Predators and relations with man

No information available.

## Prehistory

No information available. Some Pliocene genera are known from Asia and Miocene examples from Europe, but no fossil remains of dormice are known from the African continent.

## Taxonomy

Roberts (1951) and earlier authors place *platyops* in the genus *Gliriscus* and in association with *eastwoodae* and *rupicola* as distinct species. Ellerman *et al.* (1953), however, interpret *Gliriscus* as a synonym of *Claviglis*. The species *platyops* is interpreted by them as a nominate subspecies together with *eastwoodae* (Woodbush), *rupicola* (Karibib), *montosus* (near Berseba), *kaokoensis* (Kamanjab) and *australis* (Eenriet). In addition to these named forms, they also list *angolensis* (southwestern Angola), *jordani* (northeastern Zambia) and *parvulus* (southern Angola) as races of *Graphiurus platyops* (extralimital to this book). Misonne (1974) incudes all these forms in *Graphiurus (Claviglis) platyops*.

# *Graphiurus (Claviglis) murinus* (Desmarest, 1822)

### Woodland Dormouse
### Boswaaierstertmuis

For the derivation of *Graphiurus* and *Claviglis*, see *Graphiurus ocularis* and *Graphiurus (Claviglis) platyops* respectively. The species name is derived from the Latin *mus, muris* and is given because of the rather mouse-grey colour of the original specimen.

This species was first described as *Myoxus murinus* by Desmarest in 1822 in the *Encyclopédie Méthodique, Mammalogie* from specimens obtained at the Cape of Good Hope. The type specimen, initially collected by Delalande is now in the Natural History Museum in Paris (Sclater 1901). It has a wide distribution from the Cape to the southern fringes of the Sahara Desert; it thus occurs in a great variety of vegetation types. The external appearance of the species is also variable.

## Outline of synonymy

1822 *Myoxus murinus* Desmarest, *Encycl. Méth. Mamm.* Suppl.: 542. Cape of Good Hope.
1825 *M. lalandianus* Schinz, Cuvier's *Thierreich* 4:393. (Renaming of *murinus*.)
1829 *M. erythrobronchus* A. Smith, *Zool. Journ.* 4:438. South Africa.
1842 *M. cineraceus* Rüppel, *Mus. Senckenb.* 3:136. 'Port Natal' = Durban.
1845 *M. cinerascens* Schinz, *Synops. Mamm.* 2:80 (emendation).
1875 *Graphiurus urinus* Alston, *Proc. zool. Soc. Lond.*: 317.
1887 *Eliomys microtis* Noack, *Zool. Jahrb. Jena*: 248.

1890 *E. kelleni* Reuvens, *Die Myoxidae oder Schläfer:* 35. Damaraland. SWA/Namibia. (Mossamedes, according to Hill & Carter, 1942, *Bull. Amer. Mus. N.H.* 89:186.)

1896 *Myoxus (Eliomys) nanus* De Winton, *Proc. zool. Soc. Lond.:* 799. Mazoe, Zimbabwe.

1939 *Claviglis murinus murinus* (Desmarest). *In* Allen, *Bull. Mus. comp. Zool.* 83:308.

1940 *Graphiurus murinus murinus* (Desmarest). *In* Ellerman, *The families and genera of living rodents.*

1951 *Claviglis murinus murinus* (Desmarest). *In* Roberts, *The mammals of South Africa.*

1953 *Graphiurus murinus* (Desmarest). *In* Ellerman *et al., Southern African mammals.*

1964 *Graphiurus murinus* (Desmarest). *In* Meester *et al., An interim classification of southern African mammals.*

1974 *Graphiurus (Claviglis) murinus* (Desmarest). Misonne, *in* Meester & Setzer (eds), *The mammals of Africa: an identification manual.*

TABLE 66
Measurements of male and female *Graphiurus (Claviglis) murinus*

| | Parameter | Value (mm) | N | Range (mm) | CV (%) |
|---|---|---|---|---|---|
| Males | HB | 93 | 27 | 82–113 | 8,3 |
| | T | 73 | 26 | 58–92 | 11,4 |
| | HF | 17 | 27 | 15–20 | 7,6 |
| | E | 15 | 27 | 10–19 | 10,8 |
| | Mass: | 27,9 g | 10 | 24–34 g | 12,2 |
| Females | HB | 95 | 31 | 78–110 | 8,2 |
| | T | 76 | 27 | 62–94 | 9,1 |
| | HF | 17 | 28 | 15–20 | 7,2 |
| | E | 16 | 27 | 13–20 | 11,3 |
| | Mass: | 27,6 g | 13 | 23–34 g | 11,8 |

## Identification

The Woodland Dormouse is a small species with a head-body length of less than 125 mm. The general colour dorsally is mouse-grey. The fur is soft and thick, dark slaty at the base with ashy-brown tips. From the whiskers to the eyes and around each eye there may be a darker stripe and ring, not always well marked. The ears are large, rounded and not profusely haired. The tail is about as long as the head-body length, bushy, covered with long hairs which may become longer towards the tip of the tail. The colour is the same as the back. The lower surfaces, including the cheeks, chin and insides of the limbs, are dull white, with a slate colour to the base. Some adult specimens show a rusty red tinge on the cheeks, chin and chest. The young animals are greyer than adult specimens (Kingdon 1974) (plate 12).

The mammae consist of one axillary, one pectoral and two inguinal pairs, eight in all.

## Skull and dentition

The skull is not as flat as in *G. (C.) platyops* and is usually less than 29 mm long. The braincase is rectangular if seen from above. The lower margin of the nasals curves downwards over the incisors, giving them a spatulate shape. The interorbital constriction is marked and the rostrum broad. The palate is wide. The bullae are moderately inflated.

The molars are like those found in the closely related *G. (C.) platyops* and in contrast to *G. (G.) ocularis*, the Pm$^4$ is not quite as small (fig. 111). Like the other species, the upper incisors of these dormice are diagnostic: the front surfaces are usually warped inwards towards each other so that they and their cutting edges form a backwardly directed angle, resulting in a sharp point on the external surface. The incisors are plain and usually orthodont.

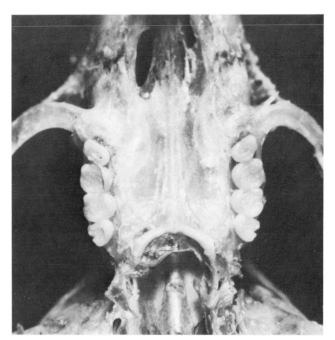

**Figure 111** Cheekteeth of the Woodland Dormouse *Graphiurus (Claviglis) murinus*. The Pm$^4$ is not as reduced as in *G. (G.) ocularis*.

## Distribution

The Woodland Dormouse is widely distributed in Africa south of the Sahara, ranging from Somalia and Senegal to the Cape. In southern Africa it occurs in the southern, eastern and northern Cape Province, the Kalahari and southern Botswana, the Transvaal, Lesotho, Natal and KwaZulu into Mozambique, Zimbabwe and northern SWA/Namibia (map 73).

## Habitat

This species is found in more wooded areas, especially in

251

the east and central districts of southern Africa. In Botswana it occurs where large trees provide holes for shelter (Smithers 1971). It also occurs there in dry areas where stands of *Acacia erioloba* and *Zizyphus mucronata* occur. In the Okavango Delta, woodland dormice are also associated with *Combretum imberbe*, and *Acacia* spp., as well as *Colophospermum mopane*. They have been taken in a certain species of *Aloe* growing in the rocky habitat in southern Mozambique (Ressano Garcia) (Smithers & Lobão Tello 1976).

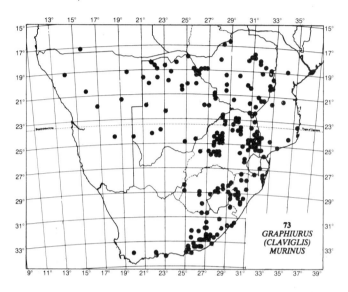

**73**
**GRAPHIURUS**
**(CLAVIGLIS)**
**MURINUS**

## Diet

Kingdon (1974) reports that they occupy inhabited beehives where they feed on dead bees, honey and wax. They also take termites. King (*in* Shortridge 1934) has stated that they also consume the wattle 'bagworm', found in Natal, while it feeds on the large millipede *Doratogonus flavifilis* in the Kruger National Park (Pienaar *et al.* 1980). Apart from these insects and millipedes, woodland dormice are probably mainly vegetarian, taking seeds of *Zizyphus mucronata* (Smithers 1971). Misonne (1974) also says that fruits and seeds are eaten.

The species of plants and insects eaten are, however, little known. It is known that *Graphiurus (C.) murinus* may have a special association with earwigs, which it eats. Chapin (Hatt 1940) found one nest of this dormouse overlain by a layer of several hundred earwigs and that the associated rodents were '. . . stained a dark mahogany . . .', a colour readily extracted from these insects (Rosevear 1969). This may explain the reddish tinge found on the chests, cheeks and chins of some southern African specimens as well. Bird flesh is an effective bait for trapping these animals (Kingdon 1974).

## Habits

Woodland dormice are swift and agile, able climbers, arboreal, partly terrestrial and nocturnal. It is a fairly common species, living in trees, bushes or rocky places

(Misonne 1974). *Graphiurus (Claviglis) murinus* has been recorded by Marshall (*in* Sclater 1901) as living in the nests of the large sociable spider *Stegodyphus* sp. These spiders build structures varying from a single chamber about as large as the egg of a domestic fowl, to a mass of tough silk as large as a man's head, intersected throughout by passages and chambers. Sclater (1901) and Roberts (1951) describe these aspects more fully. A chamber of suitable size is hollowed out and this is lined by the dormice with feathery grassheads, downy seeds and even feathers. It is related that the dormice drive out these spiders (which belong to a family normally living on the ground when they occupy the nests. Certain populations of *G. (C.) murinus* often take refuge in nests of various species of weaver *(Ploceus* spp.) (Roberts 1951), while others have been encountered in the nest of the Hamerkop *Scopus umbretta* (Shortridge 1934).

In captivity they tend to be territorial (Kingdon 1974) which may be the case under natural conditions too. They are inclined to bite when handled (Shortridge 1934).

An interesting phenomenon has been reported by Mohr (1941) on the regenerative powers of tails, which are lost. This unique feature (i.e. their ability to lose and regenerate their tails) is not often encountered in mammals. If the tip is broken off, a broad brush of hair is grown. In the European forms regeneration of a caudal vertebra has been observed, as well as the regeneration of a disc between the vertebrae (Kingdon 1974).

## Reproduction

The mammae have been described as axillary, pectoral and inguinal. Other authorities describe them as two pectoral pairs, one abdominal pair and one inguinal pair. The mammae number eight altogether. The animal builds rounded nests of leaves and grass (Sclater 1901) while Kingdon (1974) says that they will use any available material ranging from leaves, grass, bark, feathers and cotton to cloth and paper. It can also be situated more or less anywhere – from hollow coconuts to beehives.

Four young make up a litter (Sclater 1901) born from November to January in nests which are often built in holes of large tree trunks. The gestation period lasts for about 24 days and the birthweight per pup is approximately 3,5 g (Kingdon 1974).

## Parasites

Three families of mites are known to be associated with *G. (C.) murinus*. These include the laelaptid *?Macronyssus galagus*, the ereynetid *Speleognathopsis bakeri* and the following trombiculids: *Trombicula claviglia, T. claviglicola, T. jadini, Y. panieri, T. rodhaini, Schoutedenichia andrei, S. penetrans* and *Acomatacarus claviglis* (Zumpt 1961a).

The fleas associated with this rodent are fairly well known. The Pulicidae include *Xenopsylla piriei, X. versuta, X. brasiliensis, X. cornigera* and *X. hamula*. The

Hystrichopsyllidae include *Ctenophthalmus acanthurus, C. cophurus, Dinopsyllus ellobius, D. kempi, D. lypusus, D. pringlei* and *Listropsylla dolosa*. The Ceratopsyllidae include *Nosopsyllus incisus* while the Chimaeropsyllidae are represented by *Chiastopsylla octavii* (Zumpt 1966).

Finally, a rhipicephalid tick *Rhipicephalus simus* has been recorded in its immature form (Theiler 1962).

### Relations with man

The Woodland Dormouse is often found under the roofs of houses in the northwestern Transvaal *(pers. obs.)*. They are common in cultivated areas and, according to Kingdon (1974) the species is a familiar commensal in houses and huts. Kingdon also points out that they are unpopular with aviculturists. Eggs and nestlings of birds are eaten and they may even kill the sleeping birds at night.

### Predators and prehistory

No information available.

### Taxonomy

Roberts (1951) and earlier authors place *murinus* in the genus *Claviglis* and subdivide the genus into several species *(alticola, woosnami, schneideri, griselda, littoralis, zuluensis, vandami, kelleni* and *tasmani* apart from *murinus)*. Ellerman *et al.* (1953) retain *Claviglis* only as a subgenus and regard all the species previously recognised as at the most subspecifically identifiable. More than one species *(murinus)* may, however, be present (Ansell 1960). The subspecies alluded to above include *kelleni* (SWA/Namibia), *griselda* (Kuruman), *woosnami* (Botswana), *tzaneenensis* (Tzaneen), *pretoriae* (Pretoria), *streeteri* (Hectorspruit), *tasmani* (Gwelo), *littoralis* (Mozambique), *vandami* (Mozambique), *alticola* (Wakkerstroom), *zuluensis* (Ubombo), *selindensis* (Mount Selinda), and *schneideri* (SWA/Namibia). To this list is to be added the extralimital forms of G. (C.) *murinus* which are also accepted as subspecies: *microtis* (Tanzania), *johnstoni* (Malaŵi), *ansorgei* (Angola) and *cuanzensis* (Angola).

A final remark: two other species of *Graphiurus (Claviglis)*, G. (C.) *crassicaudatus* and G. (C.) *dorothae* (sic) (= *dorotheae*) occurring in West Africa, are perhaps not separable from G. (C.) *murinus*, according to Misonne (1974).

# Glossary

*adaptation* A genetically determined characteristic enhancing the ability of an organism to cope with its environment.

*aestivate* The adaptation of an animal that enables it to survive a hot, dry summer in a torpor.

*allo-grooming* Grooming directed at another individual.

*allopatric* In separate and mutually exclusive geographic regions.

*altricial* Young animals who are unable to move and run about soon after birth.

*alveolar* Pertaining to an alveolus designating sockets in the mandible or maxilla in which the roots of the teeth are attached.

*amphibious* Capable of living both in water and on land.

*angular process* A term applied to the posterior part of the mandibular ramus.

*anoplurid* An order of insects (sucking lice), characterised by the absence of wings.

*antorbital* Situated in front of the orbit or eye.

*anterior* To the front; furthest from the tail.

*anterior palatine foramina* The two foramina seen in the ventral aspect of the skull situated in the bony palate which lies between the incisors and the cheekteeth.

*antero-internal cusp* The frontmost cusp of a cheektooth situated on the lingual side.

*arboreal* Inhabiting or attached to trees.

*argasid tick* The Argasidae (superfamily Ixodiodea) comprise the soft ticks, distinguished from the hard-bodied ticks (Ixodidae) by absence of the scutum.

*arthropod* An animal belonging to the Arthropoda, an invertebrate phylum, having a hard, jointed exoskeleton and paired, jointed legs and including among other classes the arachnids and insects.

*ascaidid mite* A small family of mesostigmatic mites of which *Myonyssoides* is a well-characterised genus from South Africa.

*auditory bulla* Paired bones on the ventral side of the skull, housing middle and inner ear structures.

*auditory meatus* The external opening of the ear.

*axillary* Pertaining to the space between the thorax and medial side of the upper arm.

*baculum* A bone in the penis of insectivores, bats, rodents, carnivores, many primates except man.

*basioccipital* Pertaining to the basal portion of the occipital bone.

*bifurcate* Forked; divided into two parts.

*biogeography* The scientific study of geographic distribution of living organisms.

*biome* The recognisable community unit of a given region, usually designated according to the characteristic climax vegetation e.g. woodland, grassland, etc.

*biotope* Habitat; the kind of locality in which an animal lives.

*bipedal* Having two feet.

*brachyodont* Teeth with low crowns.

*braincase* Part of the skull actually housing the brain.

*breccia* A composite rock consisting of angular fragments of stone cemented by lime.

*buccal* Pertaining to or directed towards the cheek.

*bunodont* Cheekteeth with low, rounded cusps on the occlusal surface of the crown.

*caecum* The first part of the large intestine forming a dilated pouch into which open the ileum and colon.

*calliphorid fly* Scavenger flies which deposit their eggs in decaying matter, on wounds or in openings of the body.

*carnivore* An animal that eats flesh.

*caudal* Denoting a position more towards the tail.

*ceratophyllid flea* Fleas which are members of a large family, in which most of the world's bird fleas are to be found.

*cestode* Any tapeworm or platyhelminth of the class Cestoidea which have a head (scolex) and segments (proglottides).

*cheekteeth* The premolars and molars together.

*chimaeropsyllid flea* A small subfamily of fleas containing two genera (*Chimaeropsylla* and *Hypsophthalmus*). The former does not occur in southern Africa in contrast to the latter.

*chitinous* Composed of chitin, the principal constituent of the exoskeleton of arthropods.

*chromosome* Structures in the nucleus of cells transmitting genetic information.

*circumorbital* Situated around the orbit.

*coleopterous* Pertaining to an order of insects comprising the beetles.

*commensal* Animals which live as tenants and share their food.

*concave* The hollowed inner surface of a segment of a sphere.

*condylar process* A rounded projection on a bone. In this case, the process at the extreme posterior corner of the ramus of the mandible, articulating with the glenoid fossa.

*condylobasal length* The distance between the gnathion to the most caudid portion of the left condyle (condylion).

*conspecific* Pertaining to the same species.

*convergent evolution* The appearance of similar forms and/or functions in two or more lines not sufficiently related phylogenetically to account for the similarity.

*convex* Resembling the segment of the external surface of a sphere.

*coprophagy* The ingestion of dung or faeces.

*cosmopolitan* Belonging to all parts of the world, not restricted to any one country.

*coronoid process* A fingerlike projection on the upper margin of the mandibular ramus.

*crepuscular* Appearing or active in the twilight.

*cribriform plate* Having the form or appearance of a sieve.

*cricetid* Mouse-like rodents whose occlusal surfaces of the molars are based on a pattern of five crests of re-entrant enamel folds.

*cryptic* Concealed, hidden.

*cusp* An elevation on the occlusal surface of a tooth, usually a premolar or a molar.

*cytology* The study of cells, their origin, structure, function and pathology.

*dental formula* A way of expressing the number of each category of tooth occurring in a species.

*dentary* The bone forming the lower jaw in mammals.

*dentine* The chief substance or tissue of the teeth which surrounds the pulp cavity and is covered by enamel on the crown and by cementum on the roots of the teeth.

*diastema* A general term for a space or a cleft. In rodents it refers to the gap between the incisor and the frontmost cheektooth.

*digit* Finger or toe. Digits on each hand or foot are numbered from the inside outwards, the first being the thumb or big toe (pollex and hallux respectively, D1), the fifth being the little finger or little toe (D5).

*diphyodont* Having two dentitions, a deciduous and permanent set successively.

*diploid* An individual or cell having two full sets of homologous chromosomes.

*distal* Further from the medial axis (as opposed to proximal).

*diurnal* Active during the day.

*dorsal* On or pertaining to the back.

*ecology* The science of organisms as affected by factors of their environments.

*ecosystem* The fundamental unit in ecology, comprising the living organisms and the non-living elements interacting in a certain defined area.

*ectoparasite* A parasite that lives on the outside of the body of the host.

*enamel* The white, compact and very hard substance that covers and protects the dentine of the crown of a tooth.

*endemic* Peculiar to a certain region or country; native to a restricted area.

*endoparasite* A parasite which lives within the body of its host.

*enzootic* A disease present in an animal community at all times usually occurring in small numbers of cases.

*epidemic* A season of extensive prevalence of any particular disease.

*epididymis* The elongated structure along the posterior border of the testes in the ducts of which the spermatozoa are stored.

*ereynetid mite* A family of trombidiform mites.

*erythrism* Redness of the hair and pelage.

*ethology* The scientific study of animal behaviour.

*exoccipital* Those parts of the occipital bone which form the sides of the foramen magnum and support the condyles.

*family* A taxonomic subdivision subordinate to an order (or suborder) and superior to a tribe (or subfamily).

*femur* The thigh or proximal leg bone.

*feral* Living in the wild state after having been domesticated.

*fetus* The unborn offspring of any viviparous animal.

*fibula* The outer and smaller of the two bones of the lower leg.

*fossorial* Having a faculty of digging, burrowing.

*frontal* One of a pair of bones forming in rodents the upper part of the skull between the orbits.

*frontoparietal suture* The suture between the frontal and parietal bones of the skull.

*fusiform* Spindle shaped, i.e. tapering at each end.

*genotype* 1. The fundamental hereditary constitution of an individual (cf. *phenotype*). 2. The type of species of a genus.

*genus* A taxonomic category subordinate to a tribe (or subtribe) and superior to a species (or subgenus).

*gestation* The action or process of carrying young in the womb from conception to birth.

*glenoid fossa* A shallow cavity in the squamosal process articulating with the condyle of the lower jaw.

*graminivorous* Eating or feeding on grass.

*granivorous* Eating or feeding on grain.

*gregarious* Animals given to live and associate with others of the same species.

*habitat* The natural abode or home of an animal or plant species.

*haemoglobin* The oxygen-carrying pigment of erythrocytes.

*hallux* The great toe or first digit of the hindfoot.

*hamular* Shaped like a hook.

*herbivorous* Subsisting upon grasses and herbs.

*hibernate* The dormant state in which certain animals pass the winter.

*hirsute* Shaggy; having long hair.

*hoard* To amass and put away for preservation or future use.

*holotype* A single type specimen used and designated as such by an author in describing a new species or subspecies.

*homology* Structural similarity due to descent from a common ancestor.

*homonym* A name identical in orthography with another and based on a different type.

*humurus* The bone extending from the shoulder to the elbow.

*hypsodont* Teeth with high crowns.

*hystricognath* Rodents with a specialised lower jaw, showing an outwards distortion of the angular process.

*hystricomorph* A term designating porcupines and their kin.

*hystricopsyllid flea* A very large flea family.

*implantation* Attachment of the blastocyst to the epithelial lining of the uterus after fertilisation of the ovum.

*incisive foramina* Anterior palatine foramina (q.v.)

*incisor* The front tooth in each jaw, with wedge-like crowns adapted to cutting.

*indigenous* Native, not exotic, to a particular place or country.

*infraorbital foramen* Lying ventral to the orbit.

*inguinal* Pertaining to the groin.

*insectivorous* Subsisting upon an insect diet.

*interorbital constriction* A constriction situated between the orbits.

*interparietal* A separate bone situated partly between and partly posterior to the parietal bones in the skull.

*intradermal* Within the skin.

*intranasal* Within the nose.

*invertebrate* One of the Invertebrata, a division of the animal kingdom, including all forms that have no spinal column.

*iridescence* The condition of gleaming with bright and changing colours.

*ivory* The bonelike substance of the tusks of elephants (modified dentine).

*ixodid tick* A member of the Acarina, containing the hard ticks which are characterised by the presence of a scutum.

*jugal* The bone joining the maxillary process to the squamosal process, thus forming the middle portion of the zygomatic (or jugal) arch.

*K-selection* See *r-selection*.

*karyotype* Characteristic chromosomes of a particular species.

*lachrymal* A small bone situated at the anterior rim of the orbit containing the lachrymal foramen.

*laelaptid mite* A family of mesostigmatic mites.

*lambdoid* The suture connecting the two parietal bones with the occipital in the skull.

*lapsus calami* A mistake, usually of spelling, made by an author.

*lateral* Denoting a position further from the median plane or midline of the body or of a structure.

*leptopsyllid flea* A large family, mostly infesting rodents.

*lingual* Pertaining to the tongue.

*listrophorid mite* Typical ectoparasites of small and medium-sized mammals.

*litter* The offspring produced at one birth by a multiparous animal.

*lophodont* Cheekteeth in which the cusps have become connected to form ridges.

*lumbar* Pertaining to the loins.

*mammae* Glandular structures secreting milk for nourishment of the young.

*mammal* An individual belonging to the class Mammalia, vertebrate animals possessing hair and which suckle their young.

*mammalogy* The study of mammals.

*mandible* The lower jaw.

*manus* The distal portion of the arm, or hand, including the carpus, metacarpus, and fingers.

*masseter knob* A small bony process for the attachment of part of the masseter muscle usually situated anterior to and often just below the maxillary process.

*masseter muscle* A strong muscle, subdivided in parts, joining the mandible to the skull and used for the act of chewing.

*mastication* The chewing of food.

*mastoid* A bone situated posterior to the auditory meatus, sometimes greatly inflated in rodents.

*maxilla* The irregularly shaped bone that with its fellow forms the upper jaw lodging the canines and cheekteeth.

*maxillary process* A projection of bone from the maxilla constituting the anterior root of the zygoma.

*meatus* An opening; anatomically used as a general term to designate some passageway in the body.

*medial* Pertaining to the middle; closer to the midline of the body.

*melanism* Excessive pigmentation or blackening of the integument.

*mesopterygoid fossa* An opening immediately posterior to the bony palate, bounded by the pterygoids.

*molar* The most posterior teeth on either side in each jaw, not preceded by milk teeth.

*molariform* Shaped like a molar tooth.

*monotypic* Species which have no subspecies.

*morphology* The science of form and structure of biological organisms.

*moult* To shed or cast hair or pelage.

*multimammate* The condition of having many mammary glands.

*multiparous* A female who has had two or more pregnancies which resulted in viable offspring whether or not the offspring were alive at birth.

*murine* Pertaining to or affecting mice or rats.

*myobiid mite* Parasitic mites, having the first pair of legs modified for hair-clasping.

*myomorph* A term designating rats, mice and their kin.

*nares interni* The opening between the nasal cavity and the nasopharynx.

*nasals* A pair of bones forming the anterior upper part of the rostrum.

*nemathelminthes* The phylum of roundworms which includes Acanthocephala and Nematoda.

*nematode* An endoparasite or species belonging to the Nematoda.

*niche* All the components of the environment with which the organism or population interacts.

*nidicolous* Referring to young birds that hatch naked, with their eyes closed and are unable to move about (see *altricial* q.v.).

*nidifugous* Referring to young birds that hatch covered with down and with their eyes open and are able to move about (see *precocial* q.v.).

*nocturnal* Pertaining to the night.

*nomen nudem* A scientific name published with no (or inadequate) description of the relevant animal, making it impossible to identify, rendering the proposed name taxonomically inadmissible.

*nominate race* A race (subspecies) bearing the name of a species, only brought into recognition by the naming of other races.

*nuchal crest* A crest pertaining to the nape or scruff of the neck.

*occipital* Pertaining to the occiput, the back part of the skull.

*occiput* The posterior part of the skull above the foramen magnum.

*occlusal surface* The horizontal surface of the cheekteeth which comes into direct contact with that of the opposing jaw.

*oestrus* The recurrent restricted period of sexual receptivity in female mammals leading to changes in the genital tract as a result of hormonal activity.

*olfactory* Pertaining to the sense of smell.

*omnivorous* Subsisting upon food of every kind.

*omnivore* Eating food of every kind indiscriminately.

*opisthodont* Of the incisors, directed posteriorly.

*orbit* The bony cavity containing the eyeball.

*order* A taxonomic category subordinate to the class and superior to a family (or suborder).

*orthodont* Of the incisors, directed in a more or less vertical plane.

*palaeontology* The study of early forms of life on earth.

*palate* The partition separating the nasal and oral cavities, i.e. the roof of the mouth.

*parasitology* The science or study of parasites.

*paratype* All the specimens of a type series used by an author in the original description of a species or subspecies except that designated as holotype, are paratypes.

*parietal* One of a pair of bones forming part of the side of the braincase.

*paroccipital process* Processes occurring in proximity to the occipital bone.

*parturition* The act or process of giving birth.

*pathogen* Any disease producing microorganism or material.

*pectoral* Pertaining to or situated upon the chest.

*pelage* The fur or hairy covering of a mammal.

*perennial* Lasting through the year or for several years.

*perforate* Bored or pierced through.

*pes* The terminal organ of the leg or hind limb.

*phenotype* The outward, visible expression of the hereditary constitution of an organism. (cf. *genotype*).

*phylogeny* The developmental and evolutionary history of a group of animals.

*physiology* The science which studies the functions of the living organism and its parts.

*plantar* Pertaining to the sole of the foot.

*plantigrade* Characterised by walking on the full sole of the foot.

*platyhelminth* One of the Platyhelminthes, a phylum of flatworms.

*pneumatic* Filled or fitted to contain air.

*pollex* The first digit of the hand or thumb.

*porocephalid arachnids* Legless, blood-sucking arthropods.

*polygamy* The concurrent marriage to more than one mate.

*population* Individuals collectively constituting a category or inhabiting a specified geographic area.

*polymorphism* The quality or character of being in several or alternative forms.

*polytypic* A category containing two or more immediately subordinate categories.

*postauricular* Behind the ear.

*posterior* To the back; furthest from the head.

*posterior palatine foramina* The two foramina seen on the ventral aspect of the skull situated in the bony palate which lies between the two rows of cheekteeth.

*postorbital process* A bony process above and slightly posterior to the orbit.

*postpartum oestrus* An oestrus occurring after birth.

*precocial* Young animals who are able to move and run about soon after birth.

*prehensile* Adapted for grasping or seizing.

*premaxilla* Also referred to as the *os incisivum*, these separate paired elements bear the upper incisor teeth.

*premolar* Situated in front of the molar teeth and generally (but not in rodents) preceded by a corresponding milktooth.

*proodonty* Of the incisors, directed anteriorly.

*proximal* Nearest; closer to any point of reference such as the medial longitudinal axis; opposed to distal.

*pterygoid processes* One of a pair of narrow vertical processes immediately posterior to the palate and forming the lateral walls of the depression known as the mesopterygoid fossa. (q.v.)

*pulicid flea* A large family of fleas, with a global distribution.

*pygiopsyllid flea* A family occurring predominantly in the Australian Region.

*Q-fever* A rickettsial infection caused by *Coxiella burnettii*.

*r- and K-selection* Alternative expressions of selection on traits that determine fecundity and survivorship to favour rapid population growth at low population density *(r)* or competative ability at densities near the carrying capacity *(K)*.

*radius* The bone on the medial (thumb) side of the forearm.

*ramus of mandible* The quadrilateral process projecting superiorly from the posterior part of either side of the jaw bone.

*raptors* Birds of prey.

*Rassenkreis* A polytypic species composed of several subspecies.

*regeneration* The natural renewal of a substance, such as lost tissue or part.

*rhizome* A subterranean stem sending forth shoots at its upper end and decaying at the other.

*riparian* Along the bank of a river or lake.

*rostrum* Part of the skull that lies anterior to a line joining the front margins of the orbits.

*sagittal crest* A ridge of bone situated along the medial axis of the skull, forming a deeper or shallower erect plate.

*saltation* The action of leaping.

*sarcoptiformid mite* A suborder comprising soil-dwelling mites (Oribatei) and weaker-skinned mites (Acaridei).

*scalpriform* Shaped like a chisel.

*scansorial* Climbing or adapted to climbing.

*scatter-hoarding* Each individual load of food is separately concealed.

*schistosome* An individual of the genus *Schistosoma*, a trematode parasite or blood fluke.

*schizomycete* Any organism belonging to the Schizomycetes, a taxonomic class of typically unicellular organisms, considered plants, often causing disease in plants or animals.

*sciurognath* Rodents with an unspecialised lower jaw, with the angular portion never distorted outwards.

*sciuromorph* A term designating squirrels and their kin.

*serology* The study of antigen-antibody reactions *in vitro*.

*sibling* Another offspring of the same parents as the individual of reference.

*species* A taxonomic category subordinate to a genus (or subgenus) and superior to a subspecies or variety.

*spirochaete* A general term for any micro-organism of spiral shaped bacteria of the order Spirochaetales.

*squamosal* One of a pair of bones forming part of the walls of the braincase.

*squamosal process* A projection of bone from the squamosal forming the posterior root of the zygoma.

*squamosoparietal suture* The suture between the squamosal and parietal bones in the skull.

*subauricular* Below the pinna of the ear.

*subgenus* A taxonomic category sometimes established, subordinate to a genus and superior to a species.

*suborder* A taxonomic category sometimes established, subordinate to an order and superior to a family.

*subspecies* A taxonomic category below the rank of species. Names of subspecies are trinomial.

*subterranean* At some depth in the earth.

*succession* Replacement of populations in a habitat through a regular progression to a stable state.

*superfamily* A taxonomic category sometimes established, subordinate to an order and superior to a family.

*supraorbital* Situated above the orbit.

*sympatric* Having the same or overlapping regions of distribution.

*synonym* One of two or more names applied to the same taxon.

*symphysis* A site or line of unions as between the left and right lower jaws.

*tarsus* The region of articulation between the foot and the leg., *taxon* A particular category into which related organisms are classified, such as species, genus, family, order, or class.

*taxonomy* The orderly classification of organisms into appropriate taxa on the basis of relationships among them, with the application of suitable and correct names.

*temperate* Not liable to excessive heat (a *temperate* climate).

*terrestrial* Pertaining to the earth.

*territory* A restricted area preempted by an animal or pair of animals, usually for breeding purposes and guarded from other individuals of the same species.

*terminal heel* A medially placed posterior supplementary cusp on the $M_1$ and $M_2$.

*thoracic* Pertaining to the chest.

*tibia* The inner and larger bone of the hindleg, below the knee, articulating with the femur proximally and the talus distally.

*tragus* A cartilaginous projection anterior to the external openings of the ear.

*trematode* Any parasitic animal organism belonging to the class Trematoda, platyhelminths which includes the flukes.

*tribe* A taxonomic category subordinate to a family (or subfamily) and superior to a genus (or subtribe).

*trombiculid mite* Mites of some medical and veterinary importance.

*trypanosome* An organism of the genus *Trypanosoma,* a sporozoan parasite found in the blood plasma of man and animals, causing trypanosomiasis. These parasites are conveyed by tsetse flies, *Glossina palpalis* and *G. morsitans.*

*tuber* A swollen stem or root in which reserves are stored.

*tympanic bulla* Auditory bulla (q.v.).

*type* Holotype (q.v.).

*type locality* The exact place from which an original type specimen came.

*ulna* A bone of the forearm on the opposite side of that of the thumb.

*vector* A carrier, especially the animal which transfers an infective agent from one host to another.

*vegetarian* One whose food is exclusively of vegetable origin.

*ventral* On the underside of an animal, as opposed to the back.

*vertebra* Any one of the bones of the spinal column.

*vertebrate* An animal having a vertebral column.

*viviparous* Bearing live young which derive nutrition directly from the maternal organism through the placenta.

*xiphiopsyllid flea* A small family, endemic in East Africa.

*zoogeography* The study of the distribution of animal life on earth.

*zoonosis* A disease of animals that may secondarily be transmitted to man.

*zygomatic arch* The arched bones on each side of the skull, comprising the maxillary and squamosal processes as well as the jugal.

*zygomatic plate* The expanded and flattened lower branch of the maxillary process typically found in the Gerbillinae and Murinae.

# References

ACOCKS, J.P.H. 1953. Veld types of South Africa. *Mem. bot. Surv.* 28:1–192.

AJAYI, A. 1971. Wildlife as a resource of protein in Nigeria: Some priorities for development. *Niger. Fld.* 36(3):115–127.

AJAYI, S.S. 1974. Giant rats for meat – and some taboos. *Oryx* 12(3):379–380.

ALEXANDER, Anne J. 1956. Bone carrying by a porcupine. *S. Afr. J. Sci.* 52:257–258.

ALEXANDER, R.A. 1936. The horse sickness problem in South Africa. *J. S. Afr. Vet. Med. Ass.* 7:211–220.

ALLANSON, M. 1958. Growth and reproduction in the males of two species of gerbils. *Proc. zool. Soc. Lond.* 130:373–396.

ALLEN, G.M. 1939. A checklist of African mammals. *Bull. Mus. comp. Zool. Harv.* 83:1–763.

ALSTON, E.R. 1876. On the classification of the Order Glires. *Proc. zool. Soc. Lond.*: 61–98.

AMTMANN, E. 1966. Zur Systematik Afrikanischer Streifenhörnchen der Gattung *Funisciurus*. Ein Beitrag zur Problematik Klimaparalleler Variation und Phänetik. *Bonn. zool. Beitr.* 17(1/2):1–44.

AMTMANN, E. 1975. Part 6,1. Family Sciuridae. *In* MEESTER, J. & H.W. SETZER (eds), *The mammals of Africa: an identification manual.* Washington DC: Smithsonian Institution Press.

ANDERSON, S. 1967. Introduction to the rodents. *In* ANDERSON, S. & J.K. JONES Jr (eds), *Recent mammals of the world – a synopsis of families.* New York: Ronald Press.

ANDERSON, S. & J.K. JONES Jr (eds). 1967. *Recent mammals of the world – a synopsis of families.* New York: Ronald Press.

ANDREWES, C.H. & J.R. WALTON. 1977. *Viral and bacterial zoonoses.* London: Bailliere Tindall.

ANSELL, W.F.H. 1958. Four new African rodents. *Ann. Mag. nat. Hist.* (13) 1:337–344.

ANSELL, W.F.H. 1960. *Mammals of northern Rhodesia.* Lusaka: Government Printer.

ANSELL, W.F.H. 1963. Additional breeding data on northern Rhodesian mammals. *Puku* 1:9–28.

ANSELL, W.F.H. 1964. Addenda and corrigenda to 'Mammals of Northern Rhodesia'. *Puku* 2:14–52.

ANSELL, W.F.H. 1969. Addenda and corrigenda to 'Mammals of Northern Rhodesia', No. 3. *Puku* 5:43–46.

ANSELL, W.F.H. & P.D.H. ANSELL 1973. Mammals of the northeastern montane areas of Zambia. *Puku* 7:21–69.

ARATA, A.A. 1975. The importance of small mammals in public health. *In* GOLLEY, F.B., K. PETRUSEWICZ & L. RYSZ-KOWSKI (eds), *Small mammals, their productivity and population dynamics.* Cambridge: Cambridge University Press.

ASTLEY MABERLY, C.T. 1963. *The game animals of southern Africa.* London: Nelson.

AVERY, D.M. 1977. Past and present distribution of some rodents and insectivore species in the southern Cape Province, South Africa: new information. *Ann. S. Afr. Mus.* 74(7):201–209.

BATE, D.M.A. 1947. An extinct Reed-rat (*Thryonomys arkelli*) from the Sudan. *Ann. Mag. nat. Hist.* (11) 14:65–71.

BAUER, K. & J. NIETHAMMER 1959. Uber eine kleine Säugetierausbeute aus Südwest-Afrika. *Bonn. zool. Beitr.* 10(1959):236–260.

BAYLIS, H.A. 1935. Note on the cestode *Moniezia (Fuhrmanella) transvaalensis* (Baer, 1925). *Ann. Mag. nat. Hist.* (10)15:673–675.

BAYLIS, H.A. 1949. A new human cestode infection in Kenya. *Inermicapsifer arvicanthidis*, a parasite of rats. *Trans. R. Soc. trop. Med. Hyg.* 42:531–542.

BEDFORD, G.A.H. 1932. A synoptic checklist and hostlist of the ectoparasites on South African mammals, aves and reptilia. *Rep. vet. Res. Un. S. Afr.* 18:223–523.

BELL, C. 1978. Observations on a breeding pair of Black-shouldered Kites. *Honeyguide* 95:26–27.

BELLIER, L. 1968. Contribution a l'étude d' *Uranomys ruddi* Dollman. *Mammalia* 32:419–446.

BENSON, C.W. 1965. The Grass-owl and the Marsh-owl. *Puku* 3:175–176.

BERRY, R.J., Josephine PETERS & R.J. VAN AARDE. 1978. Subantarctic Housemice: colonization, survival and selection. *J. zool., Lond.* 184:127–141.

BIGALKE, R. 1939. *Animals and zoos today.* London: Cassel & Co.

BIRKENSTOCK, P.J. & J.A.J. NEL. 1977. Laboratory and field observations on *Zelotomys woosnami* (Rodentia: Muridae). *Zool. afr.* 12(2):429–443.

BOHMANN, L. 1942. Die Gattung *Dendromus* A. Smith. Versuch einer natürlichen Gruppierung. *Zool. Anz.* 139:33–53.

BOLWIG, N. 1958. Aspects of animal ecology in the Kalahari. *Koedoe* 1:115–135.

BOOTH, A.H. 1970. *Small mammals of West Africa.* London: Longmans.

BOTHMA, J. du P. 1965. Random observations of the food habits of certain carnivora (Mammalia) in southern Africa. *Fauna Flora, Pretoria.* 16:16–32.

BOURLIÈRE, F. 1948. Sur la réproduction et la croissance de *Cricetomys gambianus.* *Terre Vie*: 45–48.

BRANDT, J.F. 1855. Beitrage zur nähern kentniss der Saugethiere Russlands. *Mem. Acad. Imp. Sci. St. Petersbourg* 6(9):1–365.

BROOKS, P.M. 1974. The ecology of the four-striped field mouse *Rhabdomys pumilio* (Sparrman, 1784) with particular reference to a population on the Van Riebeeck Nature Reserve. DSc dissertation, University of Pretoria, Pretoria.

BROOM, R. 1930. The age of *Australopithecus. Nature, Lond.*: 814.

BROOM, R. 1934. On the fossil remains associated with *Australopithecus africanus. S. Afr. J. Sci.* 31:471–480.

BROOM, R. 1937. On some new Pleistocene mammals from limestone caves of the Transvaal. *S. Afr. J. Sci.* 33:750–768.

BROOM, R. 1939. The fossil rodents of the limestone cave at Taungs. *Ann. Transv. Mus.* 19:315–317.

BROOM, R. 1948a. Some South African Pliocene and Pleistocene mammals. *Ann. Transv. Mus.* 21:1–38.

BROOM, R. 1948b. The giant rodent mole *Gypsorhychus. Ann. Transv. Mus.* 21:47–49.

BROOM, R. & G.W.H. SCHEPERS. 1946. The South African fossil ape-men, the Australopithecinae. *Memoirs Transvaal Museum* 2:1–272.

BUFFON, Comte de, 1774–1789. *Histoire Naturelle.* Paris.

BURTON, J. 1963. My mice were zebras. *Animals* 2: 34–37.

BUTYNSKI, T.M. 1973. Life history and economic value of the springhare (*Pedetes capensis* Forster) in Botswana. *Botsw. Notes & Records* 3:209–213.

CHANLETT, E.T. 1973. *Environmental protection.* New York: McGraw-Hill.

CHAPIN, J. 1932. The birds of the Belgian Congo. *Bull. Amer. Mus. nat. Hist.* 65:1–756.

CHOATE, T.S. 1971. Research on captive wild mammals – with special reference to *Rhabdomys pumilio. Rhod. Sci. News* 5(2):47–51.

CHOATE, T.S. 1972. Behavioural studies on some Rhodesian rodents. *Zool. afr.* 7(1):103–118.

COCKRUM, E.L. 1962. *Introduction to mammalogy.* New York: Ronald Press.

COE, M.J. 1967. Preliminary notes on the eastern Kenya Springhare, *Pedetes surdaster larvalis* Hollister. *E. Afr. Wildl. Jnl* 5:174.

COE, M.J. 1969. The anatomy of the reproductive tract and breeding in the springhare. *Pedetes surdaster larvalis* Hollister. *J. Reprod. Fert.*, Suppl. 6:159–174.

COETZEE, C.G. 1965. The breeding season of the Multimammate Mouse *Praomys (Mastomys) natalensis* (A. Smith) in the Transvaal highveld. *Zool. afr.* 1(1):29–39

COETZEE, C.G. 1970. The relative tail length of striped mice *Rhabdomys pumilio* (Sparrman 1784), in relation to climate. *Zool. afr.* 5(1):1–6.

COETZEE, C.G. 1972. The identification of southern African small mammal remains in owl pellets. *Cimbebasia* Ser. A. 2(4):53–62.

COETZEE, C.G. 1977. Part 6, 8. Genus *Steatomys. In* MEESTER, J. & H.W. SETZER (eds), *The mammals of Africa: an identification manual.* Washington, DC: Smithsonian Institution Press.

COLBERT, E.H. 1969. *Evolution of the vertebrates.* 2nd ed. New York: John Wiley.

COLLINS, H. Maria. 1972. Cestodes from rodents in the Republic of South Africa. *Onderstepoort J. vet. Res.* 39(1):25–50.

COOKE, H.B.S. 1941. A preliminary account of the Wonderwerk Cave, Kuruman District. Section II. The fossil remains. *S. Afr. J. Sci.* 37:303–312.

COPLEY, H. 1950. *Small mammals of Kenya.* Nairobi: Highway Press.

CORBET, G.B. & L.A. JONES. 1965. The specific characters of the crested porcupines, subgenus *Hystrix. Proc. zool. Soc. Lond.* 144(2):285–300.

COX, J.M. 1978. Auditory communication in the Cane Rat *Thryonomys swinderianus* Temminck. *S. Afr. J. Sci.* 74:144–145.

CURSON, H.H. 1928. Nagana in Zululand. *Rept. (13/14) Dir. Vet. Educ. U. of S. Africa*: 309–414.

DART, R.A. 1925. *Australopithecus africanus,* the man-ape of South Africa. *Nature, Lond.*:195–199.

DART, R.A. 1957. The osteodontokeratic culture of *Australopithecus prometheus. Memoirs Transvaal Museum* 10:1–105.

DAVIS, D.H.S. 1942. Rodent damage in plantations and its prevention. *Jl. S. Afr. For. Ass.* 8:64–69.

DAVIS, D.H.S. 1948. Ecological studies of rodents in relation to plague control. *Proc. 4th Int. Congr. on Trop. Med. & Malaria* 1:250–256.

DAVIS, D.H.S. 1950. Notes on the status of the American Grey Squirrel *(Sciurus carolinensis* Gmelin) in the south-western Cape (South Africa) *Proc. zool. Soc. Lond.* 120:265–268.

DAVIS, D.H.S. 1953. Plague in Africa from 1935–1949. A survey of wild rodents in African territories. *Bull. Wld. Hlth. Org.* 9:665–700.

DAVIS, D.H.S. 1959. The Barn Owl's contribution to ecology and palaeoecology. *Proc. 1st pan-African ornith. Congr. Ostrich* Suppl. 3:144–153

DAVIS, D.H.S. 1962. Distribution patterns of southern African Muridae, with notes on some of their fossil antecedents. *Ann. Cape. Prov. Mus.* 2:56–76.

DAVIS, D.H.S. 1963. Wild rodents as laboratory animals and their contribution to medical research in South Africa. *S. Afr. J. med. Sci.* 28:53–69.

DAVIS, D.H.S. 1965. Classification problems of African Muridae. *Zool. afr.* 1(1):125–145.

DAVIS, D.H.S. 1966. The small rodents of the Transvaal: some taxonomic, biogeographic, economic and health problems. *Fauna Flora, Pretoria* 17:4–12.

DAVIS, D.H.S. 1974. The distribution of some small southern African mammals (Mammalia: Insectivora, Rodentia). *Ann. Transv. Mus.* 29(9):135–184.

DAVIS, D.H.S. 1975 a. Part 6,4. Genera *Tatera* and *Gerbillurus. In* MEESTER, J. & H.W. SETZER (eds), *The mammals of Africa: an identification manual.* Washington DC: Smithsonian Institution Press.

DAVIS, D.H.S. 1975 b. Part. 6,6. Genus *Aethomys. In* MEESTER, J. & H.W. SETZER (eds), *The mammals of Africa: an identification manual.* Washington DC: Smithsonian Institution Press.

DAVIS, D.H.S. & X. MISONNE 1964. Gazetteer of collecting localities of African rodents. *Mus. r. Afr. cent., Tervuren. Zool.* 7:100.

DAVIS, J.A. 1959. A giant pouched rat in the zoo. *Anim. Kingd.* 62:124–125.

DAVIS, R.M. 1973. The ecology and lifehistory of the Vlei Rat *Otomys irroratus* (Brants, 1827) on the Van Riebeeck Nature Reserve, Pretoria. DSc. dissertation, University of Pretoria, Pretoria.

DEAN, W.R.J. 1977. The ecology of owls at Barberspan, Transvaal.

Proc. symp. *African Predatory Birds,* pp.25–45. Pretoria: Northern Transvaal Ornithological Society.

DEAN, W.R.J. 1978. Conservation of the White-tailed Rat in South Africa. *Biol. conserv.* (13) 2:133–140.

DE BEER, J.J. 1972. The feeding habits of two rodents, *Tatera brantsii* and *Praomys (Mastomys) natalensis* in the Transvaal, based on a quantative analysis of their stomach contents. BSc (Hons) project, University of Pretoria, Pretoria.

DE GRAAFF, G. 1960 a. A preliminary investigation of the mammalian microfauna in Pleistocene deposits of caves in the Transvaal system. *Palaeont. afr.* 7:79–118.

DE GRAAFF, G. 1960 b. 'n Ontleding van uilklonte van die nonnetjiesuil *Tyto alba. Ostrich* 31:1–5.

DE GRAAFF, G. 1961. On the fossil mammalian microfauna collected at Kromdraai by Draper in 1895. *S. Afr. J. Sci.* 56(9):259–260.

DE GRAAFF, G. 1962. On the nest of *Cryptomys hottentotus* in the Kruger National Park. *Koedoe* 5:157–161.

DE GRAAFF, G. 1964. On the parasites associated with the Bathyergidae. *Koedoe* 7:113–123.

DE GRAAFF, G. 1965. A systematic revision of the Bathyergidae (Rodentia) of southern Africa. DSc dissertation, University of Pretoria, Pretoria.

DE GRAAFF, G. 1972. On the Mole Rat *(Cryptomys hottentotus damarensis)* (Rodentia) in the Kalahari Gemsbok National Park. *Koedoe* 15:25–35.

DE GRAAFF, G. 1975. Part 6,9. Family Bathyergidae. *In* MEESTER, J. & H.W. SETZER (eds), *The mammals of Africa: an identification manual.* Washington DC: Smithsonian Institution Press.

DE GRAAFF, G. 1979. Molerats (Bathyergidae, Rodentia) in South African National Parks: Notes on their taxonomic 'isolation' and hystricomorph affinities of the family. *Koedoe* 22:89–107.

DE GRAAFF, G. & J.A.J. NEL. 1965. On the tunnel system of Brants' Karroo Rat, *Paratomys brantsi* in the Kalahari Gemsbok National Park. *Koedoe* 8:136–139.

DELANY, M.J. 1964. An ecological study of the small mammals in the Queen Elizabeth Park, Uganda. *Revue Zool. Bot. afr.* 70:129–147.

DELANY, M.J. 1971. The biology of small rodents in Mayanja Forest, Uganda. *J. Zool., Lond.* 165:85–129.

DELANY, M.J. 1972. The ecology of small rodents in tropical Africa. *Mammal Rev.* 2(1):1–42.

DELANY, M.J. 1975. *The rodents of Uganda.* London: British Museum (Natural History).

DE MEILLON, B., D.H.S. DAVIS & Felicity HARDY. 1961. *Plague in southern Africa.* Vol. I, The Siphonaptera (excluding the Ischnopsyllidae). South African Institute for Medical Research, Johannesburg. Pretoria: Government Printer.

DE MOOR, P.P. 1969. Seasonal variation in local distribution, age classes and population density of the gerbil *Tatera brantsii* on the South African highveld. *J. zool., Lond.* 157:399–411.

DE WINTON, W.E. 1896. On collections of rodents made by Mr J. ffolliott Darling in Mashonaland and Mr. C.F. Selous in Matabeleland, with short field notes by the collectors. *Proc. zool. Soc. Lond:* 798–808.

DIETERLEN, F. 1961. Beiträge zur Biologie der Stachelmaus, *Acomys cahirinus dimidiatus* Cretzschmar. *Z. Säugetierk.* 26:1–13.

DIETERLEN, F. 1967. Jahreszeiten und Fortpflanzungperioden bei den Muriden des Kivusee-Gebietes (Congo). Teil I. Ein Beitrag zum Problem der Populationsdynamik in Tropen: *Z. Säugetierk.* 32:1–44.

DIETERLEN, F. 1969. Aspekte zur Herkunft und Verbreitung der Muriden. Bedeutung der systematischen Stellung der Otomyinae. *Z. zool. Syst. Evol.* 7:237–242.

DIETERLEN, F. 1971. Beiträge zur Systematik, Ökologie und Biologie der Gattung *Dendromus* (Dendromurinae, Cricetidae, Rodentia), insbesondere ihrer zentral afrikanischen Formen. *Säugetierk. Mitt.* 19:91–132.

DIETRICH, W.O. 1941. Die Säugetierpaläontologische Ergebnisse der Kohl-Larsen'schen Expedition 1937–1939 in nördlicher Deutsch Ost-Afrika. *Cbl. Min. Geol. Paläont.* Stuttgart 1941B:217–223.

DIXON, J.E.W. 1966. Notes on the mammals of Ndumu Game reserve. *Lammergeyer* 6:24–40.

DOBRORUKA, L.J. 1970. Behaviour in the Bush Squirrel *Paraxerus cepapi* (A. Smith, 1836). *Revue Zool. Bot. afr.* 82:131–141.

DOLLMAN, G. 1909. New mammals from British East Africa. *Ann. mag. nat. Hist.* (8) 4:549–553.

259

DOLLMAN, G. 1910. A list of the mammals obtained by Mr. R.B. Woosnam during the expedition to Lake Ngami, with field notes by the collector. *Ann. Mag. nat. Hist.* (8) 6:388–401.

DOLLMAN, G. 1914. Notes on a collection of east African mammals presented to the British Museum by Mr G.P. Cosens. *Proc. zool. Soc. Lond.:* 307–318.

DREYER, T.F. 1910. South African moles. *Agric. J. Cape G.H.* Reprint 79:1–6.

DREYER, T.F. & A. LYLE. 1931. New fossil mammals and man from South Africa. Dept. of Zoology, Grey Univ. Coll., Bloemfontein.

DU TOIT, C.A. 1965. Current and future research on African mammals. *Zool. afr.* 1:263–265.

DU TOIT, R.M. & J. GOOSEN. 1949. Onderstepoort records. Unpublished observations.

EARL. Z. 1977. Female *Saccostomus campestris* carrying young in cheek pouches. *J. Mammal.* 58:242.

EARL, Z. 1978. Postnatal development of *Saccostomus campestris*. *Afr. Small Mammal Newsletter* 2:10–12.

EARL, Z. & J.A.J. NEL. 1976. Climbing behaviour in three African rodent species. *Zool. afr.* 11(1):183–192.

ELLERMAN, J.R. 1940. *The families and genera of living rodents.* Vol. 1. London: British Museum (Natural History).

ELLERMAN, J.R. 1941. *The families and genera of living rodents.* Vol. 2. London: British Museum (Natural History).

ELLERMAN, J.R., T.C.S. MORRISON-SCOTT & R.W. HAYMAN. 1953. *Southern African Mammals 1758 to 1951: a reclassification.* London: British Museum (Natural History).

ELOFF, G. 1952. Sielkundige aangepastheid van die mol aan onderaardse leefwyse en sielkundige konvergensie. *Tydskr. Wet. Kuns.* 12:210–225.

ELOFF, G. 1954. The Free State Mole *(Cryptomys). Fmg. S. Afr.* Reprint 45:1–3.

ELOFF, G. 1958. The functional and structural degeneration of the eye in the South African rodent moles, *Cryptomys bigalkei* and *Bathyergus maritimus. S. Afr. J. Sci.* 54:293–302.

EWER, R.F. 1967. The behaviour of the African Giant Rat *(Cricetomys gambianus* Waterhouse). *Z. Tierpsychol.* 24:6–79.

FINLAYSON, M.H. 1944. Some characters exhibited by a strain of *P. whitmori* isolated from a case of chronic melioidosis. *S. Afr. med. J.* 18:113–115.

FINDLAYSON, M.H. & J.M. GROBLER. 1940. A study of South African typhus strains and the protection afforded by Zinnser-Casteneda vaccine against infection with these strains. *S. Afr. med. J.* 14:129–134.

FITZSIMONS, F.W. 1919/1920. *The natural history of South Africa.* 4 vols. London: Longmans Green.

FITZSIMONS, V.F.M. 1962. *The snakes of southern Africa.* Cape Town: Purnell.

FISCHER VON WALDHEIM, G. 1803. *Das Nationalmuseum der Naturgeschichte zu Paris.*

FRENCH, N.R., D.M. STODDART & B. BOBEK. 1975. Patterns of demography in small mammal populations. *In:* GOLLEY, F.B., K. PETRUSEWICZ & L. RYSZKOWSKI (eds), *Small mammals, their productivity and population dynamics.* Cambridge: Cambridge University Press.

FREYE, H.-A. 1969. Die Nagetiere. *In* GRZIMEK, G. (ed.), *Grzimeks Tierleben.* Zürich: Kindler Verlag.

FRIPP, P.J. 1978. Rodents as laboratory hosts for schistosomes. *S. Afr. Vet. Assn.* 49(3):233–234.

GARGETT, V. 1977. Mackinder's Eagle Owl in the Matopos, Rhodesia. *Proc. symp. African Predatory Birds* pp. 46–61. Pretoria: Northern Transvaal Ornithological Society.

GARGETT, V. & J.H. GROBLER 1976. Prey of the Cape Eagle Owl *Bubo capensis mackinderi* Sharpe 1899, in the Matopos, Rhodesia. *Arnoldia Rhod.* 8(7):1–7.

GEAR, J. 1954. The rickettsial diseases of southern Africa. *S. Afr. J. clin. Sci.* 5:158–175.

GEAR, J.H.S. & D.H.S. DAVIS. 1942. The susceptibility of South African gerbils (Genus *Tatera*) to rickettsial diseases and their use in the preparation of anti-typhus vaccine. *Trans. Roy. Soc. trop. Med. Hyg.* 36:1–7

GEAR, J., P. ROUX & C. DE V. BEVAN. 1942. An account of recent food infection outbreaks caused by organisms of the *Salmonella* group investigated at the South African Institute of Medical Research. *S. Afr. med. J.* 16:125–126.

GEAR, J, B. DE MEILLON & D.H.S. DAVIS. 1944. Typhus fever in the Transkei. *S. Afr. med. J.* 18:368–369.

GEAR, J. *et al.* 1951–1957. Rickettsial diseases. *Ann. Rpts. S. Afr. Inst. Med. Res.*

GENEST-VILLARD, H. 1967. Revision du genre *Cricetomys* (Rongeurs, Cricetidae). *Mammalia* 31:390–455.

GEOFFROY. 1803. *Cat. Mamm. Mus. Nat. Hist. Paris.*

GOLLEY, F.B., K. PETRUSEWICZ & L. RYSZKOWSKI. 1975. *Small mammals: their productivity and population dynamics.* Cambridge: Cambridge University Press.

GRAFTON, R.N. 1965. Food of the Black-backed Jackal: a preliminary report. *Zool. afr.* 1(1):41–43.

GRASSÉ P.P. (ed.) 1955. *Traité de Zoologie.* Paris: Masson et Cie.

GRAY, J.E. 1837. Description of some new or little known mammalia principally in the British Museum collection. *Mag. nat. Hist.* 1:586.

GREENWOOD, Marjorie, 1955. Fossil Hystricoidea from the Makapan Valley, Transvaal. *Palaeont. afr.* 3:77–85

HAAGNER, A. 1920. *South African mammals.* London: H.F. & G. Witherby.

HAIG, D.A. & A.G. LUND. 1948. Transmission of the South African strain of dourine to laboratory animals. *Onderstepoort J. vet. Res. An. Ind.* 23:59–62.

HALL, A. III, R.L. PERSING, D.C. WHITE & R.T. RICKETTS Jr. 1967. *Mystromys albicaudatus* (the African White-tailed Rat) as a laboratory species. *Lab. Anim. Care* 17:180–188.

HALLETT, A.F. & J. MEESTER. 1971. Early postnatal development of the South African hamster, *Mystromys albicaudatus. Zool. afr.* 6(2):221–228.

HAMILTON, W.R. & J.A. VAN COUVERING. 1977. Lower Miocene mammals from South West Africa. *Bull. Desert Ecol. Res. Unit, Gobabeb* 2:9–11.

HANNEY, P. 1965. The Muridae of Malaŵi (Africa: Nyasaland). *J. Zool., Lond.* 146:577–633.

HANNEY, P.W. 1975. *Rodents. Their lives and habits.* London: Davis and Charles.

HARRINGTON, C. & E.A.E. YOUNG. 1944. Murine typhus in the district of Port Elizabeth. *S. Afr. med. J.* 18:368–369.

HATT, R.T. 1934. Fourteen hitherto unrecognised African rodents. *Amer. Mus. Nov.* 708:1–15.

HATT, R.T. 1940. Lagomorpha and Rodentia other than Sciuridae, Anomaluridae and Idiuridae, collected by the American Museum Congo Expedition. *Bull. Amer. Mus. nat. Hist.* 76:457–604.

HEBEL, R. & M.W. STROMBERG. 1976. *Anatomy of the laboratory rat.* Baltimore: Williams & Wilkens.

HECHTER-SCHULZ, K. 1951. Plan for the control of rodents causing damage to conifer forests. *Rept. S. Afr. For. Inv. Ltd.* Sabie.

HEDIGER, H. 1950. Gefangenschaftsgeburt eines afrikanischen Springhasen *Pedetes caffer. Zool. Gart.* Leipzig. 17:166.

HENDEY, Q.B. 1978. Preliminary report on the Miocene vertebrates from Arrisdrift, South West Africa. *Ann. S. Afr. Mus.* 76(1):1–41.

HENNING, M.W. 1937. On the variation of the specific phase of *Salmonella amersfoort* n.sp. *J. Hyg. Camb.* 37:561–570.

HEROLD, W. & J. NIETHAMMER. 1963. Zur systematische Stellung des Südafrikanischen *Gerbillus paeba* Smith, 1834 (sic) (Rodentia: Gerbillinae) auf Grund seines Alveolenmusters. *Säugetierk. Mitt.* 11:49–58.

HICKMAN, G.C. 1979. Burrow system structure of the bathyergid *Cryptomys hottentotus* in Natal, South Africa. *Z. Säugetierk.* 44:153–162.

HILL, J.E. 1942. A supposed adaptation against sunstroke in African diurnal rodents. *J. Mammal.* 23:210.

HINTON, M.A.C. 1919. Notes on the genus *Cricetomys* with descriptions of four new forms. *Ann. Mag. nat. Hist.* (9)4:282–289.

HOBSON, W. (ed.), 1969. *The theory and practice of public health.* London: Oxford University Press.

HOLLISTER, N. 1916. The generic names *Epimys* and *Rattus. Proc. biol. Soc. Wash.* 29:126.

HOLLISTER, N. 1919. East African Mammals in the United States National Museum. *Smithson. Instn. Bull.* 99 (2):1–184. Washington: Government Printer.

HOPWOOD, A.T. 1931. Preliminary report on the fossil Mammalia.

*In* LEAKEY, L.S.B., *The Stone Age cultures of Kenya Colony.* Cambridge: Cambridge University Press.

ILLIGER, J.C.W. 1811. *Prodromus Syst. Mamm. et Avium.* Berlin: Salfeld.

INGOLDBY, C.M. 1927. Some notes on the African squirrels of the genus *Heliosciurus. Proc. zool. Soc. Lond.*: 471–487.

JACOBSEN, N.H.G. 1977. An annotated checklist of the amphibians, reptiles and mammals of the Nylsvley Nature Reserve. *S. Afr. Nat. Sci. Prog. Rpt.* 21:1–65.

JARVIS, J.U.M. & J.B. SALE. 1971. Burrowing and burrow patterns of East African mole-rats, *Tachyoryctes, Heliophobius* and *Heterocephalus. J. Zool., Lond.* 163:451–479.

JARVIS, M.J.F. & J. CRICHTON. 1977. Notes on Long-crested eagles in Rhodesia. *Proc. Symp. African Predatory Birds*: pp. 17–24. Pretoria: Northern Transvaal Ornithological Society.

JOHNSON, P.T. 1960. The Anoplura of African rodents and insectivores. *U.S. Dept. Agr. tech. Bull.* (1211):1–116.

JOHNSTON, H.J. & W.D. OLIFF. 1954. The oestrus cycle of the female *Rattus (Mastomys) natalensis* (Smith) as observed in the laboratory. *Proc. zool. Soc. Lond.* 124:605–613.

JONES, F.W. 1941. The external characters of a neonatal *Pedetes. Proc. zool. Soc. Lond.* 110:119–206.

JOUBERT, C.J. 1967. Total nutritive requirements for small laboratory rodents (including rodents indigenous to South Africa). *In* M.L. CONALTY (ed.), *Husbandy of laboratory animals.* London: Academic Press.

JOYTEUX, C. & J.G. BAER. 1927. Etude de quelques cestodes provenant des colonies francaises d'Afrique et de Madagascar. *Ann. Parasit. hum. comp.* 5:27–36.

KASCHULA, V.R. 1961. The propagation and modification of strains of Rift Valley Fever virusses in embryonated eggs, and their use as immunising agents for domestic ruminants. D.V.Sc. dissertation, University of Pretoria, Pretoria.

KEMP, M.I. & A.C. KEMP. 1977. *Bucorvus* and *Sagittarius*: two modes of terrestrial predation. *Proc. Symp. African Predatory Birds.*: pp. 13–16. Pretoria: Northern Transvaal Ornithological Society.

KEOGH, H.J. & M. ISAÄCSON. 1978. Wild rodents as laboratory models and their part in the study of diseases. *J. S. Afr. vet. Assn.* 49(3):229–231.

KERN, N.G. 1977. The influence of fire on populations of small mammals of the Kruger National Park. MSc thesis, University of Pretoria, Pretoria.

KINGDON, J. 1974. *East African Mammals.* Vol. II, Pt B (Hares and Rodents). London: Academic Press.

KLEVEZAL, G.A. & S.E. KLEINENBERG. 1967. *Age determination of mammals by layered structure in teeth and bone.* Moscow: Izdatelstvo Nauka.

LANDRY, S.O. 1957. The inter-relationships of the New and Old world hystricomorph rodents. *Univ. Calif. Publs. Zool.* 56:1–118.

LAVOCAT, R. 1956a. La faune de rongeurs des grottes australopithé-ques. *Palaeont. afr.* 4:69–75.

LAVOCAT, R. 1956b. Réflections sur la classification des rongeurs. *Mammalia.* 20:49–56.

LAVOCAT, R. 1957. Sur l'âge des grottes à faunes de rongeurs des australopithéques. *In* J. DESMOND CLARKE & SONIA COLE (eds), *Proc. 3rd pan-African congress on prehistory. Livingstone 1955.* London: Chatto & Windus.

LAVOCAT, R. 1959. Origine et afnités des rongeurs de la sous-famille des Dendromurinés. *C. R. Acad. Sci. Paris.* 248:1375–1377.

LAVOCAT, R. 1962. Etudes systématiques sur la dentition des Muridés. *Mammalia.* 26:107–127.

LAVOCAT, R. 1964. On the systematic affinities of the genus *Delanymys* (Hayman) *Proc. Linn. Soc. Lond.* 175:183–185.

LAVOCAT, R. 1967. À propos de la dentition des rongeurs et du probleme de l'origine des Muridés. *Mammalia* 31:205–216.

LAVOCAT, R. 1973. Les rongeurs du Miocene d'Afrique orientale. I. Miocene inférieur. *Mem. trav. Inst. Montpellier* 1:1–284.

LAVOCAT, R. 1974. What is an hystricomorph? *In* I.W. ROWLANDS & Barbara WEIR (eds), *The biology of hystricomorph rodents.* Zoological Society of London: Academic Press.

LAWRENCE, B. 1941. Incisor tips of young rodents. *Publs. Field Mus. nat. Hist.* 27:313–317.

LEISTNER, O.A. & J.W. MORRIS. 1976. Southern African place names. *Ann. Cape Prov. Mus.* 12:1–565.

LE ROUX, P.L. 1929a. On an oesophagostome *(Oesophagostomum susannae* sp. nov.) from the Spring Hare *(Pedetes caffra),* together with remarks on closely related species. *Ann. Rep. Dir. Vet. Serv. Un. S. Afr.* 15:465–479.

LE ROUX, P.L. 1929b. Note on the life cycle of *Schistosoma mattheei. Rept. Dir. vet. Serv., U. of S. Africa:* 407–438.

LE ROUX, P.L. 1930a. The generic position of *Oxyuris polyoon* von Linstow, 1909 in the subfamily Oxyurinae Hall, 1916. *Ann. Rept. Dir. Vet. Serv. Un. S. Afr.* 16:205–210.

LE ROUX, P.L. 1930b. A Spirurid *(Streptopharagus geosciuri* sp. nov.) from the stomach of a Cape Ground Squirrel *(Geosciurus capensis). Ann. Rept. Dir. Vet. Serv. Un. S. Afr.* 16:201–204.

LE ROUX, P.L. 1930c. A new nematode *(Rictularia aethechini* sp. nov.). *Ann. Rept. Dir. Vet. Serv. Un. S. Afr.* 16:217–227.

LINNAEUS, C. 1758. *Systema Naturae.* 10th ed. Stockholm.

LUNDHOLM, B.G. 1949. An account of the mammals collected on the Bernard Carp expedition to eastern Caprivi, 1949. Unpublished typescript, Transvaal Museum, Pretoria.

LUNDHOLM, B.G. 1955a. Remarks on some South African Murinae. *Ann. Transv. Mus.* 22:321–329.

LUNDHOLM, B.G. 1955b. Descriptions of new mammals. *Ann. Transv. Mus.* 22:279–303.

LURIE, H.J. & B. DE MEILLON. 1956. Experimental bilharziasis in laboratory animals. III Comparison of the pathogenicity of *S. bovis,* South African and Egyptian strains of *S. mansoni* and *S. haematobium. S. Afr. Med. J.* 30:79–82.

LYNCH, C.D. 1975. The distribution of mammals in the Orange Free State (South Africa). *Nov. nas. Mus. Bloemfontein.* 3:109–139.

MACINNES, D.G. 1957. A new Miocene rodent from East Africa. *Fossil Mammals Afr.* 12:1–35.

MACKELLAR, A.J. 1952. Report on rodent damage in plantations in eastern Transvaal, 1950–1951. *Rep. Dep. For. S. Afr.*

MAGUIRE, J.M. 1978. Southern African fossil porcupines. *S. Afr. J. Sci.* 74:144.

MAJOR, FORSYTH, C.I. 1893. On some Miocene squirrels, with remarks on the dentition and classification of the Sciurinae. *Proc. zool. Soc. Lond.:* 179–215.

MARINIER, S. 1978. Ecology of the Cane Rat *Thryonomys swinderianus. S. Afr. J. Sci.* 74:143.

MASON, J.H. & R.A. ALEXANDER. 1938. Studies of the rickettsias of the Typhus-Rockey Mountain-Spotted Fever group in South Africa. *Onderstepoort J. vet. Sc. An. Ind.* 13:41–65.

MASON, J.H. & C.R. AMIES. 1948. Plague. *Ann. rept. S. Afr. Inst. Med. Res.* 19–20.

MATTHEWS, L. HARRISON. 1971. *The life of mammals.* 2 vols. London: Weidenfeld & Nicolson.

MATTHEY, R. 1958. Les chromosomes et la position systématique de quelques Murinae africaines. *Acta trop.* 15(2):97–117.

MATTHEY, R. 1959. Formules chromosomiques de Muridae et Spalacidae. La question de polymorphisme chromosomique chez les Mammifères. *Revue suisse Zool.* 66:175–209.

MATTHEY, R. 1964. Analyse caryologique de cinq especes de Muridae Africains (Mammalia, Rodentia). *Mammalia* 28:403–418.

MATTHEY, R. 1966. Le polymorphisme chromosomique des *Mus* africains du sous-genre *Leggada. Revue suisse Zool.* 73:585–607.

MEESTER, J. 1954. Research on the mammals of Africa. *Nature, Lond.* 174:1149.

MEESTER, J. 1958. A litter of *Rattus namaquensis* born in captivity. *J. Mammal.* 51:703–711.

MEESTER, J. 1960. Early post-natal development of multimammate mice *Rattus (Mastomys) natalensis* (A. Smith). *Ann. Transv. Mus.* 24:25–35.

MEESTER, J. 1970. Collecting small mammals. *Spectrum* 8(2):403–408.

MEESTER, J.A.J. 1976. South African Red Data Book – Small Mammals. *S. Afr. Nat. Sci. Prog. Rpt.* 10. CSIR.

MEESTER, J., D.H.S. DAVIS & C.G. COETZEE. 1964. *An interim classification of southern African mammals.* Distributed by the Zoological Society of southern Africa and the CSIR, Roneographed.

MEESTER, J. & A.F. HALLETT. 1970. Notes on early post-natal development in certain southern African Muridae and Cricetidae. *J. Mammal.* 51:703–711.

MEESTER, J.A.J., C.N.V. LLOYD & D.T. ROWE-ROWE. 1979. A note on the ecological role of *Praomys natalensis. S. Afr. J. Sci.* 75(4):183–184.

MILLER, G.S. 1910. The generic name of the House-rats. *Proc. biol. Soc. Wash.* 23:57–60.

MILLER, G.S. 1923. List of North American recent mammals. *Bull. U.S. Mus.* 128 (1924):1–673.

MILLER, G. & J.W. GIDLEY. 1918. Synopsis of the supergeneric groups of rodents. *Jour. Wash. Acad. Sci.* 8:431–448.

MILLS, M.G.L. & M.E.J. MILLS. 1978. The diet of the Brown Hyaena *Hyaena brunnea* in the southern Kalahari. *Koedoe* 21:125–149.

MISONNE, X. 1963. Les rongeurs du Ruwenzori et des régions voisines. *Explor. Parc natn. Albert* (2) 14:1–164.

MISONNE, X. 1969. African and Indo-Australian Muridae. Evolutionary trends. *Annls Mus. r. Afr. cent., Tervuren, Zool.* 172:1–219.

MISONNE, X. 1974. Part 6,0. Rodentia, main text. *In* J. MEESTER & H.W. SETZER (eds), *The mammals of Africa: an identification manual.* Washington DC: Smithsonian Institution Press.

MITCHELL, J.A. 1927. Plague in South Africa: historical summary (up to June 1926). *Publ. S. Afr. Inst. Med. Res.* 3:89–108.

MOHR, E. 1941. Schwanzverlust und Schwanzregeneration bei Nagetieren. *Zool. Anz.* 135:49–65.

MÖNNIG, H.O. 1927. On a new *Physaloptera* from an eagle and a trichostrongyle from the cane rat, with notes on *Polydelphis quadricornis* and the genus *Spirostrongylus*. *Trans. R. Soc. S. Afr.* 14:261–265.

MÖNNIG, H.O. 1931. A second species of the nematode genus *Acanthoxyuris*. *Rept. Dir. Vet. Serv. Anim. Ind. Un. S. Afr.* 17:269–271.

MOORE, J.C. 1959. Relationship among living squirrels of the Sciurinae. *Bull. Amer. Mus. nat. Hist.* 118:153–206.

MOREAU R.E. 1952. Africa since the Mesozoic: with particular reference to certain biological problems. *Proc. zool. Soc. Lond.* 121:769–913.

MORRIS, B. 1963. Notes on the Giant Rat *(Cricetomys gambianus)* in Nyassaland. *Afr. Wild Life.* 17(2):103–107.

MOSSMAN, H.W. 1957. Endotheliochorial placentation in the rodents *Castor* and *Pedetes*. *Proc. zool. Soc., Calcutta.* Mookerjee Memor. Vol.: 183–186.

MYLLYMAKI, A. 1975. Control of field rodents. *In* F.B. GOLLEY, K. PETRUSEWICZ & L. RYSZKOWSKI (eds), *Small mammals, their productivity and population dynamics.* Cambridge: Cambridge University Press.

NEITZ, W.O. 1965. A checklist and hostlist of the zoonoses in mammals and birds in South and South West Africa. *Onderstepoort J. vet. Res.* 32(2):189–376.

NEITZ, W.O. & K.C.A. SCHULZ. 1949. Diagnoses of rabies in South Africa. *Rept. Dir. Vet. Serv. An. Ind. Un. S. Afr.*: 71–98.

NEL, J.A.J. 1967. Burrow systems of *Desmodillus auricularis* in the Kalahari Gemsbok National Park. *Koedoe* 10:118–121.

NEL, J.A.J. 1969. The interrelationship of the genera of the subfamily Dendromurinae. DSc. dissertation, University of Pretoria, Pretoria.

NEL, J.A.J. 1978. Habitat heterogeneity and changes in small mammal community structure and resource utilization in the southern Kalahari. *In* D.A. SCHLITTER (ed.), Ecology and taxonomy of African small mammals. *Bull. Carnegie Mus. nat. Hist.* 6:118–131.

NEL, J.A.J. & H. NOLTE. 1965. Notes on the prey of owls in the Kalahari Gemsbok National Park. *Koedoe* 8:75–81.

NEL, J.A.J. & I.L. RAUTENBACH. 1974. Notes on the activity patterns, food and feeding behaviour of *Parotomys brantsi* (Smith, 1840). *Mammalia* 38(1):7–15.

NEL, J.A.J. & I.L. RAUTENBACH. 1977. Body temperatures of some Kalahari rodents (Mammalia: Muridae, Cricetidae). *Ann. Transv. Mus.* 30(17):207–210.

NEL, J.A.J. & C.J. STUTTERHEIM. 1973. Notes on early post-natal development of the Namaqua gerbil, *Desmodillus auricularis.* *Koedoe* 16:117–125.

NELSON, G.S., C.TEESDALE & R.B. HIGHTON. 1962. The role of animals as reservoirs of bilharziasis in Africa. *In* C.E.W. WOLSTENHOLME & M. O'CONNOR (eds), *Bilharziasis.* London: Churchill.

NIESCHULZ, O. 1932. The infection of mice with African horse sickness. *Tijdschr. Dieregeneesk.* 59:1433–1445.

OLIFF, W.D. 1953. The mortality, fecundity and intrinsic rate of natural increase of the Multimammate Rat *Rattus (Mastomys) natalensis* (Smith) in the laboratory. *J. Anim. Ecol.* 22:217–226.

ORTLEPP, R.J. 1922a. A new species of oesophagostomum *(Oesophagostomum xeri* sp.n.) from a rodent *Xerus setosus.* *Proc. zool. Soc. Lond.*: 461–469.

ORTLEPP, R.J. 1922b. The nematode genus *Physaloptera* Rud. *Proc. zool. Soc. Lond.*: 999–1107.

ORTLEPP, R.J. 1937. Some undescribed species of the nematode genus *Physaloptera* Rud., together with a key to the sufficiently known forms. *Onderstepoort J. Vet. Sci. Anim. Ind.* 9(1):71–84.

ORTLEPP, R.J. 1938a. South African Helminths. Pt.III. Some avian and mammalian helminths. *Onderstepoort J. Vet. Sci. Anim. Ind.* 11(1):23–50.

ORTLEPP, R.J. 1938b. South African Helminths, Pt. V. Some avian and mammalian helminths. *Onderstepoort J. Vet. Sci. Anim. Ind.* 11(1):63–104.

ORTLEPP, R.J. 1939. South African Helminths. Pt. VI. Some helminths chiefly from rodents. *Onderstepoort J. Vet. Sci. Anim. Ind.* 12(1):75–101.

ORTLEPP, R.J. 1961. A record of three cases of human infection in southern Africa with a common tapeworm of rats. *S. Afr. med. J.* 35:837–839.

ORTLEPP, R.J. 1962a. On two new *Catenotaenia* tapeworms from a South African rat, with remarks on the species of the genus. *Onderstepoort J. Vet. Res.* 29(1):11–19.

ORTLEPP, R.J. 1962b. On *Trachypharynx natalensis* sp. nov. and some associated genera of nematodes. *Onderstepoort J. Vet. Res.* 29(2):159–163.

PACKER, J.T., K.L. KRANER, D.S. ROSE, R.A. STUHLMAN & L.R. NELSON. 1970. Diabetes Mellitus in *Mystromys albicaudatus.* *Archs. Path.* 89:410–415.

PALLAS, P.S. 1779. *Novae Species Quadrupedum e Glirium Ordine.* Erlangen: Wolfgang Walther.

PARRIS, R. 1976. A study of the major components of the Kalahari pan ecosystem. MSc (Wildlife Management) thesis. University of Pretoria, Pretoria.

PERRIN, M.R. *(in press,* a). The breeding strategies of two co-existing rodents, *Rhabdomys pumilio* and *Otomys irroratus.*

PERRIN, M.R. *(in press,* b). The feeding habits of two co-existing rodents, *Rhabdomys pumilio* and *Otomys irroratus* in relation to rainfall and reproduction.

PETERS, W. 1852. *Naturwissenschaftliche Reise nach Mossambique. Säugeth.* Berlin: George Reimer.

PETTER, F. 1963. Contribution à la connaissance des souris africaines. *Mammalia* 27:602–607.

PETTER, F. 1964. Affinités du genre *Cricetomys.* Une novelle sousfamille de Rongeurs Cricetidae, les Cricetomyinae. *C. r. Acad. Sci. Paris.* 258:6516–6518.

PETTER, F. 1966a. L'origine des Muridés: plan cricetin et plan murins. *Mammalia* 30:205–225.

PETTER, F. 1966b. Affinités des genres *Beamys, Saccostomus* et *Cricetomys* (Rongeurs, Cricetomyinae). *Annls. Mus. r. cent. Tervuren. Serv. 8. vo. Sci. Zool.* 144:13–25.

PETTER, F. 1975. Part 6,3 Subfamily Gerbillinae. *In* J. MEESTER & H.W. SETZER (eds), *The mammals of Africa: an identification manual.* Washington DC: Smithsonian Institution Press.

PETTER, F., A. CHIPPAUX & C. MONMIGNAUT. 1964. Observations sur la biologie, la réproduction et la croissance de *Lemniscomys striatus* (Rongeurs, Muridés). *Mammalia* 28:620–627.

PETTER, F. & H. GENEST. 1964. Spécialisation laceale des incisives de jeunes rongeurs Muridés d'Afrique. *Science et Nature* 65:1–3.

PETTIFER, H.L. & J.A.J. NEL. 1977. Hoarding in four southern African rodent species. *Zool. afr.* 12(2):409–418.

PIANKA, E.R. 1970. On r and K-selection. *Am. Nat.* 104:592–597.

PIENAAR, U. DE V. 1964. The small mammals of the Kruger National Park – a systematic list and zoogeography. *Koedoe* 7:1–25.

PIENAAR, U. DE V., I.L. RAUTENBACH & G. DE GRAAFF. 1980. *The small mammals of the Kruger National Park.* Pretoria: National Parks Board of Trustees.

PIJPER, A. & H. DAU. 1933. A preliminary note on a typhus-like virus in South African rats. *S. Afr. med. J.* 7:715–716.

PIRIE, J.H.H. 1927. Observations on the comparative susceptibility to plague of various veld rodents and associated animals. *S. Afr. Inst. Med. Res.* 3:119–137.

PITCHFORD, R.J. 1959. Natural schistosome infections in South African rodents. *Trans. roy. Soc. trop. Med. Hyg.* 53:213.

PITMAN, C.R.S. & J. ADAMSON. 1978. Notes on the ecology and ethology of the Giant Eagle Owl *Bubo lacteus* (Temminck). Pt. 2. *Honeyguide* 96:25–52.

POCOCK, T.N. 1974. New mammal record for genus *Mus* for southern Africa. *S. Afr. J. Sci.* 70(10):315.

POCOCK, T.N. 1976. Pliocene mammalian microfauna from Langebaanweg: a new fossil genus linking the Otomyinae with the Muridae. *S. Afr. J. Sci.* 72:58–60.

POWELL, W. 1925. Rodents. Description, habits and methods of destruction. Pamphlet 321. Pretoria: Department of Health.

QUINN, P.J. 1959. *Food and feeding habits of the Pedi.* Johannesburg: Witwatersrand University Press.

RAFINESQUE. 1814. *Préc. des Découv. et Trav. Somiologiques:* 13.

RAHM, U. 1967. Les Muridés des environs du Lac Kivu et des régions voisines (Afrique Centrale) et leur écologie. *Revue suisse Zool.* 74:439–520.

RANDERIA, J.D. 1978. The inbreeding of the Y and the Z strains of *Praomys natalensis* with special reference to the laboratory uses of the *Mastomys. J. S. Afr. vet. Assn.* 49(3):197–199.

RANGER, G. 1927. Notes on some small mammals found near Kei Road. *Rec. Albany Mus.* 3:487–490.

RAUTENBACH, I.L. 1978a. A numerical re-appraisal of the southern African biotic zones. *In* D.A. SCHLITTER (ed.), Ecology and taxonomy of African small mammals. *Bull. Carnegie Mus. nat. Hist.* 6:175–187.

RAUTENBACH, I.L. 1978b. The mammals of the Transvaal. PhD dissertation, University of Natal, Pietermaritzburg.

ROBBINS, C.B. 1978. Ecology and systematics of the Multimammate Mouse *Mastomys* and its relationship to the epidemiology of Lassafever. *African Small Mammal Newsletter* 1:8.

ROBERTS, A. 1913. The collection of mammals in the Transvaal Museum registered up to the 31st March 1913, with descriptions of new species. *Ann. Transv. Mus.* 4:65–107.

ROBERTS, A. 1917. Fourth supplementary list of mammals in the collection of the Transvaal Museum. *Ann. Transv. Mus.* 5:263–278.

ROBERTS, A. 1926. Some new south African mammals and some changes in nomenclature. *Ann. Transv. Mus.* 11:245–263.

ROBERTS, A. 1929. New forms of African Mammals. *Ann. Transv. Mus.* 13:82–121.

ROBERTS, A. 1938. Descriptions of new forms of mammals. *Ann. Transv. Mus.* 19:231–245.

ROBERTS, A. 1944. Andrew Smith's early descriptions of animals. *Sthn. Afr. Mus. Ass. Bull.:* 238–241.

ROBERTS, A. 1946. Descriptions of numerous new subspecies of mammals. *Ann. Transv. Mus.* 20:303–328.

ROBERTS, A. 1951. *The mammals of South Africa.* Johannesburg: Trustees of 'The Mammals of South Africa' Book Fund.

ROBINSON, E.M. 1929. Notes on a few outbreaks of botulism in domestic animals and birds. *Rep. Dir. Vet. Serv. An. Ind., U. of S. Africa* 15:111–117.

ROBINSON, E.M. 1930a. A note on the serological diagnosis of *Trypanosoma congolense* infection. *Rep. Dir. Vet. Serv. An. Ind. U. of S. Africa:* 61–65.

ROBINSON, E.M. 1930b. The bacteria of the *Clostridium botulinum* C and D types. *Rep. Dir. Vet. Serv. An. Ind., U. of S. Africa:* 16:107–142.

ROMER, A.S. 1971. *Vertebrate Paleontology.* (3rd ed., third impr.) Chicago: University of Chicago Press. (1958: 2nd ed. seventeenth impr.)

ROMER, A.S. & P.H. NESBIT. 1930. An extinct Cane-rat (*Thryonomys logani* sp.n.) from the Central Sahara. *Ann. mag. nat. Hist.* (10)6:687–690.

ROSEVEAR, D.R. 1963. On the West African forms of *Heliosciurus* Trouessart. *Mammalia.* 27:177–185.

ROSEVEAR, D.R. 1969. *The rodents of West Africa.* London: British Museum (Natural History).

SANDERSON, I.T. 1940. The mammals of the North Cameroons forest area. Being the result of the Percy Sladen Expedition to the Mamfe division of the British Cameroons. *Trans. zool. Soc. Lond.* 24(7):623–725.

SANDON, H. 1941. Studies on South African endozoic ciliates. II. *Meiostoma georychi* gen. nov., sp. nov. from the caecum of *Georychus capensis. S. Afr. J. med. Sci.* 6:128–135.

SCHAUB. S. 1953a. Remarks on the distribution and classification of the Hystricomorpha. *Verh. Naturf. Ges. Basel.* 64:389–400.

SCHAUB, S. 1953b. La trigonodontie des rongeurs simplicidentés. *Annls Paléont.* 39:29–57.

SCHLITTER, D.A. 1973. A new species of gerbil from South West Africa with remarks on *Gerbillus tytonis* Bauer & Niethammer, 1959 (Rodentia: Gerbillinae). *Bull. sthn Calif. Acad. Sci.* 72(1):13–18.

SCHULZ, K.C.A. 1978. Aspects of the burrowing system of the Cape Dune Mole *Bathyergus suillus. S. Afr. J. Sci.* 74:145–146.

SCHWANN, H. 1906. A list of the mammals obtained by Messrs R.B. Woosnam and R.E. Dent in Bechuanaland. *Proc. zool. Soc. Lond.:* 101–111.

SCLATER, W.L. 1901. *The mammals of South Africa. In: The Fauna of South Africa.* London: Porter.

SETZER, H.W. 1956. Mammals of the Anglo-Egyptian Sudan. *Proc. U.S. natn. Mus.* 106:447–587.

SETZER, H.W. 1975. Part 6,5. Genus *Acomys. In* J. MEESTER & H.W. SETZER (eds), *The mammals of Africa: an identification manual.* Washington DC: Smithsonian Institution Press.

SHORTRIDGE, G.C. 1934. *The mammals of South West Africa.* 2 vols. London: Heinemann.

SHORTRIDGE, G.C. 1942. Field notes on the First and Second Expeditions of the Cape Museums' Mammal Survey of the Cape Province; and descriptions of some new subgenera and subspecies. *Ann. S. Afr. Mus.* 36:27–100.

SILBERBAUER. G.B. 1965. *Report of the Government of Bechuanaland on the Bushman Survey.* 1–138. Mafeking: Bechuanaland Press.

SIMPSON, G.G. 1945. The principles of classification and a classification of mammals. *Bull. Amer. Mus. nat. Hist.* 85:1–350.

SKEAD, C.J. 1973. Zoo-historical gazetteer. *Ann. Cape Prov. Mus.* 10:1–259.

SMITH, A. 1826. *A descriptive catalogue of the South African Museum.*

SMITH, A. 1834. *South African Quarterly Journal* 2:156–157.

SMITH, A. 1845. *Zoological illustrations: South African mammals.*

SMITHERS, R.H.N. 1966. *The mammals of Rhodesia, Zambia and Malaŵi.* London: Collins.

SMITHERS, R.H.N. 1971. The mammals of Botswana. *Mem. natn. Mus. Rhod.* 4:1–340.

SMITHERS, R.H.N. 1975. *Guide to the rats and mice of Rhodesia.* Salisbury: Trustees of National Museums and Monuments.

SMITHERS, R.H.N. & J.L. LOBÃO TELLO. 1976. Checklist and atlas of the mammals of Moçambique. *Mem. natn. Mus. Rhod.* 8:1–184.

SNELL, K.C. & H.L. STEWART. 1967. Neoplastic and nonneoplastic renal disease in *Praomys (Mastomys) natalensis. J. natn. Cancer Inst.* 39:95.

SOKOLOFF, L., K.C. SNELL & H.L. STEWART. 1967. Degenerative joint disease in *Praomys (Mastomys) natalensis. Ann. Rheum. Diseases* 26:146.

SPARRMAN, A. 1785. *Voyage au Cap de Bonne-Esperance... et le pays des Hottentots et des Caffres.* Paris.

STEHLIN, H.G. & S. SCHAUB. 1951. Die Trigonodontie der Simplicidentaten Nager. *Schweiz. paläont. Abhandl.* 67:1–384.

STEVENSON-HAMILTON, J. 1950. *Wild Life in South Africa.* (2nd ed.) London: Cassel.

STIEMIE, S. & J.A. J. NEL. 1973. Nest-building behaviour in *Aethomys chrysophilus, Praomys (Mastomys) natalensis* and *Rhabdomys pumilio. Zool. afr.* 8(1):91–100.

ST LEGER, J. 1933. A new species of *Myomys* from South West Africa. *Proc. zool. Soc. Lond.:* 411.

STRASCHIL, B. 1974. Albinism in the Cape Ground Squirrel *Xerus inauris* (Zimmermann, 1870 (sic)). *S. Afr. J. Sci.* 70(10):315.

STRASCHIL, B. 1975. Sandbathing and marking in *Xerus inauris* (Zimmerman (sic), 1870 (sic)). *S. Afr. J. Sci.* 71:215–216.

STROMER, E. 1922. Erste mitteilung über Tertiäre Wirbeltier-Reste aus Deutsch-Südwestafrika. *Sber. bayer. akad. Wiss. München:* 331–340.

STROMER, E. 1924. Ergebnisse der Bearbeitung mitteltertiër Wirbeltier-Reste aus Deutsch-Südwestafrikas. *Sber. bayer. akad. Wiss. München* 253–270.

STROMER, E. 1926. Reste Land-und Süsswasserbewohnender Wirbeltiere aus den Diamantenfeldern Deutsch-Südwestafrikas. *In* E. KAISER (ed.). *Die Diamantenwüste Südwestafrikas* 2:107–153. Berlin: Reimer.

SUNDEVALL, C.J. 1846. *Oefvers. Vetensk. akad. Förh. Stockholm* 3:120.

SWANEPOEL, P. 1972. The population dynamics of rodents at Pongola, northern Zululand, exposed to dieldrin cover spraying. MSc thesis. University of Pretoria, Pretoria.

TATE, G.H.H. 1936. Some Muridae of the Indo-Autralian Region. *Bull. Amer. Mus. nat. Hist.* 72:512–580.

TAYLOR, K.D. & M.G. GREEN. 1976. The influence of rainfall on diet and reproduction in four African rodent species. *J. Zool., Lond.* 180:367–389.

TAYLOR, P. & D.H. GORDON. 1978. Sibling species of *Praomys (Mastomys) natalensis:* identification, distribution and sympatric population ecology. *S. Afr. J. Sci.* 74:143.

THEILER, A. & E.M. ROBINSON. 1927. Der Botulismus der Haustiere. *Zschr. f. Infek.-Krh.* 31:165–220.

THEILER, G. 1962. The Ixodoidea parasites of vertebrates in Africa south of the Sahara (Ethiopian Region). *Rep. Dir. Vet. Services, Onderstepoort.* Project S 9958.

THOMAS, O. 1882. On a small collection of rodents from south-western Africa. *Proc. zool. Soc. Lond.:* 265–267.

THOMAS, O. 1894, On the mammals of Nyasaland: third contribution. *Proc. zool. Soc. Lond.:* 136–146.

THOMAS, O. 1896. On the genera of the rodents: an attempt to bring up to date the current arrangement of the Order. *Proc. zool. Soc. Lond.:* 1012–1028.

THOMAS, O. 1897a. Exhibition of small mammals collected by Mr Alexander Whyte during his expedition to the Nyika plateau and the Masuku Mountains, n. Nyasa. *Proc. zool. Soc. Lond.:* 430–436.

THOMAS, O. 1897b. On a new dormouse from Mashunaland. *Ann. Mus. nat. Hist.* (6)19:388–389.

THOMAS, O. 1904. On mammals from northern Angola collected by Mr. W.J. Ansorge. *Ann. Mag. nat. Hist.* (7) 13:405–421.

THOMAS, O. 1907. On further new mammals obtained by the Ruwenzori Expedition. *Ann. Mag. nat. Hist.* (7) 19:118–123.

THOMAS, O. 1909. New African small mammals in the British Museum collection. *Ann. Mag. nat. Hist.* (8) 4:98–112.

THOMAS, O. 1911. New African mammals. *Ann. Mag. nat. Hist.* (8) 8:375–378.

THOMAS, O. 1915a. New African rodents and insectivores, mostly collected by Dr C. Christy for the Congo Museum. *Ann. Mag. nat. Hist.* (8) 16:146–152.

THOMAS, O. 1915b. List of mammals (exclusive of Ungulata) collected on the Upper Congo by Dr Christy for the Congo Museum, Tervuren. *Ann. Mag. nat. Hist.* (8) 16:465–481.

THOMAS, O. 1916a. On the rats usually included in the genus *Arvicanthis. Ann. Mag. nat. Hist.* (8) 18:67–70.

THOMAS, O. 1916b. On small mammals obtained in Sankuru, south Congo, by Mr H. Wilson. *Ann. Mag. nat. Hist.* (8) 18:234–239.

THOMAS, O. 1918. New species of *Gerbillus* and *Taterillus. Ann. Mag. nat. Hist.* (9) 2:146–151.

THOMAS, O. 1919. Scientific results from the mammal survey of India, No. 19. A synopsis of the groups of true mice found within the Indian Empire. *J. Bombay nat. Hist. Soc.* 26:417–421.

THOMAS, O. 1920. The generic positions of 'Mus' *nigricauda*, Thos., and *woosnami*, Schwann. *Ann. Mag. nat. Hist.* (9) 5:140–142.

THOMAS, O. 1922. On the animals known as 'Ground hogs' or 'Cane Rats' in Africa. *Ann. Mag. nat. Hist.* (9) 9:389–392.

THOMAS, O. 1926a. The generic position of certain African Muridae, hitherto referred to *Aethomys* and *Praomys. Ann. Mag. nat. Hist.* (9) 17:174–179.

THOMAS, O. 1926b. On mammals from Ovamboland and the Cunene River, obtained during Capt. Shortridge's Third Percy Sladen and Kaffrarian Museum expedition into South West Africa. *Proc. zool. Soc. Lond.:* 285–312.

THOMAS, O. 1927. On the mammals of the Gobabis district, eastern Damaraland, South West Africa, obtained during Capt. Shortridge's Fourth Percy Sladen and Kaffrarian Museum Expedition. With field notes by the collector. *Proc. zool. Soc. Lond.:* 371–398.

THOMAS, O. 1929. On mammals from the Kaokoveld, South West Africa, obtained during Captain Shortridge's Fifth Percy Sladen and Kaffrarian Museum Expedition. *Proc. zool. Soc. Lond.:* 99–111.

THOMAS, O. & M.A.C. HINTON. 1925. On mammals collected in 1923 during the Percy Sladen and Kaffrarian museum expedition to South-West Africa. *Proc. zool. Soc. Lond.:* 221–246.

THOMAS, O. & H. SCHWANN. 1904. On mammals from British Namaqualand. *Proc. zool. Soc. Lond.:* 171–183.

THOMAS, O. & H. SCHWANN. 1906. The Rudd Exploration of South Africa IV. List of mammals obtained by Mr Grant at Knysna. *Proc. zool. Soc. Lond.* 1:159–168.

THOMAS, O. & R.C. WROUGHTON. 1908a. The Rudd Exploration of South Africa. IX. List of mammals obtained by Mr Grant on the Gorongoza Mountains, Portugese southeast Africa. *Proc. zool. Soc. Lond.:* 164–173.

THOMAS, O. & R.C. WROUGHTON. 1908b. The Rudd Exploration of South Africa. X. List of mammals collected by Mr Grant near Tette, Zambesia. *Proc. zool. Soc. Lond.:* 535–553.

THUNBERG, C.P. 1788. *Resa uti Europa, Africa, Asia,* etc. I.

TILL, W.M. 1957. Mesostigmatic mites living as parasites of reptiles in the Ethiopian region (Acarina: Laelaptidae). *J. ent. Soc. S. Afr.* 20:120–143.

TROUESSART, E.L. 1880. Révision du genre écureuil *(Sciurus). Le Naturaliste:* 37.

TROUESSART, E.L. 1881. Catelogue des mammifères vivants et fossilés. *Bull. soc. Etud. scient. Angers.:* 177–209.

TULLBERG, T. 1899. Uber das System der Nagetiere: eine Phylogenetische Studie. *Nova Acta R. Soc. Scient. upsal.* 3(18):1–514.

TWIGG, G.I. 1978. The role of rodents in plague dissemination: a worldwide review. *Mammal Review* 8(3):77–110.

VANDEBROEK, G. 1961. The comparative anatomy of the teeth of the lower and non-specialised mammals. Intern. Colloq. Evol. Mammals. *Kon. vlaams. Acad. Wetensch. Lett. Sch. Kunsten,* Brussels.

VANDEBROEK, G. 1966. Plans dentaires fondamentaux chez les rongeurs. Origine des Muridés. *Annls. Mus. r. Afr. cent. Ser. 8 Sci. zool.* 144:115–152.

VAN DER HORST, C.J. 1935. On the reproduction of the Springhare, *Pedetes caffer. Pamph. S. Afr. biol. Soc.* 8:47.

VAN ROOYEN, R.J. 1955. The systematics of the genus *Thallomys.* MSc Thesis, University of Pretoria, Pretoria.

VAUGHAN, T.A. 1972. *Mammalogy.* Philadelphia: W.B. Saunders.

VEENSTRA, A.J.F. 1958. The behaviour of the Multimammate Mouse *Rattus (Mastomys) natalensis* (A. Smith). *Anim Behav.* 6:195–206.

VERMEIREN, L. 1979. Pan-african taxonomical revision of the genus *Leggada. Afr. Small Mamm. Newsletter* 3:32–34.

VERNON, C.J. 1972. An analysis of owl pellets collected in southern Africa. *Ostrich* 43:109–124.

VERSTER, A.J.M. 1960. *Trichuris* species from South African rodents and a hyracoid. *Onderstepoort J. vet. Res.* 28(3):465–471.

VERSTER, A.M.J. & D. BROOKER. 1970. Helminth parasites of small laboratory animals at the veterinary research institute Onderstepoort. *J. S. Afr. Med.* 41:183–184.

VESEY-FITZGERALD, D.F. 1966. The habits and habitats of small rodents in the Congo River catchment region of Zambia and Tanzania. *Zool. afr.* 2(1):111–122.

VILJOEN, P.R. 1921. Das Vorkommen von Sarcosporidien in Südafrikanischen Tieren (Haustieren und Wild). Dr. Med. Vet. Thesis, Univ. Bern.

VILJOEN, S. 1973. Competition between crested barbets and squirrels. *N. Tvl. Ornith. Soc. Newsletter* 13:2.

VILJOEN, S. 1975. Aspects of the ecology, reproductive physiology and ethology of the Bush Squirrel *Paraxerus cepapi cepapi* (A. Smith, 1836). MSc thesis, University of Pretoria, Pretoria.

VILJOEN, S. 1977a. Feeding habits of the Bush Squirrel *Paraxerus cepapi cepapi* (Rodentia, Sciuridae). *Zool. afr.* 12(2):459–467.

VILJOEN, S. 1977b. Ectoparasites of the Bush Squirrel *Paraxerus cepapi cepapi* in the Transvaal (Rodentia: Sciuridae). *Zool. afr.* 12(2):498–500.

VILJOEN, S. 1978a. Tree Squirrels. *Afr. wild Life* 31(6):36–39.

VILJOEN, S. 1978b. Comparative eco-ethology of southern African Tree squirrels. *Afr. Small. Mamm. Newsletter* 2:21–22.

VON LEHMANN, E. 1955. Neue Säugetierrassen aus Südwestafrika. *Bonn. zool. Beitr.* 6:171–172.

VON LEHMANN, E. 1960. Die Unterarten von *Desmodillus auricularis* (A. Smith, 1834) in Südwestafrika. *Säugetierk. Mitt.* 8:161.

WALKER, E.P. (ed.) 1975. *Mammals of the world.* 2 vols. (3rd ed.) Baltimore: The John Hopkins Press.

WEBER, M. 1928. *Die Säugetiere.* Jena: Gustav Fischer Verlag.

WELLINGTON, J.H. 1955. *Southern Africa. A geographical study.* 2 vols. Cambridge: Cambridge University Press.

WERGER, M.J.A. (ed.) 1978. *Biogeography and ecology of southern Africa.* 2 vols. The Hague: W. Junk.

WILLAN, K. & MEESTER 1978. Breeding biology and postnatal development of the African Dwarf Mouse. *Acta theriol.* 23(3):55–73.

WILSON, V.J. 1975. Mammals of the Wankie National Park, Rhodesia. *Mem. natn. Mus. Rhod.* 5:1–147.

WINGE, H. 1941. *The Interrelationships of the Mammalian genera.* 3 vols. Copenhagen: C.A. Reitzels Forlag.

WINTERBOTTOM, J.M. 1973. The relationships of the avifauna of the South West Arid area of Africa. *Zool. afr.* 8:83–90.

WOLSTENHOLME, B. 1952. Rickettsial diseases. *Ann. Rept. S. Afr. Inst. Med. Res.* 34–35.

WOLSTENHOLME, B. 1960. Rickettsial diseases. *Ann. Rept. S. Afr. Inst. Med. Res.:* 125.

WOLSTENHOLME, B. & HARWIN. 1951. Rickettsial diseases. *Rept. S. Afr. Inst. Med. Res.:* 36–37.

WOLSTENHOLME, B & HARWIN. 1952. Rickettsial diseases. *Rept. S. Afr. Inst. Med. Res.:* 34–35.

WOOD, A.E. 1955. A revised classification of the rodents. *J. Mammal.* 36:165–187.

WOOD, A.E. 1958. Are there rodent suborders? *Syst. Zool.* 7: 169–173.

WOOD, A.E. 1959. Eocene radiation and phylogeny of the rodents. *Evolution* 13:354–361.

WOOD, A.E. 1974. The evolution of the Old world and New World hystricomorphs. *In* I.W. ROWLANDS & Barbara J. WEIR (eds), *The biology of hystricomorph rodents.* Zoological Society of London: Academic Press.

WOOD, A.E. & B. PATTERSON. 1959. The rodents of the Deseadan Oligocene of Patagonia and the beginnings of South American rodent evolution. *Bull. Mus. comp. Zool. Harv.* 120:279–428.

WORDEN, C.J. & HALL. 1978. Observations on the White-faced Owl *Otis leucotis* at Cleveland dam, Salisbury. *Honeyguide* 94:31–37.

WRANGHAM, R.W. 1969. Captivity, behaviour and post-natal development of the Cape Pouched Rat *Saccostomus campestris* Peters. *Puku* 5:207–210.

WROUGHTON, R.C. 1907. On a collection of mammals made by Mr S.A. Neave in Rhodesia, north of the Zambezi, with field notes by the collector. *Mem. Proc. Manchr. lit. phil. Soc.* 51(5):1–39.

WROUGHTON, R.C. 1908. Three new African species of *Mus. Ann. Mag. nat. Hist:* (8) 1:255–257.

ZEALLY, E.A.V. 1916. A breccia of mammalian bones at Bulawayo Waterworks Reserve. *Proc. Rhod. Sci. Assn.* 15:5–16.

ZINSSER, H. 1942. *Rats, lice and history.* London: Routledge.

ZUMPT, F. 1959. Is the multimammate rat a natural reservoir of *Borrelia duttoni? Nature, Lond.* 184:793–794.

ZUMPT, F. (ed.). 1961a. The arthropod parasites of vertebrates in Africa south of the Sahara (Ethiopian Region). Vol. I. (Chelicerata). *Publ. S. Afr. Inst. Med. Res.* 9(1):1–457.

ZUMPT, F. 1961b. A case of traumatic myiasis in a wild rat (Diptera: Calliphoridae). *J. ent. Soc. S. Afr.* 24:350.

ZUMPT, F. (ed.). 1966. The arthropod parasites of vertebrates in Africa south of the Sahara (Ethiopian Region). Vol. III. (Insecta, excluding Phthiraptera.) *Pub. S. Afr. Inst. Med. Res.* 13(52):1–283.

# Index

This Index is divided into three parts: (A) Scientific names, (B) English vernacular names, and (C) Afrikaans vernacular names.

## (C) Afrikaans vernacular names